# CISM International Centre for Mechanical Sciences

Courses and Lectures

Volume 587

The series presents lecture notes, monographs, edited works and proceedings in the field of Mechanics, Engineering, Computer Science and Applied Mathematics. Purpose of the series is to make known in the international scientific and technical community results obtained in some of the activities organized by CISM, the International Centre for Mechanical Sciences.

More information about this series at http://www.springer.com/series/76

Sinisa Mesarovic • Samuel Forest • Hussein Zbib
Editors

# Mesoscale Models

## From Micro-Physics to Macro-Interpretation

Springer

*Editors*
Sinisa Mesarovic
Department of Mechanical and Materials Engineering
Washington State University
Pullman
Washington, USA

Samuel Forest
Centre des Matériaux
Mines ParisTech
Evry, France

Hussein Zbib
Department of Mechanical and Materials Engineering
Washington State University
Pullman
Washington, USA

ISSN 0254-1971 ISSN 2309-3706 (electronic)
CISM International Centre for Mechanical Sciences
ISBN 978-3-319-94185-1 ISBN 978-3-319-94186-8 (eBook)
https://doi.org/10.1007/978-3-319-94186-8

Library of Congress Control Number: 2018952930



This Springer imprint is published by the registered company Springer Nature Switzerland AG
The registered company address is: Gewerbestrasse 11, 6330 Cham, Switzerland

# Preface

In the past decades, significant advances have been made in computational modeling of materials on multiple length and time scales. In particular, the finest scale (ab initio, atomistic level) and coarsest (classic continua) are well understood and are now fine-tuned for numerous applications. From numerous studies, the realization has emerged that most macroscopic properties of materials depend critically on the mesoscale: the evolving microstructure and interactions which occur on the length and time scales much larger than atomic, yet much smaller than macroscopic observables. These lecture notes follow the 1-week advanced course presented in the International Centre for Mechanical Sciences (CISM), Udine, in May 2017.

The topics of the course are chosen with the following criteria in mind:

- Significant progress has been made in the past decades
- Some outstanding problems remain as a challenge for future researchers.
- The past efforts have been multidisciplinary, and future efforts are expected to remain such, involving researchers in engineering, physics and mathematics.

Three broad areas representative of the field are selected: dislocation plasticity and creep of crystals and polycrystals, interfaces in solids and fluids, and granular deformation and flow.

Geometrical complexity of dislocation assemblies—a mixture of order and disorder arising from highly constrained, yet rich, kinematics, combined with the long-range interactions (inherent non-locality)—results in a very challenging problem. Experimentally observed size effects indicate a size-dependent mesoscale model, but the consensus about the correct formulation has not been reached.

The structure of crystalline interfaces, mechanisms of their motion, and their role in the overall deformation of polycrystalline assemblies have been the subject of intense research but many questions remain open, among which are dislocation-interface interactions. Fluid interfaces in capillary flows exhibit topological discontinuities (breaking and coalescence of subdomains). These are effectively handled with the phase field formulations, so that questions arise regarding the appropriate phase field model.

Depending on the packing density, the mechanical behavior of granular materials may resemble that of solids or fluids, but also exhibits unique features (dilatancy, critical state, shear localization with a characteristic width, vortex flow). Microscopically, the geometry is strongly disordered and interparticle forces strongly inhomogeneous. A continuum theory capable of describing all the salient features is yet to be formulated.

The lecture notes are intended to provide answer to the following general questions. How far have the understanding and mesoscale modeling advanced in recent decades? What are the key open questions that require further research? What are the mathematical and physical requirements for a mesoscale model intended to provide either insight or a predictive engineering tool?

The book begins and ends with two perspectives on continuum theory. In chapter "Physical Foundations of Mesoscale Continua" (*Mesarovic*), the relationship between microscale physics and mesoscale continua is examined, with a goal of providing a method for deriving the latter from the former. Chapter "Generalized Continua and Phase-Field Models: Application to Crystal Plasticity" (*Forest*) presents a comprehensive overview of higher order continua and phase field models.

Chapters "Multiscale Dislocation-Based Plasticity" and "Statistical Theory of Dislocation" are dedicated to dislocation plasticity. In chapter "Multiscale Dislocation-Based Plasticity" (*Zbib*), computational discrete dislocation mechanics and its relationship to continuum computational models are reviewed. In chapter "Statistical Theory of Dislocation", *Istvan Groma* brings the power of modern statistical mechanics to bear on discrete dislocation dynamics, with the aim of formulating a continuum model.

In chapter "Granular Materials: Micromechanical Approaches of Model Systems", *Jean-Noël Roux* gives a comprehensive review of computational experiments on granular materials (discrete element method) and the advances in understanding of granular deformation/flow that follows from such experiments.

In chapter "Multiscale Modeling of Interfaces, Dislocations, and Dislocation Field Plasticity", *David McDowell* provides a tour de force through the variety of length and time scales from first principle models to macroscale continua, with a broad scope of applications.

The editors are indebted to the CISM for funding and organizing the lectures and to the guest lecturers for their contributions.

Pullman, WA, USA S. Dj. Mesarovic
Pullman, WA, USA H.M. Zbib
Evry, France S. Forest

# Contents

# Physical Foundations of Mesoscale Continua

**Sinisa Dj. Mesarovic**

**Abstract** Assume that the physics on the microscale (interactions between atoms, molecules, defects in crystals . . . ) is understood. What is the appropriate mesoscale continuum theory for the problem? What are the assumptions involved and how do they define the limitations of the continuum model? To answer these questions, we begin with the definition of mesoscale continuum kinematics from the microscale kinematics. The geometry of micro-structure (e.g., order vs. disorder) has a decisive role in defining the continuum kinematics. We thus arrive at three kinematic formulations: mass continuum, lattice continuum and granular continuum. Then, upon formulating the power balance, we use the principle of virtual power to arrive at a variety of mathematical formulations: simple continuum with moving boundaries, phase field formulation, and, a higher order, size-dependent continuum. The problems considered include: mixing of fluids and capillary flows, granular flow/deformation, and, polycrystalline diffusional/dislocation creep accompanied by dislocation plasticity.

## 1 Introduction

Continuum mechanics is an old discipline beginning (arguably) with Euler's contributions in the eighteenth century (Euler 1757), and much of its fundamentals developed in the nineteenth century by, among others, Cauchy and Navier. The first part of the twentieth century has been marked by the adoption of tensor analysis formalism and development of the thermodynamic foundations of constitutive

S. Dj. Mesarovic (✉)
School of Mechanical and Materials Engineering, Washington State University, Pullman, WA, USA
e-mail: smesarovic@wsu.edu

 
S. Mesarovic et al. (eds.), *Mesoscale Models*, CISM International Centre for Mechanical Sciences 587, https://doi.org/10.1007/978-3-319-94186-8_1

theories.[1] Important developments in the last half century include: computational methods (e.g. Hughes 2000), treatment of defects (e.g., cracks and dislocations) as discontinuities,[2] expansion of constitutive theories for simple continua (e.g., Rice 1971), rapid progress in polar, size-dependent continua motivated by experimental observations, further alignment with Hamiltonian mechanics through the concept of configurational forces (Gurtin 1995), and—most importantly for the present lectures—a growing need to connect the continuum models to models relevant at finer scales (atomistic, dislocations, particles), motivated by an increasing ability to manipulate matter on small scales and the availability of computational power.

Continua are approximations to the discrete structure of matter and necessarily involve loss of information about the details of motion and interaction between discrete components. This is particularly important in mesoscale continua which are often expected to interface with microscale models, so that a precise definition of continuum variables is required to formulate boundary conditions. Consequently, the mathematical structure of a mesoscale continuum is an open question; the possibilities of a non-simple continuum or additional independent variables arising from the analysis, must be considered.

In this set of lectures, we aim to provide an algorithm for formulating a mesoscale continuum from a well-understood microscale physics. We begin with kinematics in Sect. 2, by addressing the fundamental question: What is the material (point, element)? Or, in mathematical terms, what is it that moves with the continuum velocity field? We first consider the standard *mass* continuum using the example of two-component fluid flow (Sect. 2.1). The motion of crystalline solids is better described by the *lattice* continuum (Sect. 2.2), which, in case of dislocation-mediated deformation, is naturally extended to a size-dependent version (Sect. 2.3). Finally, the continuum kinematics arising for dense granular matter is very specific (Sect. 2.4), although loose granular matter behaves as fluid and is better described by the mass continuum.

In Sect. 3, we address thermodynamics and governing equations for a set of representative problems. To keep the derivations simple, we avoid heat transfer issues by limiting the discussion to isothermal processes, whereby any heat generated by the deformation/flow is assumed to be transported instantaneously out of the system. The method employed for deriving governing equations is the principle of virtual power (Germain 1973a, b). This method emphasizes the importance of a rigorous definition of kinematic variables. The proper kinematic analysis, aided by the thermodynamic restrictions to constitutive models, fully determines the mathematical structure of the problem. "…it provides the shortest way to obtain the required equations and to avoid any ambiguity or error." (Germain 1973b). In Sect. 3.1, the kinematics of fluid mixing is used to develop the phase field model for capillary flows with diffusive motion of the triple line. In Sect. 3.2, the

[1]E.g., Truesdell and Toupin (1960), Eringen (1962).

[2]Eshelby (1956), Rice (1968), Hirth and Lothe (1992).

lattice continuum kinematics is used to derive the sharp interface formulation for diffusional and dislocation creep. Finally, a size dependent theory of dislocation plasticity is derived in Sect. 3.3, on the basis of the kinematics of dislocation pile-ups, discussed previously. Where appropriate, short paragraphs suggesting future research directions are included.

It is assumed that the reader is equipped with standard continuum mechanics concepts at the level of a first graduate course (e.g., Malvern 1969). Specifically, familiarity with the following concepts is expected: tensor analysis, Eulerian (material derivative, transport theorem) and Lagrangean (deformation gradient) kinematics, basic thermodynamics, and, simplest constitutive laws (elasticity and viscosity). We will limit the discussion to Cartesian systems, so that the gradient operator can be treated formally as a vector, and the comma in index notation can be used for partial derivatives without ambiguity. The dyadic product is given without any operation symbol. For example, a vector field $\mathbf{u}(\mathbf{x})$ has left and right gradients and curls:

$$(\nabla \mathbf{u})_{ij} = u_{j},_{i}; \quad (\mathbf{u}\nabla)_{ij} = u_{i},_{j}; \quad (\nabla \times \mathbf{u})_{m} = \in_{mij} u_{j},_{i} = -(\mathbf{u} \times \nabla)_{m}.$$

# 2 Kinematics

The first step in formulating a continuum theory is the definition of primary kinematic variables. While the simplest problems require only velocity or displacement vector fields, additional independent kinematic variables may be required (diffusional fluxes, slip fields, phase fields...). We start with the velocity: At instant $t$, *material* point (element) currently at the position $\mathbf{x}$ is moving with the velocity $\mathbf{v}(\mathbf{x}, t)$. But what is the physical meaning of the *material*? In this section, we provide three different answers to this question for three continuum theories: mass continuum, lattice continuum, and, granular continuum. We also illustrate definitions of additional primary variables: diffusional fluxes and plastic slip and climb rates.

## 2.1 *Mass Continuum: Mixing of Fluids*

Even in a single component fluid, the material point (element) cannot be identified with any specific set of atoms. As illustrated in Fig. 1, the continuous mixing of atoms in a flow prevents any such definition. Instead, the material is identified with mass density field. Such continuum will be called *mass continuum*. Specifically, the mass density field $\rho(\mathbf{x}, t)$ is convected with the velocity $\mathbf{w}(\mathbf{x}, t)$.

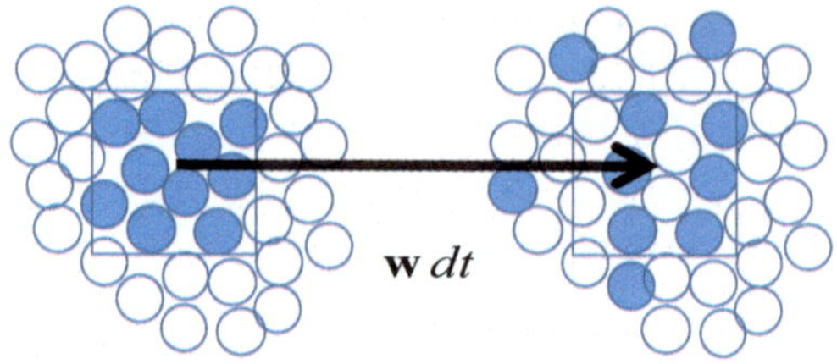


**Fig. 1** Uniform velocity flow of a single component fluid. Atoms belonging to the material element at time $t$ are colored. At time $t + dt$, the material element contains a different set of atoms

Consider now a mix of two fluids.[3] In the current instant, two species, $A$ and $B$, occupy an elementary volume $V$. The respective masses in $V$ are $M^A$ and $M^B$, and the average velocities[4] of respective atoms are $\mathbf{w}^A$ and $\mathbf{w}^B$. We define the partial mass densities $\rho^A$ and $\rho^B$, such that their sum is the total mass density $\rho$:

$$\rho = \frac{M}{V} = \frac{M^A}{V} + \frac{M^B}{V} = \rho^A + \rho^B. \tag{1}$$

The continuum velocity field $\mathbf{w}(\mathbf{x})$ is defined so that the total momentum in unit volume is $\rho\mathbf{w}$ (barycentric velocity):

$$\rho\mathbf{w} = \rho^A\mathbf{w}^A + \rho^B\mathbf{w}^B. \tag{2}$$

From the mass balance of the two species,

$$\frac{\partial\rho^{A(B)}}{\partial t} + \nabla\left(\rho^{A(B)}\mathbf{w}^{A(B)}\right) = 0, \tag{3}$$

we derive the mass balance for the mix:

$$\frac{D\rho}{Dt} = -\rho\nabla\cdot\mathbf{w}, \tag{4}$$

where $D/Dt$ represents the material derivative. For a material field $Y(\mathbf{x}, t)$ (in the present case, a field given per unit mass):

$$\frac{DY}{Dt} = \frac{\partial Y}{\partial t} + \mathbf{w}\cdot\nabla Y. \tag{5}$$

[3]Here, we follow the simplest mixture theory outlined by Lowengrub and Truskinovsky (1998) with additional clarifications given by Dehsara et al. (2017). For a more elaborate models of mixing, see Atkin and Crane (1976) and Joseph and Renardy (1993).

[4]The assumption underlying any continuum formulation is the separability of thermal motion of atoms and atomic drift, which includes convection (velocity) and mixing (diffusional flux).

The transport theorem for the material volume $V(t)$, can be written as

$$\frac{D}{Dt}\int\limits_{V(t)} \rho Y dV = \int\limits_{V(t)} \rho \frac{DY}{Dt} dV. \tag{6}$$

We assume that pure components $A$ and $B$ are incompressible and have different mass densities. Upon defining the mass fraction of $A$: $\phi = \rho^A/\rho$, the mass density in an isothermal process is a function of the mass fraction only: $\rho = \rho(\phi)$. The mass flux of $A$ is then

$$\mathbf{q} = \rho^A \left(\mathbf{w}^A - \mathbf{w}\right). \tag{7}$$

The diffusion law then follows (Dehsara et al. 2017):

$$\rho \frac{D\phi}{Dt} = -\nabla \cdot \mathbf{q}. \tag{8}$$

The above discussion is effectively a simple change of variables. We have started with two vector and two scalar fields $\{\mathbf{w}^{A(B)}, \rho^{A(B)}\}$, related through the two balance laws (Eq. 3), then replaced them with different pairs of vector and scalar fields $\{\mathbf{w}, \mathbf{q}, \rho, \phi\}$, related through the balance laws (Eqs. 4 and 8). While the choice of the barycentric velocity ensures that the linear momentum density of the mix is simply $\rho\mathbf{w}$, the kinetic energy density is more complicated:

$$\frac{1}{2}\rho \mathbf{w} \cdot \mathbf{w} + \frac{1}{2} \frac{1}{\rho\phi\,(1-\phi)} \mathbf{q} \cdot \mathbf{q}. \tag{9}$$

The second term (kinetic energy of mixing) is typically assumed to be negligibly small. This is certainly justified in the context of phase field models for immiscible fluids, where $\mathbf{q} \neq 0$ only in a thin interface layer.

The kinematic variables $\{\mathbf{w}, \mathbf{q}, \rho, \phi\}$ are not independent. The additional constraint follows from the functional dependence $\rho = \rho(\phi)$:

$$\frac{D\rho}{Dt} = \frac{d\rho}{d\phi}\frac{D\phi}{Dt} = -\rho \nabla \cdot \mathbf{w} = -\frac{1}{\rho}\frac{d\rho}{d\phi} \nabla \cdot \mathbf{q}. \tag{10}$$

Although the pure components are incompressible, enforcement of incompressibility for the mix ($\nabla \cdot \mathbf{w} = 0$) would effectively freeze any diffusion. Thus, this definition of continuum kinematics does not support incompressible flow.

In the phase field models of immiscible fluids, mixing occurs only in the thin layer (the phase field representation of the interface between the two phases). In such case, the *compositionally compressible*[5] kinematics developed above results in

[5]Lowengrub and Truskinovsky (1998) use the expression *quasi-incompressible*.

an unnecessarily complicated mathematical model. Moreover, it has been shown (Dehsara et al. 2017) that the compositional compressibility of the thin layer is inconsistent with incompressibility of the rest of the domain, as any change of interface layer density requires motion of a large mass of pure (incompressible) components. We will address that question in more detail in Sect. 3.1. At the moment, we remain focused on kinematics and seek an alternative definition of continuum velocity which will allow the convenient assumption of incompressibility.

**Incompressible flow as an approximation to compositionally compressible flow** Upon rewriting Eq. (10) as

$$\nabla \cdot \mathbf{w} - \frac{1}{\rho^2}\frac{d\rho}{d\phi}\nabla \cdot \mathbf{q} = 0, \tag{11}$$

it is evident that the mixing law $\rho(\phi)$ such that

$$\frac{1}{\rho^2}\frac{d\rho}{d\phi} = \alpha = \text{const.}, \tag{12}$$

will allow definition of alternative continuum velocity field

$$\mathbf{v}(\mathbf{x}) = \mathbf{w}(\mathbf{x}) - \alpha\mathbf{q}(\mathbf{x}), \tag{13}$$

which is divergence-free, i.e., satisfies the "incompressibility" condition:

$$\nabla \cdot \mathbf{v} = 0. \tag{14}$$

If mixing/diffusion is limited to a thin interface region $I$, then in the reminder of the volume $V$:

$$\mathbf{v}(\mathbf{x}) \equiv \mathbf{w}(\mathbf{x}); \qquad \text{for } \mathbf{x} \in V\backslash I. \tag{15}$$

Thus, the error of this approximation can be made arbitrarily small for a sufficiently small thickness of the phase field interface.

Specifically, the mixing law requiring that atomic volume of a component remains the same in pure component and in the mix,

$$\frac{1}{\rho(\phi)} = \frac{\phi}{\rho^{A0}} + \frac{1-\phi}{\rho^{B0}}; \qquad \phi \in [0,1], \tag{16}$$

where $\rho^{A0}$ and $\rho^{B0}$ are the mass densities of pure components, satisfies Eq. (12):

$$\frac{1}{\rho^2}\frac{d\rho}{d\phi} = \frac{\rho^{A0} - \rho^{B0}}{\rho^{A0}\rho^{B0}} = \alpha = \text{const.} \tag{17}$$

Upon replacing the mass fraction $\phi$ with the new phase variable $c$, such that

$$\rho(c) = c\rho^{A0} + (1-c)\,\rho^{B0}; \qquad c \in [0, 1]\,, \tag{18}$$

we obtain the replacement for diffusion equation (8):

$$\frac{Dc}{Dt} = -\nabla \cdot \mathbf{J}, \tag{19}$$

where the new flux vector is related to the old one as

$$\mathbf{J} = \frac{\rho}{\rho_{0A}\rho_{0B}}\mathbf{q}, \tag{20}$$

and the material derivative now has the meaning different from the one implied in Eq. (5):

$$\frac{DY}{Dt} = \frac{\partial Y}{\partial t} + \mathbf{v} \cdot \nabla Y. \tag{21}$$

Details of the derivation are given in Dehsara et al. (2017).

To complete the structure of the new (approximate) continuum, the mass density $\rho$ should be convected with the new velocity $\mathbf{v}$:

$$\frac{D\rho}{Dt} = -\rho\nabla \cdot \mathbf{v} = 0. \tag{22}$$

Clearly, this new continuum has an intrinsic inconsistency; Eq. (22) is in direct contradiction to Eq. (18), which implies

$$\frac{D\rho}{Dt} = \frac{d\rho}{dc}\frac{Dc}{Dt} \neq 0. \tag{23}$$

Nevertheless, a mathematically consistent model can be formulated if proper care is taken in derivation, as we will show in Sect. 3.1.

Finally, we note that all of the above refers to the Eulerian (spatial) description of the continuum. The failure to identify the material with a physical entity renders any Lagrangean (material) description meaningless.

## 2.2 *Lattice Continuum: Diffusion, Dislocations, Lattice Growth*

In solid mechanics problems Lagrangean kinematics is essential. In crystalline solids, the relevant deformation mechanisms are referred to the crystalline lattice.

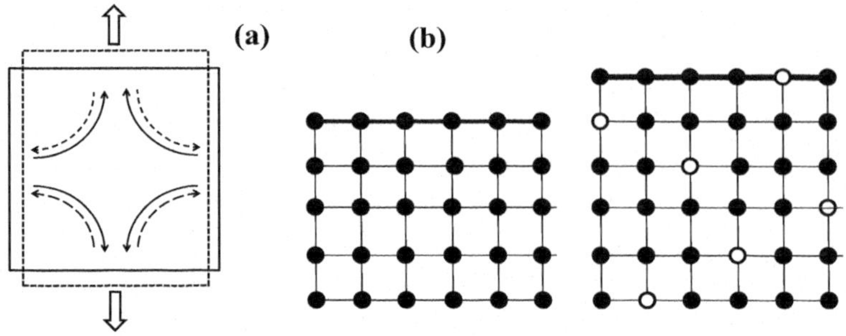


**Fig. 2** (**a**) Schematic representation of vacancy diffusion (dashed arrows) and atom diffusion (solid arrows) within a crystalline grain leading to the creation of new lattice planes at the boundary under tension and disappearance of lattice planes at other boundaries. (**b**) Schematic illustration of the creation of a new lattice plane. Thick solid line represents crystal boundary, solid and hollow circles are atoms and vacancies. Initial vacancy-free crystal (left) absorbs vacancies from the boundary. The new configuration (right) has an extra lattice plane

For example, plastic slip is produced by the motion of dislocations with respect to the lattice, while elastic deformation is associated with lattice stretching. Thus, the standard crystal elasto-plasticity[6] is, in fact, based on the *lattice continuum* but without diffusion. In this case, the lattice continuum is equivalent to the mass continuum but with material element identified with specific atoms at their lattice positions, so that the distinction between lattice and mass is unnecessary.

In the solid mechanics problems with diffusion, the importance of crystalline lattice was noted early by Larché and Cahn (1973, 1978, 1985) but in the form of lattice constraint imposed on the otherwise mass-based continuum. Berdichevsky et al. (1997) noted the absence of Lagrangean description in the mass continuum. Elements of the lattice continuum were introduced by Garikipati et al. (2001), but it was only recently that full formulation of the lattice continuum was developed for Nabarro-Herring creep (Mesarovic 2016) and dislocation creep (Mesarovic 2017).

At high temperatures, vacancy diffusion takes place by the vacancy-atom exchange mechanism, thus producing the complementary fluxes of atoms and vacancies (Fig. 2). Crystal boundaries serve as sources and sinks of vacancies resulting in lattice growth/loss on the boundary faces under tension/compression.

**Continuity equation and the transport theorem** We consider two species occupying lattice sites: atoms and vacancies. At any instant $t$, the lattice velocity field $\mathbf{v}(\mathbf{x}, t)$ is defined in the crystal domain. The lattice density (number of lattice sites per unit volume) $N(\mathbf{x}, t)$ represents the material; it is convected with continuum velocity $\mathbf{v}(\mathbf{x}, t)$, and a material field $Y(\mathbf{x}, t)$ is now given per lattice site. The material derivative has the same form as before but its physical meaning has changed; it now represents the time derivative following the lattice.

[6]Lee (1969), Hill and Havner (1982), Asaro (1983), Bassani (1994).

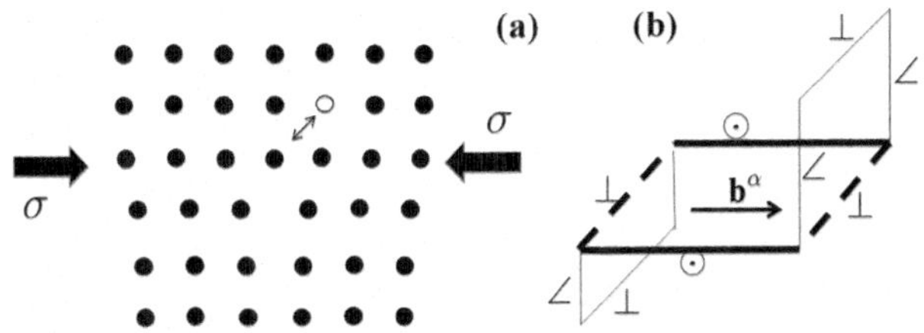


**Fig. 3** (**a**) Edge dislocation and vacancy (hollow circle). The vacancy-atom exchange takes place. Dislocation climbs and a lattice point disappears. (**b**) Illustration of climb starting from a planar dislocation loop with the Burgers vector $\mathbf{b}^{\alpha}$ (thick lines). The screw components (⊙) don't climb. The climb of initial edges (⊥) produces parallel edges as well as out-of-plane edge segments (∠) which cannot glide. As both parallel and out-of-plane edges continue to climb, the loop expands and operates as climb analogue of the Frank-Read source

The mechanism of dislocation climb, illustrated in Fig. 3a, consists of vacancy absorption by the dislocation core. As the vacancy is absorbed, the core of edge dislocation moves one atomic space and one lattice position is lost. Thus, the dislocation climb mechanism represents the lattice sink, as well as the vacancy sink. The continuity equation, i.e., the lattice balance law, must now include—not only the convection of lattice sites, but also the local sink strength. Let $\dot{s}\,(\mathbf{x}, t)$ be the fraction of lattice sites lost per unit time. The lattice site balance law is then:

$$\frac{DN}{Dt} = \frac{\partial N}{\partial t} + \mathbf{v} \cdot \nabla N = -N\nabla \cdot \mathbf{v} - N\dot{s}. \tag{24}$$

The transport theorem with lattice sink must also take into account the loss of lattice sites. Let the quantity $Y(\mathbf{x}, t)$ be given per lattice site. Consider the lattice volume $V(t)$, i.e., the volume that follows the prescribed set of lattice sites. The transport theorem has the following form:

$$\frac{D}{Dt}\int\limits_{V(t)} NY dV = \int\limits_{V(t)} N\left[\frac{DY}{Dt} - \dot{s}Y\right] dV. \tag{25}$$

**Climb and glide** The slip system$\alpha$ is described by a triad of unit orthogonal vectors: the slip direction $\mathbf{s}^{\alpha} = \mathbf{b}^{\alpha}/\mid \mathbf{b}^{\alpha}\mid$, the slip plane normal $\mathbf{m}^{\alpha}$, and $\mathbf{t}^{\alpha} = \mathbf{s}^{\alpha} \times \mathbf{m}^{\alpha}$. The evolution of climbing edge dislocations, starting from a rectangular dislocation loop on the slip system $\alpha$ is illustrated in Fig. 3b. Evidently, the climb of original edges produces, not only the parallel edge segments (⊥), but also the out-of-plane edge segments (∠) which cannot glide. Further climb by all edges results in a multiplication mechanism analogous to the Frank-Read source for glide (Hirth and Lothe 1992). Let $\theta^{\alpha}$ be the density (length per volume) of all edge dislocations (⊥ and ∠) associated with the slip system $\alpha$, and let $v^{C\alpha}$ be the average climb velocity for edges belonging to the slip system. Then, in analogy with the Orowan equation for the slip rate, the strength of the lattice sink associated with the slip system $\alpha$ is:

$\dot{s}^{\alpha} = \theta^{\alpha} v^{C\alpha} \mid \mathbf{b}^{\alpha} \mid$, and total lattice sink strength is obtained by summation over all slip systems:

$$\dot{s} = \sum\nolimits_{\alpha} \dot{s}^{\alpha}. \tag{26}$$

When dislocation climb operates, we must distinguish between the *true slip* field $\hat{\gamma}^{\alpha}(\mathbf{x}, t)$, which is the one related to dislocation glide through the Orowan equation and on which the power associated with dislocation glide is expanded on, by the Peierls (1940)–Nabarro (1948) stress, and the *apparent slip* field $\gamma^{\alpha}(\mathbf{x}, t)$, which corresponds to the geometric intuition. This is illustrated in Fig. 4. The true slip field is the slip occurring without any climb. The loss of lattice planes parallel to the slip plane affects the apparent slip. The relation between true slip rate $\dot{\hat{\gamma}}^{\alpha}$ and apparent slip rate $\dot{\gamma}^{\alpha}$ is (Mesarovic 2017):

$$\dot{\gamma}^{\alpha} - \dot{\hat{\gamma}}^{\alpha} = \gamma^{\alpha} \sum\nolimits_{\beta} \phi_{\alpha\beta}^{2} \dot{s}^{\beta}; \quad \phi_{\alpha\beta} = \mathbf{m}^{\alpha} \cdot \mathbf{s}^{\beta}. \tag{27}$$

**Reference configuration and elastic-plastic decomposition** The Eulerian kinematics is formally similar to the kinematics of mass continuum, except that the velocity field is now the lattice velocity. The deformation rate tensor $\mathbf{D}(\mathbf{x}, t)$ is the symmetric part of velocity gradient $\mathbf{L}(\mathbf{x}, t)$:

$$\mathbf{L} = \mathbf{v}\nabla; \quad L_{ij} = \partial v_i / \partial x_j = v_{i,j}; \quad \mathbf{D} = \frac{1}{2}(\mathbf{v}\nabla + \nabla\mathbf{v}). \tag{28}$$

When diffusion is present, Lagrangean description poses a challenge (Berdichevsky et al. 1997), but only if one insists that the reference configuration represents real physical configuration (i.e., the initial configuration). But that is not necessary. The reference configuration can be fictitious, as long as the difference between the current and reference configurations accurately represents the relevant components of deformation. To that end, we start from the current configuration

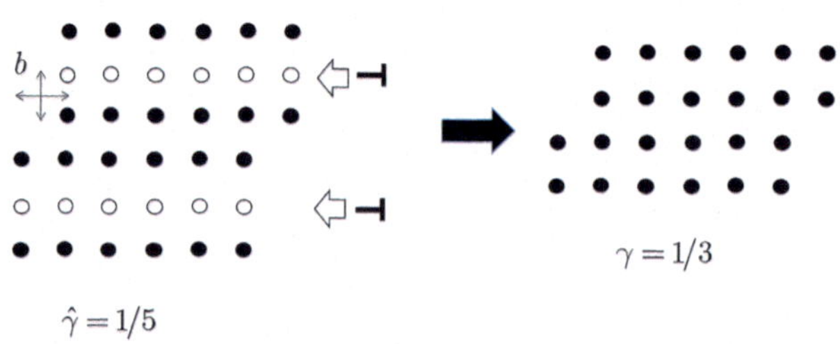


**Fig. 4** **Left:** Initial lattice volume $5b \times 5b$ is traversed by an edge dislocation resulting in the true slip $\hat{\gamma} = 1/5$. Two edge dislocations on an orthogonal plane are poised to climb (arrows) and destroy the lattice planes indicated by hollow circles. **Right:** After the climb, the apparent slip is $\gamma = 1/3$

and define the reference configuration suitable for elastic-plastic decomposition, as follows.

In addition the lattice density $N(\mathbf{x}, t)$ and the vacancy concentration $c(\mathbf{x}, t)$, the material (lattice) point currently located at $\mathbf{x}$, carries the deformation gradient $\mathbf{F}(\mathbf{x}, t)$, which, for the existing lattice, evolves as

$$\frac{D}{Dt}\mathbf{F} = \mathbf{L} \cdot \mathbf{F}. \tag{29}$$

To extend the definition of lattice-advected variables to the lattice being currently created at a boundary, it is sufficient to assert the continuity of deformation gradient, vacancy concentration and lattice density, within the crystal. The key elements of the Lagrangean description are thus defined. The reference configuration $\mathbf{X}$ can, in principle, be obtained from the current configuration, but it will be fictitious for the lattice grown during the process. For an assembly of grains, the mapping $\mathbf{x} \rightarrow \mathbf{X}$ is multiple-valued, with gaps/overlaps at grain boundaries, as illustrated in Fig. 5.

The deformation gradient tensor can be decomposed into the plastic deformation gradient $\mathbf{F}^p$ and the elastic-compositional deformation gradient $\mathbf{F}^*$. The decomposition is multiplicative and formally identical to the standard (glide-only) elastic-plastic decomposition (Lee 1969):

$$\mathbf{F} = \mathbf{F}^* \cdot \mathbf{F}^p. \tag{30}$$

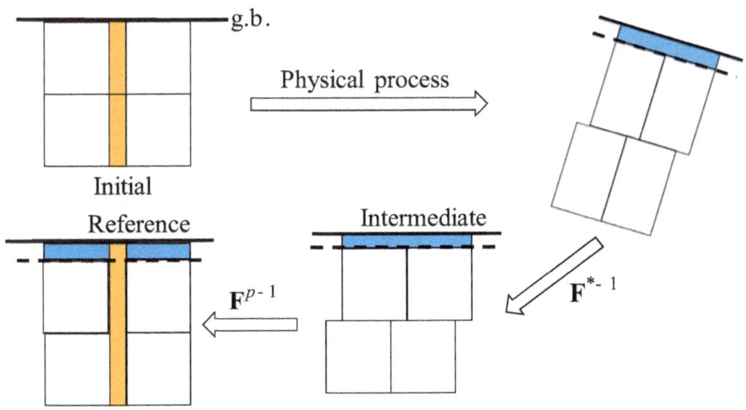


**Fig. 5** Illustration of elastic-plastic decomposition when the reference configuration is not the initial configuration. The initial element (upper left) at the grain boundary (g.b.) is transformed into the current configuration (upper right). The physical deformation process includes: loss of lattice planes through dislocation climb (yellow), dislocation glide along the horizontal pane, creation of new lattice planes at the boundary (blue), elastic-compositional lattice stretching, and, rotation. The reference configuration is defined by inverse mappings, which include all deformation mechanisms except creation/destruction of lattice planes at the grain boundary

We envision a sequential application of the two gradients (Fig. 5). First, the plastic deformation gradient is applied to the reference configuration to produce the intermediate (isoclinic) configuration $\mathbf{F}^p : \mathbf{X} \rightarrow \tilde{\mathbf{x}}$. In contrast to standard crystal plasticity based on dislocation glide, here $F^p = \det \mathbf{F}^p \neq 1$; lattice planes are lost/created, resulting in the translational motion of the parts of the lattice, but lattice symmetries and orientation are preserved. Then, the intermediate configuration is transformed into the current configuration, $\mathbf{F}^* : \tilde{\mathbf{x}} \rightarrow \mathbf{x}$. The elastic-compositional deformation gradients accounts for lattice stretching (compositional and elastic strains) and rotation, but the connectivity of lattice points is preserved.

The Eulerian deformation rates are additive:

$$\mathbf{L} = \mathbf{L}^* + \mathbf{L}^p = \dot{\mathbf{F}}^* \cdot \mathbf{F}^{*-1} + \mathbf{F}^* \cdot \tilde{\mathbf{L}}^p \cdot \mathbf{F}^{*-1}; \qquad \tilde{\mathbf{L}}^p = \dot{\mathbf{F}}^p \cdot \mathbf{F}^{p-1}. \tag{31}$$

The plastic velocity gradient on intermediate configuration $\tilde{\mathbf{L}}^p$ accounts for changes in lattice topology and is defined from the elementary slip system state variables: true slip rates[7] $\dot{\hat{\gamma}}^\alpha$ and climb rates $\dot{s}^\alpha$:

$$\tilde{\mathbf{L}}^p = \sum_\alpha \dot{\hat{\gamma}}^\alpha \tilde{\mathbf{s}}^\alpha \tilde{\mathbf{m}}^\alpha - \sum_\alpha \dot{s}^\alpha \tilde{\mathbf{s}}^\alpha \tilde{\mathbf{s}}^\alpha; \qquad \mathrm{tr}\tilde{\mathbf{L}}^p = -\dot{s}, \tag{32}$$

where the vectors $\tilde{\mathbf{s}}^\alpha$ and $\tilde{\mathbf{m}}^\alpha$ are unit and orthogonal in the intermediate configuration.[8] We note that neither $\mathbf{F}^*$ and $\mathbf{F}^p$, nor $\mathbf{L}^*$ and $\mathbf{L}^p$, are compatible. While they are assumed continuous within the grain, they are not gradients of single-valued vector fields. Their incompatibility will be the key component of the size-dependent crystal continuum discussed in the next subsection.

Let the lattice density and elementary volume in intermediate configuration be $\tilde{N}$ and $d\tilde{\Omega}$. Then, elastic-compositional changes between intermediate and current configurations can be expressed as

$$F^* N = \tilde{N}; \quad d\Omega = F^* d\tilde{\Omega}; \quad F^* = \det \mathbf{F}^*. \tag{33}$$

For any material field, we write $Y(\mathbf{x}) = Y(\tilde{\mathbf{x}}) \; (\mathbf{F}^* : \tilde{\mathbf{x}} \rightarrow \mathbf{x})$, and switch the domains of integration between the current and intermediate configurations:

$$\int_\Omega Y(\mathbf{x})\, d\Omega = \int_{\tilde{\Omega}} F^* Y(\tilde{\mathbf{x}})\, d\tilde{\Omega}. \tag{34}$$

**Diffusional creep rate tensor, diffusion and mass balance** The lattice velocity field $\mathbf{v}(\mathbf{x}, t)$ is defined on an open domain $\Omega$ (crystalline grain), its closure $\partial\Omega$ (grain boundary) having the outer normal $\mathbf{n}$. The growth/disappearance of lattice planes at

[7]In Mesarovic (2017), the apparent slip rate is erroneously used in Eq. (32). Later, in thermodynamics analysis, this error produces an additional dissipation term.

[8]The choice between intermediate and current configurations as the basis for definition of slip has been extensively discussed in literature (Hill and Havner 1982; Asaro 1983; Bassani 1994).

the boundary requires an additional independent scalar field: the normal boundary velocity $V^n$, defined on the closure set $\partial\Omega$. Then, the boundary lattice growth rate is

$$g = V^n - \mathbf{n} \cdot \mathbf{v}. \tag{35}$$

In contrast to the locally-defined (fields defined at each point) deformation rates (elastic-compositional and plastic), the diffusional creep deformation rate is defined as constant in each grain domain $\Omega$, from the boundary lattice growth rate[9]:

$$\mathbf{C} = \frac{1}{\Omega} \int_{\partial\Omega} g\mathbf{nn} d\partial\Omega. \tag{36}$$

The total average deformation rate for the grain $\Omega$ can then be defined as

$$\langle \mathbf{D} \rangle_\Omega^{total} = \langle \mathbf{D} \rangle_\Omega^{intrinsic} + \mathbf{C}, \tag{37}$$

where the intrinsic deformation rate is

$$\langle \mathbf{D} \rangle_\Omega^{intrinsic} = \frac{1}{\Omega} \int_\Omega \mathbf{D} d\Omega = \frac{1}{2\Omega} \int_{\partial\Omega} (\mathbf{vn} + \mathbf{nv}) \, d\partial\Omega. \tag{38}$$

Let the mass of an atom be $m$. The mass density is $\rho = mN(1 - c)$. If the average velocity of atoms and vacancies are $\mathbf{v}^a$ and $\mathbf{v}^v$, the vacancy flux can be defined as

$$\mathbf{J} = -\mathbf{J}^{atom} = -\frac{1 - c}{F^*} \left( \mathbf{v}^a - \mathbf{v} \right) = \frac{c}{F^*} \left( \mathbf{v}^v - \mathbf{v} \right). \tag{39}$$

The local mass balance results in the vacancy diffusion equation:

$$\frac{Dc}{Dt} = -F^* \nabla \cdot \mathbf{J} - (1 - c) \, \dot{s}. \tag{40}$$

The transport theorem for a variable crystal domain $\Omega(t)$, with lattice growing/vanishing at the boundary $\partial\Omega(t)$ with the rate $g$, will have the form:

$$\frac{D}{Dt} \int\limits_{\Omega(t)} NY d\Omega = \int\limits_{\Omega(t)} N \left[ \frac{DY}{Dt} - \dot{s}Y \right] d\Omega + \int\limits_{\partial\Omega(t)} NYg d\partial\Omega. \tag{41}$$

Applying Eq. (41) to the mass density, the mass of a crystal grain will change as

$$\frac{D}{Dt} \int\limits_{\Omega(t)} \rho d\Omega = m\tilde{N} \int\limits_{\partial\Omega(t)} \left[ \frac{1 - c}{F^*} g + J_n \right] d\partial\Omega; \quad J_n = \mathbf{n} \cdot \mathbf{J}. \tag{42}$$

[9]For boundary representation of the average strain in a domain, see Mesarovic and Padbidri (2005).

Thus, mass conservation for the grain $\Omega$ requires:

$$\int_{\partial\Omega} \left[ \frac{1-c}{F^*} g + J_n \right] d\partial\Omega = 0. \tag{43}$$

The weak constraint (43) requires a model for boundary diffusion. In the absence of boundary diffusion, the stronger, local condition:

$$\frac{1-c}{F^*} g + J_n = 0, \tag{44}$$

requires that lattice growth/disappearance at the boundary be directly related to the normal vacancy flux. The creep deformation rate (Eq. 36) can then be written as

$$\mathbf{C} = -\frac{1}{\Omega} \int_{\partial\Omega} \frac{F^* J_n}{1-c} \mathbf{nn} d\partial\Omega. \tag{45}$$

## 2.3 *Linearized Kinematics, Size-Dependence*

Observations of size effects in plasticity of metals and Ashby's (1970) qualitative explanation of size effects arising from accumulation (pile-up) of dislocations at the obstacles, have motivated development of numerous phenomenological theories,[10] which, together with the early works,[11] contributed much to the understanding of the possible mathematical structures of size-dependent continua.

Size-dependent continuum requires a material length scale. Investigations of the physical nature of that length scale are less numerous that the proposed size-dependent formulations. Groma et al. (2006) consider the case of completely disordered set of dislocation in 2D; they associate the material length scale with the average spacing between dislocations. Mesarovic et al. (2010) consider the perfectly ordered configurations—periodically stacked dislocation pile-ups, and conclude that the material length scale is the stacking distance, i.e., the distance between the active slip planes.

In the absence of diffusion (lattice growth on the boundaries, climb, lattice sinks), distinction between lattice and mass continua is lost. Moreover, the resulting material element consist of atoms, leading to the direct interpretation of the reference configuration.

**Linearized kinematics of a crystal** As before, dislocation glide in the slip system $\alpha$ is described by the continuum field $\gamma^\alpha(\mathbf{x}, t)$. Without climb, the true and apparent

[10] E.g., Fleck and Hutchinson (1993, 1997), Gurtin (2002).

[11] Kondo (1952), Nye (1953), Kröner (1958), Mindlin (1964), Eringen and Suhubi (1964).

slips (Eq. 27) are identical. To simplify the mathematical developments, we limit the discussion to the linearized kinematics, whereby the geometric distinction between reference, intermediate and current configuration is lost. The linearized analogue of the plastic velocity gradient is the slip tensor:

$$\boldsymbol{\gamma}(\mathbf{x}, t) = \sum\nolimits_{\alpha} \gamma^{\alpha} \mathbf{s}^{\alpha} \mathbf{m}^{\alpha}. \tag{46}$$

Linearization results in the additive decomposition of displacement gradients:

$$\mathbf{F} = (\mathbf{I} + \mathbf{e} + \boldsymbol{\omega}) \cdot (\mathbf{I} + \boldsymbol{\gamma}) \approx \mathbf{I} + \mathbf{e} + \boldsymbol{\omega} + \boldsymbol{\gamma}; \qquad \mathbf{u}\nabla = \mathbf{e} + \boldsymbol{\omega} + \boldsymbol{\gamma}, \tag{47}$$

where $\mathbf{e}$ is the elastic strain, while the plastic strain tensor is the symmetric part of the slip tensor $(\boldsymbol{\gamma} + \boldsymbol{\gamma}^T)/2$. The small rotation tensor is $\boldsymbol{\omega} + (\boldsymbol{\gamma} - \boldsymbol{\gamma}^T)/2$. As before, the tensors $(\mathbf{e} + \boldsymbol{\omega})$ and $\boldsymbol{\gamma}$ are incompatible. Kröner's (1958) incompatibility tensor is based on elastic strains:

$$\boldsymbol{\eta} = \mathrm{Inc}(\mathbf{e}) \equiv \nabla \times \mathbf{e} \times \nabla; \qquad \eta_{ij} = \epsilon_{ipm} \epsilon_{jnq} e_{mn,pq}. \tag{48}$$

The Nye's (1953) dislocation density tensor, $\mathbf{A}$, quantifies the accumulation of dislocations, i.e. the non-vanishing net Burgers vector in the volume element:

$$\mathbf{A} = -\boldsymbol{\gamma} \times \nabla; \qquad A_{ij} = -\epsilon_{jkp} \gamma_{ik,p}. \tag{49}$$

The in-plane components of the slip gradient define the partial densities of edge ($\perp$) and screw ($\odot$) dislocations:

$$g_{\perp}^{\alpha} = \mathbf{s}^{\alpha} \cdot \nabla \gamma^{\alpha}; \qquad g_{\odot}^{\alpha} = \mathbf{t}^{\alpha} \cdot \nabla \gamma^{\alpha}; \qquad \mathbf{g}^{\alpha} = g_{\odot}^{\alpha} \mathbf{t}^{\alpha} + g_{\perp}^{\alpha} \mathbf{s}^{\alpha}. \tag{50}$$

Nye's dislocation density tensor can be defined from the partial densities:

$$\mathbf{A}(\mathbf{x}, t) = -\sum\nolimits_{\alpha} \mathbf{s}^{\alpha} \left( \mathbf{m}^{\alpha} \times \mathbf{g}^{\alpha} \right) = \sum\nolimits_{\alpha} \left( -g_{\odot}^{\alpha} \mathbf{s}^{\alpha} \mathbf{s}^{\alpha} + g_{\perp}^{\alpha} \mathbf{s}^{\alpha} \mathbf{t}^{\alpha} \right), \tag{51}$$

where the summation is over all currently or previously active slip systems. The two representations of incompatibility are related through

$$\boldsymbol{\eta} = \frac{1}{2} \left[ (\nabla \times \mathbf{A}) + (\nabla \times \mathbf{A})^T \right]. \tag{52}$$

The field $\mathbf{A}(\mathbf{x}, t)$ can be discontinuous and can represent an assembly of discrete dislocations.[12] For example, consider the thin film configuration under shear, illustrated in Fig. 6, with active slip planes stacked periodically at distance $h$.

[12]Indeed, many of the solutions to discrete dislocation problems (Hirth and Lothe 1992) have been obtained using Kröner's (1958) formalism.

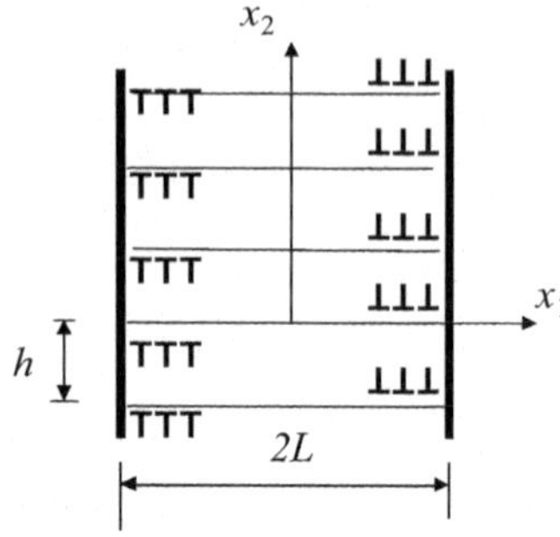


**Fig. 6** Elastic-plastic thin film with one slip system, sandwiched between elastic half-spaces. Active slip planes, orthogonal to the boundary, are periodically stacked to infinity. Dislocations pile-up against the boundaries

The relevant Nye's tensor component can represent discrete, semi-discrete and continuous fields:

$$\begin{aligned}
&\hat{A}_{13}(x_1, x_2) = -b \sum_j \sum_{k=-\infty}^{\infty} \operatorname{sign}\left(x_1^j\right) \delta\left(x_1^j\right) \delta(x_2 - kh);\\
&\overline{A}_{13}(x_1, x_2) = B(x_1) \sum_{k=-\infty}^{\infty} \delta(x_2 - kh);\\
&A_{13}(x_1) = B(x_1)/h,
\end{aligned} \tag{53}$$

where $b$ is the magnitude of the Burgers vector of individual dislocations, $\delta(\cdot)$ is the Dirac delta function, and $B(x_1)$ is the continuum representation of Burgers vectors within one slip plane.

The key argument for the physical nature of length scale is thermodynamic. The length scale must be embedded into a size-dependent energy, which is the result of elastic dislocation interactions, and is therefore represented by a quadratic form of $\mathbf{g}^\alpha$ (or, a quadratic form of $\mathbf{A}$, under certain conditions). However, it is easily shown[13] that all the components of energy corresponding to the continuous fields $\gamma^\alpha$ and $\mathbf{g}^\alpha$ (and hence, a continuous $\mathbf{A}$) are already accounted for in the energy of the simple continuum, which has no length scale. The size-dependent energy is the error in energy arising from the kinematic coarsening illustrated in Eq. (53). Baskaran et al. (2010) showed that the energy error arising from in-plane coarsening ($\hat{\mathbf{A}} \to \overline{\mathbf{A}}$) is much smaller than the one arising from the next step ($\overline{\mathbf{A}} \to \mathbf{A}$), which is characterized by the length scale $h$.

While the all the kinematics discussed above equally applies to the simple and the size-dependent elastic-plastic crystal continua, the role of the slip fields in the mathematical structure of the problem changes. This will be discussed in the thermodynamics section (3.3). An inconvenient result of the kinematic coarsening exemplified in Eq. (53) is that the gradient of slip fields are singular at the boundaries. The order of singularity is the same as for crack tip singularity; dislocation pile-ups are mathematically equivalent to cracks (mode II for edges and mode III for screws), as shown by Fleck (1991). This singularity is particularly

[13]Mesarovic et al. (2010, 2015), Mesarovic (2010). Detailed discussion of the absence of length scale in the energy associated with continuous $\mathbf{A}$ is given in Mesarovic (2005).

inconvenient for the continuum description of dislocation-boundary interactions. Moreover, computational treatment of a singular continuum field requires high mesh density, which defeats the purpose of a continuum theory. An approximation is required such that the field is smooth inside the crystal, and the discontinuities are placed at the boundaries. Such approximation is illustrated in Fig. 7, where singular slip gradient at the boundary is approximated by a finite jump in slip. The slowly-varying slip distribution in the interior remains. Similar approximation was used by Yassar et al. (2007) to model inclusion hardening in metals.

To replace the singular portion of dislocation densities $\mathbf{g}^\alpha$ by the appropriate boundary values of slip jumps at the boundary, the geometry at the intersection of slip plane and the boundary must be defined. Let $\mathbf{n}$ be the outer unit normal to the boundary (Fig. 8). The slip plane is bounded by the intersection line with the boundary characterized by the unit tangent $\boldsymbol{\xi}^\alpha$ and unit outer in-plane normal $\mathbf{N}^\alpha$.

The geometry is easily expressed in terms of the system triad $(\mathbf{s}^\alpha, \mathbf{m}^\alpha, \mathbf{t}^\alpha)$ and the surface normal $\mathbf{n}$, by defining the slip system projection operator $\mathbf{P}^\alpha = \mathbf{I} - \mathbf{m}^\alpha\mathbf{m}^\alpha$, which extracts the in-plane component of an arbitrary vector:

$$| \mathbf{P}^\alpha \cdot \mathbf{n} |= \cos\theta^\alpha \neq 0; \quad \mathbf{N}^\alpha \cos\theta^\alpha = \mathbf{P}^\alpha \cdot \mathbf{n} = \cos\varphi^\alpha \mathbf{s}^\alpha + \sin\varphi^\alpha \mathbf{t}^\alpha, \tag{54}$$

where $\theta^\alpha$ is the angle between the trace normal $\mathbf{N}^\alpha$ and the surface normal $\mathbf{n}$, while $\varphi^\alpha$ determines edge/screw nature of dislocations piled-up against the boundary.

Densely packed dislocations at the boundary are represented by the step in the slip field, or equivalently, the Dirac delta function in the slip gradient field. In the discrete representation, this signifies a wall of super-dislocations at the boundary.

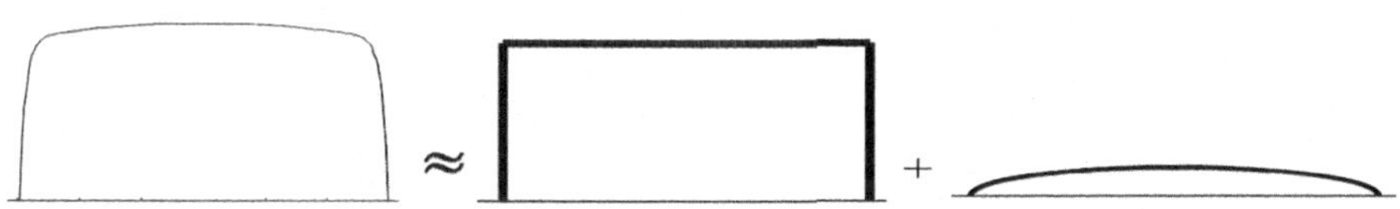


**Fig. 7** **Left:** Typical slip distribution between impenetrable boundaries using the semi-discrete problem formulation. At the boundaries, the slip gradient is singular. **Right:** Approximation to the actual slip distribution. Singular gradients at the boundaries are approximated by jumps (step functions), so that the approximate slip field consists of slowly varying portion with non-vanishing boundary values

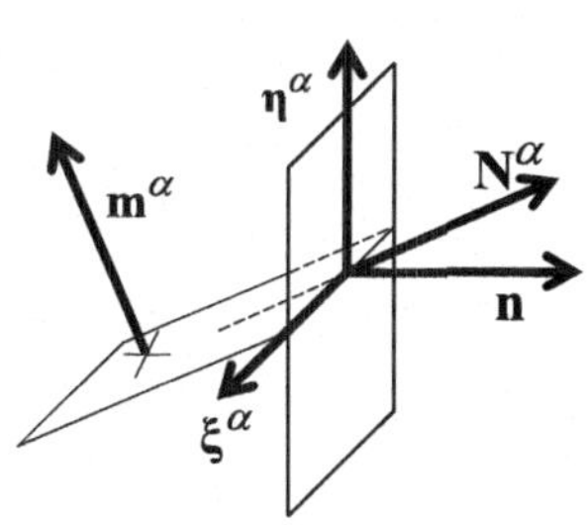


**Fig. 8** Geometry of an interface intersected by a slip plane

The slip step at the boundary is numerically equal to the negative value of the slip as the boundary is approached from the inside of the crystal: $\Delta\gamma^{\alpha} = -\gamma^{\alpha}(\partial V^{-})$. It represents the line density of Burgers vector, given per unit length of in the direction $h^{\alpha}$ (Fig. 8, orthogonal to the boundary-slip plane intersection):

$$\mathbf{B}^{\alpha} = \Delta\gamma^{\alpha}\cos\theta^{\alpha}\mathbf{s}^{\alpha}. \tag{55}$$

The factor $\cos\theta^{\alpha}$ arises from the relation between the normal spacing $h^{\alpha}$ between the stacked slip planes and the spacing of boundary-slip plane intersections. To compute the energy, we need the description that distinguishes between screw and edge pile-ups:

$$\mathbf{\Gamma}^{\alpha} = \mathbf{B}^{\alpha}\cdot\left(\mathbf{s}^{\alpha}\mathbf{N}^{\alpha}\right) = \Delta\gamma^{\alpha}\cos\theta^{\alpha}\mathbf{N}^{\alpha}. \tag{56}$$

With such kinematic description, the coarsening error in energy can be represented in terms of continuum fields $\mathbf{g}^{\alpha}$, defined on the open set $V$ (the bulk of the crystal), and $\mathbf{\Gamma}^{\alpha}$, defined on the closure $\partial V$. To formalize the relationship between the two fields, defining the surface Dirac delta function $\delta(\partial V)$ as infinite on $\partial V$, vanishing elsewhere, its line integral across $\partial V$ being equal 1. The definition of the slip gradient can be formally extended to the closed set $\overline{V} = V \cup \partial V$:

$$\mathbf{g}^{\alpha}\left(\overline{V}\right) = \mathbf{g}^{\alpha}(V) + \delta\left(\partial V\right)\mathbf{\Gamma}^{\alpha}. \tag{57}$$

Note that while $\mathbf{g}^{\alpha}$ have dimensions (length)$^{-1}$, the boundary fields $\mathbf{\Gamma}^{\alpha}$, $\mathbf{B}^{\alpha}$ are non-dimensional.

The characteristic lengths are the (average) spacing of slip planes $h^{\alpha}$. These lengths evolves during deformation; dislocations spread to new parallel slip planes by means of the double cross-slip mechanism. The initial values for an annealed crystal are large, of the order of density of Frank-Read sources. As dislocations spread onto new planes, the average spacing will be weighted average of two lengths: the initial spacing of sources and the characteristic double cross-slip distance, the latter being the saturation value. Multiplicity of characteristic lengths is particularly important at the boundaries, where different slip system interact, as illustrated in Fig. 9.

**Linearized kinematics of grain boundaries** The dislocation-boundary relaxation represents an important component of polycrystalline deformation. Dislocations arriving at the boundary may[14]: (i) promote nucleation of, or react with, dislocations on the opposite or the same side of the boundary, or, (ii) decompose into mobile and sessile boundary dislocation without interacting with any other system. If there are mobile products, the boundary plane acts as a slip plane. The sessile products form

[14]Poulat et al. (1998, 2001), Priester (2001), Spearot et al. (2007), Tschopp and McDowell (2008), Guleryuz and Mesarovic (2016).

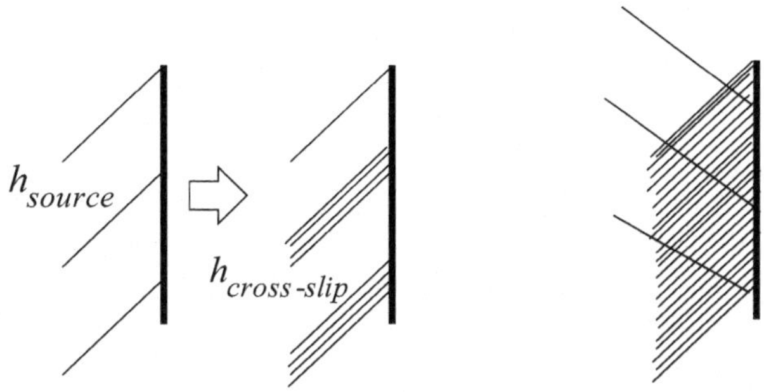


**Fig. 9** **Left:** Evolution of characteristic length on the primary slip system. **Right:** The secondary and primary slip systems may have very different characteristic lengths

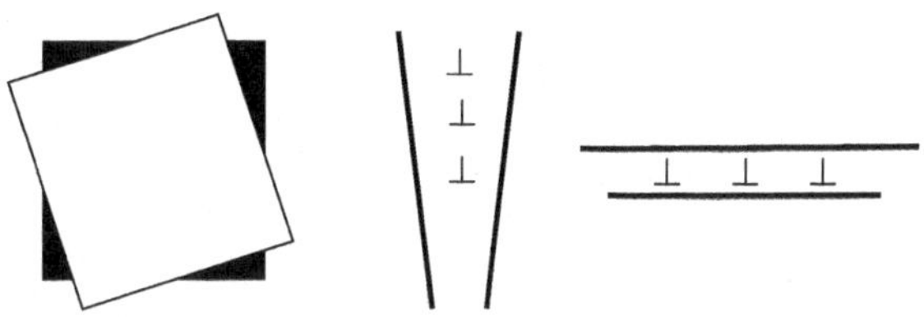

**Fig. 10** Effect of additional intrinsic boundary dislocations created by relaxation of pile-ups from the two adjacent crystals on the initially matching surface elements. **Left:** Parallel screws create additional twist. **Center:** Parallel edges with the Burgers vector normal to the interface create additional tilt. **Right:** Parallel edges with the Burgers vector in the boundary plane create mismatch strains

a network of intrinsic boundary dislocations effectively changing the tilt angle, twist angle and the mismatch of the boundary (Fig. 10).

The dislocation wall at the boundary, characterized by dual representation $\mathbf{\Gamma}^\alpha$ and $\mathbf{B}^\alpha$, partially relaxes, so that we can decompose the vectors into the unrelaxed $\overline{\mathbf{\Gamma}}^\alpha$ and relaxed $\hat{\mathbf{\Gamma}}^\alpha$ parts:

$$\mathbf{\Gamma}^\alpha = \overline{\mathbf{\Gamma}}^\alpha + \hat{\mathbf{\Gamma}}^\alpha; \qquad \mathbf{B}^\alpha = \overline{\mathbf{B}}^\alpha + \hat{\mathbf{B}}^\alpha. \tag{58}$$

To understand the distinction between the unrelaxed $\overline{\mathbf{B}}^\alpha$ and relaxed $\hat{\mathbf{B}}^\alpha$ Burgers vector density, it may be helpful to consider the unrelaxed part as being placed infinitesimally close to the boundary, but still inside the crystal. On the other hand, the relaxed Burgers vector $\hat{\mathbf{B}}^\alpha$ changes the nature of the grain boundary. While the geometry of interfaces, albeit complex, is well understood (Sutton and Balluffi 1995), their plasticity, including sliding and relaxation is, at present, an open area of research. Nevertheless, we can at least identify the unknowns and sketch the shape of the future theory, which awaits the solution to the interface relaxation problem.

Consider the interface $k$ with the unit normal $\mathbf{n}^k$ and the crystal into which the normal is pointing denoted by $+$ and the other by $-$. Let the Burgers vector density of created intrinsic mobile interface dislocations be $\mathbf{B}^{k,\,mob}$. The relaxed Burgers vectors must add up to the net interface Burgers vector:

$$\mathbf{B}^{k} + \mathbf{B}^{k,mob} = \sum\nolimits_{\alpha(-)} \hat{\mathbf{B}}^{\alpha} - \sum\nolimits_{\alpha(+)} \hat{\mathbf{B}}^{\alpha}. \tag{59}$$

The remaining, sessile part of intrinsic Burgers vector density must decompose into twist, tilt and two mismatch components representing the two principal mismatch strains, one of which is shown in Fig. 10:

$$\mathbf{B}^{k} = \mathbf{B}^{k,tws} + \mathbf{B}^{k,tlt} + \mathbf{B}^{k,m1} + \mathbf{B}^{k,m2}. \tag{60}$$

At present, a general algorithm for decompositions embodied in Eqs. (59) and (60) does not exist; Eq. (59) requires knowledge of slip direction for the particular boundary, which is only known for few special boundaries, while Eq. (60) requires detailed knowledge of which slip systems on the two sides of the interface interact, which is, in principle, feasible for specific geometries, but no general rule is available. However, once decompositions (Eqs. 59 and 60) are accomplished, we can write the boundary conditions for the two crystals in terms of in-plane displacement gradients, as follows.

Define the in-plane projection operator for the boundary $k$ and the in-plane gradient operator: $\mathbf{P}^{k} = \mathbf{I} - \mathbf{n}^{k}\mathbf{n}^{k}$ and $\nabla^{k} = \mathbf{P}^{k} \cdot \nabla$. Let the displacement discontinuity at the boundary be $\Delta\mathbf{u}^{k} = \mathbf{u}^{+} - \mathbf{u}^{-}$. Then,

$$\begin{aligned} \mathbf{B}^{k,tlt} \times \boldsymbol{\xi}^{k} &= \nabla^{k}\left(\mathbf{n}^{k} \cdot \Delta\mathbf{u}^{k}\right); \\ \mid \mathbf{B}^{k,tws} \mid \mathbf{n}^{k} &= \nabla^{k} \times \left(\mathbf{P}^{k} \cdot \Delta\mathbf{u}^{k}\right); \\ \mathbf{B}^{k,m1}\mathbf{B}^{k,m2} &= \tfrac{1}{2}\left[\left(\mathbf{P}^{k} \cdot \Delta\mathbf{u}^{k}\right)\nabla^{k} + \nabla^{k}\left(\mathbf{P}^{k} \cdot \Delta\mathbf{u}^{k}\right)\right], \end{aligned} \tag{61}$$

where $\boldsymbol{\xi}^{k}$ is the dislocation line direction (unit vector) associated with $\mathbf{B}^{k,\,tlt}$, i.e., the tilt axes, obtained as part of decomposition (Eq. 60).

The utility of dual representation $(\boldsymbol{\Gamma}^{\alpha}, \mathbf{B}^{\alpha})$ should now be apparent from Eqs. (57 and 61). $\boldsymbol{\Gamma}^{\alpha}$ is convenient for kinematic analysis inside the crystal; $\mathbf{B}^{\alpha}$ is convenient for formulation of kinematic boundary conditions for the crystal grain.

## 2.4 Granular Continuum

Dense granular materials exhibit peculiar mechanical behavior: (i) When sheared under constant pressure, material can dilate or compact, depending on the initial porosity and pressure, until it reaches the *critical state* (Schofield and Wroth 1968), whereupon the material shears at constant volume, shear rate and shear stress, effectively transitioning from solid state to fluid state, (ii) Localization into persistent shear bands with apparently invariant characteristic width (10–20 average particle diameters) is ubiquitous,[15] and, (iii) Vortex flow appears in the bulk and in the shear bands, at very low Reynolds numbers.[16]

Some granular phenomena also characterize different material systems: dilatancy and glass transition in simple and polymer fluids and shear localization in solidifying metals.[17]

Particles in a granular assembly are disordered as atoms in fluids or amorphous solids, but their motion is heavily constrained. They differ from atoms in the following important aspects: (1) They do not possess the equivalent of thermal atomic vibration; any vibrational energy is quickly dissipated, (2) The particle motion is constrained by friction, (3) The interactions are strictly local: contact forces between nearest neighbors, (4) The normal interactions forces are either strictly compressive, or predominantly compressive, and, (5) A granular assembly typically includes a range of particle sizes and shapes. For these reasons the granular mechanics has been one of the most intractable problems that has resisted efforts of engineers, physicist and mathematicians.[18]

The collective behavior in a granular assembly arises on different length scales. Understanding of such behavior requires multiple levels of modeling. Here, we outline the first level of continuum models, defined on the scale of particle size, but lacking continuous velocity gradients.

The geometry of the granular assembly is described using the elements of graph theory (Satake 1993; Bagi 1996). For an assembly of identical spherical particles, the first step is the standard Voronoi tessellation.[19] The particles sharing the face of a Voronoi cell are nearest neighbors. The dual graph is the Delaunay graph (Fig. 11) obtained by connecting the centers of nearest neighbors. Not all nearest neighbors are in contact, i.e., the set of contacts is a subset of the Delaunay set. Delaunay cells are simplexes (tetrahedra in 3D, triangles in 2D).

Bagi (1996) generalized the tessellation to assembly of arbitrarily shaped dissimilar convex particles by defining the dual graphs: material and space cells. The material cell of a particle consist of points whose distance to the (closest point of the) particle is shorter than the distance to any other particle. The material graphs are polyhedral whose faces are not necessarily flat. The dual graph (space cells) is obtained by connecting the pairs of particle centers which share a face of a material cell. As before, the lines of the space graph connect nearest neighbors which may not be in contact, and space cells are simplexes. For non-spherical

[15]Rechenmacher (2006), Alshibli and Sture (1999), Saada et al. (1999), Finno et al. (1997).

[16]Tordesillas et al. (2008), Abedi et al. (2012), Radjai and Roux (2002).

[17]Klein and Kumacheva (1995), Zou et al. (2009), Gourlay and Dahl (2007).

[18]The difficulty of the problem is perhaps best illustrated by its longevity; the questions that puzzle us today are not very different from those considered by Coulomb (1773) and Reynolds (1885).

[19]The algorithm for Voronoi tessellation: (1) connect all particle centers with lines, (2) bisect each line with a plane, (3) for each particle select the smallest enclosing polyhedron, this is the Voronoi cell. Voronoi cells cover the space. For unequal spheres, Dirichlet tessellation (Bagi 1996) produces analogous results.

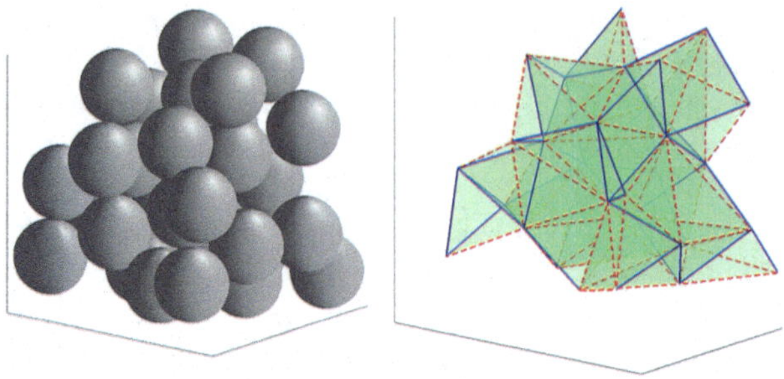

**Fig. 11** An assembly of spherical particles and its Delaunay graph connecting nearest neighbors. Blue, solid lines represent contacts. Red, dashed lines connect nearest neighbors which are not in contact

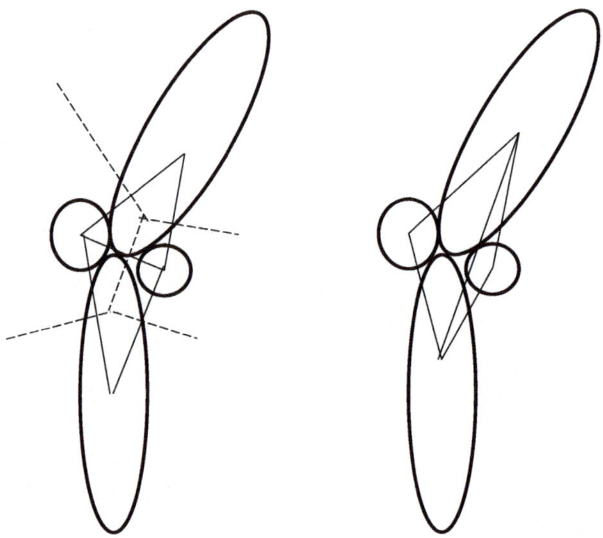

**Fig. 12** 2D assembly of two elongated and two circular particles. **Left:** Voronoi tessellation (dashed lines) and Delaunay graph (solid lines). The Delaunay graph does not include the contact between the two elongated particles. **Right:** Space cell graph includes all contacts

particles, Delaunay graph produced from Voronoi (or Dirichlet) tessellations will not always include true contacts as its subset, but Bagi's space cell graph will, as illustrated in Fig. 12.

Let the velocities of the centers of four particles forming the space cell be $\mathbf{v}^k$ $(k = 1, 2, 3, 4)$. The complete description requires specification of angular velocities $\boldsymbol{\omega}^k$ $(k = 1, 2, 3, 4)$. In dynamic deformation/flow, the rotational inertia is negligible compared to the translational inertia (Padbidri and Mesarovic 2011); in quasi-static[20] deformation the rotation is a secondary degree of freedom induced by the intergranular friction. The frictional resistance to particle rearrangement is minimized if the particles roll over each other instead of sliding,[21] but the rolling is constrained in a dense packing. This has been suggested as the physical mechanisms which determines the shear band width (Padbidri et al. 2012).

If the velocities of particles are known, then linear interpolation within each cell produces a continuous velocity field. Within each cell

$$\hat{\mathbf{v}}(\mathbf{x}, t) = \mathbf{v}^0(t) + \sum_{k=1}^{4} \hat{\mathbf{L}}(t) \cdot \mathbf{x}. \tag{62}$$

The 12 unknown scalar components of $\mathbf{v}^0(t)$ and $\hat{\mathbf{L}}(t)$ can be obtained by solving 12 scalar equations: $\hat{\mathbf{v}}\left(\mathbf{x}^k, t\right) = \mathbf{v}^k$ $(k = 1, 2, 3, 4)$. The vector field $\mathbf{v}^*(\mathbf{x}, t)$, obtained by concatenating the fields $\hat{\mathbf{v}}(\mathbf{x}, t)$ for each cell, is $C^0$ continuous[22] throughout the domain. The velocity gradient field,

$$\mathbf{L}^*(\mathbf{x}, t) = \mathbf{v}^* \nabla, \tag{63}$$

is piecewise constant, with discontinuities across cell boundaries.

Thus far, we have considered a fixed topology of the space cell graph; nearest neighbors remain nearest neighbors throughout the deformation, at any time the space cell (or Delaunay) graph is topologically equivalent to the initial one. This limits the analysis to very small strains (Roux and Combe 2010), which are of little interest. To describe the larger strains we must consider the mechanism of topological changes in the space cells (or Delaunay) graph.

For a simplex graph, there are few generic mechanisms for topological changes—*flips*, two in 3D, and one in 2D (Edelsbrunner 2000). The 2-2 flip in 2D is illustrated in Fig. 13. The generic 2-3 flip in 3D transforms two tetrahedra into three; the 3-2 flip is the reverse transformation. The flipping cluster (two or three cells) gradually deforms in an interval $\Delta t$ to reach the intermediate configuration in Fig. 13a. This is an unstable configuration and the reminder of the flip is dynamic and requires negligible time (Fig. 13b).

The distribution of contact forces in a granular material is strongly non-homogeneous (Drescher and De Josselin de Jong 1972), with forces organized in force chains. This implies strong inhomogeneities in the elastic strain energy (typically, we consider normal mutual indentation of particles as elastic). Flips represent relaxation of highly stressed contacts; at the end of the flip we expect

[20]In the literature on granular materials, the word quasi-static is used with different meanings. The inertia number (GDR MiDi 2004, da Cruz et al. 2005) distinguishes between solid and fluid phases of a granular material, while the quasi-static coefficient (Padbidri and Mesarovic 2011) follows the standard continuum mechanics meaning.

[21]Oda et al. (1982), Oda and Kazama (1998), Kuhn and Bagi (2004).

[22]$C^0$ means that the field is continuous but its gradient is not. This field is identical to the one produced by linear tetrahedral (constant strain) finite elements.

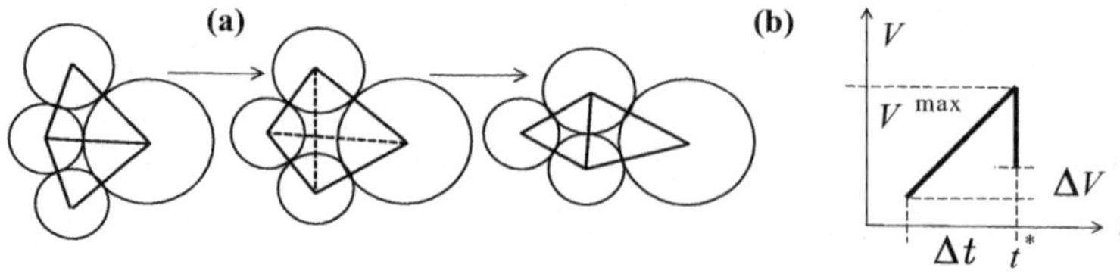


**Fig. 13** (**a**) The generic 2-2 flip in 2D. (**b**) Volume of the pair of flipping cells as function of time

$\Delta V > 0$ in Fig. 13b. Simulations indicate that the dilatancy is the result of both, the relaxed configuration at the end of the flip, and the increasing rate of flipping in a unit volume, since each flipping cluster spend most of its time on the raising part of the curve in Fig. 13b (Mesarovic et al. 2012).

Owing to intergranular friction, deformation of a dense granular material is never strictly elastic; the frictional sliding between grains (and thus, dissipation) is always present. In this context, it is opportune to decompose the deformation on the basis of topology of the space cell graph into *isotopologic* (no flips) and *heterotopologic* (flips). The velocity gradient defined above is isotopologic. It is defined only between the flips but not in the instants $t^*$ when flips occur. The total velocity gradient is the sum of the isotopologic (*) and heterotopologic ($H$) parts:

$$\mathbf{L}(\mathbf{x}, t) = \mathbf{L}^*(\mathbf{x}, t) + \mathbf{L}^H(\mathbf{x}, t) \tag{64}$$

The total deformation gradient is $C^0$ in space but discontinuous in time; the heterotopologic part $\mathbf{L}^H$is represented by a sequence of Dirac delta functions in time.

**Future directions** At present, the question of a useful Lagrangean representation of the granular deformation remains open. We have seen earlier (lattice continuum) that the reference configuration need not correspond to any physical configuration. We can also handle multiple-valued mappings (elastic-plastic decomposition).

The strongly non-homogeneous distribution of elastic energy implies that unloading of a granular assembly will necessarily include flipping of cells. If Lagrangean description is to be useful, it should, at least describe elasticity. One possibility is to forgo the Lagrangean description of heterotopologic deformation, which is in any case lost when transition to fluid state occurs. The isotopologic deformation can then be measured with respect to the intermediate configuration which is defined on the cell-by-cell basis (unloading of a cell). The inverse mapping (current to intermediate) is multiple valued (unladed cells will overlap).

Although the formal definition of time-discontinuous velocity gradients has been proven useful in explaining the mechanisms of dilatancy, the next step must be the coarsening and definition of a smooth velocity gradient field. As in the case of size-dependent plasticity, the kinematic analysis is likely to be inseparable from thermodynamic arguments. Possibly, the critical step may be the definition of variable(s) which describes the non-homogeneity of the strain energy density.

The transition from solid to fluid behavior has been modeled as a phase field phenomenologically (Aranson and Tsimring 2005), but the components of the phase field model are not understood on the particle interaction level. The physical mechanisms controlling the intrinsic length scale have been proposed (Tordesillas 2007; Padbidri et al. 2012), but it is not yet clear how they should enter the continuum theory.

# 3 Thermodynamics and Governing Equations

## *3.1 Phase Field Models for Capillary Flows*

Flows controlled by capillary forces have long been a subject of interest (de Gennes 1985; Bonn et al. 2009) as a fundamental physical problem, as well as related to industrial and natural processes, including: materials joining,[23] painting (Patton 1979) and coating (Marrion 2004), biological phenomena such as super-hydrophilic leafs in plant biology (Koch et al. 2008), and geology such as molten lava flows (Lezzi and Ventura 2005).

**Phase field models** Sharp interface models applied to capillary flows face two key problems. The first is the singularity of the solutions to fluid dynamics equations with no-slip boundary condition in the vicinity of the triple line (Durbin 1988; Dussan et al. 1991). The second is computational treatment of topological discontinuities (breakup and coalescence of domains) within the interface tracking algorithm. The phase field (or diffuse interface) models naturally accommodate the solution to both problems. Multiple materials in multiple domains are treated as single material/domain with phases distinguished by the value of the phase field variable, which removes the need for interface tracking algorithms. The phase field model allows for distinction between convective (slip) and diffusive motion of the triple line. While the convective motion is still subject to the no-slip boundary condition, the contact line moves by surface diffusion mechanism, as indicated by molecular dynamics studies (Thompson and Robins 1990; Koplik and Banavar 1998) and molecular kinetic theory (Blake and Haynes 1969; Seveno et al. 2009). Based on Seppecher's (1996) analysis, the mathematical formulation of the diffusive boundary condition has been developed by Jacqmin (2000); it was implemented into the finite element framework for incompressible flow (Yue and Feng 2011) and for compositionally compressible flow (Dehsara et al. 2017).

The interface between immiscible fluids is represented by a conserved phase field leading to a nonlinear, fourth order diffusion equation (Cahn and Hilliard 1958).[24]

[23]Liu and Sekulic (2011), Eustathopoulos et al. (2013), Fu et al. (2016).

[24]Both conserved and non-conserved phase fields (Cahn 1961), applied to fluids and solids, have been subject of review articles and monographs (Anderson et al. 1998; Chen 2002; Emmerich 2003; Provatas and Elder 2010).

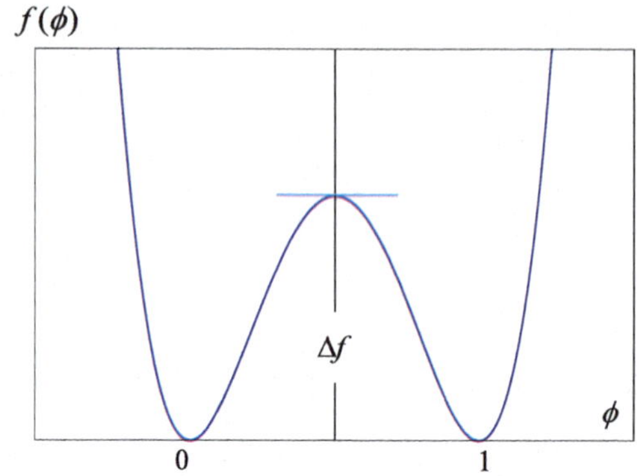


**Fig. 14** Free energy density of a homogeneous fluid as function of the phase variable

It is useful to think of the phase field formulation as the mathematical technique for representing a multi-domain problem with moving interfaces as a single domain. Therefore, the parameters of the phase field model must be firmly rooted in the sharp interface thermodynamics, in the sense of Gibbs (1878) excess quantities, and the solutions to the phase field equations must converge to the sharp interface solutions in the limit of vanishing interface thickness. The key components of a phase field model are as follows.

The energy density $f(\phi)$ (or specific energy, if given per unit mass) of a homogeneous fluid is function of the phase field parameter $\phi$, which takes values 0 and 1 in the two phases. It is a double-well potential with minima corresponding to the two phases (Fig. 14).

Although few systematic studies are available, experience and intuition suggests that the detailed shape of the double-well potential is of little importance as long as the key parameter $\Delta f$ is fixed ($\phi$will be different from 0 or 1 only in a thin interface layer). The free energy density (or specific energy if given per unit mass) of a non-homogeneous fluid is written as

$$\omega\left(\phi, \nabla\phi\right) = f\left(\phi\right) + \frac{1}{2}\kappa(\nabla\phi)^2. \tag{65}$$

The parameter $\kappa$ and $\Delta f$ are directly related to the interface energy—a physical parameter, and, to the width of the diffuse interface—a computational parameter chosen with some latitude.

**The paradox of the compositionally compressible phase field** Much of fluid mechanics has been developed for incompressible fluids, and with a good reason; with notable exceptions (large temperature changes, sound waves, liquid-gas phase change, polymeric fluids), the volumetric deformation rates are often negligible compared to the deviatoric ones. Capillary flow of immiscible fluids is not an exception. Here, too, we assume that the two phase are incompressible. However, the kinematic analysis in Sect. 2.1 leads to the conclusion that compositional compressibility is inevitable, although it is manifested (in the context if phase field

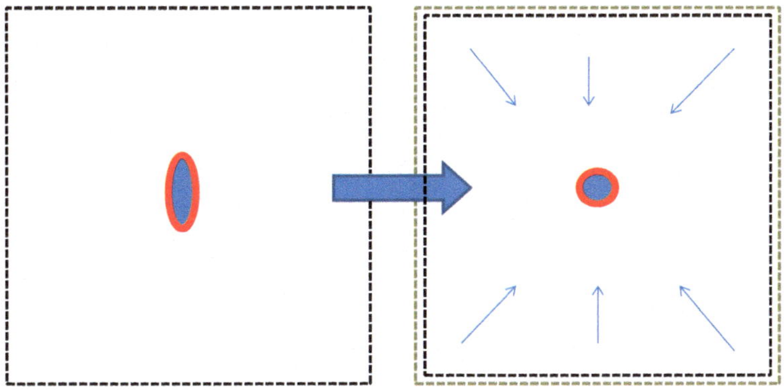

**Fig. 15** Ellipsoidal interface evolves into a spherical one and changes its volume

models) only in a thin interface region. This is then reflected in thermodynamic analysis (Lowengrub and Truskinovsky 1998). The compositional compressibility of a thin interface layer, combined with incompressibility of pure phases, produces a paradox best illustrated on an example. Consider a droplet (Fig. 15) of a blue fluid immersed into the white fluid and evolving from initial (ellipsoidal) shape to the equilibrium (spherical) shape. The interface layer (red) is compositionally compressible, while the pure phases (white and blue) are incompressible. Any change of volume (say, shrinking) in the small region around and including the interface layer requires (inward) displacement of a large mass of white fluid (arrows). This is neither physical (large masses of fluid do not move because interfaces change shape), nor computationally tractable (consider periodic or rigid boundary conditions).

On the other hand, to keep the interface layer at constant volume, any change in interface area will have to be accompanied by the corresponding change in interface thickness. However, the shape of the phase field across the interface (and hence the interface thickness) is strongly constrained by Cahn-Hilliard equations, although it can be somewhat changed by introducing elasticity (Asle-Zaeem et al. 2011).

In computational treatments of fluids with different mass densities with compositionally compressible model, the results exhibit anomalous bands of velocity divergence (Guo et al. 2014; Dehsara et al. 2017), shadowing the interface but somewhat removed from it, and persisting for a wide range of mesh densities. At present, it is not clear if these effects represent a true (albeit non-physical) solution to the mathematical problem, or, are they simply the spurious solutions to the discrete numerical problem. The existence results are only available for the incompressible flow (Abels 2009).

In the end, it appears that in the context of phase field models, compositional compressibility is an unnecessary complication; it makes a difference only in a negligibly thin layer (which itself is a mathematical construct), but creates numerous theoretical and computational difficulties.[25] In Sect. 2.1, an alternative, divergence-free velocity field is derived as an approximation to the rigorous kinematics of mixing. This velocity field can be used as the basis for continuum kinematics to derive the governing equations for incompressible capillary flow.

**Incompressible phase field model** The independent kinematic variables are: the divergence-free velocity field $\mathbf{v}(\mathbf{x}, t)$ and the flux field $\mathbf{J}(\mathbf{x}, t)$. We consider a computational cell with volume$V$, with boundary $\partial V$ consisting of the solid boundary $S$, and fluid boundary $A$: $\partial V = S \cup A$. We will assume that on the boundary normal flux vanishes:

$$\mathbf{n} \cdot \mathbf{J} = 0 \qquad \text{on } \partial V. \tag{66}$$

This implies that no diffusion occurs across the fluid boundary, which will be the case if the phase field interfaces are orthogonal to the boundary. This will enable us to reduce the number of independent variables; instead of the vector $\mathbf{J}$, only the scalar $Dc/Dt = -\nabla \cdot \mathbf{J}$ will appear in the final equations. The fluid boundary is also subject to the usual fluid mechanics inlet/outlet conditions (Panton 1996; Ranacher 2007).

The first step is the analysis of the power balance. For an isothermal process:

$$\int_{\partial V} \mathbf{t} \cdot \mathbf{v} d\partial V = \frac{D\mathcal{P}}{Dt} + \frac{D\mathcal{K}}{Dt} + \mathcal{D}, \tag{67}$$

where $\mathcal{P}$ and $\mathcal{K}$ are the total potential and kinetic energy, and $\mathcal{D}$ is the total dissipation rate (i.e., the heat production, in isothermal processes assumed to be instantaneously taken out of the system). The power input arises only from boundary tractions $\mathbf{t}$ (gravity is a conservative force and is included in the potential energy).

*Potential energy* The total potential energy of the system can be written as

$$\mathcal{P} = \int_V [\omega(c, \nabla c) - \rho\psi(\mathbf{x})] \, dV + \int_S \gamma(c) d\partial V, \tag{68}$$

where $\omega(c, \nabla c)$ is given by Eq. (65), $\psi(\mathbf{x})$ is the gravitational potential ($\nabla\psi = \mathbf{g} = const.$), and $\gamma(c)$ is a monotonic interpolation between the values of interface energies between the solid and the two fluids, $\gamma^{SA}$ and $\gamma^{SB}$, corresponding to $c = 0$ and $c = 1$.

To derive the rate of change of potential energy, we use the transport theorem. The internal inconsistency of the model (Eqs. 22 and 23) is resolved by neglecting

[25] Here, we present the derivation of the incompressible model in some detail. For intricacies of the compositionally compressible model theory, the reader is referred to Dehsara et al. (2017).

the power expanded on $(D\rho/Dt)$.[26] The result is

$$\frac{D\mathcal{P}}{Dt} = \int_V \left\{ \mathcal{M}^0 \frac{Dc}{Dt} - \kappa\left(\nabla c \nabla c\right) : (\mathbf{v}\nabla) - \rho\mathbf{g}\cdot\mathbf{v} \right\} dV + \int_S \theta^0 \frac{Dc}{Dt} d\partial V;$$
$$\mathcal{M}^0 = \frac{df}{dc} - \kappa\nabla^2 c; \qquad \theta^0 = \frac{d\gamma}{dc} + \kappa\mathbf{n}\cdot\nabla c. \tag{69}$$

The capillary equilibrium at the solid surface, $\theta^0 = 0$, is the phase field equivalent of the Young equation for equilibrium of surface tensions at the triple line:

$$\gamma^{SB} - \gamma^{SA} = \Gamma\cos\alpha, \tag{70}$$

where $\Gamma$ is the interface energy between two fluids, and $\alpha$ is the contact angle. Equation (70) can be obtained by integrating $\theta^0 = 0$ over the interface thickness $h$, and using the fact that the equilibrium solution to the problem of flat infinite interface between fluids $A$ and $B$ gives the one-to-one relationship between $(\Gamma, h)$ and the phase field parameters $(\kappa, \Delta f)$ (Eq. 65; Fig. 14):

$$\kappa = \frac{3}{2}\Gamma h; \quad \Delta f = \frac{3}{4}\Gamma/h. \tag{71}$$

*Dissipation* Three different physical process produce dissipation. For linear isotropic viscosity, the viscous stress tensor is given as

$$\boldsymbol{\tau} = \mu(c)\left(\mathbf{v}\nabla + \nabla\mathbf{v}\right), \tag{72}$$

where $\mu(c)$ is monotonic interpolation between the viscosities of the two pure fluids. We assume that both, diffusion in the bulk and diffusion on the solid surface, are proportional to the rate of change of the phase variable:

$$\mathcal{D} = \int_V \left[\boldsymbol{\tau} : (\mathbf{v}\nabla) - \mathcal{M}\frac{Dc}{Dt}\right] dV - \int_{\partial V} \theta\frac{Dc}{Dt} d\partial V. \tag{73}$$

The bulk diffusion is subject to the conservation law (Eq. 19), so that the simplest linear constitutive law is

$$\mathbf{J} = -B\nabla\mathcal{M}; \qquad \text{in } V. \tag{74}$$

[26]The argument used here is from Dehsara et al. (2017). Different arguments, with more or less rigor, have been used to derive incompressible flow models (Hohenberg and Halperin 1977; Gurtin et al. 1996; Ding et al. 2007; Abels et al. 2012).

The surface diffusion is not subject to a conservation. We assume

$$\frac{Dc}{Dt} = -b\theta; \qquad \text{on } S. \tag{75}$$

The second law of thermodynamics for isothermal processes requires $\mathcal{D} \geq 0$, which is satisfied if: $\mu(c) > 0$, $B > 0$ and $b > 0$.

*Principle of virtual power and governing equations* We can now write the power balance (Eq. 67) in the form which emphasizes the treatment of inertial forces as external body forces equivalent to gravity:

$$\begin{aligned} &\int_{\partial V} \mathbf{t} \cdot \mathbf{v} d\partial V + \int_{V} \rho \left[ \mathbf{g} - \frac{D\mathbf{v}}{Dt} \right] \cdot \mathbf{v} dV = \\ &= \int_{V} \left\{ \boldsymbol{\sigma} : (\mathbf{v}\nabla) - \left( \mathcal{M} - \mathcal{M}^{0} \right) \frac{Dc}{Dt} \right\} dV - \int_{S} \left( \theta - \theta^{0} \right) \frac{Dc}{Dt} d\partial V. \end{aligned} \tag{76}$$

Internally, the symmetric stress tensor $\boldsymbol{\sigma} = \boldsymbol{\tau} - \kappa(\nabla c \nabla c)$ expands power on the Eulerian deformation rate $\mathbf{D} = (\mathbf{v}\nabla + \nabla\mathbf{v})/2$, while $\left(\mathcal{M} - \mathcal{M}^{0}\right)$ and $(\theta - \theta^{0})$ expand power on $\nabla \cdot \mathbf{J} = -Dc/Dt$, in volume and on the a solid surface. Thus, the external forces expand power on independent "velocities" $(\mathbf{v}, \mathbf{J})$; internally, the power is expended on the "gradients" of the same velocities, by the corresponding power conjugates. With the power balance structured as in Eq. (76), we can formulate the principle of virtual power.

At instant $t$, we consider virtual velocity fields $\left(\tilde{\mathbf{v}}, \tilde{\mathbf{J}}\right)$. These are arbitrary everywhere, except that they must vanish where the velocities are prescribed (essential boundary conditions). Thus, $\tilde{\mathbf{v}}$ vanishes on $S$, while $\tilde{\mathbf{J}}$ vanishes on $\partial V$. Moreover, $\left(\mathbf{v} + \tilde{\mathbf{v}}, \mathbf{J} + \tilde{\mathbf{J}}\right)$ must satisfy all the additional constraints for $(\mathbf{v}, \mathbf{J})$. Thus, in the present case: $\nabla \cdot \tilde{\mathbf{v}} = 0$. The principle of virtual power can then be stated as follows.

*The power expanded by external forces on virtual velocities equals the power expanded internally by power conjugates on virtual gradients.*

An important caveat to the simple statement of the principle of virtual power is that it is strictly true only if all components of velocities and their gradients are independent, i.e., if both external and internal power are written as linear combination of independent scalar fields. If not, the constraints can be included in the statement with the corresponding Lagrange multipliers. In the present case, the constraint (on gradients) is the incompressibility condition (Eq. 14).[27] The mathematical expression of the principle follows directly from Eq. (76), with

[27] In the compositionally compressible flow, the constraint is given by Eq. (11).

addition of the constraint enforced by the Lagrange multiplier $\pi(\mathbf{x}, t)$:

$$\int_{\partial V} \mathbf{t} \cdot \tilde{\mathbf{v}} d\partial V + \int_{V} \rho \left[ \mathbf{g} - \frac{D\mathbf{v}}{Dt} \right] \cdot \tilde{\mathbf{v}} dV =$$
$$= \int_{V} \left\{ \boldsymbol{\sigma} : (\tilde{\mathbf{v}}\nabla) - \pi \nabla \cdot \tilde{\mathbf{v}} - \left( \mathcal{M} - \mathcal{M}^0 \right) \frac{D\tilde{c}}{Dt} \right\} dV - \int_{S} \left( \theta - \theta^0 \right) \frac{D\tilde{c}}{Dt} d\partial V. \tag{77}$$

The standard procedure involving the generalized Gauss-Ostrogradsky theorem and the argument based on arbitrariness of $\tilde{\mathbf{v}}$ and $\tilde{c}$, yields the strong form of the initial/boundary value problem, for the unknowns $\mathbf{v}(\mathbf{x}, t)$, $\pi(\mathbf{x}, t)$, $c(\mathbf{x}, t)$ and $\mathcal{M}(\mathbf{x}, t)$. In addition to the field equations (Navier-Stokes equation modified by capillary stresses, incompressibility condition and Cahn-Hilliard equations)[28]:

$$\begin{gathered} \rho \frac{D\mathbf{v}}{Dt} = \rho \mathbf{g} + \nabla \cdot [\mu (\mathbf{v}\nabla + \nabla\mathbf{v}) - \kappa \nabla c \nabla c] - \nabla \pi; \qquad \nabla \cdot \mathbf{v} = 0; \\ \frac{Dc}{Dt} = \nabla \cdot (B \nabla \mathcal{M}); \qquad \mathcal{M} = \frac{df}{dc} - \kappa \nabla^2 c, \end{gathered} \tag{78}$$

the analysis also yields the diffusive boundary condition:

$$\frac{Dc}{Dt} = -b \left( \kappa \mathbf{n} \cdot \nabla c + \frac{d\gamma}{dc} \right); \qquad \text{on } S. \tag{79}$$

Thus, although the velocity is subject to the no-slip/no-separation boundary condition:$\mathbf{v} = \mathbf{v}^{solid}$ on $S$, the triple line moves by diffusion. The Lagrange multiplier $\pi(\mathbf{x})$ is recognized as the non-capillary pressure. The total pressure $p$ includes the capillary pressure:

$$p = \pi - \frac{1}{3} \, tr \left[ -\kappa \left( \nabla\phi \nabla\phi \right) \right] = \pi + \frac{1}{3} \, \kappa (\nabla\phi)^2. \tag{80}$$

The present incompressible model has several versions, which are *mathematically equivalent*, but may produce numerical differences. Different orders of derivatives of the main variables may have an effect on the finite element computations, particularly if low order interpolation is used. One alternative is obtained by noting that $\nabla \cdot (-\kappa \nabla c \nabla c) = \mathcal{M} \nabla c - \nabla \omega$. Next, we note that division of pressure into capillary and non-capillary parts is arbitrary and replace the variable

[28]The chemical potential can clearly be eliminated from the equations, resulting in fourth order diffusion equation, which requires continuous ($C^1$) Galerkin finite elements (Asle Zaeem and Mesarovic 2009, 2010). The form (Eq. 78) is consistent with the discontinuous ($C^0$) Galerkin formulation (Key and Welford 2006).

$\pi$ with $\hat{\pi} = \pi + \omega$. The alternative modified Navier-Stokes equation is

$$\rho \frac{D\mathbf{v}}{Dt} = \rho \mathbf{g} + \nabla \cdot [\mu (\mathbf{v}\nabla + \nabla \mathbf{v})] + \mathcal{M}\nabla c - \nabla \hat{\pi}, \tag{81}$$

and the total pressure is now $p = \hat{\pi} - f(c) - \kappa(\nabla c)^2/6$.

*Parameter identification and computations* The functions $\rho(c)$, $\mu(c)$ and $\gamma(c)$ are produced by monotonic interpolation between the corresponding values of mass density, viscosity and interface energies between the solid and the two fluids. The computational interface thickness $h$ is chosen so that it is significantly smaller than the anticipated smallest radius of curvature of the interface during the process. This may require iterations. The parameters $(\kappa, \Delta f)$are then computed from Eq. (71). The remaining phase field parameters are the bulk mobility $B$ and triple line mobility $b$.

The triple line mobility $b$ determines the relationship between the capillary forces and triple line velocity. It could be, in principle, determined from molecular dynamics simulations, or from the combination of molecular statics (energy landscape) simulations and activation rate theory. The most practical solution appears to be matching of experimental kinetics in the regime where the capillary forces dominate (Fu et al. 2016).

For immiscible fluids, the bulk mobility $B$ has no physical significance; it is a computational parameter that enables diffusional rearrangements within the phase field interface layer. Intuitively, it should be large enough not to obstruct the motion of the interface. However, the computational studies (Dehsara et al. 2017) indicate that the changes in $B$ by four orders of magnitude have negligible effect on the kinetics of capillary flow.

The capillary stress tensor, $-\kappa(\nabla\phi\nabla\phi)$, has non-zero values only across the interface. It is the phase field description of the interface tension and the pressure jump across the interface (Davis and Scriven 1982). For the interface with principal curvatures $k^1$ and $k^2$ the pressure jump is $\Delta p = \Gamma(k^1 + k^2)$. In Fig. 16, we compare the computational (phase field) and analytical (sharp interface) solutions to the problem of equilibrium capillary rise between parallel plates. Note the sharp negative peak in pressure and the jump between the values on the two sides of the interface. Integral of the negative pressure represents the interface tension.

The triple line mobility can be extracted from the T-joint experiments, shown in Fig. 17 (Fu et al. 2016). The fluid rests on the horizontal glass plate. The vertical glass plate is brought gradually into contact and the capillary rise follows. In the initial stage of the experiment, capillary forces dominate; the triple line mobility parameter, $b$, is fitted to the initial slope of the kinetic curve in Fig. 17.

## 3.2 Diffusional and Dislocation Creep

We aim to derive the governing equations for a single crystalline grain, based on the lattice kinematics (Sect. 2.2) and encompassing: diffusion, dislocation glide and

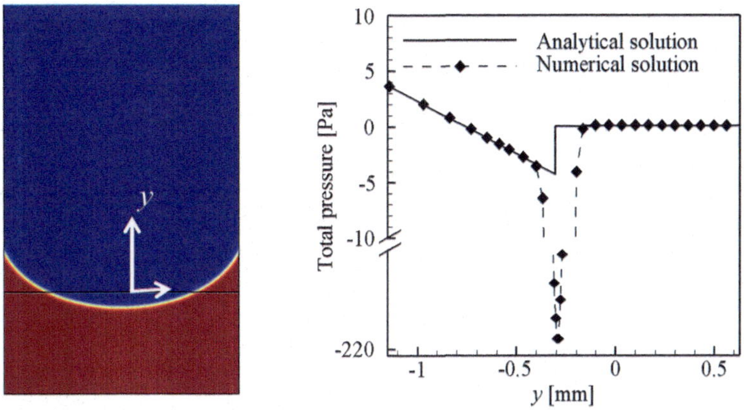


**Fig. 16** Capillary rise between parallel plates. **Left:** Equilibrium phase field contours. Blue is low-density gas, red is high density fluid. **Right:** Pressure distribution across the interface in the centerline $x = 0$

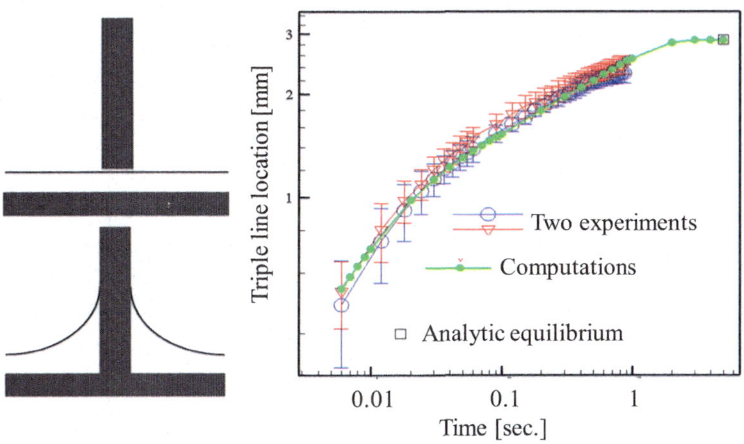


**Fig. 17** **Left:** Schematic of the T-joint experimental setup. **Right:** Triple line kinetics: experimental, computational and analytic equilibrium solution

climb, as well as elastic and compositional deformation. As before, we begin with the power balance for a grain occupying the current domain $\Omega$, bounded by $\partial\Omega$ with outer normal $\mathbf{n}$. The boundary tractions expand power on boundary velocities whose normal component differs from the lattice velocity $\mathbf{v}(\mathbf{x}, t)$ by the lattice growth rate $g(\mathbf{x} \in \partial\Omega, t)$ (Eq. 35). With the potential energy $\mathcal{P}$ and the dissipation rate $\mathcal{D}$:

$$\int_{\partial\Omega} (t_n g + \mathbf{t} \cdot \mathbf{v})\, d\partial\Omega = \frac{D\mathcal{P}}{Dt} + \mathcal{D}; \qquad t_n = \mathbf{t} \cdot \mathbf{n}. \tag{82}$$

With Eq. (44), we rewrite Eq. (82) as

$$\int_{\partial\Omega} \mathbf{t}\cdot\mathbf{v}d\partial\Omega = \frac{D\mathcal{P}}{Dt} + \mathcal{D} + \int_{\partial\Omega} \frac{F^*}{1-c} t_n J_n d\partial\Omega. \tag{83}$$

**Potential energy** At a constant temperature, the free energy density, given per unit volume in intermediate configuration $\tilde{\Omega}$, is function of concentration and elastic strains: $\Phi(\mathbf{F}^*, c)$. Using Eqs. (33) and (34), the total free energy of a grain with current volume $\Omega$ can be written as

$$\mathcal{P} = \int_{\tilde{\Omega}} \Phi\left(\mathbf{F}^*, c\right) d\tilde{\Omega} = \int_{\Omega} \frac{1}{F^*} \Phi\left(\mathbf{F}^*, c\right) d\Omega = \int_{\Omega} \frac{N}{\tilde{N}} \Phi\left(\mathbf{F}^*, c\right) d\Omega. \tag{84}$$

By means of the transport theorem for a grain with growing/disappearing boundary (Eq. 41), the material rate of the potential energy is

$$\begin{aligned} \frac{D\mathcal{P}}{Dt} = &\int_{\Omega} N\left[\frac{1}{\tilde{N}}\left(\left(\frac{\partial\Phi}{\partial\mathbf{F}^*}\right)^T : \frac{D\mathbf{F}^*}{Dt} - \frac{\Phi}{\tilde{N}}\frac{D\tilde{N}}{Dt}\right) + \frac{\partial\Phi}{\partial c}\frac{Dc}{Dt} - \Phi\dot{s}\right] d\Omega \\ &+ \int_{\partial\Omega} \frac{\Phi g}{F^*} d\partial\Omega. \end{aligned} \tag{85}$$

Now note, with reference to Fig. 5, that while dislocation climb changes the total number of lattice points, it does not change the lattice density; the translation of the lattice compensates so that the lattice density in intermediate (isoclinic) configuration is constant. Using Eq. (32):

$$\frac{D\tilde{N}}{Dt} = -\tilde{N}\mathrm{tr}\tilde{\mathbf{L}}^p - \tilde{N}\dot{s} = 0. \tag{86}$$

Upon introducing the Cauchy stress $\boldsymbol{\sigma}$ and the non-symmetric intermediate configuration stress $\tilde{\mathbf{S}}$ (Bassani 1994):

$$\boldsymbol{\sigma} = \frac{1}{F^*}\mathbf{F}^* \cdot \left(\frac{\partial\Phi}{\partial\mathbf{F}^*}\right)^T = \frac{1}{F^*}\left(\frac{\partial\Phi}{\partial\mathbf{F}^*}\right)\cdot\mathbf{F}^{*T}; \quad \tilde{\mathbf{S}} = F^*\left(\mathbf{F}^{*-1}\cdot\boldsymbol{\sigma}\cdot\mathbf{F}^*\right), \tag{87}$$

the first term in Eq. (85) can be written as

$$\begin{aligned} \int_{\Omega} \frac{1}{F^*}\left(\frac{\partial\Phi}{\partial\mathbf{F}^*}\right)^T : \frac{D\mathbf{F}^*}{Dt} d\Omega &= \int_{\Omega} \boldsymbol{\sigma} : \mathbf{L}^* d\Omega \\ &= \int_{\Omega} \boldsymbol{\sigma} : (\mathbf{v}\nabla)\, d\Omega - \int_{\tilde{\Omega}} \tilde{\mathbf{S}} : \tilde{\mathbf{L}}^p d\tilde{\Omega}. \end{aligned} \tag{88}$$

Finally, upon substitution of od (Eqs. 32, 40, and 88) into Eq. (85), we obtain

$$
\begin{aligned}
\frac{D\mathcal{P}}{Dt} &= \int\limits_{\Omega} \left[ \boldsymbol{\sigma} : (\mathbf{v}\nabla) - \frac{\partial \Phi}{\partial c} \nabla \cdot \mathbf{J} \right] d\Omega + \int\limits_{\partial\Omega} \frac{\Phi g}{F^*} d\partial\Omega \\
&\quad - \int\limits_{\tilde{\Omega}} \sum_{\alpha} \left[ \hat{\tau}^{\alpha} \dot{\hat{\gamma}}^{\alpha} + \hat{\Lambda}^{\alpha} \dot{s}^{\alpha} \right] d\tilde{\Omega}; \\
\hat{\Lambda}^{\alpha} &= \hat{\sigma}^{\alpha} + \frac{\partial \Phi}{\partial c} (1 - c) + \Phi; \quad \hat{\sigma}^{\alpha} = -\tilde{\mathbf{s}}^{\alpha} \cdot \tilde{\mathbf{S}} \cdot \tilde{\mathbf{s}}^{\alpha}; \\
&\qquad \hat{\tau}^{\alpha} = \tilde{\mathbf{m}}^{\alpha} \cdot \tilde{\mathbf{S}} \cdot \tilde{\mathbf{s}}^{\alpha}.
\end{aligned}
\tag{89}
$$

**Dissipation** Local diffusional dissipation rate in the lattice is modeled simply as proportional to the divergence of flux, while the dissipation rates associated with glide and climb of dislocations are proportional to $\dot{\hat{\gamma}}^{\alpha}$ and $\dot{s}^{\alpha}$:

$$
\begin{aligned}
\mathcal{D}_{diff} &= \int\limits_{\Omega} \mathcal{M} \nabla \cdot \mathbf{J} d\Omega = \int\limits_{\partial\Omega} \mathcal{M} J_n d\partial\Omega - \int\limits_{\Omega} \nabla \mathcal{M} \cdot \mathbf{J} d\Omega; \\
\mathcal{D}_{climb} &= \int\limits_{\tilde{\Omega}} \sum_{\alpha} \Lambda^{\alpha} \dot{s}^{\alpha} d\tilde{\Omega}; \quad \mathcal{D}_{glide} = \int\limits_{\tilde{\Omega}} \sum_{\alpha} \tau^{\alpha} \dot{\hat{\gamma}}^{\alpha} d\tilde{\Omega}.
\end{aligned}
\tag{90}
$$

In addition to the bulk dissipation, the vacancy creation and absorption at the boundaries must be a dissipative process. The dissipation rate at a boundary point can be equivalently assumed to be proportional either to the lattice growth rate $g$, or to the normal vacancy flux:

$$
\mathcal{D}_{bdr} = \int\limits_{\partial\Omega} \theta J_n d\partial\Omega. \tag{91}
$$

The total dissipation rate,

$$
\begin{aligned}
\mathcal{D} &= \int\limits_{\partial\Omega} (\mathcal{M} + \theta) J_n d\partial\Omega - \int\limits_{\Omega} \nabla \mathcal{M} \cdot \mathbf{J} d\Omega \\
&\quad + \int\limits_{\tilde{\Omega}} \sum_{\alpha} \Lambda^{\alpha} \dot{s}^{\alpha} d\tilde{\Omega} + \int\limits_{\tilde{\Omega}} \sum_{\alpha} \tau^{\alpha} \dot{\hat{\gamma}}^{\alpha} d\tilde{\Omega},
\end{aligned}
\tag{92}
$$

will satisfy the second law of thermodynamics ($\mathcal{D} \geq 0$) if we assume linear constitutive laws with positive mobilities $B$, $\beta$ and $H$:

$$
\begin{aligned}
&\mathbf{J} = -B \nabla \mathcal{M}, \qquad \dot{s}^{\alpha} = \beta \Lambda^{\alpha}, \qquad \beta, B > 0, \quad \text{in } \Omega; \\
&J_n = H (\theta + \mathcal{M}), \qquad\qquad\qquad\quad H > 0, \quad \text{on } \partial\Omega,
\end{aligned}
\tag{93}
$$

and any one of the slip constitutive laws (Bassani 1994) which satisfy $\tau^{\alpha} \dot{\hat{\gamma}}^{\alpha} \geq 0$.

**Principle of virtual power and governing equations** Upon rewriting the power balance (Eq. 82) in the form

$$\begin{aligned}\int_{\partial\Omega} \mathbf{t}\cdot\mathbf{v}d\partial\Omega &= \int_{\Omega}\left[\boldsymbol{\sigma}:(\mathbf{v}\nabla)+\left(\mathcal{M}-\frac{\partial\Phi}{\partial c}\right)\nabla\cdot\mathbf{J}\right]d\Omega \\ &\quad+\int_{\partial\Omega}\left[\theta+\frac{F^{*}t_n-\Phi}{1-c}\right]J_n d\partial\Omega \\ &\quad+\int_{\tilde{\Omega}}\sum_{\alpha}\left(\Lambda^{\alpha}-\hat{\Lambda}^{\alpha}\right)\dot{s}^{\alpha}d\tilde{\Omega}+\int_{\tilde{\Omega}}\sum_{\alpha}\left(\tau^{\alpha}-\hat{\tau}^{\alpha}\right)\dot{\hat{\gamma}}^{\alpha}d\tilde{\Omega},\end{aligned} \tag{94}$$

we formulate the principle of virtual power, as follows.

In addition to independent velocities, $\mathbf{v}(\mathbf{x},t)$ and $\mathbf{J}(\mathbf{x},t)$, the rates, $\dot{s}^{\alpha}(\mathbf{x},t)$ and $\dot{\hat{\gamma}}^{\alpha}(\mathbf{x},t)$, are subject to independent variations. Thus, we consider the set of independent virtual velocities/rates (test functions): $\tilde{\mathbf{v}}(\mathbf{x})$, $\tilde{\mathbf{J}}(\mathbf{x})$, $\tilde{\dot{s}}^{\alpha}(\mathbf{x})$ and $\tilde{\dot{\hat{\gamma}}}^{\alpha}(\mathbf{x})$, arbitrary except that each must vanish at the points where the corresponding velocity/rate is prescribed. From Eq. (94), the principle of virtual power is written as

$$\begin{aligned}\int_{\partial\Omega} \mathbf{t}\cdot\tilde{\mathbf{v}}d\partial\Omega &= \int_{\Omega}\left[\boldsymbol{\sigma}:(\tilde{\mathbf{v}}\nabla)+\left(\mathcal{M}-\frac{\partial\Phi}{\partial c}\right)\nabla\cdot\tilde{\mathbf{J}}\right]d\Omega \\ &\quad+\int_{\partial\Omega}\left[\theta+\frac{F^{*}t_n-\Phi}{1-c}\right]\tilde{J}_n d\partial\Omega \\ &\quad+\int_{\tilde{\Omega}}\sum_{\alpha}\left(\Lambda^{\alpha}-\hat{\Lambda}^{\alpha}\right)\tilde{\dot{s}}^{\alpha}d\tilde{\Omega}+\int_{\tilde{\Omega}}\sum_{\alpha}\left(\tau^{\alpha}-\hat{\tau}^{\alpha}\right)\tilde{\dot{\hat{\gamma}}}^{\alpha}d\tilde{\Omega}.\end{aligned} \tag{95}$$

Equation (95) must be satisfied for arbitrary allowable variations. The usual manipulation and the argument based on independence of variations yields:

- Stress (quasi-) equilibrium with standard boundary conditions:

$$\nabla\cdot\boldsymbol{\sigma}=0 \quad \text{in } \Omega(t); \qquad \mathbf{n}\cdot\boldsymbol{\sigma}=\mathbf{t}^{0} \text{ or } \mathbf{v}=\mathbf{v}^{0} \quad \text{on } \partial\Omega(t), \tag{96}$$

- The diffusion potential $\mathcal{M}=\partial\Phi/\partial c$ and $\theta=-(F^{*}t_n-\Phi)/(1-c)$, resulting in the coupled, moving-boundary diffusion problem with an unusual boundary condition on the moving boundaries:

$$\begin{aligned}&\frac{Dc}{Dt}=BF^{*}\nabla^{2}\frac{\partial\Phi}{\partial c}-(1-c)\dot{s} && \text{in } \Omega(t);\\ &\frac{1-c}{F^{*}}g=B\mathbf{n}\cdot\nabla\frac{\partial\Phi}{\partial c}=H\left(\frac{F^{*}t_n-\Phi}{1-c}-\frac{\partial\Phi}{\partial c}\right) && \text{on } \partial\Omega(t),\end{aligned} \tag{97}$$

- The power conjugate of the true slip and climb rates:

$$\tau^{\alpha} = \hat{\tau}^{\alpha} = \tilde{\mathbf{m}}^{\alpha} \cdot \tilde{\mathbf{S}} \cdot \tilde{\mathbf{s}}^{\alpha}; \qquad \Lambda^{\alpha} = \hat{\Lambda}^{\alpha} = \hat{\sigma}^{\alpha} + (1-c)\,\partial\Phi/\partial c + \Phi. \tag{98}$$

With constitutive equations for elasticity with compositional strains, and plasticity, this completes the initial-boundary value problem. The resolved shear stress $\tau^{\alpha}$ is the standard concept in crystal plasticity. The climb power conjugate $\Lambda^{\alpha}$requires some attention. Since slip and climb are defined consistently in the intermediate configuration, $\hat{\sigma}^{\alpha}$ is the normal (compressive) component of the stress tensor $\tilde{\mathbf{S}}$ on the plane with normal $\tilde{\mathbf{s}}^{\alpha}$. Compression on a lattice plane favors the loss of lattice planes, while the tension favors creation of new planes. The expression $(1-c)\partial\Phi/\partial c$ is the osmotic stress (Lothe and Hirth 1967; Gao and Cocks 2009). For small elastic-compositional deformation gradient and small vacancy concentration, with the regular solution model, it can be written as

$$(1-c)\frac{\partial\Phi}{\partial c} \approx \frac{\partial\Phi}{\partial c} \approx kT\tilde{N}\ln\frac{c}{c_0} - 3\eta p \approx kT\tilde{N}\ln\frac{c}{c^{p,eq}} - p, \tag{99}$$

where $k$ is the Boltzman constant, $T$ is the temperature, $c_0$ and $c^{p,eq}$ are the equilibrium vacancy concentrations at the temperature $T$—the first at zero pressure, the second at the current pressure $p$. The Vegard's law coefficient $\eta$ describes the volumetric compositional strain as $-\eta(c-c_0)$. The last term, $\Phi$, accounts for the fact that each lattice site carries internal energy. When a lattice site is lost, its internal energy is dissipated.

**Elastic-compositional deformation, Nabarro-Herring creep** At low stresses and high temperature, polycrystals creep. Diffusion of vacancies through the bulk of a grain produces lattice growth/disappearance at the grain boundaries (Nabarro 1967; Herring 1950). In the absence of dislocation glide and climb $\mathbf{F} \equiv \mathbf{F}^{*}$. The elastic and compositional deformation is small and we can partially linearize the governing equations (96 and 99). Define the small displacement $\mathbf{u}$ in a crystalline grain by subtracting the rigid body motion of the crystal. Then,

$$\mathbf{F} = (\mathbf{I}+\boldsymbol{\omega})\cdot(\mathbf{I}+\boldsymbol{\varepsilon}) \approx \mathbf{I}+\boldsymbol{\omega}+\boldsymbol{\varepsilon}; \qquad \boldsymbol{\varepsilon} = \frac{1}{2}(\mathbf{u}\nabla + \nabla\mathbf{u}). \tag{100}$$

The energy density is now function of strain $\Phi(\boldsymbol{\varepsilon}, c)$. The total strain is the sum of elastic strain $\mathbf{e}$ and compositional volumetric strain:

$$\boldsymbol{\varepsilon} = \mathbf{e} - \eta\,(c-c_0)\,\mathbf{I}; \qquad \eta > 0 \text{ for vacancies.} \tag{101}$$

The energy density may be expressed as the sum of elastic energy and mixing energy with the regular solution model (Haasen 1986):

$$\begin{gathered}\Phi(\boldsymbol{\varepsilon}, c) = \phi(\boldsymbol{\varepsilon}, c-c_0) + \psi(c-c_0);\\ \phi(\boldsymbol{\varepsilon}, c-c_0) = \tfrac{1}{2}\mathbf{e}:\mathbf{E}:\mathbf{e} = \tfrac{1}{2}(\boldsymbol{\varepsilon}+\eta(c-c_0)\mathbf{I}):\mathbf{E}:(\boldsymbol{\varepsilon}+\eta(c-c_0)\mathbf{I});\\ \psi \approx \tfrac{1}{2}\kappa(c-c_0)^2; \qquad \kappa = kTN/c_0.\end{gathered} \tag{102}$$

where $c_0$ is the equilibrium vacancy concentration for unstressed crystal at a given temperature $T$, $\mathbf{E}$ is the elastic stiffness tensor, $k$ is the Boltzman constant. Thus, the compositional strain plays the role of the eigenstrain (stress-free strain analogous to thermal dilation/contraction). Further,

$$\sigma_{ij} = \frac{\partial \Phi}{\partial \varepsilon_{ij}}; \quad p = -\frac{1}{3}\mathrm{tr}\boldsymbol{\sigma}; \quad \Phi' = \frac{\partial \Phi}{\partial c} = -3\eta p + \kappa\,(c - c_0)\,. \tag{103}$$

Then, the equations governing diffusion in the variable domain take form

$$\begin{aligned} &\frac{\partial c}{\partial t} = B\nabla^2 \Phi' && \text{in } \Omega(t); \\ &g = B\mathbf{n}\cdot\nabla\Phi' = b\left[t_n - \Phi'\right] && \text{on } \partial\Omega(t). \end{aligned} \tag{104}$$

The motion of the boundaries is described by the lattice growth rate $g$, which also defines the creep strain rate (Eqs. 36 and 45). The presence of pressure in the field equation and normal tractions at the boundaries implies that the diffusion problem is coupled to the elastic-compositional equilibrium governed by

$$\nabla\cdot(\mathbf{E} : (\mathbf{u}\nabla)) + \eta\,(\nabla c)\cdot\mathbf{E} : \mathbf{I} = 0 \qquad \text{in } \Omega(t), \tag{105}$$

and the appropriate boundary conditions. For a certain class of problems (Mesarovic 2016), the diffusion p.d.e. can decouple from stress but the boundary coupling cannot be eliminated.

**Future directions** The lattice continuum theory has been developed recently. Although some approximate asymptotic results have been obtained for Nabarro-Herring creep, the appropriate numerical method for the sharp interface lattice continuum model is yet to be developed. Given the nature of the problem (moving boundaries), a phase field formulation[29] is indicated. Such models should provide important insight into cavity nucleation process and transition from secondary to tertiary creep. The process of vacancy nucleation/absorption at boundaries is poorly understood at present; the current models are phenomenological.

The role of interface energy and interface stresses has been briefly addressed in Mesarovic (2016). It appears that interface orientation dependence of interface energy is as important as the interface energy itself. Related to the interface energy are: the question of the model for grain boundary diffusion (Coble creep), and, the question of vacancy nucleation without subsequent absorption by the lattice. The dependence of surface energy of the vacancy content (excess volume) appears to be the key component of such process.

---

[29]Phase field models have been constructed (Garikipati et al. 2001; Mishin et al. 2013; Villani et al. 2015) but without the full lattice continuum sharp interface model.

Another open problem related to the grain boundaries is the relaxation of gliding and climbing dislocation at the grain boundary. In the next section we take up the problem of glide and boundary relaxation. To describe the relaxation of a gliding dislocation at the boundary, the continuum slip field must become an independent variable. It appears that this naturally leads to a size-dependent continuum.

## 3.3 Size-Dependent Plasticity of Crystals and Polycrystals

In order to construct a size-dependent continuum theory, we begin with thermodynamic analysis of the standard size-invariant crystal plasticity formulation. Specifically, we analyze the free energy of an elastic-plastically crystal, as might be computed from different descriptions of dislocations. Consider the finite volume $V$, bounded by the surface $\partial V$. The free energy of the system consists of the elastic strain energy. For linear elasticity (and ignoring for the moment boundary conditions), the free energy in the simple continuum:

$$\Phi^0 = \int_V \varphi^0 (\mathbf{e})\, dV; \qquad \varphi^0 (\mathbf{e}) = \frac{1}{2}\mathbf{e} : \mathbf{E} : \mathbf{e}. \tag{106}$$

The strain energy density $\varphi^0$is function of the elastic strain $\mathbf{e}$, $\mathbf{E}$is the elastic stiffness tensor and the stress is $\boldsymbol{\sigma} = \partial\phi^0/\partial\mathbf{e} = \mathbf{E} : \mathbf{e}$. When plastic deformation is present, the elastic strain $\mathbf{e}$ is incompatible: $\boldsymbol{\eta} \neq 0$ in Eq. (48). In general, there is infinity of partitions into the compatible strain $\boldsymbol{\varepsilon}^c$and the incompatible strain $\boldsymbol{\varepsilon}^{inc}$:

$$\mathbf{e} = \boldsymbol{\varepsilon}^c + \boldsymbol{\varepsilon}^{inc} : \quad \mathrm{Inc}\left(\boldsymbol{\varepsilon}^c\right) = 0 \ \text{ and } \ \mathrm{Inc}\left(\boldsymbol{\varepsilon}^{inc}\right) = \mathrm{Inc}\,(\mathbf{e}) = \boldsymbol{\eta} \neq 0. \tag{107}$$

However, the orthogonality condition:

$$\int_V \boldsymbol{\varepsilon}^c : \mathbf{E} : \boldsymbol{\varepsilon}^{inc}\, dV = 0, \tag{108}$$

(the stress corresponding to one does no work on the other), renders a unique partition (Mesarovic et al. 2010). Orthogonality implies that the strain energies corresponding to the two fields are additive:

$$\Phi^0 = \Phi^c + \Phi^{inc} = \frac{1}{2}\int_V \boldsymbol{\varepsilon}^c : \mathbf{E} : \boldsymbol{\varepsilon}^c\, dV + \frac{1}{2}\int_V \boldsymbol{\varepsilon}^{inc} : \mathbf{E} : \boldsymbol{\varepsilon}^{inc}\, dV. \tag{109}$$

In crystal plasticity, incompatibility, $\boldsymbol{\eta} \neq 0$, is equivalent to a non-vanishing Nye's tensor or slip gradient vectors: $\mathbf{A} \neq 0$ or $\mathbf{g}^\alpha \neq 0$ (Eq. 51), so that the incompatible strain energy $\Phi^{inc}$ is, in fact, identical to the elastic interaction energy between dislocation segments which contribute to the net Burgers vector (Mesarovic 2005), which can be expressed as a double convolution of the quadratic form of either Nye's tensor or slip gradient vectors:

$$\begin{aligned}\Phi^{inc} &= \frac{1}{2}\int_V dx^3 \int_V dy^3 \mathbf{A}(\mathbf{x}) : \mathbf{L}(\mathbf{x}-\mathbf{y}) : \mathbf{A}(\mathbf{y}) \\ &= \frac{1}{2}\int_V dx^3 \int_V dy^3 \sum_\alpha \sum_\beta \mathbf{g}^\alpha(\mathbf{x}) \cdot \mathbf{\Delta}^{\alpha\beta}(\mathbf{x}-\mathbf{y}) \cdot \mathbf{g}^\beta(\mathbf{y}).\end{aligned} \tag{110}$$

Clearly, if the continuous fields $\mathbf{A}$ and $\mathbf{g}^\alpha$ are used in Eq. (110), the incompatible energy, say $\Phi^{inc}(\mathbf{g}^\alpha)$, is included in the energy of the simple continuum $\Phi^0$ (Eq. 106), which contains no length scale. However, the smoothing out of the discrete dislocations $\hat{\mathbf{g}}^\alpha$, as in Eq. (53), results in error in energy:

$$\Delta\Phi^{inc} = \Phi^{inc}\left(\hat{\mathbf{g}}^\alpha\right) - \Phi^{inc}\left(\mathbf{g}^\alpha\right). \tag{111}$$

The free energy of the size-dependent continuum can then be written as

$$\Phi = \Phi^0 + \Delta\Phi^{inc}. \tag{112}$$

While the total incompatible energy (Eq. 110) is irreducibly non-local, it can be shown that the error (Eq. 111) is localized to the neighborhood of the size $(h^\alpha h^\beta)^{1/2}$, so that, with some approximation (Mesarovic et al. 2010; Mesarovic 2010):

$$\Delta\Phi^{inc} = \frac{1}{2}\int_V dV \sum_\alpha \sum_\beta \mathbf{g}^\alpha \cdot \mathbf{D}^{\alpha\beta} \cdot \mathbf{g}^\beta + \frac{1}{2}\int_{\partial V} \sum_\alpha \sum_\beta \mathbf{\Gamma}^\alpha \cdot \mathbf{F}^{\alpha\beta} \cdot \mathbf{\Gamma}^\beta d\partial V, \tag{113}$$

where the constitutive tensors have the form

$$\begin{aligned}\mathbf{D}^{\alpha\beta} &= h^\alpha h^\beta \left[ d_{\perp\perp}^{\alpha\beta} \mathbf{s}^\alpha \mathbf{s}^\beta + d_{\perp\odot}^{\alpha\beta} \mathbf{s}^\alpha \mathbf{t}^\beta + d_{\odot\perp}^{\alpha\beta} \mathbf{t}^\alpha \mathbf{s}^\beta + d_{\odot\odot}^{\alpha\beta} \mathbf{t}^\alpha \mathbf{t}^\beta \right]; \\ \mathbf{F}^{\alpha\beta} &= \sqrt{h^\alpha h^\beta} \left[ f_{\perp\perp}^{\alpha\beta} \mathbf{s}^\alpha \mathbf{s}^\beta + f_{\perp\odot}^{\alpha\beta} \mathbf{s}^\alpha \mathbf{t}^\beta + f_{\odot\perp}^{\alpha\beta} \mathbf{t}^\alpha \mathbf{s}^\beta + f_{\odot\odot}^{\alpha\beta} \mathbf{t}^\alpha \mathbf{t}^\beta \right].\end{aligned} \tag{114}$$

The constitutive relations (114) are the result of interactions between dislocations embedded in an elastic medium. The coefficients $d^{\alpha\beta}$ and $f^{\alpha\beta}$ have dimensions of elastic moduli. They are easily derived for isotropic elasticity but require numerical treatment for anisotropic crystals and interfaces between dissimilar media (Mesarovic et al. 2015).

The additive nature of Eq. (112) is the direct consequence of the orthogonality condition (108), while the additive decomposition between the bulk and the boundary (Eq. 113) is possible because of the localized nature of $\Delta\Phi^{inc}$. These additive decompositions are the key requirements for the formulation of the principle of virtual power (or work): independent velocities and their gradients.[30]

**Single crystal with rigid boundaries** We first consider the problem where the crystal has rigid boundaries, i.e., dislocation piled-up against the boundaries cannot relax by any mechanisms so that dissipation is limited to the bulk of the crystal. Denote the time derivative with a dot, e.g., $d\mathbf{u}/dt = \dot{\mathbf{u}}$. The power balance for such crystal can be written as

$$\int_{\partial V} \mathbf{t} \cdot \dot{\mathbf{u}} d\partial V = \dot{\Phi} + \mathcal{D}. \tag{115}$$

The dissipated power is expanded on slip rates:

$$\mathcal{D} = \int_V \sum_\alpha \tau^\alpha \dot{\gamma}^\alpha dV. \tag{116}$$

Using the slip field representation shown in Fig. 7 and Eqs. (55 and 56), the rate of change of potential energy is conveniently expressed as

$$\dot{\Phi} = \int\limits_V \left[ \sigma : \dot{\mathbf{e}} + \sum_\alpha \mathbf{p}^\alpha \cdot \dot{\mathbf{g}}^\alpha \right] dV + \int\limits_{\partial V} \sum_\alpha \mathbf{q}^\alpha \cdot \dot{\boldsymbol{\Gamma}}^\alpha d\partial V, \tag{117}$$

where the power conjugates are defined with respect to the bulk and surface energy density functions $\phi(\mathbf{e}, \mathbf{g}^\alpha)$ and $\psi(\boldsymbol{\Gamma}^\alpha)$, both arising from linear elasticity and therefore represented by quadratic forms, so that

$$\mathbf{p}^\alpha = \frac{\partial \phi}{\partial \mathbf{g}^\alpha} = \sum_\beta \mathbf{D}^{\alpha\beta} \cdot \mathbf{g}^\beta; \quad \mathbf{q}^\alpha = \frac{\partial \psi}{\partial \boldsymbol{\Gamma}^\alpha} = \sum_\beta \mathbf{F}^{\alpha\beta} \cdot \boldsymbol{\Gamma}^\beta. \tag{118}$$

The elastic *boundary constitutive law* with coefficients $\mathbf{F}^{\alpha\beta}$ is the key feature that makes the theory applicable to problems with a variety of boundary conditions. The higher order boundary conditions for $\boldsymbol{\Gamma}^\alpha$ and $\mathbf{q}^\alpha$ (or equivalently, $\gamma^\alpha$ and $\nabla\gamma^\alpha$) are intrinsically difficult to formulate.[31] The boundary constitutive law (Eq. 118), combined with Eq. (57), bypasses that problem as will be shown below.

The power conjugates $\mathbf{p}^\alpha$ and $\mathbf{q}^\alpha$ in Eq. (118) do not necessarily lie in the slip plane $\alpha$. It is convenient to define the in-plane gradient operator $\nabla^\alpha = \mathbf{P}^\alpha \cdot \nabla$. The

[30] An insightful discussion on this issue is given by Gurtin (2000, 2002) who showed (among other things) that the power conjugate to slip rate (Eq. 116) cannot be the resolved shear stress (as is the case in the simple crystal continuum).

[31] Gurtin (2000, 2002, 2003), Gurtin and Needleman (2005).

power balance (Eq. 115) can now be written as

$$\begin{aligned}\int_{\partial V} \mathbf{t}\cdot\dot{\mathbf{u}}d\partial V = \int_V \left[\boldsymbol{\sigma} : (\dot{\mathbf{u}}\nabla) + \sum_\alpha \left(\tau^\alpha - \tau_R^\alpha\right)\dot{\gamma}^\alpha + \sum_\alpha \mathbf{p}^\alpha \cdot \nabla^\alpha \dot{\gamma}^\alpha\right] dV \\ + \int_{\partial V} \sum_\alpha \mathbf{q}^\alpha \cdot \dot{\boldsymbol{\Gamma}}^\alpha d\partial V,\end{aligned} \tag{119}$$

where the resolved shear stress is

$$\tau_R^\alpha = \boldsymbol{\sigma} : \left(\mathbf{s}^\alpha \mathbf{m}^\alpha\right). \tag{120}$$

The independent primary rates are $\dot{\mathbf{u}}(\mathbf{x})$ and $\dot{\gamma}^\alpha(\mathbf{x})$; we recall from Sect. 1.3 that $\dot{\boldsymbol{\Gamma}}^\alpha$ is but a representation of the boundary value of $\dot{\gamma}^\alpha$.

The *principle of virtual work* is the analog of the principle of virtual power, but without reference to time. Instead of virtual velocities $\tilde{\mathbf{v}}$, we think of virtual displacements $\delta\mathbf{u} = \tilde{\mathbf{v}}\Delta t$. The statement of the principle of virtual work is identical to Eq. (119); we only replace the rates virtual fields $\delta\mathbf{u}$ and $\delta\gamma^\alpha$, arbitrary everywhere except where the corresponding boundary values are prescribed. The standard manipulation then yields

$$\begin{aligned}\int_{\partial V} (\mathbf{t} - \mathbf{n}\cdot\boldsymbol{\sigma})\cdot\delta\mathbf{u}d\partial V - \int_{\partial V}\sum_\alpha \mathbf{N}^\alpha \cdot \left(\mathbf{q}^\alpha - \mathbf{p}^\alpha\right)\delta\Delta\gamma^\alpha \cos\theta^\alpha d\partial V \\ = \int_V \left[-(\nabla\cdot\boldsymbol{\sigma})\cdot\delta\mathbf{u} + \sum\nolimits_\alpha \left(\tau^\alpha - \tau_R^\alpha - \nabla^\alpha\cdot\mathbf{p}^\alpha\right)\delta\gamma^\alpha\right] dV.\end{aligned} \tag{121}$$

Arbitrariness and independence of virtual fields implies the strong form of the boundary value problem: In addition to the stress equilibrium and the usual boundary conditions:

$$\nabla\cdot\boldsymbol{\sigma} = 0 \quad \text{in } V, \quad \mathbf{u} \text{ or } \mathbf{t} \text{ prescribed on } \partial V, \tag{122}$$

we obtain additional differential equations and boundary conditions for each active slip system, $\alpha = 1, 2, \ldots$:

$$\nabla^\alpha\cdot\mathbf{p}^\alpha = \tau^\alpha - \tau_R^\alpha \ \text{ in } V; \quad \mathbf{N}^\alpha\cdot\mathbf{p}^\alpha = \mathbf{N}^\alpha\cdot\mathbf{q}^\alpha \text{ on } \partial V. \tag{123}$$

To understand the nature of the equations (123), it is useful to write them explicitly in terms of slips

$$\begin{aligned}&\sum_\beta \mathbf{D}^{\alpha\beta} : \left(\nabla\nabla\gamma^\beta\right) = \tau^\alpha - \tau_R^\alpha && \text{in } V; \\ &\sum_\beta \left(\mathbf{n}\cdot\mathbf{D}^{\alpha\beta}\cdot\nabla\gamma^\beta + \mathbf{n}\cdot\mathbf{F}^{\alpha\beta}\cdot\mathbf{n}\gamma^\beta\right) = 0 && \text{on } \partial V^-,\end{aligned} \tag{124}$$

where $\gamma^{\beta}$ and $\nabla\gamma^{\beta}$ in the boundary condition are the values immediately inside the boundary ($\partial V^{-}$), not including the $\gamma^{\beta}$ drop to zero at the boundary (nor the Dirac delta function for the gradient). Thus, the constitutive law (Eq. 118) provides sufficient higher order boundary conditions for the second order p.d.e's in slip.

We now re-interpret the field equations (123) as differential yield conditions:

$$| \tau_R^{\alpha} + \nabla^{\alpha} \cdot \mathbf{p}^{\alpha} | < \tau_Y^{\alpha} \Rightarrow \dot{\gamma}^{\alpha} = 0. \tag{125}$$

Otherwise, the slip system is active, $\dot{\gamma}^{\alpha} \neq 0$, and the field equations (123) apply. The dissipative power conjugate $\tau^{\alpha}$ interpreted as the yield stress with the appropriate sign, and the yield stress evolves according to the prescribed hardening function:

$$\tau^{\alpha} = \mathrm{sign}\left(\dot{\gamma}^{\alpha}\right) \tau_Y^{\alpha}; \quad \dot{\tau}_Y^{\alpha} = \sum\nolimits_{\beta} H^{\alpha\beta} \mid \dot{\gamma}^{\beta} \mid . \tag{126}$$

**Polycrystals, boundary relaxation and dissipation** Boundary relaxation/dissipation of dislocation pile-ups is the key component of the plasticity of polycrystals, and even of constrained single crystals as dislocations on different slip systems in the crystal relax at the boundary. This has been clearly demonstrated by discrete dislocation dynamics simulations (Shu et al. 2001) and continuum model (Mesarovic and Baskaran 2011) of the sheared thin film sandwiched between elastic half spaces, which can be summarized as follows.

Without boundary relaxation, the plasticity of a thin film is practically indistinguishable from elasticity; the energy required to produce slip by building a dislocation pile-up against the boundary is so high that it is comparable to elastic energy for the same value of elastic shear. Only when relaxation at the boundary is allowed between dislocations in the primary and secondary slip systems, a clear distinction between elastic and plastic deformation is observed.

Consider the crystalline grain occupying volume $V$ whose boundary consist of planar facets corresponding to interfaces with another grains: $\partial V = S^1 \cup S^2 \cup S^3 \ldots$ The near-boundary dislocation densities partially relax: $\mathbf{\Gamma}^{\alpha} = \overline{\mathbf{\Gamma}}^{\alpha} + \hat{\mathbf{\Gamma}}^{\alpha}$, so that the corresponding energy density in Eq. (118) is now the quadratic form of the unrelaxed parts, with the corresponding power conjugate $\mathbf{q}^{\alpha}$:

$$\psi = \frac{1}{2} \sum_{\alpha} \sum_{\beta} \overline{\mathbf{\Gamma}}^{\alpha} \cdot \mathbf{F}^{\alpha\beta} \cdot \overline{\mathbf{\Gamma}}^{\beta}; \quad \mathbf{q}^{\alpha} = \frac{\partial \psi}{\partial \overline{\mathbf{\Gamma}}^{\alpha}} = \sum_{\beta} \mathbf{F}^{\alpha\beta} \cdot \overline{\mathbf{\Gamma}}^{\beta}. \tag{127}$$

Dissipation (Eq. 116) now includes boundary relaxation:

$$\mathcal{D} = \int_V \sum\nolimits_{\alpha} \tau^{\alpha} \dot{\gamma}^{\alpha} dV + \int_V \sum\nolimits_{\alpha} \mathbf{r}^{\alpha} \cdot \dot{\hat{\mathbf{\Gamma}}}^{\alpha} dV. \tag{128}$$

Upon performing the same operations as before, the new power balance will read as Eq. (119) with the additional term on the right hand side:

$$-\int_{\partial V}\sum_{\alpha}\left(\mathbf{q}^{\alpha}-\mathbf{r}^{\alpha}\right)\cdot\dot{\hat{\boldsymbol{\Gamma}}}^{\alpha}\,d\partial V, \tag{129}$$

where only the component $\mathbf{N}^{\alpha}\cdot\mathbf{r}^{\alpha}$ has physical meaning, so we take $\mathbf{r}^{\alpha}=r^{\alpha}\mathbf{N}^{\alpha}$.

To formulate the principle of virtual power, in addition to virtual fields $\delta\mathbf{u}$ and $\delta\gamma^{\alpha}$, defined in $V\cup\partial V$, we now must introduce the additional independent virtual field $\delta\hat{\boldsymbol{\Gamma}}^{\alpha}$, defined on $\partial V$. The resulting field equations are identical to the previously derived ones (Eqs. 122–126), except for the additional equations:

$$r^{\alpha}=q^{\alpha}=\mathbf{N}^{\alpha}\cdot\mathbf{q}^{\alpha}\quad\text{on }\partial V. \tag{130}$$

We then interpret Eq. (130) as the boundary yield condition:

$$r^{\alpha}=\operatorname{sign}\left(\mathbf{N}^{\alpha}\cdot\dot{\hat{\boldsymbol{\Gamma}}}^{\alpha}\right)r_{Y}^{\alpha};\quad |\,q^{\alpha}\,|<r_{Y}^{\alpha}\Rightarrow\dot{\hat{\boldsymbol{\Gamma}}}^{\alpha}=0, \tag{131}$$

i.e., no relaxation for this slip system is allowed until the boundary yield condition is satisfied:

$$|\,q^{\alpha}\,|=r_{Y}^{\alpha}\Rightarrow\dot{\hat{\boldsymbol{\Gamma}}}^{\alpha}\neq 0. \tag{132}$$

Finally, the boundary conditions accompanying the stress equilibrium equations (122) must change. The new boundary conditions include the traction matching between the grains:

$$\mathbf{t}^{-}=-\mathbf{t}^{+}\quad\text{on }S^{k},\,k=1,2,3,\ldots, \tag{133}$$

and the conditions which include linear combinations of gradients of displacement discontinuity on the boundary (Eqs. 59–61), with:

$$\hat{\mathbf{B}}^{\alpha}=\hat{\boldsymbol{\Gamma}}^{\alpha}\cdot\left(\mathbf{N}^{\alpha}\mathbf{s}^{\alpha}\right). \tag{134}$$

**Future directions** The major unsolved problem in size-dependent theory of polycrystals (or composites) is the boundary relaxation. The key elements of the theory that require input are the boundary decompositions (Eqs. 59–61) and the boundary yield condition (Eqs. 131 and 132).

Although we have used the linearized kinematics here, the major elements of the full kinematics have been developed for the bulk of a crystal (Gurtin 2002; Forest 2012), but not for the boundaries.

Numerical methods have been developed for various higher order theories, but not for the mathematical structure that includes boundary constitutive laws (Eqs. 118

and 132) and the specific boundary conditions (Eqs. 59–61 and 133). The existence and uniqueness of solutions with such boundary conditions is intuitively plausible, but it remains to be proven.

# Bibliography

S. Abedi, A.L. Rechenmacher, A.D. Orlando, Vortex formation and dissolution in sheared sands. Granul. Matter **14**(6), 695–705 (2012)

H. Abels, Existence of weak solutions for a diffuse interface model for viscous, incompressible fluids with general densities. Commun. Math. Phys. **289**, 45–73 (2009)

H. Abels, H. Garcke, G. Grün, Thermodynamically consistent, frame indifferent diffuse interface models for incompressible two-phase flows with different densities. Math. Models Methods Appl. Sci. **22**, 1150013 (2012)

K.A. Alshibli, S. Sture, Sand shear band thickness measurements by digital imaging techniques. J. Comput. Civ. Eng. **13**(2), 103–109 (1999)

D.M. Anderson, G.B. McFadden, A.A. Wheeler, Diffuse-interface methods in fluid mechanics. Annu. Rev. Fluid Mech. **30**, 139–165 (1998)

I.S. Aranson, L.S. Tsimring, Patterns and collective behavior in granular media: theoretical concepts. Rev. Mod. Phys. **78**(2), 641–692 (2005)

R.J. Asaro, Micromechanics of crystals and polycrystals. Adv. Appl. Mech. **23**, 1–115 (1983)

M.F. Ashby, The deformation of plastically non-homogeneous materials. Philos. Mag. **21**, 399–424 (1970)

M. Asle Zaeem, S.Dj. Mesarovic, Investigation of phase transformation in thin film using finite element method. Solid State Phenom. **150**, 29–41 (2009)

M. Asle Zaeem, S.Dj. Mesarovic, Finite element method for conserved phase fields: stress-mediated diffusional phase transformation. J. Comput. Phys. **229**, 9135–9149 (2010)

M. Asle Zaeem, H. El Kadiri, S.Dj. Mesarovic, M.F. Horstemeyer, P.T. Wang, Effect of the compositional strain on the diffusive interface thickness and on the phase transformation in a phase-field model for binary alloys. J. Phase Equilibria Diffus. **32**, 302–308 (2011)

R.J. Atkin, R.E. Crane, Continuum theories of mixtures: basic theory and historical development. Q. J. Mech. Appl. Math. **29**, 209–244 (1976)

K. Bagi, Stress and strain in granular materials. Mech. Mater. **22**, 165–177 (1996)

R. Baskaran, S. Akarapu, S.Dj. Mesarovic, H.M. Zbib, Energies and distributions of dislocations in stacked pile-ups. Int. J. Solids Struct. **47**, 1144–1153 (2010)

J.L. Bassani, Plastic flow of crystals. Adv. Appl. Mech. **30**, 191–257 (1994)

V. Berdichevsky, P. Hazzledine, B. Shoykhet, Micromechanics of diffusional creep. Int. J. Eng. Sci. **35**(10/11), 1003–1032 (1997)

T.D. Blake, J.M. Haynes, Kinetics of liquid/liquid displacement. J. Colloid Interface Sci. **30**, 421–423 (1969)

D. Bonn, J. Eggers, J. Indekeu, J. Meunier, E. Rolley, Wetting and spreading. Rev. Mod. Phys. **81**, 739–805 (2009)

J.W. Cahn, On spinodal decomposition. Acta Metall. **9**, 795801 (1961)

J.W. Cahn, J.E. Hilliard, Free energy of a nonuniform system. I. Interfacial free energy. J. Chem. Phys. **28**, 258–267 (1958)

L.Q. Chen, Phase-field models for microstructure evolution. Annu. Rev. Mater. Res. **32**, 113–140 (2002)

C.A. Coulomb, Sur une application des règles, de maximis et minimis à quelque problèmes de statique, relatifs à l'architecture. Mém. Math. Phys. **7**, 343–382 (1773). Acad. Royal des Sciences, Paris

F. da Cruz, S. Emam, M. Prochnow, J.-N. Roux, F. Chevoir, Rheophysics of dense granular materials: discrete simulation of plane shear flow. Phys. Rev. E **72**, 021309 (2005)
H.T. Davis, L.E. Scriven, Stress and structure in fluid interfaces. Adv. Chem. Phys. **49**, 357–454 (1982)
P.G. de Gennes, Wetting: statics and dynamics. Rev. Mod. Phys. **57**, 827–863 (1985)
M. Dehsara, H. Fu, S.Dj. Mesarovic, D.P. Sekulic, M. Krivilyov, (In)compressibility and parameter identification in phase field models for capillary flows. Theor. Appl. Mech. (2017). https://doi.org/10.2298/TAM170803009D
H. Ding, P.D.M. Spelt, C. Shu, Diffuse interface model for incompressible two-phase flows with large density ratios. J. Comput. Phys. **226**, 2078–2095 (2007)
A. Drescher, G. De Josselin de Jong, Photoelastic verification of a mechanical model for the flow of a granular material. J. Mech. Phys. Solids **20**, 337–340 (1972)
P. Durbin, Considerations on the moving contact-line singularity, with application to frictional drag on a slender drop. J. Fluid Mech. **197**, 157–169 (1988)
V. Dussan, E. Rame, S. Garoff, On identifying the appropriate boundary conditions at a moving contact line: an experimental investigation. J. Fluid Mech. **230**, 97–116 (1991)
H. Edelsbrunner, Triangulations and meshes in computational geometry. Acta Numerica **9**, 133–213 (2000)
H. Emmerich, *The Diffuse Interface Approach in Materials Science*, Lecture Notes in Physics (Springer, New York, 2003)
A.C. Eringen, *Nonlinear Theory of Continuous Media* (McGraw-Hill, New York, 1962)
A.C. Eringen, E.S. Suhubi, Nonlinear theory of simple microelastic solids: I. Int. J. Eng. Sci. **2**, 189–203 (1964)
J.D. Eshelby, The continuum theory of lattice defects, in *Solid State Physics*, vol. III, ed. by F. Seitz, D. Turnbull (Academic Press, New York, 1956)
L. Euler, Principes généraux de movement des fluides. Académie Royale des Sciences et des Belles-Lettres de Berlin. Mémoires **11**, 274–315 (1757)
N. Eustathopoulos, F. Hodaj, O. Kozlova, The wetting process in brazing, in *Advances in Brazing*, ed. by D.P. Sekulic (Woodhead Publishing, Oxford, 2013), pp. 3–30
R.J. Finno, W. Harris, M.A. Mooney, G. Viggiani, Shear bands in plane strain compression of loose sand. Geotechnique **47**(1), 149–165 (1997)
N.A. Fleck, Brittle fracture due to an array of microcracks. Proc. R. Soc. Lond. A **432**, 55–76 (1991)
N.A. Fleck, J.W. Hutchinson, A phenomenological theory for strain gradient effects in plasticity. J. Mech. Phys. Solids **41**, 1825–1857 (1993)
N.A. Fleck, J.W. Hutchinson, Strain gradient plasticity. Adv. Appl. Mech. **33**, 295–362 (1997)
S. Forest, Generalized continuum modelling of crystal plasticity, in *Generalized Continua and Dislocation Theory*, vol. 537, CISM Courses and Lectures, ed. by C. Sansour, S. Skatulla (Springer, Wien, 2012), pp. 181–287
H. Fu, M. Dehsara, M. Krivilyov, S.Dj. Mesarovic, D.P. Sekulic, Kinetics of the molten Al-Si triple line movement during a brazed joint formation. J. Mater. Sci. **51**(4), 1798–1812 (2016)
Y. Gao, A.C.F. Cocks, Thermodynamic variational approach for climb of an edge dislocation. Acta Mech. Solida Sin. **22**, 426–435 (2009)
K. Garikipati, L. Bassman, M. Deal, A lattice-based micromechanical continuum formulation for stress-driven mass transport in polycrystalline solids. J. Mech. Phys. Solids **49**, 1209–1237 (2001)
GDR MiDi, On dense granular flows. Eur. Phys. J. E **14**, 341–365 (2004)
P. Germain, La méthode des puissances virtuelles en mécanique des milieux continus, 1ère partie: théorie du second gradient. J. de Mécanique **12**, 235–274 (1973a)
P. Germain, The method of virtual power in continuum mechanics. Part 2: Microstructure. SIAM J. Appl. Math. **25**(3), 556–575 (1973b)
J.W. Gibbs, On the equilibrium of heterogeneous substances. Trans. Conn. Acad. **3**, 108–248, 343–524 (1878). Reprinted in The Scientific Papers of J. Willard Gibbs, 55-371. Oxbow Press
C.M. Gourlay, A.K. Dahl, Dilatant shear bands in solidifying metals. Nature **445**, 70–73 (2007)

I. Groma, G. Gyorgyi, B. Kocsis, Debye screening of dislocations. Phys. Rev. Lett. **96**, 165503 (2006)

Z. Guo, P. Lin, J.S. Lowengrub, A numerical method for the quasi-incompressible Cahn–Hilliard–Navier–Stokes equations for variable density flows with a discrete energy law. J. Comput. Phys. **276**, 486–507 (2014)

E.D. Guleryuz, S.Dj. Mesarovic, Dislocation nucleation on grain boundaries: low angle twist and asymmetric tilt boundaries. Crystals **6**(7), 77 (2016)

M.E. Gurtin, The nature of configurational forces. Arch. Ration. Mech. Anal. **131**, 67–100 (1995)

M.E. Gurtin, On the plasticity of single crystals: free energy, microforces, plastic strain gradients. J. Mech. Phys. Solids **48**, 989–1036 (2000)

M.E. Gurtin, A gradient theory of single-crystal viscoplasticity that accounts for geometrically necessary dislocations. J. Mech. Phys. Solids **50**, 5–32 (2002)

M.E. Gurtin, On a framework for small-deformation viscoplasticity: free energy, microforces, strain gradients. Int. J. Plast. **19**, 47–90 (2003)

M.E. Gurtin, A. Needleman, Boundary conditions in small-deformation, single crystal plasticity that account for the Burgers vector. J. Mech. Phys. Solids **53**, 1–31 (2005)

M.E. Gurtin, D. Polignone, J. Vinals, Two phase binary fluids and immiscible fluids described by an order parameter. Math. Models Methods Appl. Sci. **6**(6), 815–831 (1996)

P. Haasen, *Physical Metallurgy*, 2nd edn. (Cambridge University Press, Cambridge, 1986)

C. Herring, Diffusional viscosity of a polycrystalline solid. J. Appl. Phys. **21**, 437–445 (1950)

R. Hill, K.S. Havner, Perspectives in the mechanics of elastoplastic crystals. J. Mech. Phys. Solids **30**, 5–22 (1982)

J.P. Hirth, J. Lothe, *Theory of Dislocations* (Wiley, New York, 1992). Reprint 2nd edn. (1982)

P.C. Hohenberg, B.I. Halperin, Theory of dynamic critical phenomena. Rev. Mod. Phys. **49**, 435–479 (1977)

T.J.R. Hughes, *The Finite Element Method: Linear Static and Dynamic Finite Element Analysis* (Dover Publications, 2000)

D. Jacqmin, Contact-line dynamics of a diffuse fluid interface. J. Fluid Mech. **402**, 57–88 (2000)

D.D. Joseph, Y.Y. Renardy, *Fundamentals of two fluid mechanics* (Springer, New York, 1993)

D. Kay, R. Welford, A multigrid finite element solver for the Cahn-Hilliard equation. J. Comput. Phys. **212**, 288–304 (2006)

J. Klein, E. Kumacheva, Confinement-induced phase transitions in simple liquids. Science **269**, 816–819 (1995)

K. Koch, I.C. Blecher, G. König, S. Kehraus, W. Barthlott, The superhydrophilic and superoleophilic leaf surface of Ruelliadevosiana (Acanthaceae): a biological model for spreading of water and oil on surfaces. Funct. Plant Biol. **36**(4), 339–350 (2008)

K. Kondo, On the geometrical and physical foundations of the theory of yielding. Proc. Japan Natl. Congress Appl. Mech. **2**, 41–47 (1952)

J. Koplik, J.R. Banavar, The no-slip condition for a mixture of two liquids. Phys. Rev. Lett. **80**, 5125–5128 (1998)

E. Kröner, in *Kontinuumstheorie der Versetzungen und Eigenspannungen*. Ergeb. agnew. Math., 5. (Springer, Berlin, 1958)

M.R. Kuhn, K. Bagi, Contact rolling and deformation in granular media. Int. J. Solids Struct. **41**, 5793–5820 (2004)

F.C. Larché, J.W. Cahn, A linear theory of thermochemical equilibrium of solids under stress. Acta Metall. **21**, 1051–1063 (1973)

F.C. Larché, J.W. Cahn, Thermochemical equilibrium of multiphase solids under stress. Acta Metall. **26**, 1579–1589 (1978)

F.C. Larché, J.W. Cahn, Interaction of composition and stress in crystalline solids. Acta Metall. **33**(3), 331–357 (1985)

E.H. Lee, Elastic–plastic deformation at finite strains. J. Appl. Mech. **36**, 1–6 (1969)

G. Lezzi, G. Ventura, The kinematics of lava flows inferred from the structural analysis of enclaves: a review. Geol. Soc. Am. Spl. Pap. **396**, 15–28 (2005)

W. Liu, D.P. Sekulic, Capillary driven molten metal flow over topographically complex substrates. Langmuir **27**, 6720–6730 (2011)

J. Lothe, J.P. Hirth, Dislocation climb forces. J. Appl. Phys. **38**, 845–849 (1967)

J. Lowengrub, L. Truskinovsky, Quasi-incompressible Cahn-Hilliard fluids and topological transition. Proc. R. Soc. Lond. **A454**, 2617–2654 (1998)

L.E. Malvern, *Introduction to the mechanics of a continuous medium* (Prentice-Hall, Englewood Cliffs, NJ, 1969)

A.R. Marrion, *The Chemistry and Physics of Coatings* (The Royal Society of Chemistry, Cambridge, 2004)

S.Dj. Mesarovic, Energy, configurational forces and characteristic lengths associated with the continuum description of geometrically necessary dislocations. Int. J. Plast. **21**, 1855–1889 (2005)

S.Dj. Mesarovic, Plasticity of crystals and interfaces: From discrete dislocations to size-dependent continuum theory. Theor. Appl. Mech. **37**(4), 289–332 (2010). http://www.ssm.org.rs/WebTAM/journal.html

S.Dj. Mesarovic, Lattice continuum and diffusional creep. Proc. R. Soc. A **472**, 20160039 (2016)

S.Dj. Mesarovic, Dislocation creep: climb and glide in the lattice continuum. Crystals **7**(8), 243 (2017)

S.Dj. Mesarovic, R. Baskaran, in *Interfaces in Size-Dependent Crystal Plasticity*. Proceedings of the 3rd International Conference on Heterogeneous Material Mechanics (ICHMM-2011), Shanghai, 2011

S.Dj. Mesarovic, J. Padbidri, Minimal kinematic boundary conditions for simulations of disordered microstructures. Philos. Mag. **85**(1), 65–78 (2005)

S.Dj. Mesarovic, R. Baskaran, A. Panchenko, Thermodynamic coarse-graining of dislocation mechanics and the size-dependent continuum crystal plasticity. J. Mech. Phys. Solids **58**(3), 311–329 (2010)

S.Dj. Mesarovic, J.M. Padbidri, B. Muhunthan, Micromechanics of dilatancy and critical state in granular matter. Geotech Lett **2**, 61–66 (2012)

S.Dj. Mesarovic, S. Forest, J.P. Jaric, Size-dependent energy in crystal plasticity and continuum dislocation models. Proc. R. Soc. A **471**, 20140868 (2015). https://doi.org/10.1098/rspa.2014.0868

R.D. Mindlin, Micro-structure in linear elasticity. Arch. Ration. Mech. Anal. **16**, 51–78 (1964)

Y. Mishin, J.A. Waren, R.F. Sekerka, W.J. Boettinger, Irreversible thermodynamics of creep in crystalline solids. Phys. Rev. B **88**, 184303 (2013)

F.R.N. Nabarro, in *Deformation of Crystals by the Motion of Single Ions*. Report of a Conference on Strength of Solids (The Physical Society, London, 1948), pp. 75–90

F.R.N. Nabarro, Steady-state diffusional creep. Philos. Mag. **16**, 231–237 (1967)

J.F. Nye, Some geometrical relations in dislocated crystals. Acta Metall. **1**, 153–162 (1953)

M. Oda, H. Kazama, Microstructure of shear bands and its relation to the mechanisms of dilatancy and failure of dense granular soils. Geotechnique **48**(4), 465–481 (1998)

M. Oda, J. Konishi, S. Nemat-Nasser, Experimental micromechanical evaluation of strength of granular materials: effects of particle rolling. Mech. Mater. **1**(4), 269–283 (1982)

J. Padbidri, S.Dj. Mesarovic, Acceleration of DEM algorithm for quasistatic processes. Int. J. Numer. Methods Eng. **86**, 816–828 (2011)

J.M. Padbidri, C.M. Hansen, S.Dj. Mesarovic, B. Muhunthan, Length scale for transmission of rotations in dense granular materials. J. Appl. Mech. **79**, 031011 (2012)

R.L. Panton, *Incompressible Flow* (Wiley-Interscience, New York, 1996)

T.C. Patton, *Paint Flow and Pigment Dispersion, A Rheological Approach to Coating and Ink Technology* (Wiley-Interscience, New York, 1979)

R.E. Peierls, The size of a dislocation. Proc. Phys. Soc. **52**, 34–38 (1940)

S. Poulat, B. Decamps, L. Priester, Weak-beam transmission electron microscopy study of dislocation accommodation processes in nickel $\Sigma = 3$ grain boundaries. Phil. Mag. A **77**(6), 1381–1397 (1998)

S. Poulat, B. Decamps, L. Priester, J. Thibault, Incorporation processes of extrinsic dislocations in singular, vicinal and general grain boundaries in nickel. Mater. Sci. Eng. A **309–310**, 483–485 (2001)
L. Priester, "Dislocation-interface" interaction – stress accommodation processes at interfaces. Mater. Sci. Eng. A **309–310**, 430–439 (2001)
N. Provatas, K. Elder, *Phase Field Methods in Materials Science and Engineering* (Wiley-VCH, Weinheim, 2010)
F. Radjai, S. Roux, Turbulentlike fluctuations in quasistatic flow of granular media. Phys. Rev. Lett. **89**(6), 064302 (2002)
R. Ranacher, in *Methods for Numerical Flow Simulations*, ed. by G.P. Galdi, R. Rannacher et al. Lecture Notes, Oberwolfach Seminar Hemodynamical Flows: Aspects of Modeling, Analysis and Simulation (Birkhäuser, Basel, 2007)
A.L. Rechenmacher, Grain-scale processes governing shear band initiation and evolution in sands. J. Mech. Phys. Solids **54**, 22–45 (2006)
O. Reynolds, On the dilatancy of media composed of rigid particles in contact. Philos. Mag. **20**(127), 469–481 (1885)
J.R. Rice, Mathematical analysis in the mechanics of fracture, in *Fracture, An Advanced Treatise*, vol. II, ed. by H. Liebowitz (Academic Press, New York, 1968), pp. 191–311
J.R. Rice, Inelastic constitutive relations for solids: an internal-variable theory and its application to metal plasticity. J. Mech. Phys. Solids **19**(6), 433–455 (1971)
J.-N. Roux, G. Combe, in *How Granular Materials Deform in Quasistatic Conditions*, ed. J. Goddard, J.T. Jenkins, P. Giovine. IUTAM-ISIMM Symposium on Mathematical Modeling and Physical Instances of Granular Flow (American Institute of Physics, College Park, MD, 2010), pp. 260–270
A.S. Saada, L. Liang, J.L. Figueroa, C.T. Cope, Bifurcation and shear band propagation in sands. Geotechnique **49**(3), 367–385 (1999)
M. Satake, New formulation of graph-theoretical approach in the mechanics of granular materials. Mech. Mater. **16**, 65–72 (1993)
A.N. Schofield, C.P. Wroth, *Critical State Soil Mechanics* (McGraw-Hill, London, 1968)
P. Seppecher, Moving contact lines in the Cahn-Hilliard theory. Int. J. Eng. Sci. **34**, 977–992 (1996)
D. Seveno, A. Vaillant, R. Rioboo, H. Adao, J. Conti, J. De Coninck, Dynamics of wetting revisited. Langmuir **25**(22), 13034–13044 (2009)
J.Y. Shu, N.A. Fleck, E. Van der Giessen, A. Needleman, Boundary layers in constrained plastic flow: comparison of non-local and discrete dislocation plasticity. J. Mech. Phys. Solids **49**, 1361–1395 (2001)
D.E. Spearot, M.A. Tschopp, K.I. Jacob, D.L. McDowell, Tensile strength of <100> and <110> tilt bicrystal copper interfaces. Acta Mater. **55**, 705–714 (2007)
A.P. Sutton, R.W. Balluffi, *Interfaces in crystalline materials* (Clarendon Press, Oxford, 1995)
P.A. Thompson, M.O. Robbins, Shear flow near solids: epitaxial order and flow boundary solutions. Phys. Rev. A **41**, 6830–6837 (1990)
A. Tordesillas, Force chain buckling, unjamming transitions and shear banding in dense granular assemblies. Philos. Mag. **87**(32), 4987–5016 (2007)
A. Tordesillas, M. Muthuswamy, S.D.C. Walsh, Mesoscale measures of nonaffine deformation in dense granular assemblies. J. Eng. Mech. **134**(12), 1095–1113 (2008)
C. Trusdell, R.A. Toupin, The classical field theories, in *Encyclopedia of Physics*, vol. III(1), ed. by S. Flugge (Springer, Berlin, 1960)
M.A. Tschopp, D.L. McDowell, Dislocation nucleation in $\Sigma 3$ asymmetric tilt grain boundaries. Int. J. Plast. **24**, 191–217 (2008)
A. Villani, E.P. Busso, S. Forest, Field theory and diffusion creep predictions in polycrystalline aggregates. Model. Simul. Mater. Sci. Eng. **23**, 055006 (2015)
R.S. Yassar, S.Dj. Mesarovic, D.P. Field, Micromechanics of hardening of elastic-plastic crystals with elastic inclusions: I. Dilute concentration. Int. J. Plast. **23**, 1901–1917 (2007)

P. Yue, J.J. Feng, Wall energy relaxation in the Cahn–Hilliard model for moving contact lines. Phys. Fluids **23**, 012106 (2011)
L.-N. Zou, X. Cheng, M.L. Rivers, H.M. Jaeger, S.R. Nagel, The packing of granular polymer chains. Science **326**, 408–410 (2009)

# Multiscale Dislocation-Based Plasticity

**Hussein M. Zbib, Mehdi Hamid, Hao Lyu, and Ioannis Mastorakos**

**Abstract** This chapter, outlines a multiscale dislocation-based plasticity framework coupling discrete dislocation dynamics (DDD) with continuum dislocation-based plasticity. In this framework, and guided by DDD, a continuum dislocation dynamics (CDD) plasticity model involving a set of spatio-temporal evolution equations for dislocation densities representing mobile and immobile species is developed. The evolution laws consist of a set of components each corresponding to a physical mechanism that can be explicitly evaluated and quantified from DDD analyses. In this framework, stochastic events such as cross-slip of screw dislocations and uncertainties associated with initial microstructural conditions are explicitly incorporated in the continuum theory based on probability distribution functions defined by activation energy and activation volumes. The result is a multiscale dislocation-based plasticity model which can predict not only the macroscopic material mechanical behavior but also the corresponding microscale deformation and the evolution of dislocation patterns, size and gradient-dependent deformation phenomena, and related material instabilities at various length and time scales.

## 1 Introduction

This chapter deals with the issue of plastic deformation in metals with focus on the development of models that are based on underlying fundamental mechanisms that control the process. Particularly, the focus is on recent developments by Zbib and

H. M. Zbib (✉) · M. Hamid · H. Lyu
School of Mechanical and Materials Engineering, Washington State University, Pullman, WA, USA
e-mail: zbib@wsu.edu

I. Mastorakos
Department of Mechanical & Aeronautical Engineering, Clarkson University, Potsdam, NY, USA

 
S. Mesarovic et al. (eds.), *Mesoscale Models*, CISM International Centre for Mechanical Sciences 587, https://doi.org/10.1007/978-3-319-94186-8_2

co-workers (Zbib et al. 1998; Zbib and Diaz de la Rubia 2002; Alankar et al. 2013; Li et al. 2014; Mastorakos and Zbib 2014; Askari et al. 2015; Lyu et al. 2015a, b, c; Taheri-Nassaj and Zbib 2015; Lyu et al. 2016a, b, c; Taheri-Nassaj and Zbib 2016; Lyu et al. 2017) in the field of discrete dislocation dynamics and multiscale dislocation-based plasticity theory. The principal idea in the multiscale approach is to utilize and pass information from small scale (say microscale) to larger scale (say mesoscale) and thereby develop physically-based and continuum models that can accurately predict deformation phenomena. The emphasis in this chapter is on the development of models that can predict plastic deformation in metals and its relation to underlying dislocation mechanisms, size effects and microstructure.

Plastic deformation in metals is a very complex phenomenon originating from highly nonlinear dynamical processes associated with small-scale defects, such as dislocations, vacancies, interstitials, cracks, voids, etc. Dislocations, in particular, and their mutual interaction and interaction with other defects determine material strength and ductility. Moreover, and because of their long-range and short range-interactions, dislocation tend to self-organize in spatial patterns at the microscale, resulting in heterogeneous deformation field and thereby affecting macroscopic properties. The stability of plastic deformation and patterning was examined by (Holt 1970) who investigated the stability of a uniformly dense dislocation distribution, and showed that an initially homogenous distribution is unstable, and a stable heterogeneous distribution may emerge. The idea was further investigated by (Walgraef and Aifantis 1985) who viewed dislocation patterning as a self-organization phenomenon driven by the dynamics of dislocations. They developed a nonlinear system of equations and discussed the role and nature of various gradient coefficients, appearing in the form of diffusion-like coefficients for different dislocation populations, and reaction-like terms modeling dislocation production or annihilation. Similarly, (Kratochvil 1988) analyzed dislocation patterning by considering a net dipole drift to stress concentrations created by glide dislocations under an applied stress state, including two sets of mobile dislocations with opposite Burgers vectors. These are only a few examples of numerous works that followed on this topic and there are many papers on the formation and stability of dislocation patterns, e.g. (Kubin and Estrin 1988; Arsenlis and Parks 1999; Groma and Balogh 1999; Morita et al. 2001; Arsenlis et al. 2004; Ananthaktishna 2007; Hallber and Ristinmaa 2013; Pontes et al. 2015). Zbib and co-workers (Shizawa and Zbib 1999a, b; Morita et al. 2001) extended this idea to three-dimensions within a continuum mechanics framework using the concept of Nye's and incompatibility tensors to represent geometrically necessary dislocations and dislocation dipoles, enabling the prediction of dislocation cell wall formation as well as size effects.

Size effects play an important role in plastic deformation, especially at small length scales. Plasticity and strength at small length can be strongly dependent on both the size of the underlying microstructure (grain size, particle size, etc.) and the specimen/component size when the dimensions are small. In polycrystalline materials, yield strength increases with decreasing grain size. This effect is captured by the classical Hall-Petch (Hall 1951) relationship which describes the yield stress of a polycrystal as $\sigma_y = K/\sqrt{d} + \sigma_0$, where $K$ is a material constant, $d$ is the grain size, and $\sigma_0$ is the constant yield stress. This relationship is based on the mechanism of dislocation pileup against grain boundaries when dislocations emit from a dislocation source under uniform applied stress. However, under non uniform states of stress this relationship is no longer valid since the dislocation pileup differs depending on the gradients of the stress field. In this case an alternative relationship emerges leading to the so called stress-gradient plasticity theory (Hirth 2006a, b; Chakravarthy and Curtin 2011; Taheri-Nassaj and Zbib 2015). The Hall-Petch relation is thought of as a special case of the stress-gradient theory (Hirth 2006a, b; Chakravarthy and Curtin 2011; Forest and Sab 2012; Liu et al. 2014; Taheri-Nassaj and Zbib 2015). The stress-gradient theory is based on the classical problem of dislocation pileups against grain boundaries as natural obstacles under an inhomogeneous state of stress. The problem of dislocation pileup relates to earlier works by Hall (1951), Petch (1953) Eshelby et al. (1951). Stress gradient plasticity theories lead to a dependence of flow stress on higher-order stress-gradients in addition to the obstacle spacing, e.g. grain size. The well-known Hall-Petch relation is a special case in the stress-gradient theory (Hirth 2006a, b; Chakravarthy and Curtin 2011; Taheri-Nassaj and Zbib 2015).

Size effects at small scales also emerge when the deformation field includes torsional and bending mode that lead to the formation of so-called geometrically necessary dislocations (GNDs). Theories to model size effects resulting from GNDs are based on the concept of dependence of strength on higher order strain gradient that relate the gradient of plastic strain to the density of GNDs (Zbib and Aifantis 1992; Fleck et al. 1994; Zbib 1994; Fleck and Hutchinson 1997; Zhu et al. 1997; Aifantis 1999; Huang et al. 2000, 2004; Gurtin and Anand 2005; Mesarovic 2005a, b; Ohashi 2005; Forest and Sab 2012; Lyu et al. 2015a, b, c; Wulfinghoff and Böhlke 2015; Bayerschen et al. 2016). This is in contrast to the dislocation pileup mechanism and the stress-gradient theory discussed above, where hardening is due to pile up of dislocations against obstacles. Moreover, strain-gradient theories contain some critical complexities. The intrinsic length scale is not well defined and can be experiment-dependent. In contrast, for stress-gradient theory, the stress-gradient coefficients are given explicitly in terms of obstacle spacing.

A combination of strain-gradient and stress-gradient theory is proposed in (Lyu et al. 2016a, b, c) which can capture the full spectrum of size effect. It is argued that both strain-gradient and stress-gradient plasticity are complementary rather than competing theories. Although these two theories represent two completely different mechanisms and length scale effects, a combined theory can more accurately describe the size-dependency phenomenon in plastic deformation over a wide range of length scales. This theory is reviewed and implemented in the multi-scale model discussed in this chapter based on a discrete dislocation dynamic and a continuum dislocation theory, coupled with crystal plasticity. The effect of the GNDs density is integrated into the model for the mean free path of dislocations. This was found to contribute to strain hardening. A combined stress/strain-gradient plasticity theory [8] is implemented into the multiscale framework as discussed in the next section.

## 2 Multiscale Dislocation-Based Plasticity Framework

The physical features and phenomena associated with plastic deformation can span over a length scale ranging from 10s of nm to 10s of mm, and involve a number of underlying deformation mechanisms: dislocations, grain boundaries, interfaces, localization, size effect, and damage. While discrete methods, such as discrete dislocation dynamics and molecular dynamics, are physically based and thus can provide rigorous understating of small scale phenomena, continuum methods, such as crystal plasticity, are phenomenological but provide a means to address large scale problems. The central idea in the multiscale framework is to bridge these scales and so that a given model for a given length scale can be more accurate if it uses information passed into it from a smaller scale (Zbib and Diaz de la Rubia 2002; Ohashi et al. 2007; Akasheh et al. 2009; Groh et al. 2009; Zbib et al. 2011)

The framework depicted in Fig. 1 is based on dislocation mechanisms and models that are derived from molecular dynamics (MD) and discrete dislocation dynamic (DDD), and integrated in continuum dislocation dynamics (CDD) which is, in turn, coupled with crystal plasticity (CP) (Li et al. 2014; Lyu et al. 2015a, b, c; Taheri-Nassaj and Zbib 2015; Lyu et al. 2016a, b, c; Taheri-Nassaj and Zbib 2016). Furthermore, impeded in this framework, and in order to capture size effect, two gradient models, strain-gradient and stress-gradient, which are based on different dislocation mechanisms as discussed in Sect. 1.

In this approach, mesoscale analyses are based at the scale of a single grain, while macroscale analyses are for large polycrystalline scale. There are different types of mesoscale analyses that can be performed. In this chapter, continuum

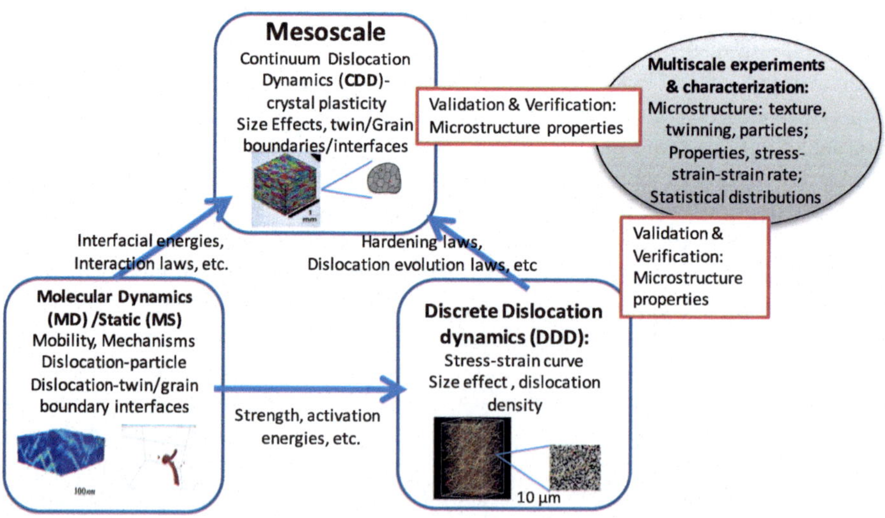


**Fig. 1** Multiscale framework coupling discrete dislocation dynamics, continuum dislocation dynamics, crystal plasticity within input form molecular dynamics and multiscale experiments

dislocation dynamics occurring at the scale of the grain is presented. Also, crystal plasticity, which is a mix of discrete and continuum concepts, starts at the grain scale and can be used for polycrystalline analysis with the use of averaging schemes. The mesoscale analysis in this chapter is discussed in the context of internal state variables, which are rooted in continuum level. Because the discrete dislocation dynamics and crystal plasticity formulations require a large capacity of computing power, they are not generally used to solve large scale boundary value problems of structural components or systems. It is the mesoscale internal state variable continuum theory that is often employed in solving practical engineering problems. It is noted that particular discrete details of dislocation nucleation, motion, and interaction can be quite clearly captured in the discrete dislocation dynamics (DDD) formulation. However, in DDD the degrees of freedom required to solve an engineering problem are prohibitive. Likewise, the crystal plasticity formulation employs discrete crystals but treats the dislocation effects in a phenomenological manner. Hence, it loses some of the details but can capture a larger scale boundary value problem. In this chapter the focus is on the details of a continuum dislocation dynamics (CDD) model coupled with CP as it sets at the core of the multiscale framework. The details of the discrete dislocation dynamics model can be found in a number of articles by Zbib and co-workers, e.g. (Hirth et al. 1996, 1998; Zbib et al. 1998, 2000; Rhee et al. 1998; Zbib and Diaz de la Rubia 2002)

Within the continuum mechanics framework, the governing equations of the material response are developed based on a representative volume element (RVE) over which the deformation field is assumed to be homogeneous. In this approach, the effect of internal defects, such as dislocations, voids, microcracks, etc., on material behavior and the manner they influence material properties is modeled through a set of internal variables and corresponding phenomenological evolution equations. Generally, the material response is measured in terms of the macroscopic strain rate (also called stretching) tensor, and spin tensor $\boldsymbol{W}$ and their relations to the stress tensor $\boldsymbol{\sigma}$. The tensors $\boldsymbol{D}$ and $\boldsymbol{W}$ are the symmetric and antisymmetric parts of the velocity gradient $\boldsymbol{L}$, respectively, i.e.

$$\boldsymbol{D} = \frac{1}{2}\left[\nabla \mathbf{v} + \nabla \mathbf{v}^T\right], \boldsymbol{W} = \frac{1}{2}\left[\nabla \mathbf{v} - \nabla \mathbf{v}^T\right], \tag{1}$$

where $\mathbf{v}$ is the material particle velocity and $\nabla$ (.) is the spatial gradient operator. For elasto-viscoplastic behavior, $\boldsymbol{D}$ can be decomposed into elastic part $\boldsymbol{D}^e$ and plastic part $\boldsymbol{D}^p$, and $\boldsymbol{W}$ into lattice spin tensor $\boldsymbol{\omega}$ and plastic spin $\boldsymbol{W}^p$, such that

$$\boldsymbol{D} = \boldsymbol{D}^e + \boldsymbol{D}^p, \boldsymbol{W} = \boldsymbol{\omega} + \boldsymbol{W}^p \tag{2}$$

For most metals the elastic response is linear and can be expressed by the rate form of Hooke's law for large deformation and material rotation as follows.

$$\overset{o}{\sigma} = \left[\mathbf{C}^e\right]\left[\mathbf{D} - \mathbf{D}^p\right]; \overset{o}{\sigma} = \dot{\sigma} - \omega\sigma + \sigma\omega \tag{3}$$

where $\mathbf{C}^e$ is fourth order elasticity tensor. The main challenge in the plasticity theory is the development of proper constitutive laws for $\boldsymbol{D}^p$ and $\boldsymbol{W}^p$ (Bammann 1984; Voyadjis and Mohammad 1991; Breunig et al. 1992; Bammann and Dawson 1997; Forest and Sievert 2003, McDowell 2005, 2010). More importantly, these laws should be based on the underlying microstructure, mainly dislocations. Nonetheless, this task is perhaps formidable, especially when bridging two scales orders of magnitude apart, i.e. the continuum scale and the discrete dislocation scale. Here it is emphasized that the "*assumed*" constitutive nature of $\boldsymbol{D}^p$ and flow stress and their dependence on internal variables and gradients of internal variables is very critical, since they dictate, among other things, the length scale of the problem and the phenomena that can be predicted by the model. In this respect, it goes without question that the most rigorous and physically based approach of computing the plastic strain and strain hardening in metals is DDD, with all relevant length scales, is through explicit evaluation of interaction, motion and evolution of all individual discrete dislocations and all relevant other defects in the crystal. However, within the framework of crystal plasticity (CP), constitutive equations based on DDD can be developed and incorporated into the expression for $\boldsymbol{D}^p$ and $\boldsymbol{W}^p$ which are given by the following constitutive relations.

$$\boldsymbol{D}^p = \sum_{\alpha=1}^{N} \dot{\gamma}^\alpha \frac{\left(\boldsymbol{n}^\alpha \otimes \boldsymbol{m}^\alpha + \boldsymbol{m}^\alpha \otimes \boldsymbol{n}^\alpha\right)}{2}, \quad \boldsymbol{W}^p = \sum_{\alpha=1}^{N} \dot{\gamma}^\alpha \frac{\left(\boldsymbol{n}^\alpha \otimes \boldsymbol{m}^\alpha - \boldsymbol{m}^\alpha \otimes \boldsymbol{n}^\alpha\right)}{2}, \tag{4}$$

where $N$ is the total number of slip systems, $\boldsymbol{n}^\alpha$ and $\boldsymbol{m}^\alpha$ are unit vectors representing slip direction and slip plane normal of slip system $\alpha$, respectively, and $\dot{\gamma}^\alpha$ is the crystallographic slip rate on slip system $\alpha$ and is usually given by phenomenological strain-rate-dependent power law of the following form.

$$\dot{\gamma}^\alpha = \dot{\gamma}_0 \left| \frac{\tau^\alpha}{\tau^{*\alpha}} \right|^{1/m} \operatorname{sign}\left(\tau^\alpha\right) \tag{5}$$

where $\dot{\gamma}_0$ is reference strain rate, $\tau^{*\alpha}$ is the slip resistance of slip system $\alpha$, $\tau^\alpha$ is the resolved shear stress on slip system $\alpha$, and $m$ is the strain rate sensitivity exponent. The hardening law that defines the evolution of the critical resolved shear stress for each slip system is given by a phenomenological equation of the following form (Asaro 1983).

$$\dot{\tau}^{*\alpha} = \sum_{\alpha} h_{\alpha\beta} \dot{\gamma}^\beta \tag{6}$$

where $h_{\alpha\beta}$ is the hardening coefficient matrix consisting of self-hardening and latent hardening components which are obtained from experimental results using curve fitting.

Alternative to Equation (6) and within the continuum mechanics framework, one can introduce a more physically-based relationship relating shear strain rate to dislocation density through the Orowan equation (Orowan 1940), i.e.

$$\dot{\gamma}^{\alpha} = \rho_m^{\alpha} \mathrm{b} \overline{\upsilon}_g^{\alpha} \tag{7}$$

where $b$ is the magnitude of the Burgers vector, $\rho_m^{\alpha}$ is the total mobile dislocation density on slip system $\alpha$ and $\overline{\upsilon}_g^{\alpha}$ is the average dislocation glide velocity. In turn, the dislocation velocity can be expressed by a power law relationship of the form,

$$\overline{\upsilon}_g^{\alpha} = \upsilon_0 \left| \frac{\tau^{\alpha}}{\tau^{*\alpha}} \right|^{1/m} \mathrm{sign}\left(\tau^{\alpha}\right) \tag{8}$$

In this equation, $v_0$ is the reference velocity, and $\tau^{*\alpha}$ is slip resistance to dislocation motion for each slip system $\alpha$. $\tau^{*\alpha}$ can be divided into three additive terms, internal friction term $\tau_0$, dislocation hardening term $\tau_{dh}^{\alpha}$and size-dependent term $\tau_s^{\alpha}$, such that

$$\tau^{*\alpha} = \tau_0 + \tau_{dh}^{\alpha} + \tau_s^{\alpha} \tag{9}$$

The dislocation hardening term is the slip resistance to the total dislocation density $\rho_T^{\alpha}$ (mobile and immobile) on all slip systems and against the moving dislocation on one specific slip system $\alpha$, and is given by the Baily-Hirsch model (Ohashi 1994; Ohashi et al. 2007)

$$\tau_{dh}^{\alpha} = \alpha^* b\mu \sum \Omega^{\alpha\beta} \sqrt{\rho_T^{\beta}} \tag{10}$$

where $\mu$ is the shear modulus, $\alpha*$ is a numerical factor on the order of 0.1, $\Omega^{\alpha\beta}$ is the interaction matrix of the slip system $\alpha$ with the other slip systems such as slip system. Equations (7) through (10) above are dependent on the dislocations density and the underlying microstructure, such as grain size and distribution. The next section describes the evolution equations for the dislocation species densities and distributions through a system of dynamic equations.

## 3 Continuum Dislocation Dynamics (CDD)

In previously proposed dislocation-based crystal plasticity models, the dislocation density evolution laws was either ignored or fitted to experiments, e.g. (Kocks 1976; Lim et al. 2011). Here, a continuum dislocation dynamics (CDD) model is using an internal state variable theory along the lines discussed in (Li et al. 2013), but advanced to include stochastic and size effects. In this framework, the evolution rate of the dislocation density is based on mechanisms that can be explicitly quantified from DDD. Generally, the dislocations on each slip system can be divided into two types, mobile $\rho_m^\alpha$ and immobile $\rho_i^\alpha$. The mobile dislocation density can be further divided into two sets, one set $\rho_m^{+\alpha}$ for the dislocations of the same type but with "positive" Burgers vector$\rho_m^{+\alpha}$, and a corresponding set $\rho_m^{-\alpha}$ with negative Burgers vector, such that

$$\rho_m^\alpha = \rho_m^{+\alpha} + \rho_m^{-\alpha} \tag{11}$$

Next, and based on mechanisms that can be explicitly captured and quantified in discrete dislocation dynamics, evolution equations are derived for the three sets of dislocations, two mobile and one immobile.

### 3.1 Spatio-Temporal Evolution Equations

The evolution equations for the dislocations densities, including dislocation flux terms (first term on the right side), diffusion (climb) (second term on the right side), and sources/annihilation function (last terms), have the generalized form,

$$\dot{\rho}_m^{+\alpha} + \overline{\boldsymbol{v}}_g^\alpha . \nabla \rho_m^{+\alpha} = D\nabla^2 \rho_m^{+\alpha} + f^{+\alpha}\left(\rho_m^{+\alpha}, \rho_m^{-\alpha}, \rho_i^\alpha\right) \tag{12a}$$

$$\dot{\rho}_m^{-\alpha} - \overline{\boldsymbol{v}}_g^\alpha . \nabla \rho_m^{-\alpha} = D\nabla^2 \rho_m^{-\alpha} + f^{-\alpha}\left(\rho_m^{+\alpha}, \rho_m^{-\alpha}, \rho_i^\alpha\right) \tag{12b}$$

$$\dot{\rho}_i^\alpha = D_i \nabla^2 \rho_i^\alpha + f_i^\alpha\left(\rho_m^{+\alpha}, \rho_m^{-\alpha}, \rho_i^\alpha\right) \tag{12c}$$

where $\nabla$ is the gradient operator, $\nabla^2$ is the Laplacian operator, $D$ is the climb (diffusion) coefficients. Based on DDD, specific functional forms for the source functions $f^{+\alpha}$, $f^{-\alpha}$, $f_i^\alpha$ can be developed. Table 1 lists the basic elements in DDD and the mechanisms that can be explicitly captured and evaluated in DDD. Recent works by Zbib and co-workers proposed that the evolution rate of the dislocation density consists of the following six additive terms each corresponding to a specific physical mechanism.

1. Multiplication and growth of resident dislocations as well as the production of new dislocations from Frank-Reed sources in the slip system $\alpha$.
2. Mutual annihilation of two mobile edge or screw dislocations with opposite signs in the slip system $\alpha$ when two dislocation are at a critical distance $R_c$.
3. Trapping of mobile dislocation by immobile dislocations at critical capture radius $R_c$.
4. Mobilization of immobile dislocations due to the breakup of junctions, dipoles, pinning parts, etc., at critical stress conditions.

**Table 1** Dislocation mechanics and mechanisms

| Mechanisms and measurements from Discrete Dislocation Dynamics (SDDD) | Mesoscale: Continuum Dislocation Dynamics (CDD) –Crystal Plasticity(CP) | | Parameters that can be obtained from DDD |
|---|---|---|---|
| $N$ discrete dislocations on slip system $\alpha$, segment $j$ length $\ell_j^\alpha$, segment line sense (vector) $\zeta_j^\alpha$, segment glide (stochastic) velocity $v_{gj}^\alpha$, Burgers vector $b_j^\alpha$ Slip normal $n_j^\alpha$ | Dislocation density on each slip system $\alpha$<br>Mobile: $\rho_m^{+\alpha}$, $\rho_m^{-\alpha}$<br>Immobile: $\rho_i^\alpha$ | | |
| Plastic shearing per slip system: sum over all segments divide by volume<br>$\dot{\gamma}^\alpha = \sum_\beta b_\beta^\alpha \ell_\beta^\alpha v_\beta^\alpha / V$ | Orowan relation: $\dot{\gamma}^\alpha = \rho_m^\alpha b^\alpha v_g^\alpha$ | | |
| Plastic strain rate tensor:<br>$D_{ij}^p = \sum_\alpha \dot{\gamma}^a \left(b_i^a n_j^a\right)_{sym}$ | $D_{ij}^p = \sum_\alpha \dot{\gamma}^a \left(m_i^a n_j^a\right)_{sym}$ | | |
| Nye's dislocation density tensor:<br>$\alpha_{ij} = \sum_{all\ segments} \ell_\beta b_i \xi_{j\beta} \beta / V$ | $\alpha_{ij} = \sum_{all\ slip\ systems} \left(\rho^{+\alpha} - \rho^{-\alpha}\right) b_i^\alpha \zeta_j^\alpha$ | | |
| Long-range stress | $\sigma_{ij}^*(x) = C_{ijkl} \int_{V'} \varepsilon_{lnh} C_{pqmn} G_{kp,q} \left(x - x'\right) \alpha_{hm} \left(x'\right) dV'$ | | |
| Dislocations flux (Glide) | $\overline{v}_g^\alpha . \nabla \rho_m^{+\alpha} \& \overline{v}_g^\alpha . \nabla \rho_m^{-\alpha}$ | | |
| Multiplication and growth of mobile dislocations | $\beta_1 \rho_m^{+\alpha} \& \beta_1 \rho_m^{-\alpha}$ | $\beta_1 = \alpha_1 \frac{\overline{v}_g^\alpha}{\ell}$ | $\alpha_1$ |
| Mobile-mobile mutual annihilation | $\beta_2 \rho_m^{-\alpha} \rho_m^{+\alpha}$ | $\beta_2 = 2\alpha_2 R_c \overline{v}_g^\alpha$ | $\alpha_2$ |
| Pinning of mobile by immobile | $\beta_3 \left(\rho_i^\alpha\right)^2 \rho_m^+, \beta_3 \left(\rho_i^\alpha\right)^2 \rho_m^{(-)}$ | $\beta_3 = 2\alpha_3 A_c R_c \overline{v}_g^\alpha$ | $\alpha_3$ |
| Break-up of locks | $\beta_4 \rho_i^\alpha$ | $\beta_4 = \alpha_4 \frac{\overline{v}_g^\alpha}{\ell} \left\| \frac{\tau}{\tau_{jn}} \right\|^r$ | $\alpha_4$, $r$, $\tau_{jn}$ |
| Dipole formation | $\beta_5 \rho_m^{-\alpha} \rho_m^{+\alpha}$ | $\beta_5 = 2\alpha_5 R_c \overline{v}_g^\alpha$ | $\alpha_5$ |
| Cross-slip (Stochastic) | $\sum_\beta^N \beta_6 P^{\alpha\beta} \rho_m^{+\beta} \& \sum_\beta^N \beta_6 P^{\alpha\beta} \rho_m^{-\beta}$ | $\beta_6 = \alpha_6 \frac{\overline{v}_g^\alpha}{\ell}$ | $\alpha_6$ |
| Dislocations Climb (Diffusion) | $D\nabla^2 \rho_m^{(+)}, \quad D\nabla^2 \rho_m^{(-)} \& D\nabla^2 \rho_i$ | | - |

5. Mobile dislocations of opposite signs forming dipoles at critical capture radius $R_c$.
6. Cross-slip when it becomes energetically favorable for screw dislocation segments on one slip plane to cross to another glide plane.

Evolution equations for each of the above mentioned mechanisms have been developed by Zbib and co-workers. Here, it is assumed that the flux of gliding mobile dislocations is diffusive and is related to dislocation glide with an average value of $\overline{v}_g^\alpha$ within a three-dimensional framework. The second column in Table 1, lists the corresponding continuum variables from which the source functions in equations (11) can be constructed by adding these terms appropriately, leading to the following set of spatio-temporal evolution equations for dislocations on each slip system α.

$$\dot{\rho}_m^{+\alpha} + \overline{\boldsymbol{v}}_g^\alpha.\nabla\rho_m^{+\alpha} = D\nabla^2\rho_m^{+\alpha} + \beta_1\,\rho_m^{+\alpha} - (\beta_2+\beta_5)\,\rho_m^{+\alpha}\,\rho_m^{-\alpha} - \beta_3\left(\rho_i^\alpha\right)^2\rho_m^{+\alpha}$$
$$+\beta_4\rho_i^\alpha + \beta_6\sum_\beta^N P^{\alpha\beta}\rho_m^{+\beta} \tag{13a}$$

$$\dot{\rho}_m^{-\alpha} - \overline{\boldsymbol{v}}_g^\alpha.\nabla\rho_m^{-\alpha} = D\nabla^2\rho_m^{-\alpha} + \beta_1\rho_m^{-\alpha} - (\beta_2+\beta_5)\,\rho_m^{-\alpha}\rho_m^{+\alpha} - \beta_3\left(\rho_i^\alpha\right)^2\rho_m^{-\alpha}$$
$$+\,\beta_4\rho_i^\alpha + \beta_6\sum_\beta^N P^{\alpha\beta}\rho_m^{-\beta} \tag{13b}$$

$$\dot{\rho}_i^\alpha = D_i\nabla^2\rho_i^\alpha + 2\beta_5\,\rho_m^{+\alpha}\rho_m^{-\alpha} + \beta_3\left(\rho_i^\alpha\right)^2\left(\rho_m^{+\alpha} + \rho_m^{-\alpha}\right) - 2\beta_4\rho_i^\alpha \tag{13c}$$

In this set of equations (see Table 1), is $\overline{v}_g^\alpha$ average velocity, and $\tau^{*\alpha}$ is critical resolved shear (threshold) stress which are defined in Equations (8) and (9) respectively. Also, the dislocation mean free path which is indicated by $\overline{\ell}$ in the equations is defined as inverse of square root of the total dislocation densities. Explicit expressions for the coefficient $\beta_1 - \beta_6$ are given in Table 1. These expressions are derived by assuming that the evolution, interaction and reactions occur over some characteristic time of interactions for dislocation segment of characteristic length $\overline{\ell}$ and having an average velocity of $\overline{v}_g^\alpha$. [1]

[1](i) Dislocation multiplication, term $\beta_1 = \alpha_1\,\overline{v}_g^\alpha/\overline{\ell}$.Let the rate of multiplication be $\dot{\rho}_m^{+\alpha}\big|_{(1)} = \alpha_1\rho_m^{+\alpha}/\overline{t}$, and $\overline{t}$ be some characteristic time for dislocation segment of characteristic length $\overline{\ell}$ and having an average velocity of $\overline{v}_g^\alpha$. Then $\overline{t} = \overline{\ell}/\overline{v}_g^\alpha$, which can be substituted into the expression for $\dot{\rho}_m^{+\alpha}\big|_{(1)}$, yielding $\beta_1$.
(ii) Mutual annihilation of mobile dislocations, term $\beta_2 = 2\alpha_2 R_c\overline{v}_g^\alpha$. Let $f = 1/\overline{t}$ be the frequency of which a mobile dislocation located in a circular region of radius $R_c$ (capture radius for annihilation) gets annihilated by a mobile dislocation of opposite sign sweeping through the circular region with an average velocity $\overline{v}_g^\alpha$. Then $\overline{t} = \overline{x}\,/\,\overline{v}_g^\alpha$ , where $\overline{x}$ is the average distance traveled by dislocations sweeping through the circular region. Then let the rate of annihilation be

The last term in equations (13a) and (13b) correspond to cross-slip. This mechanism is a thermal process and a stochastic phenomenon. This term is the sum of all possible cross slip of screw dislocations from slip system β to system α, with $\beta_6$ being the fraction of screw dislocations. $P^{\alpha\beta}$ is the probability of screw dislocations on β-plane to cross slip to α-plane. $P^{\alpha\beta}$ is a matrix consisting of 0 or 1 components and is determined by a Monte-Carlo (MC) type analysis. The cross-slip probability is discussed in Rhee et al. (1998) in the discrete dislocation framework and adopted here for the continuum case, i.e.

$$P = \alpha' \, \Omega \delta_t \exp\left(-\frac{\Delta W^* - \tau A}{kT}\right), \quad \Omega = \frac{\pi C_t}{L} \tag{14}$$

when $\Omega$ is the fundamental frequency of a vibrating dislocation segment of length $L$, $C_t$ is the transverse sound velocity, $\delta_t$ is the time increment, $\alpha'$ is the numerical parameter controlling frequency of cross slip, $\Delta W^*$ is the activation energy of the screw dislocation to form a kink configuration [see the model given in Rhee et al. (1998)], $\tau$ is the resolved shear stress, $A$ is the area swept by the dislocation segment, $k$ is the Boltzmann constant, and $T$ is the absolute temperature.

Stochastic effects also appear in the equation for the dislocation velocity which includes a deterministic term similar to that given in equation (6), and a stochastic term $\delta v$. Following the method developed in (Hiratani and Zbib 2002, 2003) for stochastic DDD, it is assumed that the stochastic velocity is driven by a temperature-dependent stochastic stress component, satisfying the condition of ensemble averages and follows a Gaussian process. Stochastic DDD are used to construct velocity distribution functions for use in the CDD model, where the model parameters (activation energy, frequency, and activation volume) as discussed in (Shao et al. 2014; Askari et al. 2015).

In passing it is noted that the CDD framework discussed above can be extended to include the effect of other microstructures and defects. For example, as outlined

---

$\dot{\rho}_m^{+\alpha}\big|_{(2)} = \alpha_1 N_m^{+\alpha} \, N_m^{-\alpha}/A_c \, \bar{t}$, where $A_c = \pi R_c^2$ and $N_m^{+\alpha} = \rho_m^{+\alpha} A_c$, is the number of mobile dislocations with opposite sign of burgers vector residing in $A_c$ , and $N_m^{-\alpha} = \rho_m^{-\alpha} A_c$ is the number of disloctions of opposite sign entering the area $A_c$ with velocity $\overline{v}_g^{\alpha}$. Suppose a mobile dislocation is gliding on a slip plane intersecting the circular area $A_c$ and located at distance $y < R_c$ from the center of the circle, then the distance the dislocation travels through the area is equal to $2\sqrt{R_c^2 - y^2}$. Then for all possible slip planes interesting the area A, the average distance swept through the circle by dislocations entering and exiting the circle is $\overline{x} = (1/R_c)\int_0^{R_c} 2\sqrt{R_c^2 - y^2} dy = \pi R_c/2$, substituting this result in the expression for the rate of annihilation $\dot{\rho}_m^{+\alpha}\big|_{(2)}$ leads to $\beta_2$. $\beta_3$ is derived using the same arguments.

(iii) Pinning of mobile dislocations by immobile, term $\beta_3 = \alpha_3 \pi R_c^3 \overline{v}_g^{\alpha}$. In this case it is argued that an immobile dislocation resided within the capture area of radius $R_c$ may trap equally mobile dislocations of opposite signs entering the capture area. For each mobile dislocation entering the area, say form right to left, there is a dislocation of opposite sign entering the area from left to right. Furthermore, it is assumed that since immobile dislocation (e.g. dipole or junction) is formed from two dislocations (e.g. dipole or junction), and thus the term includes $\left(\rho_i^{\alpha}\right)^2$. The derivation follows the same steps described above for term 2.

in (Li et al. 2010, 2014; Mastorakos and Zbib 2014), in irradiated materials a number of irradiation-induced defects form, and affect deformation and strength. In modeling the effect of such defects on the mechanical properties, normally the density of defects is taking as constant to save in computational effort. However, in real systems there is a continuous defect production that in a realistic multiscale model must be taken into account. The density of defects and the interaction with dislocations can be introduced in the above model in the form of additional internal state variable and corresponding evolution equations. Various defect densities can be used to account for the different types of defects encountered during the irradiation of materials. Such defects include, among others, dislocation loops, voids, helium bubbles. In this case, defects density evolution laws will be in the general form:

$$\dot{\rho}_{\mathrm{D}} = -\mathrm{v_d}\nabla\rho_{\mathrm{D}} + \mathrm{D_D}\cdot\nabla^2\rho_{\mathrm{D}} + \mathrm{f_D}\left(\rho_m^{+}, \rho_m^{+}, \rho_{\mathrm{i}}, \rho_{\mathrm{D}}, \mathrm{T}, \mathrm{dpa}, \ldots\right) \tag{15}$$

where $\rho_D$ is a general defect density, $D_D$ the defect diffusion coefficient, and *dpa* the radiation incubation dose. The unknown function $f_D$ will depend critical parameters such as temperature, dislocation densities, other defect densities, etc. These evolution laws can be derived from DDD and then put in a functional form for implementation in CDD. Within the CDD framework, the interaction of dislocations with the irradiation produced defects is in the form of extra hardening terms in equation (9), and in the equation for the dislocation mean-free path as discussed in (Li et al. 2013; Mastorakos and Zbib 2014).

## 3.2 *Size Effects*

Models based on stress-gradient and strain-gradient theories can predict size effects unlike the conventional models of crystal plasticity. As discussed above, these two theories attempt to predict size effects that arise from two different dislocation mechanisms. The main features of the two theories used in the multiscale framework discussed in this chapter are discussed below.

### 3.2.1 Strain-Gradient Theory

The main hypothesis in this theory is that strength depends on both strain and strain gradients, leading to the strain-gradient plasticity theory. This means that the gradient of strain is directly proportional to the density of dislocations. In fact, material strengthening is controlled by the density of dislocations. These can be classified as dislocations stored during uniform strengthening and those necessitated by gradient of strain, GNDs. In strain-gradient plasticity theory, the size effect in metals is attributed to the formation of GNDs leading to the dependence of flow stress on length scale, as well as higher-order strain gradients. In most strain-gradient theories (Fleck and Hutchinson 1993; Gao and Huang 2003; Huang et al.

2004), the so-called geometrically necessary dislocations (GND) are treated as the main origin of size effect (Mesarovic 2005a, b). Generally, the GNDs can be related to the spatial gradients of the plastic deformation and their effect can be directly incorporated into hardening laws. Within the dislocation model approach, the GNDs are considered as obstacles to dislocation motion and thus are incorporated in the dislocation evolution equations (7) and (8) through the mean-free path $\overline{\ell}^{\alpha}$ (Ohashi 2004; Ohashi et al. 2007).

$$\overline{\ell}^{\alpha} = \frac{c^{*}}{\sqrt{\sum w^{j\alpha}\left(\rho_T^{\alpha} + \left\|\rho_{GND}^{\alpha}\right\|\right)}} \tag{16}$$

where $c^{*}$ is a constant and $w^{j\alpha}$ is the weight matrix. Here $\left\|\rho_{GND}^{\alpha}\right\|$ denotes the density norm of the GNDs on slip system $\alpha$, which is expressed as

$$\left\|\rho_{GND}^{\alpha}\right\| = \frac{1}{b}\sqrt{\boldsymbol{\alpha} : \boldsymbol{\alpha}} \tag{17}$$

where $\boldsymbol{\alpha}$ is the Nye's tensor (see Table 1). The rate form of the Nye's tensor for large deformation, in the intermediate configuration, is given by (Shizawa and Zbib 1999a, b; Shizawa and Zbib 2000).

$$\dot{\boldsymbol{\alpha}} = \mathrm{curl}\,\left(-\boldsymbol{L}^{p}\right) \tag{18}$$

with $\boldsymbol{L}^{p}$ being the velocity gradient which is the sum of shearing from all active slip systems in the crystal lattice in the grain.

Equations (16–18) represent the main feature of the strain gradient model that is adopted in the multiscale model presented in this chapter. The equations are incorporated in CDD and crystal plasticity though the expression of the mean-free path, equation (14) appears in the expression for $\beta_1 - \beta_6$ in Equations (13a, 13b, 13c).

### 3.2.2 Stress-Gradient Theory

The stress-gradient concept originally proposed by (Hirth 2006a, b) is based on the fact that the pileup of dislocations against obstacles as they are emitted from two sources is dependent on the state of the stress field, homogenous versus inhomogeneous. Hirth applied an analytical method to solve for the pileup when the stress field is inhomogeneous by solving a singular integral equation with a kernel of Cauchy type on a finite interval. He showed that the pileup, and therefore the stress need to emit dislocations through the obstacle, is dependent not only on the grain size but also on the stress gradients. Similar approach was used for various stress gradients to researchers develop stress-gradient plasticity models for the dependence of flow stress on obstacle spacing, as in the Hall-Petch relation, as well as higher-order stress-gradients (Chakravarthy and Curtin 2011).

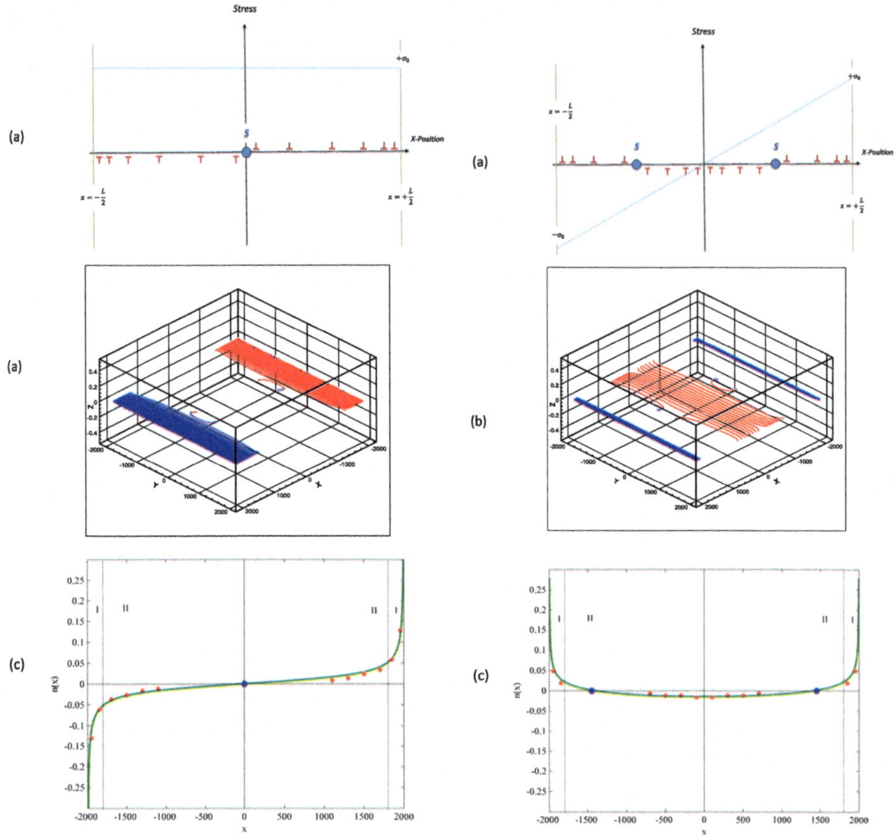


**Fig. 2** (**a**) Equilibrium distribution of discrete dislocations in a double ended pileup, (**b**) DDD-Simulation result, (**c**) Continuum and Discrete Dislocation Density versus spatial position, for the cases of uniform stress and linear stress distribution

Zbib and co-workers (Taheri-Nassaj and Zbib 2015; Lyu et al. 2017) developed a generalized stress-gradient plasticity model based on DDD and continuum dislocation theory. They analyzed the dislocation double pileup problem in the presence of various stress gradients. Then examined the stress field ahead of the pileup tip, using both discrete dislocation dynamics analyses and a continuum approach to evaluate pileup density. Figure 2 shows DDD results for the pileup of dislocations emitted from two sources when the stress is uniform versus the case of linear stress field.

The result from DDD, as can be deduced from Fig. 2, demonstrates the physical aspects of the pileup and the need to introduce more than one dislocation source to accommodate a variety of possible stress fields. Also shown in Fig. 2 the distribution of dislocation density obtained from the continuum theory. As can be deduced from these figures, analytical results from the continuum approach are in qualitative agreement with DDD results, except for singularities at the boundaries. This is because in DDD, the magnitude of the Burgers vector is constant. Results for various stress distribution (quadratic, cubic, etc..) are reported in (Taheri-Nassaj and Zbib 2015). Based on these results, Zbib and co-workers developed a stress-gradient theory (Zbib et al.), leading the following generalized equation for the stress-gradient dependent flow stress.

$$\tau_S{}^{\alpha} = \frac{K}{\sqrt{L'}}\left[1 + \frac{L'}{4\overline{\tau}}\left|\nabla\overline{\tau}\right| + \frac{L'^2}{16\overline{\tau}}\left|\nabla^2\overline{\tau}\right| + \frac{L'^3}{128\overline{\tau}}\left|\nabla^3\overline{\tau}\right| + \dots\right] \tag{19}$$

where $L'$ is the spacing between obstacles and is treated as an intrinsic material length-scale, $K$ is Hall-Petch constant, $\overline{\tau}$ is the effective stress, and $\nabla(.)$ is the first-order gradient operator, $\nabla^2(.) = \nabla.\ \nabla(.)$, $\nabla^3(.) = \nabla.\ (\nabla.\ \nabla(.))$, etc. Equation (19) is reminiscent to many strain-gradient models, e.g.(Aifantis 1986, 1987; Zbib and Aifantis 1988) [11–16].

The length scale $L'$ appearing in equation (19), is the distance between obstacles ($L'$) where dislocations pile up. In addition to the material length scale $L'$, the stress gradient term brings a length scale into the analysis depending on the problem at hand. If the local stress gradient is constant it then corresponds to a linear distribution of the local stress. In this case, the linear distribution of stress will depend on the size of the domain under consideration and the size of the local microstructure. For example, two cases were considered in details in (Taheri-Nassaj and Zbib 2015). First when the stress gradient theory is applied to continuum plasticity for homogenous media. In this case, the size of the domain and the boundary conditions determine the state of the stress and thus the local stress field and stress gradients. For the case of bending of a micro-beams (say of thickness $H$) or torsion of micro-wires (say of radius $R$) the applied stress field is linear and thus the stress gradient is constant, hence the only non-zero stress gradient term in Equation (19) becomes $L' \lceil \nabla\overline{\tau} \rceil / 4\overline{\tau} = L'/2H$ for the bending case and $L' \lceil \nabla\overline{\tau} \rceil / 4\overline{\tau} = L'/4R$ for the torsion case, illustrating the interplay between two length scales, the material length scale $L'$ and the specimen size $H$ or $R$. It was argued that the material length scale $L'$ was a constant equal to the average grain size, and, thus, as the size of the beam ($H$) or the wire ($R$) increases the stress gradient term diminishes.

## 4 Some Case Studies

The multiscale framework discussed in this chapter has been employed to address a wide range of plasticity phenomena. This includes, for example, dislocation behavior during monotonic loading (Zbib and Diaz de la Rubia 2002), dislocation patterning (Hiratani and Zbib 2002), analyses pertaining to the formation of dislocation boundaries during heavy deformation (Khan et al. 2001, 2004), the evolution of deformation and dislocation structure during nanoindentation in both bcc and fcc single crystals (Zbib et al. 2011), and dislocation interaction with shock waves during impact loading conditions (Shehadeh et al. 2005; Zbib and Shehadeh 2016). Other problems and phenomena, have also been studied by Zbib and Co-workers, including dislocation-defect interaction (Li et al. 2010, 2013; Mastorakos et al. 2010; Mastorakos and Zbib 2014) size effects, formation of deformation bands. (Akarapu et al. 2010; Shao et al. 2012, 2014; Lyu et al. 2015a, b, c; Taheri-Nassaj and Zbib 2015; Lyu et al. 2016a, b, c; Taheri-Nassaj and Zbib 2016). Results pertaining to dislocation patterning and size effect using both DDD and CDD are presented below.

## *4.1 Multiscale Modeling of Dislocation Patterning*

As discussed in Sect. 1, plastic deformation in metals is usually associated with the formation of dislocation patterning which can be viewed as a dynamic self-organization phenomenon driven by local and long-range stress interaction among dislocations. Both the DDD method and the CDD model presented above can be used to predict the formation of dislocation pattering. While the DDD method is more rigorous in its treatment of the discreteness of dislocation geometry, kinematics and kinetics, it can be more computationally extensive which limits its utility to scales in the order of micrometers, the CDD approach is based on continuum concepts but can be used to address large-scale problems.

### 4.1.1 DDD: Dislocation Patterning

The DDD method, incorporates the 3D kinematics and kinetics aspects of curved dislocations into a computational traceable framework. In the numerical implementation, three-dimensional curved dislocations are treated as a set of connected segments, whose dynamics is driven by a Newtonian equation of motion. Figure 3 shows results from DDD simulation of deformation of a single crystal cube unit cell whose size cube is 10 μm subjected to monotonic loading with low strain rate. For this case, one would construct a simulation cell with either reflected boundary conditions (Zbib et al. 1998) or periodic boundary conditions (Zbib and Diaz de la Rubia 2002), maintaining continuity and flux across the cell boundaries.

The crystal considered is bcc molybdenum with randomly distributed initial dislocation density on the <111> systems. The load is applied in the [2 9 20] direction for reasons described in (Zbib and Diaz de la Rubia 2002). Constant strain rate of 1/s is imposed in that direction. Typical results can be seen in Fig. 3b showing the formation of dislocation patterning at 0.4% strain. From the data collected, one can extract various interesting information that could be useful in many ways as discussed (Zbib and Diaz de la Rubia 2002; Hiratani and Zbib 2003). For example, by analyzing the spatial distribution of all dislocation segments, one can construct pair-distribution functions, for projections in various crystallographic directions,

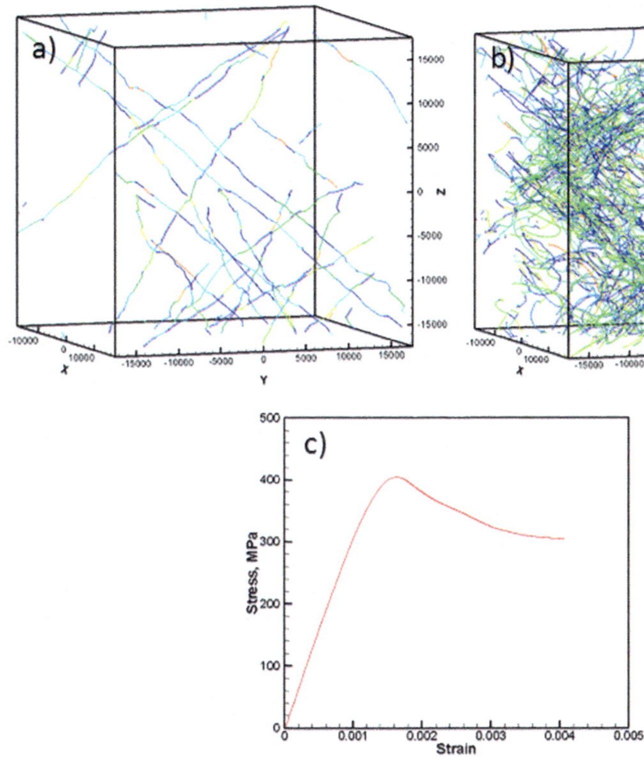


**Fig. 3** Large scale DDD simulation of dislocation patterning in molybdenum single crystal loaded under constant strain rate: (**a**) Initial dislocation structure, (**b**) dislocation structure at 0.4% strain, (**c**) calculated stress-strain curve

from which one can extract a wavelength and indication of a dislocation pattern. In this case, the analyses reveal that not only the <111>{011} systems are active but also some of the <111>{112} systems also become activated, resulting mainly from multiple cross-slip.

It is pointed out that this DDD simulation was performed under static loading conditions and the variable time step varied between $10^{-8}$s and $10^{-6}$s. The dislocation velocity and the shortest distance between two dislocations control the time step. The number of steps in this simulation was a few millions to reach a strain of about 0.4% using multiprocessors.

### 4.1.2 CDD: Linear Stability Analysis

The spatio-temporal evolution equations for the dislocation densities given by Equations (13a, 13b, 13c) are highly nonlinear and need to be solved numerically in time and space. The numerical solution to this set of equations gives the spatial distribution of dislocation patterns over time. However, first one can analyze the system of equations for equilibrium states and the stability of these states and to determine the conditions under which a homogenous state may give rise to dislocation patterning. For the sake of simplicity, a one-dimensional case is considered with constant average dislocation velocity, and no dislocation climb and cross-slip, ($D=0$, $D_i=0$, $\beta_{6=0}$). For this special case, which can be considered for predicting slip band formation, the governing equations reduce to the following set.

$$\dot{\rho}_m^{+\alpha} + \overline{\boldsymbol{v}}_g^{\alpha}\,\frac{\partial \rho_m^{+\alpha}}{\partial x} = \beta_1 \rho_m^{+\alpha} - (\beta_2+\beta_5)\,\rho_m^{+\alpha}\,\rho_m^{-\alpha} - \beta_3\left(\rho_i^{\alpha}\right)^2 \rho_m^{+\alpha} + \beta_4 \rho_i^{\alpha} \quad (20a)$$

$$\dot{\rho}_m^{-\alpha} - \overline{\boldsymbol{v}}_g^{\alpha}\,\frac{\partial \rho_m^{-\alpha}}{\partial x} = \beta_1 \rho_m^{-\alpha} - (\beta_2+\beta_5)\,\rho_m^{-\alpha}\,\rho_m^{+\alpha} - \beta_3\left(\rho_i^{\alpha}\right)^2 \rho_m^{-\alpha} + \beta_4 \rho_i^{\alpha} \quad (20b)$$

$$\dot{\rho}_i^{\alpha} = 2\beta_5\,\rho_m^{+\alpha}\rho_m^{-\alpha} + \beta_3\left(\rho_i^{\alpha}\right)^2\left(\rho_m^{+\alpha} + \rho_m^{-\alpha}\right) - \beta_4 \rho_i^{\alpha} \quad (20c)$$

It can be shown that there are five sets of steady state homogenous solutions for this set of equations:

$$\text{I} \qquad \rho_{mH}^{+\alpha} = \rho_{mH}^{+\alpha} = \rho_{iH}^{\alpha} = 0 \quad (21a)$$

II & III:

$$\rho_{mH}^{+\alpha} = \rho_{mH}^{+\alpha} = \frac{\beta_1}{\beta_2},\ \rho_{iH}^{\alpha} = \frac{\beta_2}{4\beta_1\beta_3}\left[2\beta_4 \pm \sqrt{(2\beta_4)^2 - 16\beta_3\beta_5{\beta_1}^3/{\beta_2}^3}\right] \quad (21b)$$

IV and V, $\rho_{iH}^{\alpha} = \sqrt{\beta_1/\beta_3}$, $\rho_{mH}^{-\alpha}\rho_{mH}^{+\alpha} = \frac{\beta_2\sqrt{\beta_1/\beta_3}}{[\beta_2+\beta_5]}$, and the two roots of:

$$\beta_1 \rho_{mH}^{+2} + \left[\frac{2\beta_4\,\beta_5}{\beta_2+\beta_5} - 2\beta_4\right]\sqrt{\frac{\beta_1}{\beta_3}}\rho_{mH}^{+\alpha} + \frac{\beta_1\,\beta_4}{\beta_2+\beta_5}\sqrt{\frac{\beta_1}{\beta_3}} = 0 \quad (21c)$$

Next the stability of the steady state is investigated by imposing small perturbations, $\delta\rho' s \ll 1$,from the steady state. It is assumed that the perturbations are small and the domain is infinite. Then the nonlinear differential equations are linearized around the steady and the solution to the linearized equation requires that the perturbations be harmonic in space and exponential in time, i.e.

$$\rho_m^{-\alpha} = \rho_{mH}^{-\alpha} + \delta\rho_m^{-\alpha} e^{iqx+\omega t} \quad (22a)$$

$$\rho_m^{+\alpha} = \rho_{mH}^{+\alpha} + \delta\rho_m^{+\alpha} e^{iqx+\omega t} \quad (22b)$$

$$\rho_i^{\alpha} = \rho_{iH}^{\alpha} + \delta\rho_i^{\alpha} e^{iqx+\omega t} \quad (22c)$$

where $q$ is the wave number and $\omega$ is the eigenvalue whose sign determines whether the perturbation will decay (stable homogenous steady state) or grow (unstable homogenous steady state). For a given steady state, and following standard linear stability analysis, one obtains a characteristic equation for the eigenvalues which has the of the following generalized form

$$\omega^3 + A\left(q, \beta's\right)\omega^2 + B\left(q, \beta's\right)\omega + C\left(q, \beta's\right) = 0 \tag{23}$$

The expressions for *A, B* and *C* are function of the wave number $q$ (or wavelength $\lambda = 2\pi q$) and the material parameters $\beta^{'}s$ can be easily evaluated numerically. Values for α-parameters are used from previous works for Fe Single Crystal (fcc) (Li et al. 2014) except for $\alpha_4$, and therefore $\beta_4$, which is left as a varying parameter; it is the rate of breakup of immobile dislocations (junctions) and hence is consider as a key parameter in the pattern formation. The influence of the other parameters is investigated elsewhere (Taheri-Nassaj and Zbib 2016). Initial dislocation density is set at $0.1/\mu m^2$, with constant strain rate of $10^{-4}$/sec, and $b = 2.54x10^{-4}\mu m$. All the parameters are assumed to be constant around the steady state.

For any wavelength and fixed value $\beta_4$, the characteristic equation (23) can possess three complex roots; three $\omega^{'}s$. The preferred wavelength $\lambda$ is the one that would correspond to the optimum growth rate (i.e. the root $\omega$ whose real part is maximum). The equation is solved numerically for a wide range of $q$ and $\beta_4$. Typical results are shown in Fig. 4 for two wavelengths, $\lambda = 1\ \mu m$, and $\lambda = 10\ \mu m$. There are three solutions for each case as shown in the Figures. The result for $\lambda = 1\ \mu m$ show that for $\beta_4 > 0.1$ the real part of $\omega$ is negative, indicating a stable homogeneous steady stat. However, for the case of $\lambda = 10\ \mu m$ there is always a positive $\omega$ corresponding to an unstable homogenous steady state solution and indicating that the perturbation will grow in time and will lead to pattern formation.

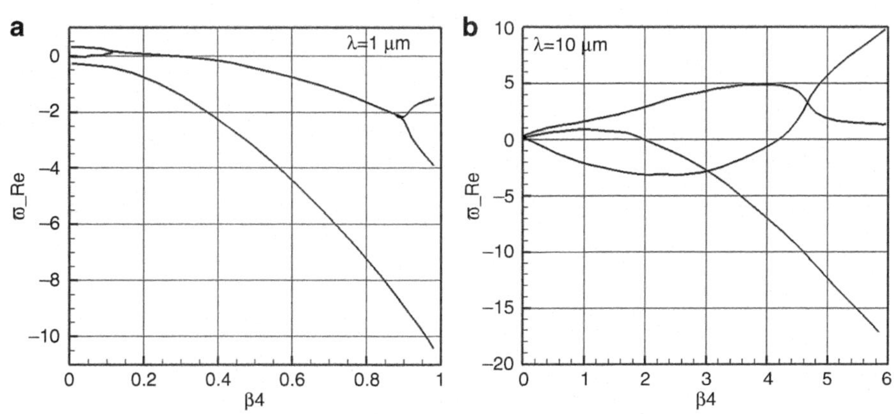


**Fig. 4** Rate of growth of he perturbation from the steady state as a function of the junction parameter, for (**a**) wavelengths, $\lambda = 1\ \mu m$, and (**b**) $\lambda = 10\ \mu m$

### 4.1.3 CDD: Dislocation Patterning (1D)

The spatio-temporal evolution equations (19a–c) are highly nonlinear and, therefore, are solved numerically using 5th order Runge-Kutta method. For the one-dimensional case with one slip activated, the space along the x-axis is discretized with mesh size equal to 0.01 $\mu m$. It is assumed that the shearing rate is constant and equal to $10^{-4}$/sec. The initial distribution of dislocations is shown in Fig. 5a and it is assumed to have a small perturbation as shown in the figure. Periodic boundary conditions are assumed. The numerical integration is performed with constant time step of $10^{-3}$ sec.

Figure 5b shows the dislocation density distribution at very low shear strain at $t = 0.1\ sec$ . As can be seen from the figure, the positive and negative mobile dislocations grow from their initial state (Fig. 5a) into periodic patterns—which is consistent with the prediction from linear stability analysis- that propagate opposite to each other (Fig. 5b), while the positive dislocations propagate to the right the negative ones propagate to the left. At this very low strain, the immobile dislocations distribution density is not very much changed from its initial condition.

With further straining, the dislocation densities evolve into the stable patterns shown in Figs. 6a and b. As can be seem from the figures two thick walls (high-density regions) of dislocation densities are formed having three distinct wavelengths: center-to-center spacing of the two walls approximately equal to 8 μm, wall width approximately equal to 3.5 μm, and a wavelength of fluctuating density within the wall approximately equal to 0.5 μm. These results clearly illustrate that the dislocation model can predict the formation of dislocation patterns at different length scales.

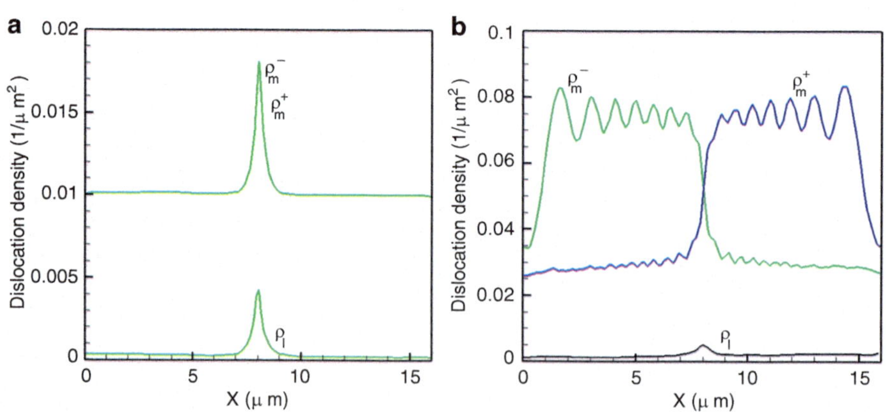


**Fig. 5** 1D-Dislocation Patterning. (**a**) Initial distribution of dislocations used in the numerical analyses; (**b**) dislocation distribution at $t = 0.1\ sec$

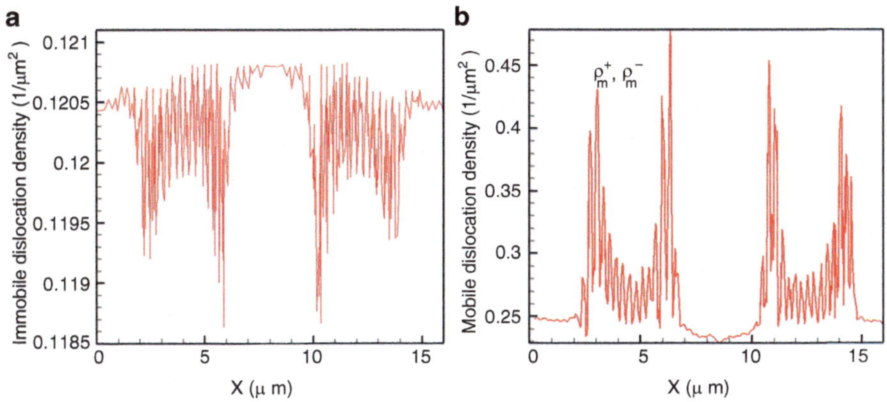


**Fig. 6** 1D-Dislocation Patterning. (**a**) spatial distribution of immobile dislocation densities and (**b**) mobile dislocation density at $t = 10$ *sec*

### 4.1.4 CDD: Dislocation Patterning (2D)

The two-dimensional version of the spatio-temporal evolution equations (13a–13c) with $D$=0, $D_i = 0$ have been implemented into the crystal plasticity framework, and solved numerically (Taheri 2016). In this case, the flux term in the equations in Cartesian coordinate system is a combination of two terms, that are mobile dislocation density fluxes in $x$ and $y$ directions. Reference velocity of 400 μm/s, reference length of 1 μm based on dislocation densities of $\rho_m^{+\alpha} = \rho_m^{-\alpha} = 0.25/\mu m^2$ and $\rho_i^{\alpha} = 0.5/\mu m^2$ and strain rate of $1.0^{-4}$/sec.

Similarly to the 1D case, initial distributions of mobile dislocations as well as immobile dislocations in two-dimension are introduced with periodic perturbations in $x$ and $y$ directions, in a simulation box of 10 μm × 10 μm, Al fcc single crystal. Figures 7 *a-d* display 2D distributions of the dislocation densities at different strains, showing pattern formation for a total strain of 2%. It can be seen for the figures that high density regions grow wider while low density regions become smaller in width over time but the pattern remains as organized as the initial one. The figures also reveal the evolution of immobile dislocation patterns for the same material and same total strain. In this case it can be seen that dislocation pattern changes over time and loose the organized initial distribution.

In passing, it is noted that preliminary results for the 3D case, to be discussed later on, show that the evolution equations produce three dimension cellular patterns. Moreover, it is also noted that the numerical solutions for all the cases presented above, in order to obtain the spatial patterns, are for one grain and are very extensive. Therefore, when solving a large-scale polycrystalline problem consisting of large number of grains, as discussed in the next section, determining the spatial distribution of dislocation density within each grain may cumbersome. Also in this case it is not necessary to know the spatial distribution within each grain in order to evaluate macroscopic properties, since after all the deformation field are homogenized over the grains, and thus the flux terms can be dropped.

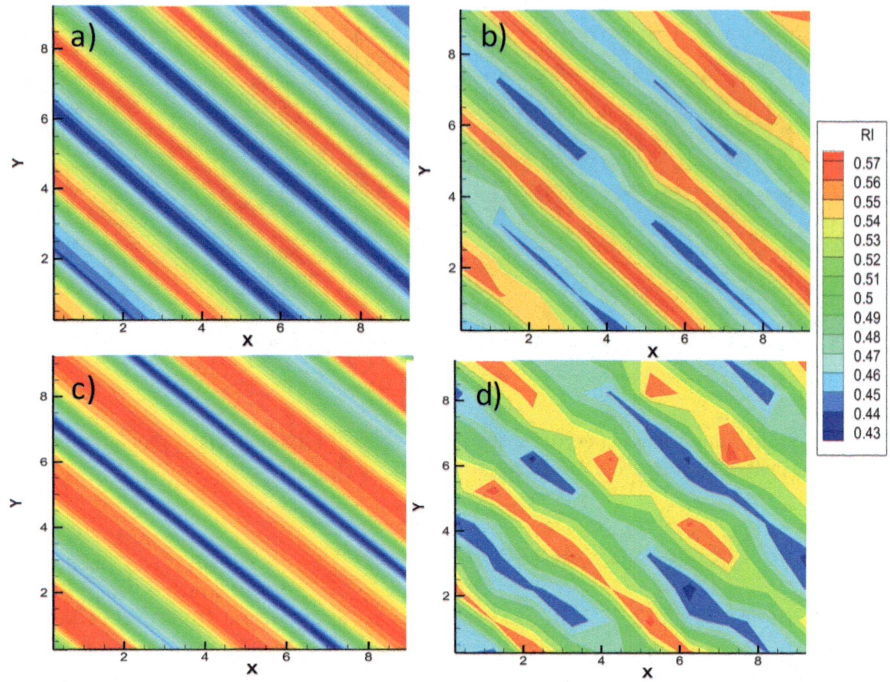


**Fig. 7** 2D- Dislocation Patterning- (**a**) and (**b**) Distribution of mobile dislocation density and immobile density respectively, at 1% strain; (**c**) and (**d**) distribution of mobile dislocation density and immobile density respectively, at 2% strain

### 4.1.5 CDD: Dislocation Patterning (3D)

The dislocation density evolution equations (13a–13c) with $D$=$0$, $D_i = 0$ have also been solved numerically using the finite element method for the 3D case. The results are presented in Fig. 8 below, and demonstrate the capability of the evolution equations in predicting three dimension cellular patterns. Furthermore, in the 3D case, and for large strains, the evolution equations exhibit hexagonal cells with wall of higher dislocation densities (the region included in the red square in Fig. 8b).

Like the 2D case, that was an extension of the 1D case, the 3D case was an extension of the 2D case. Likewise, the flux term in equations (13a–13c) is a combination of three terms, that correspond to the mobile dislocation density fluxes in $x$, $y$ and $z$ directions. The reference velocity was 40 μm/s, smaller than the 2D case which also explains why it took so high strain to produce the patterning. Like in the 2D case, dislocation densities of $\rho_m^{+\alpha} = \rho_m^{-\alpha} = 0.25/\mu m^2$ and $\rho_i^{\alpha} = 0.5/\mu m^2$ and strain rate of $1.0^{-4}$/sec have been used.

Similarly to the 1D and 2D cases, the initial distributions of mobile dislocations as well as immobile dislocations in three-dimensions had periodic perturbations in $x$, $y$ and $z$ directions, in a simulation box of 8 μm × 8 μm × 8 μm, Al fcc

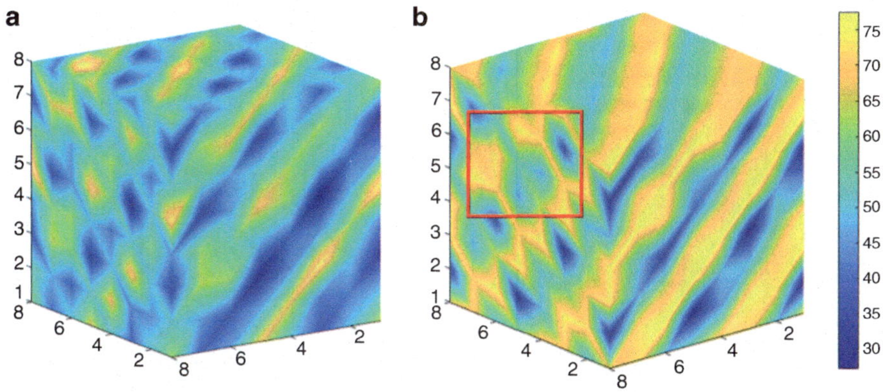


**Fig. 8** 3D- Dislocaiton patterning- (**a**) and (**b**) Distribution of immobile dislocation density and mobile density respectively at 15% strain

single crystal. Moreover, it is also noted that the numerical solutions for all the cases presented above, in order to obtain the spatial patterns, are for one grain and are very extensive. Therefore, when solving a large-scale polycrystalline problem consisting of large number of grains, as discussed in the next section, determining the spatial distribution of dislocation density within each grain may cumbersome. Also in this case it is not necessary to know the spatial distribution within each grain in order to evaluate macroscopic properties, since after all the deformation field are homogenized over the grains, and thus the flux terms can be dropped.

## 4.2 *Size Effect and Plasticity in Materials with Gradient Microstructure*

Macroscopic mechanical properties and behavior in polycrystalline metals depend critically on the nature of the underlying microstructure: grain size, texture, orientation, gain boundaries, phases, as well the spatial distribution (heterogeneous and gradient distributions) of all of these features. As discussed in the introduction, the issue of size effect has been examined extensively by many researchers. Particular, Zbib and co-workers have shown that the multiscale framework discussed in this chapter can address a wide range of size-dependent plasticity phenomena, many references are cited in this chapter and more can be found in the literature. In this section, a recently developed method using this framework for designing materials that may possess improved mechanical properties, with emphasis on strength and ductility, is presented, suggesting future utilization of such multiscale plasticity theories for the design and development of high-efficiency structural materials.

Generally, strength and ductility in a given class of materials are affected by a number of factors: phases, grain size, porosity, etc. For instance, a polycrystalline

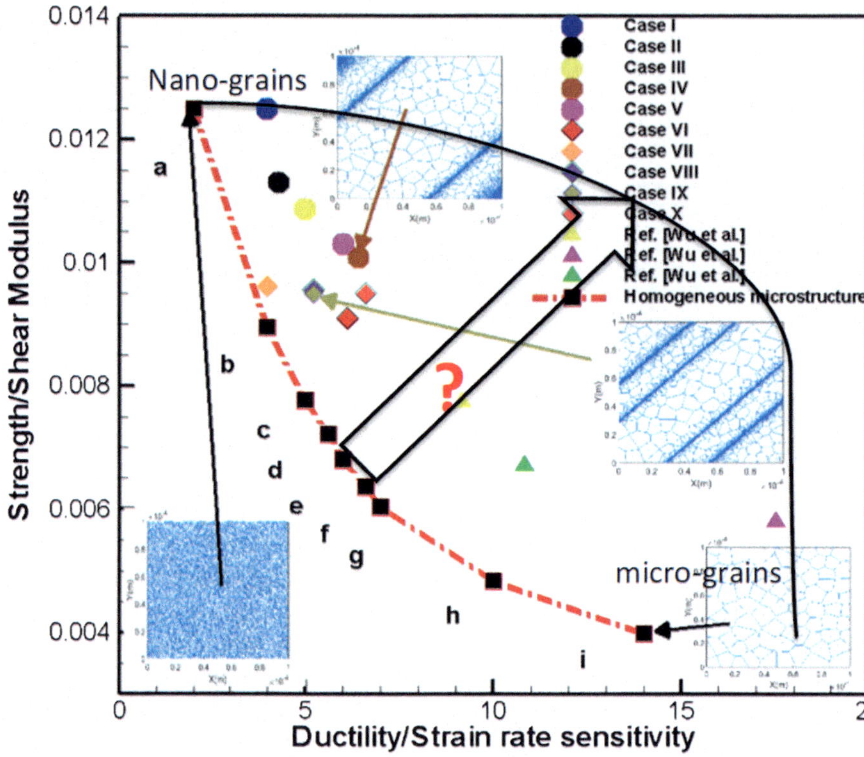


**Fig. 9** Polycrystalline metals with homogenous distribution of nano-grains have high strength and low ductility, and materials with homogenous distribution of micro-grains have low strength and high ductility (dashed curve for different grain sizes). But materials with heterogeneous nano-microstrcture have higher strength and ductility. Cases I-X are results for heterogeneous microstructures studied in (Lyu et al. 2017) and other data points are experimental results for heterogeneous microstructures from the indicated references

material with nano-size grains will have a higher strength than its counterpart with micron-size grains. However, the trade-off is that it will also become more brittle (Calcagnotto et al. 2010, 2011; Morris 2010; Lyu et al. 2016a, b, c, 2017). For a material with a homogenous microstructure, it is difficult to attain both high strength and ductility. This has been reported by many researchers, e.g. (Lu and Lu 2004; Kang et al. 2016), where it is shown that the strength versus ductility curve, for many classes of polycrystalline materials, follows a so-called 'banana curve' (Fig. 9, dashed line).

Polycrystalline materials with homogenous distribution of nano size grains can achieve very high strength but low ductility; while polycrystalline materials with homogenous distribution of micro size grains have the opposite effect. This behavior can be attributed to the activation of dislocation deformation and grain boundary sliding mechanisms. In micro-grains (large volume) a relatively large number of dislocations can be activated at relatively lower stress (yield stress) and their collective motion produces relatively large plastic strain, and thus contributes to ductility. But this is not the case in nano-grains. Dislocations in nano-grains (small volume) are scarce and thus the material becomes very brittle, leading to higher strength but cracking and failure at low strains.

Recently, there has a number of mostly experimental works that provide some insight on possible means to improve the strength-ductility trade-off, by designing and processing materials that possess heterogeneous and/or gradient microstructure and texture, e.g.(Rajagopalan and Saif 2011; Wu et al. 2014a, b; Kang et al. 2016; Yin et al. 2016). Towards this end, the multiscale framework discussed in this chapter can be used to guide in the design of microstructures that achieve this goal (Lyu et al. 2015a, b, c, 2016a, b, c, 2017). In this approach a spatial representative domain representing the polycrystalline material is defined and appropriately discretized to represent grain size and distribution. In order to investigate gradient microstructures with controlled spatial distributions, a controlled Poisson Voronoi tessellation (CPVT) model is used to generate a virtual grain structure (Zhang et al. 2011, 2012). Then each grain is assigned a crystallographic orientation randomly or for specified texture. The stress state, strain and dislocation densities in each element are computed using the multiscale model coupling CDD with crystal plasticity.

As shown by Zbib and co-workers various gradient microstructures can be generated and their mechanical response can be investigated using the multiscale framework. Figures 9a and b show examples of many microstructures that have been examined by the authors. A series of microstructures are designed and investigated. Each series consists of different gradient microstructures with grain size ranging from 100 nm to 10 μm. In order to design the layered microstructures, like the case shown in Fig. 10b, structures are produced separately and all the layers were combined to have a complete microstructure.

For example, the case shown in Fig. 10b, has several layers with different grain size. The range of grain size in layers of this microstructure is from 100 nm up to 10 μm. These gradient materials can be produced by a method called surface mechanical attrition treatment (Tao et al. 2003; Zhu et al. 2004; Balusamy et al. 2013; Yin et al. 2016). These and other studies have shown that this process can produce a surface layer with a gradient distribution of nano-grains, thereby resulting in a material with high strength and high ductility along the tensile direction (Lu et al. 2000; Wu et al. 2002; Lu and Lu 2004; Zhao et al. 2006; Yang et al. 2016). These types of gradient microstructures can also be manufactured using high-pressure torsion process (Kang et al. 2016), asymmetric rolling process, (Wu et al. 2015) and subsequest partial recrystallization process (Roumina and Sinclair 2008; Hailiang et al. 2012).

The stress-strain responses for various gradient microstructures (grain size distributions), as predicted by the multiscale model, are shown in Fig. 10c for the case of interstitial free (IF)-steel (Lyu et al. 2017). Similar analyses were performed for other materials, such as copper. From these figures one can extract both strength

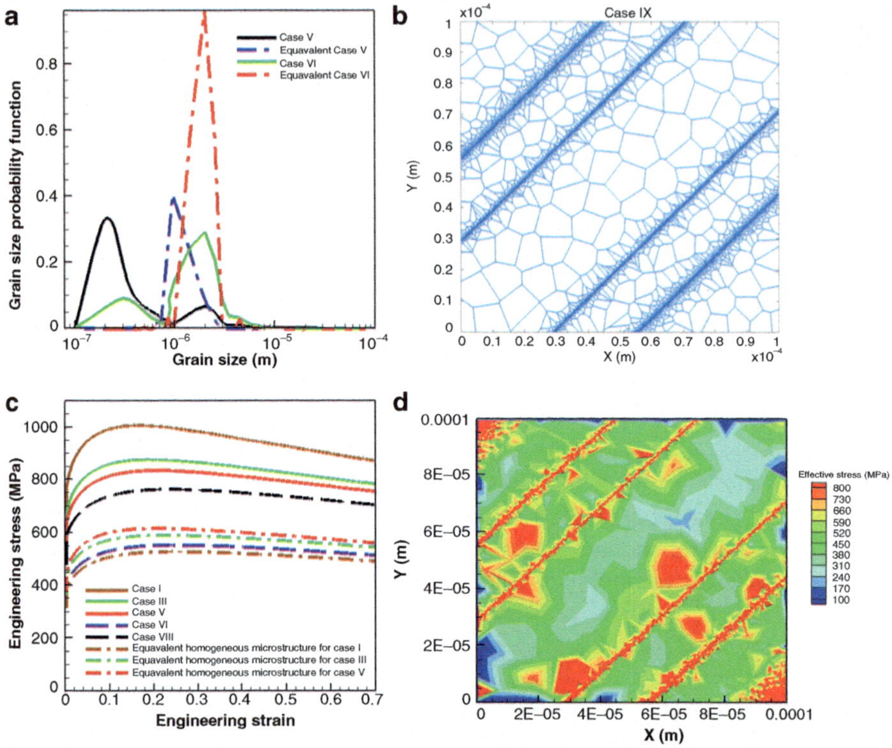


**Fig. 10** (**a**) Various Grain size distributions, (**b**) sample heterogeneous microstructure with gradients grain-size distribution, (**c**) predicted Stress-strain response of heterogeneous microstructures (solid lines) and equivalent homogeneous microstructures, (**d**) contour plots of effective strain

and ductility for each gradient microstructure case. Sample results, for a number of cases are plotted in Fig. 9, where, for generalization, the strength is normalized by the shear modulus and the ductility by the strain rate sensitivity. Also shown in the figure the results for homogenous microstructure cases, for grain size ranging from nano-meters to micrometers. As can be deduced from the figure, the predicted trend for the homogenous cases follows the traditional behavior of trade-off between strength and ductility. However, for the gradient microstructure the values for strength and ductility are both higher when compared to the homogenous cases as can be deduced from the figure.

As can be concluded from Fig. 10a, the grain size distribution in the cases of gradient microstructures have bimodal distributions as opposed to unimodal distributions as in the case of a homogeneous microstructures. This suggests that additional microstructural parameters such as the grain size distribution should also be considered when studying the material strength. This also indicates the possibility that microstructures with the same mean grain size but different grain size distribution would result in a different mechanical behavior. Hence, the traditional Hall-Petch relationship that relates strength to average grain size may not apply in the case of materials with gradient microstructures. Instead, in this case strength is dependent on both average grain size as well as a parameter representing grain-size gradient as proposed in (Lyu et al. 2017).

Gradient microstructure based on grain-size morphology is only one of other microstructural features that can also have spatial gradients and can significantly affect properties, including, texture gradients, phase gradients, and porosity gradients. Gradient textures are typically observed experimentally, but their contribution to macroscopic behavior are neglected in most modelling work. For example, initial texture distribution is a typical parameter neglected in most studies of gradient materials, which can lead to different mechanical behavior. Although different processing method may lead to the same grain size gradient, in the meanwhile it will also lead to different initial texture.

The multiscale simulation results shown in Fig. 11 are for fourteen cases all having the same grain-size gradient morphology, but each case has different initial texture distribution. For example, Case IX has a strong fiber (from BCC rolling texture) for grains with grain size <600 nm, which results in an increasing in tensilie strength but at the cost of ductility. These results indicate that there is a strong interaction between the two gradient microstructures, grain size and texture. The results show that the initial texture distribution can affect stress and ductility tremendously for a grain size gradient microstructure. Thus, simply considering one microstructural parameter is not sufficient to accurately determine the mechanical behavior. This motivates more experimental design to produce microstructures with new combintion of texture gradient and grain size gradient and optimization of mechanical behavior based on both microstructural parameters.

Phase gradients also play significant role in strength and ductility. Generally, increased strength in two-phase composites, such as DP steels, can be attributed to a number of factors including grain refinement, solid solution strengthening and precipitation hardening. In advanced high-strength steel, alloying elements are typically added so that strengthening is achieved through grain refinement. Second-phases introduce additional heterogeneity into the microstructure and various spatial distributions of second phases will affect the mechanical behavior of the material. Numerous investigations have shown the effect of second phase martensite morphology on the strength and ductility of DP steel. The spatial distribution of the second phase, the size of the second phase cluster and the shape of the cluster can lead to different mechanical behavior (Sarwar and Priestner 1996; Bag et al. 1999; Son et al. 2005; Zhang et al. 2016). A recent study by Zhang et al. (2016) showed that strength and ductility can be affected by the spatial distribution and density of ferrite grains and martensitic islands which can be produced by various heat treatment histories; more pronounced gradient multi-phase microstructure could be

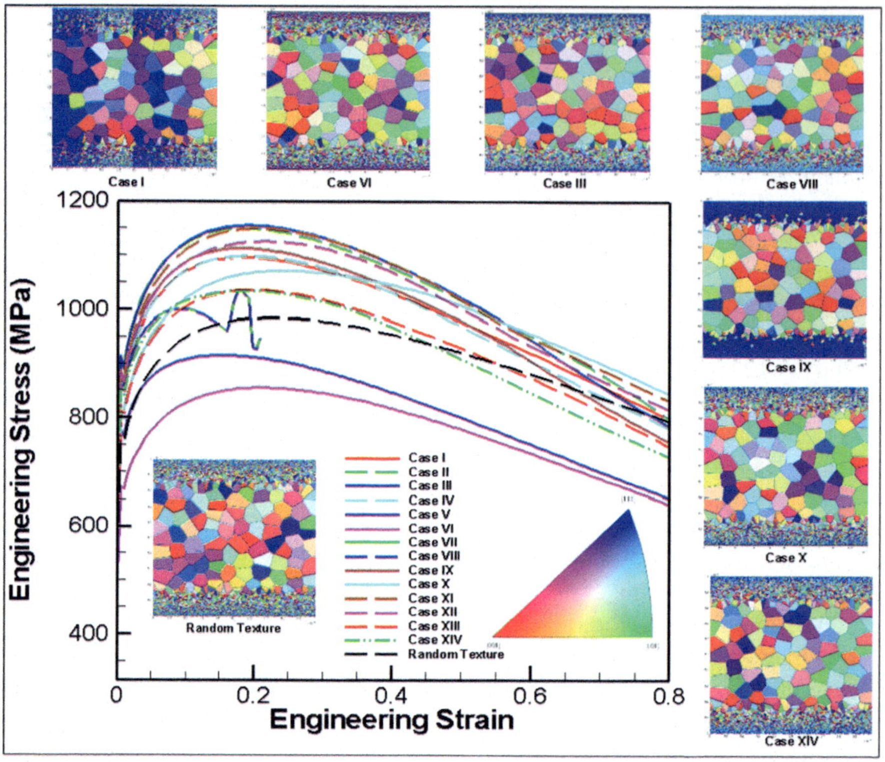


**Fig. 11** Multiscale simulation. Mechanical behavior of interstitial free (IF)-steel for various initial texture distributions. The grain-size distribution is the same for all the cases

produced by heat treatments methods (Kang et al. 2016) that yield lamella-type structures).

The multiscale framework presented in this chapter can be used to investigate gradient multi-phase microstructures, and to design phase distributions in order to achieve desired properties (Lyu et al. 2015a, b, c). The simulation results shown in Fig. 12, obtained from the multiscale framework, are for various two-phase morphologies analogous to the experimental ones given in (Zhang et al. 2016). It can be inferred from these simulation results that a higher strength can be attained with a microstructure that has martensite islands surrounded by the ferrite matrix. These results show that the mechanical behavior is not only a function of volume fractions of ferrite and maternsite, as assumed in traditional models, but also a function of their spatial distributions. This is in good agreement with the work of Bag et al. (Bag et al. 1999).

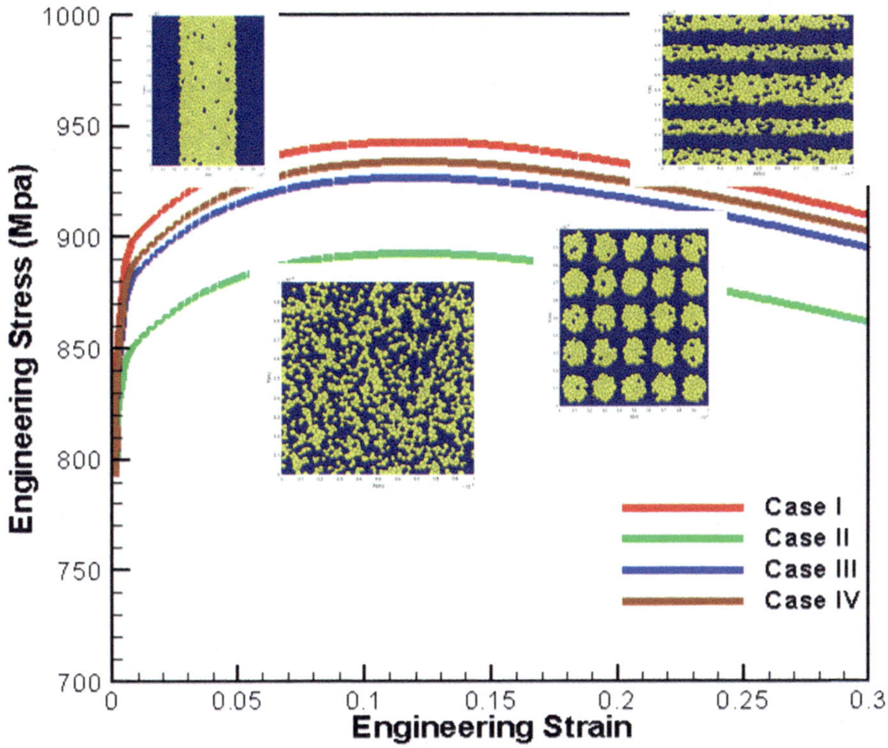


**Fig. 12** Multiscale simulation. Mechanical behavior of DP980 for various spatial phase distributions. In all the cases the volume fraction of martensite (42%) and ferrite (58%) are the same, but the spatial distributions are different

## 5 Concluding Remarks

A multiscale framework for crystal plasticity in metals, taking advantages of modeling and simulations at various length scales, has been presented in this chapter. The main idea in this framework is that a given model for a given length scale can be more accurate if it uses information passed to it from a smaller scale. Key features of the framework include a continuum dislocation-based crystal plasticity theory coupled with discrete dislocation dynamics and a stress/strain gradient theory that is based on dislocation mechanisms. The multiscale model has been implemented into a computational framework and can be used to solve a wide range of plasticity phenomena at various length scales to predict mechanical properties and their relation to underlying microstructures; e.g. dislocation patterns, grain size and gradients, providing a means for designing and manufacturing a new class of structured materials that possess improved properties. The case studies presented in this chapter are intended to highlight the significance of gradient microstructures.

The results obtained show that a combination of high strength and high ductility can be attained simultaneously in materials with gradient microstructure.

Finally, it is suggested that the multiscale methodology presented in this chapter provides a pathway not only for investigating plasticity phenomena, but also for designing a new class of high-efficiency structural materials with superior properties, a goal highlighted by a number of national reports: Materials Genome Initiative (NSTC 2011), NSF (2006) and NRC (2008). These reports highlight that material discovery and design is one of the critical components for further scientific and engineering advancements. Among other things, these reports emphasize the need for the development of new classes of materials that possess optimum mechanical properties for efficient mechanical performance for fuel efficiency and reduced emissions, while minimizing weight and material cost. Among these properties, strength and ductility are amongst the most important factors that affect component design, thus weight and cost. Ideal structural materials are the ones that can have both high strength and excellent ductility. However, current state-of-the-art materials possess a drastic tradeoff between these two properties. In order to overcome this tradeoff and to push the strength-ductility map upward as suggested in Fig. 9, new classes of multiphase gradient materials need to be developed. Nevertheless, there is still a lack of scientific understanding of key factors that may make it possible to push the strength-ductility envelope into unexplored regions (arrow in Fig. 9). The strength-ductility tradeoff, addressed in this chapter, is only one example of other tradeoffs encountered in material design, strength-density is another example. The multiscale methodology presented in this chapter can address these tradeoff phenomena, making it feasible to design microstructures where mechanisms can be synchronized to operate concurrently to yield optimum properties.

**Acknowledgement** The support provided by the National Science Foundation's CMMI program to WSU under Grant No. 1434879 is gratefully acknowledged.

## Bibliography

E.C. Aifantis, On the dynamical origin of dislocation patterns. Mater. Sci. Eng. **81**, 563–574 (1986)

E.C. Aifantis, The physics of plastic deformation. Int. J. Plast. **3**, 211–247 (1987)

E. Aifantis, Strain gradient interpretation of size effects. Int. J. Fract. **95**(1–4), 299–314 (1999)

A. Akarapu, H.M. Zbib, D.F. Bahr, Analysis of heterogeneous defromation and dislocation dynamics in single crystal micropillars under compression. Int. J. Plast. **26**, 239–257 (2010)

F. Akasheh, H.M. Zbib, S. Akarapu, S. Overman, D. Bahr, Multiscale modeling of dislocation mechanisms in nanoscale multilayered composites. Mater. Res. Soc. Symp. **1130**, W13-01 (2009)

A. Alankar, I. Mastorakos, D. Field, H.M. Zbib, Determination of dislocation interaction strengths using discrete dislocation dynamics of curved dislocations. J. Eng. Mater. Tech **134**, 4 (2013)

G. Ananthaktishna, Current theoretical approaches to collective behavior of dislocations. Phys. Rep. **440**, 113–259 (2007)

A. Arsenlis, D.M. Parks, Crystallographic aspects of geometrcially necessary and statistically-stored dislocation density. Acta Metall. **47**, 1597–1611 (1999)

A. Arsenlis, B.D. Wirth, M. Rhee, Disloction density-based constitutive model for the mechanical bahaviour of irradiated Cu. Philos. Mag. **84**(34), 3517–3635 (2004)

H. Askari, M.R. Maughan, N.S. Abdolrahim, D.F. Bahri, H.M. Zbib, A stochastic crystal plasticity framework for deformation in micro-scale polycrystalline materials. Int. J. Plast. **68**, 21–33 (2015)

A. Bag, K. Ray, E. Dwarakadasa, Influence of martensite content and morphology on tensile and impact properties of high-martensite dual-phase steels. Metall. Mater. Trans. A **30**(5), 1193–1202 (1999)

T. Balusamy, T.S. Narayanan, K. Ravichandran, I.S. Park, M.H. Lee, Influence of surface mechanical attrition treatment (SMAT) on the corrosion behaviour of AISI 304 stainless steel. Corros. Sci. **74**, 332–344 (2013)

D.J. Bammann, An internal variable model of viscoplasticity. Int. J. Eng. Sci. **22**(8–10), 1041–1053 (1984)

D.J. Bammann, P.R. Dawson, Effects of spatial gradients in hardening evolution upon localization. Physics and mechanics of finite plastic and viscoplastic deformation (1997)

E. Bayerschen, A. McBride, B. Reddy, T. Böhlke, Review on slip transmission criteria in experiments and crystal plasticity models. J. Mater. Sci. **51**(5), 2243–2258 (2016)

T.M. Breunig, S.R. Stock, S.D. Antolovich, J.H. Kinney, W.N. Massey, M.C. Nichols, A framework for relating macroscopic measures and physical processes of crack closure illustrated by a study of aluminum lithium alloy 2090, ASTM STP 1131. Fracture Mech. 22nd Sym, ASTM, Phil (1992)

M. Calcagnotto, D. Ponge, D. Raabe, Effect of grain refinement to 1 $\mu$m on strength and toughness of dual-phase steels. Mater. Sci. Eng. A **527**(29), 7832–7840 (2010)

M. Calcagnotto, Y. Adachi, D. Ponge, D. Raabe, Deformation and fracture mechanisms in fine-and ultrafine-grained ferrite/martensite dual-phase steels and the effect of aging. Acta Mater. **59**(2), 658–670 (2011)

S.S. Chakravarthy, W. Curtin, Stress-gradient plasticity. Proc. Natl. Acad. Sci. **108**(38), 15716–15720 (2011)

J. Eshelby, F. Frank, F. Nabarro, XLI. The equilibrium of linear arrays of dislocations. Lond. Edinb. Dublin Philos. Mag. J. Sci. **42**(327), 351–364 (1951)

N. Fleck, J. Hutchinson, A phenomenological theory for strain gradient effects in plasticity. J. Mech. Phys. Solids **41**(12), 1825–1857 (1993)

N. Fleck, J. Hutchinson, Strain gradient plasticity. Adv. Appl. Mech. **33**, 296–361 (1997)

N. Fleck, G. Muller, M. Ashby, J. Hutchinson, Strain gradient plasticity: theory and experiment. Acta Metallurgica et Materialia **42**(2), 475–487 (1994)

S. Forest, K. Sab, Stress gradient continuum theory. Mech. Res. Commun. **40**, 16–25 (2012)

S. Forest, R. Sievert, Elastoviscoplastic constitutive frameworks for generalized continua. Acta Mech. **160**, 71–111 (2003)

H. Gao, Y. Huang, Geometrically necessary dislocation and size-dependent plasticity. Scr. Mater. **48**(2), 113–118 (2003)

S. Groh, E.B. Marin, M.F. Horstemeyer, H.M. Zbib, Multiscale modeling of plasticity in an aluminum single crystal. Int. J. Plast. **25**, 1456–1473 (2009)

I. Groma, P. Balogh, Investigation of dislocation pattern formation in a two-dimensional self-consistent field approximation. Acta Mater. **47**, 3647–3654 (1999)

M.E. Gurtin, L. Anand, A theory of strain-gradient plasticity for isotropic, plastically irrotational materials. Part II: Finite deformations. Int. J. Plast. **21**(12), 2297–2318 (2005)

Y. Hailiang, L. Cheng, T. Kiet, L. Xianghua, S. Yong, Y. Qingbo, K. Charlie, Asymmetric cryorolling for fabrication of nanostructural aluminum sheets. Sci. Rep. **2**, 772 (2012)

E. Hall, The deformation and ageing of mild steel: III discussion of results. Proc. Phys. Soc. Sect. B **64**(9), 747 (1951)

H. Hallber, M. Ristinmaa, Microstructure evolution influenced by dislocation density gradients modeled in a reaction-diffusion system. Comput. Mater. Sci. **67**, 373–383 (2013)

M. Hiratani, H.M. Zbib, Stochastic dislocation dynamics for dislocation-defects interaction. J. Eng. Mater. Tech. **124**, 335–341 (2002)
M. Hiratani, H.M. Zbib, On dislocation-defect interactions and patterning: stochastic discrete dislocation dynamics (SDD). J. Nucl. Mater. **323**, 290–303 (2003)
J. Hirth, Dislocation pileups in the presence of stress gradients. Philos. Mag. **86**(25–26), 3959–3963 (2006a)
J.P. Hirth, Disloction pileups in the presene of stress gradients. Philos. Mag. **86**, 3959–3963 (2006b)
J.P. Hirth, M. Rhee, H.M. Zbib, Modeling of deformation by a 3D simulation of multipole, curved dislocations. J. Computer-Aided Mater. Des. **3**, 164–166 (1996)
J.P. Hirth, H.M. Zbib, J. Lothe, Forces on high velocity dislocations. Model. Simul. Mater. Sci. Eng. **6**, 165–169 (1998)
D. Holt, Dislocation cell formation in metals. J. Appl. Phys. **41**, 3197–3201 (1970)
Y. Huang, H. Gao, W. Nix, J. Hutchinson, Mechanism-based strain gradient plasticity—II. Analysis. J. Mech. Phys. Solids **48**(1), 99–128 (2000)
Y. Huang, S. Qu, K. Hwang, M. Li, H. Gao, A conventional theory of mechanism-based strain gradient plasticity. Int. J. Plast. **20**(4), 753–782 (2004)
J.Y. Kang, J.G. Kim, H.W. Park, H.S. Kim, Multiscale architectured materials with composition and grain size gradients manufactured using high-pressure torion. Sci. Rep. **6**, 26590 (2016)
A. Khan, H.M. Zbib, D.A. Hughes, *Stress patterns of deformation induced planar dislocation boundaries* (MRS, San Francisco, 2001)
A. Khan, H.M. Zbib, D.A. Hughes, Modeling planar dislocation boundaries using a multi-scale approach. Int. J. Plast. **20**, 1059–1092 (2004)
U.F. Kocks, Laws for work-hardening and low-temperature creep. ASME Trans. Ser. H. J. Eng. Mater. Technol. **98**, 76–85 (1976)
J. Kratochvil, Dislocation pattern formation in metals. Revue de physique appliquée **23**(4), 419–429 (1988)
L.P. Kubin, Y. Estrin, Strain non-uniformities and plastic instabilities. Rev. Phys. Appl. **23**, 573–583 (1988)
D.S. Li, H.M. Zbib, H.S. Garmestani, M. Khaleel, X. Sun, Modeling of irradiation hardening of polycrystalline materials. Comp. Mater. Cont. **50**, 2496–2501 (2010)
D. Li, H.M. Zbib, X. Sun, M. Khaleel, Predicting plastic flow and irradiation hardening of iron single crystal with mechanism-based continuum dislocation dynamics. Int. J. Plast. **52**, 3–17 (2013). https://doi.org/10.1016/j.ijplas.2013.01.015
D. Li, H. Zbib, X. Sun, M. Khaleel, Predicting plastic flow and irradiation hardening of iron single crystal with mechanism-based continuum dislocation dynamics. Int. J. Plast. **52**, 3–17 (2014)
H. Lim, M.G. Lee, J.H. Kim, B.L. Adams, R.H. Wagoner, Simulation of polycrystal defromation with grain and garin boundary effects. Int. J. Plast. **27**, 1328–1354 (2011)
D. Liu, Y. He, B. Zhang, L. Shen, A continuum theory of stress gradient plasticity based on the dislocation pile-up model. Acta Mater. **80**, 350–364 (2014)
K. Lu, J. Lu, Nanostructured surface layer on metallic materials induced by surface mechanical attrition treatment. Mater. Sci. Eng. A **375**, 38–45 (2004)
L. Lu, M. Sui, K. Lu, Supperplastic extensibility of nanocrystalline copper at room termperature. Science **287**(5457), 1463 (2000)
H. Lyu, A. Ruimi, H.M. Zbib, A dislocation-based model for deformation and size effect in multi-phase steels. Int. J. Plast. **72**, 44–59 (2015a)
H. Lyu, A. Ruimi, H.M. Zbib, A dislocation-based model for deformation and size effect in multiscale-phase steels. Int. J. Plast. **72**, 44–59 (2015b)
H. Lyu, A. Ruimi, F. Zhang, H.M. Zbib, A numerical investigation of the effect of texture on mechanical properties in dual phase steel using a dislocation-based crystal plasticity model. MS&T 2015 Proceedings: Multi scale Modeling of Microstructure Deformation in Material Processing (2015c)
H. Lyu, A. Ruimi, P.D. Field, H.M. Zbib, Plasticity in materials with heterogeneous microstructures. Metall. Trans. A. **47**(12), 6608–6620 (2016a)

H. Lyu, N. Taheri-Nassaj, H.M. Zbib, A multiscale gradient-dependent plasticity model for size effects. Philos. Mag. **96**, 1–26 (2016b)
H. Lyu, N. Taheri-Nassaj, H.M. Zbib, A multiscale gradient-dependent plasticity mole for size effects. Philos. Mag. **96**(18), 1883–1908 (2016c)
H. Lyu, M. Hamid, A. Ruimi, H.M. Zbib, Stress/Strain gradient plasticity model for size effects in materials with heterogeneous nano-microstructures. Int. J. Plast. **97**, 46–63 (2017)
I. Mastorakos, H. Zbib, A multiscale approach to study the effect of chromium and nickel concentration in the hardening of iron alloys. J. Nucl. Mater. **449**(1), 101–110 (2014)
I. Mastorakos, L. Le, M. Zeine, H.M. Zbib, M. Khaleel, Multiscale Modeling of irradiation induced hardening in a-Fe, Fe-Cr and Fe-Ni Systems, in *Basic Actinide Science and Materials for Nuclear Applications*, (MRS, Warrendale, 2010)
D.L. McDowell, *Internal state variable theory* (Springer, Dordrecht, 2005)
D.L. McDowell, A perspective on trends in multiscale plasticity. Int. J. Plast. **26**(9), 1280–1309 (2010)
S. Mesarovic, Energy, configurational forces and characteristic lengths associated with the continuum description of geometrically necessary dislocations. Int. J. Plast. **21**, 1855–1889 (2005a)
S.D. Mesarovic, Energy, configurational forces and characteristic lengths associated with the continuum description of geometrically necessary dislocation. Int. J. Plast. **21**, 1855–1889 (2005b)
Y. Morita, K. Shizawa, H.M. Zbib, Self-organization model and simulation of collective dislocation based on interaction between GN dislocation and dislocation dipole. Mater. Sci. Res. Int. **2**, 323–326 (2001)
D.G. Morris, The origins of strengthening in nanostructured metals and alloys (2010)
NRC, National Research Council Report. Integrated computational materials engineering. (The National Academies Press, Washington, DC, 2008), http://www.nap.edu/catalog/12199.html
NSF, Blue Ribbon Advisory Panel Report 2006 Simulation-based engineering science (2006), http://www.nsf.gov/pubs/reportssbes_final_report.pdf
NSTC, Materials Genome Initiative (MGI) for Global Competitiveness (2011), http://www.whitehouse.gov/sites/default/files/microsites/ostp/materials_genome_initiative-fina;.pdf.
T. Ohashi, Numerical modeling of plastic multislip in metal crystals of fcc type. Philos. Mag. A **70**(5), 793–803 (1994)
T. Ohashi, A new model of scale dependent crystal plasticity analysis. IUTAM Symposium on Mesoscopic Dynamics of Fracture Process and Materials Strength (Springer, 2004)
T. Ohashi, Crystal plasticity analysis of dislocation emission from micro voids. Int. J. Plast. **21**(11), 2071–2088 (2005)
T. Ohashi, M. Kawamukai, H. Zbib, A multiscale approach for modeling scale-dependent yield stress in polycrystalline metals. Int. J. Plast. **23**(5), 897–914 (2007)
E. Orowan, Problems of plastic gliding. Z. Physik **1934**, 634 (1940)
N. Petch, The cleavage strength of polycrystals. J. Iron Steel Inst. **174**, 25–28 (1953)
J. Pontes, D. Walgraef, E.C. Aifantis, On disloction patterning: multiple slip effects in the rate equations approach. Int. J. Plast. **22**, 1486–1505 (2015)
J. Rajagopalan, M.T.A. Saif, Effect of microstructural heterogeneity on the mechanical behavior of nanocrystalline metal films. J. Mater. Res. **26**(22), 2826–2832 (2011)
M. Rhee, H.M. Zbib, J.P. Hirth, H. Huang, T.D. de la Rubia, Models for long/short range interactions in 3D dislocatoin simulation. Model. Simul. Mater. Sci. Eng. **6**, 467–492 (1998)
R. Roumina, C. Sinclair, Deformation geometry and through-thickness strain gradients in asymmetric rolling. Metall. Mater. Trans. A **39**(10), 2495 (2008)
M. Sarwar, R. Priestner, Influence of ferrite-martensite microstructural morphology on tensile properties of dual-phase steel. J. Mater. Sci. **31**(8), 2091–2095 (1996)
S. Shao, H.M. Zbib, I. Mastorakos, D.F. Bahr, Deformation mechanisms, size effects, and strain hardening in nanoscale multilayerd metallic composites under nanoindentation. J. Appl. Phys. **112**, 044307 (2012)

S. Shao, N. Abdolrahim, D.F. Bahr, G. Lin, H.M. Zbib, Stochastic effects in plasticity in small volumes. Int. J. Plast. **82**, 435–441 (2014)

M. Shehadeh, H.M. Zbib, T.D. de la Rubia, Multiscale dislocation dynamics simulations of shock compressions in copper single crystal. Int. J. Plast. **21**, 2369–2390 (2005)

M.A. Shehadeh, H.M. Zbib, T.D. de la Rubia, Modeling the dynamic deformation and patterning in FCC single crystals at high strain rates: dislocation dynamic plasticity analysis. Philos. Mag. A **85**, 1667–1684 (2005)

K. Shizawa, H. Zbib, A thermodynamical theory of gradient elastoplasticity with dislocation density tensor. I: Fundamentals. Int. J. Plast. **15**(9), 899–938 (1999a)

K. Shizawa, H. Zbib, A thermodynamical theory of plastic spin and internal stress with dislocation density tensor. J. Eng. Mater. Technol. **121**(2), 247–253 (1999b)

K. Shizawa, H.M. Zbib, A strain-gradient thermodynamic theory of plasticity based on dislocation density and incompatibility tensor. Mater. Sci. Eng. A **309**, 416–419 (2001)

Y.I. Son, Y.K. Lee, K.-T. Park, C.S. Lee, D.H. Shin, Ultrafine grained ferrite–martensite dual phase steels fabricated via equal channel angular pressing: microstructure and tensile properties. Acta Mater. **53**(11), 3125–3134 (2005)

T.-N. Taheri, Dislocation-based multiscale modeling of plasticity and controlling deformation mechanisms. PhD, Washington State University, 2016

N. Taheri-Nassaj, H.M. Zbib, On dislocation pileups and stress-gradient dependent plastic flow. Int. J. Plast. **74**, 1–16 (2015)

N. Taheri-Nassaj, H.M. Zbib, A mesoscale model of plasticity: Disloction dynamics and patterning (1D). ASME J. Eng. Mater. Technol **138**(4), 1–9 (2016)

N. Tao, H. Zhang, J. Lu, K. Lu, Development of nanostructures in metallic materials with low stacking fault energies during surface mechanical attrition treatment (SMAT). Mater. Trans. **44**(10), 1919–1925 (2003)

G.Z. Voyadjis, L.M. Mohammad, Theory vs experiment for finite strain viscoplastic lagrangian constitutive model. Int. J. Plast. **7**, 329–350 (1991)

D. Walgraef, E.C. Aifantis, On the formation and stability of dislocation patterns I, II, III. Int. J. Eng. Sci. **23**, 1315–1372 (1985)

X. Wu, N. Tao, Y. Hong, B. Xu, J. Lu, K. Lu, Microstructure and evolution of mechanically-induced ultrafine grain in surface layer of Al-alloy subjected to USSP. Acta Mater. **50**(8), 2075–2084 (2002)

X. Wu, P. Jiang, L. Chen, F. Yuan, Y.T. Zhu, Extraordinary strain hardening by gradient structure. Proc. Natl. Acad. Sci. **111**(20), 7197–7201 (2014a)

X.L. Wu, P. Jiang, L. Chen, J.F. Zhnag, F.P. Youn, Y.T. Zhu, Synergetic strengthening by gradient structure. Mater. Res. Lett. **2**(4), 185–191 (2014b)

X. Wu, M. Yang, F. Yuan, G. Wu, Y. Wei, X. Huang, Y. Zhu, Heterogeneous lamella structure unites ultrafine-grain strength with coarse-grain ductility. Proc. Natl. Acad. Sci. **112**(47), 14501–14505 (2015)

S. Wulfinghoff, T. Böhlke, Gradient crystal plasticity including dislocation-based work-hardening and dislocation transport. Int. J. Plast. **69**, 152–169 (2015)

M. Yang, Y. Pan, F. Yuan, Y. Zhu, X. Wu, Back stress strengthening and strain hardening in gradient structure. Mater. Res. Lett., 1–7 (2016)

Z. Yin, X. Yang, X. Ma, J. Moering, J. Yang, Y. Gong, Y. Zhu, X. Zhu, Strength and ductility of gradient structured copper obtained by surface mechanical attrition treatment. Mater. Des. **105**, 89–95 (2016)

H. Zbib, Strain gradients and size effects in nonhomogeneous plastic deformation. Scripta metallurgica et materialia **30**(9), 1223–1226 (1994)

H.M. Zbib, E.C. Aifantis, On the localization and post localization behavior of plastic deformation-II. On the evolution and thickness of shear bands. Res. Mech. Int. J. Struct. Mech. Mater. Sci. **23**, 279–292 (1988)

A.P.H. Zbib, E. Aifantis, On the gradient-dependent theory of plasticity and shear banding. Acta Mech. **92**(1–4), 209–225 (1992)

H.M. Zbib, T. Diaz de la Rubia, A multiscale model of plasticity. Int. J. Plast. **18**(9), 1133–1163 (2002)
H.M. Zbib, M. Shehadeh, On the homogenous nucleation and propagation of dislocations under shock compression. Philos. Mag. **96**, 2752–2778 (2016)
H.M. Zbib, M. Rhee, J.P. Hirth, On plastic deformation and the dynamcis of 3D dislocations. Int. J. Mech. Sci. **40**, 113–127 (1998)
H.M. Zbib, M. Rhee, J.P. Hirth, T. Diaz de la Rubia, A 3D dislocation simulation model for plastic deformation and instabilities in single crystals. J. Mech. Behav. Mater. **11**, 251–255 (2000)
H.M. Zbib, C. Overman, F. Akasheh, D.F. Bahr, Analysis of plastic deformation in nanoscale metallic multilayers with coherent and incoherent interfaces. Int. J. Plast. **27**, 1618–1638 (2011)
P. Zhang, D. Balint, J. Lin, Controlled Poisson Voronoi tessellation for virtual grain structure generation: a statistical evaluation. Philos. Mag. **91**(36), 4555–4573 (2011)
P. Zhang, M. Karimpour, D. Balint, J. Lin, D. Farrugia, A controlled Poisson Voronoi tessellation for grain and cohesive boundary generation applied to crystal plasticity analysis. Comput. Mater. Sci. **64**, 84–89 (2012)
F. Zhang, A. Ruimi, P.C. Wo, D.P. Field, Morphology and distribution of martensite in dual phase (DP980) steel and its relation to the multiscale mechanical behavior. Mater. Sci. Eng. A **659**, 93–103 (2016)
Y.-H. Zhao, X.-Z. Liao, S. Cheng, E. Ma, Y.T. Zhu, Simultaneously increasing the ductility and strength of nanostructured alloys. Adv. Mater. **18**(17), 2280–2283 (2006)
H.T. Zhu, H. Zbib, E. Aifantis, Strain gradients and continuum modeling of size effect in metal matrix composites. Acta Mech. **121**(1–4), 165–176 (1997)
K. Zhu, A. Vassel, F. Brisset, K. Lu, J. Lu, Nanostructure formation mechanism of $\alpha$-titanium using SMAT. Acta Mater. **52**(14), 4101–4110 (2004)

# Statistical Theory of Dislocation

**István Groma**

**Abstract** The plastic deformation of materials are traditionally modeled by phenomenological crystal plasticity continuum theories. There are, however, several phenomena, like deformation size effect, hardening due to grain boundary, dislocation patter formation that cannot be described within this framework. One has to take into account that the stress-strain response of crystalline materials is determined by the collective motion of dislocations. The aim of the present chapter is to introduce a continuum theory of dislocation obtained by a systematic coarse-graining of the evolution equation of individual dislocations.

## 1 Introduction

The plastic deformation of crystalline materials is controlled by the collective motion of dislocations. So, to develop a comprehensive model for the stress-strain response of materials we have to understand the statistical properties of dislocations. Since dislocations form a complex network of line type objects, modeling the collective evolution of the dislocation network is a rather challenging problem. What make the issue even more difficult is that the dislocation motion is dissipative and the interaction between dislocations is long ranged. The statistical physics of line type dissipative systems with long range interaction is not developed.

One possibility is to study the evolution of the dislocation system with molecular dynamic (MD) simulation (Zepeda-Ruiz et al. 2016). For details see chapter "Multiscale Modeling of Interfaces, Dislocations, and Dislocation Field Plasticity". With massive parallel computers it is feasible to perform MD simulation with $1000 \times 1000 \times 1000$ atoms corresponding to a cube with about 200 nm. By such a simulation one can study complex but still "elementary" dislocation phenomena

I. Groma (✉)
Department of Materials Physics, ELTE, Eötvös Loránd University, Budapest, Hungary
e-mail: groma@metal.elte.hu

 
S. Mesarovic et al. (eds.), *Mesoscale Models*, CISM International Centre
for Mechanical Sciences 587, https://doi.org/10.1007/978-3-319-94186-8_3

like dislocation multiplication, junction formation, cross slip, etc., but macroscopic properties of the system practically cannot be obtained.

With discrete dislocation dynamics (DDD) simulations (for details see chapter "Multiscale Dislocation-Based Plasticity") one can reach the order of 1 μm sample size (Ghoniem and Sun 1999; Kubin and Canova 1992; Rhee et al. 1998; Gomez-Garcia et al. 2006; Devincre et al. 2001; Bulatov et al. 2006), but for several important problems like dislocation patter formation or size effect we would need a system size minimum of 10 μm because the characteristic size of dislocation patterns are about 1 μm.

In this chapter we discuss the statistical properties of a rather simplified dislocation system consisting of parallel edge dislocations. As it is explained below for this dislocation setup a continuum theory of the evolution of the different dislocation densities can be derived on a mathematically rigorous manner by a systematic coarse-graining of the equation of motion of dislocations. In order to get closed set of equations, however, requires some assumptions about the properties of dislocation-dislocation correlation functions. A key feature of the analysis is that these assumptions can be directly verified by DDD simulations. Moreover, the predictions of the continuum theory can be directly compared to DDD simulation results. So, the statistical continuum theory of dislocations presented is validated by DDD.

Although there are rather promising attempts to generalized the theory for more complex dislocation configurations, the 3D continuum theory of dislocations is much less developed. The 2D theory, established on a solid grounds, can help a lot in setting up the structure of the 3D continuum theory.

In the first part of the chapter the field theory of dislocations, developed by Nye, Kröner, and Kosevich, is summarized. It is explained how the stress or strain field generated by a dislocation system can be determined within the framework of a field theory. In the next part the link between the microscopic and mesoscopic descriptions of the evolution of a 2D dislocation system is established by a systematic coarse-graining. It is shown that the theory is able to predict dislocation patterning. In the last part current approaches for the 3D generalization of the theory are discussed.

## 2 Nye, Kröner, and Kosevich Field Theory of Dislocations

### *2.1 Dislocation Density Tensor*

Shortly after the concept of dislocation was introduced by Polanyi, Orovan and Taylor in 1934, it was recognized by the Burgers brothers that the elastic field generated by a straight dislocation was already determined by Vito Volterra in 1907 when he considered an elastic problem with a discontinuity on a half plane. (see for example Kovács and Zsoldos (1973)). In the 1950s Nye, Kröner and

Kosevich reconsidered the problem and developed an extremely elegant formalism to determine the elastic properties of dislocated crystals (Kröner 1981; Landau and Lifshitz 1986). In contrast to the **statistical theory of dislocations** (explained in details in this chapter) this theory does not consider the collective evolution of the dislocation network. It gives the different fields (stress, strain, etc.) generated by the dislocation system. So, it should be referred to as **field theory of dislocations**. Since, however, it is essential for the statistical theory, first it is shortly summarized. For more details the reader is referred to Kröner (1981), Landau and Lifshitz (1986), and Kosevich (1979).

The deformation of a body can be given by the transformation $\vec{R}(\vec{r})$ where $\vec{R}$ is the deformed position of the point originally located at $\vec{r}$ in the reference (undeformed) system. Assuming that the transformation is differentiable, the transformation matrix defined as

$$dR_i = F_{ij} dx_j \tag{1}$$

is

$$F_{ij} = \partial_j R_i(\vec{r}). \tag{2}$$

(Throughout this chapter a double index implies summation according to Einstein summation convention.)

By introducing the displacement field $\vec{u}(\vec{r}) = \vec{R}(\vec{r}) - \vec{r}$ the matrix $F_{ij}$ can be given as

$$F_{ij} = \delta_{ij} + \beta_{ji} \text{ with } \beta_{ij} = \partial_i u_j(\vec{r}), \tag{3}$$

where $\beta_{ij}$ is called the distortion tensor, and $\delta_{ij}$ is the unit tensor.

For an elastic body the internal stress $\sigma_{ij}$ is completely determined by the deformation tensor $\epsilon_{ij} = (F_{ik}F_{jk} - \delta_{ij})/2$. If, however, plastic deformation is involved only a certain part of the deformation defined above generates stress. To account for this, it is assumed that the total deformation of the body is reached by two subsequent steps, a plastic and an elastic deformations. The first one given by the transformation matrix $F_{ij}^p$ does not generate stress while the second one denoted by $F_{ij}^e$ is related to the stress according to the constitutive equation of the material considered. So, the "starting" point of any plasticity theory of crystalline materials is that

$$F_{ij} = F_{ik}^e F_{kj}^p, \tag{4}$$

where neither $F_{ij}^e$ nor $F_{ij}^p$ is a derivative of a vector field (for details see chapter "Generalized Continua and Phase-field Models: Application to Crystal Plasticity"). It has to be mentioned that the definitions given above do not uniquely determine $F_{ij}^e$ and $F_{ij}^p$. The issue is discussed below.

After introducing the plastic $\beta^p_{ij} = F^p_{ji} - \delta_{ij}$ and elastic $\beta^e_{ij} = F^e_{ji} - \delta_{ij}$ distortions Eq. (4) reads as

$$\partial_i u_j = \beta^e_{ij} + \beta^p_{ij} + \beta^e_{ik}\beta^p_{kj}. \tag{5}$$

In the rest of this chapter it is assumed that the distortions are small. So, the last term in the right hand side of the above equation is neglected (small deformation limit), i.e.

$$\partial_i u_j = \beta^e_{ij} + \beta^p_{ij}. \tag{6}$$

For a detailed introduction to the large deformation case see chapter "Generalized Continua and Phase-field Models: Application to Crystal Plasticity". It is important to stress, that so far, splitting $\partial_i u_j$ is just formal, we have to precisely defined $\beta^e_{ij}$ and $\beta^p_{ij}$.

Since the total distortion $\beta_{ij} = \partial_i u_j$ is a gradient of a vector field its *curl* vanishes, $e_{ikl}\partial_k \beta_{lj} = 0$ where $e_{ikl}$ is the permutation tensor. If, however, dislocations present in the crystal the plastic distortion is not *curl* free, its *curl*

$$\alpha_{ij} = -e_{ikl}\partial_k \beta^p_{lj} \tag{7}$$

is called Nye's dislocation density tensor. From Eq. (6) one can find that $\alpha_{ij} = e_{ikl}\partial_k \beta^e_{lj}$.

Taking the integral of $\alpha_{ij}$ for a surface

$$b_j = \int_A \alpha_{ij} dA_i = -\int_A e_{ikl}\partial_k \beta^p_{lj} dA_i = -\oint \beta^p_{ij} ds_i = -\oint du^p_j \tag{8}$$

gives the net Burgers vector of the dislocations crossing the surface. From this one can find that for a single dislocation

$$\alpha_{ij} = t_i b_j, \delta(\zeta) \tag{9}$$

where $l_i$ is an unit vector in the direction of the dislocation line, and $\zeta$ is the distance from the dislocation line.

In the rest of this section we assume that the dislocation density tensor is given, its evolution will be discussed in the next section. In the following we concentrate on calculating the internal stress generated by the dislocation network. Before proceeding further, there are two important issues that have to be discussed:

- Since $\alpha_{ij}$ is the *curl* of the plastic distortion, $\beta^p_{ij}$ is not uniquely defined by $\alpha_{ij}$. For giving $\beta^p_{ij}$ completely we have to put further physical input. The issue is discussed at the end of this section.
- In order to avoid that rigid body rotation generates internal stress, the stress cannot depend on the total elastic distortion $\beta^e_{ij}$, only its symmetric part, the

elastic deformation $\epsilon^e_{ij} = (\beta^e_{ij} + \beta^e_{ji})/2$. It follows that the stress state does not uniquely determine $\beta^e_{ij}$ only its symmetric part. Because of this, in the following the symmetric parts of the total, elastic, and plastic distortions are considered. From Eq. (6) they are related as

$$(\partial_j u_i + \partial_i u_j)/2 = \epsilon^e_{ij} + \epsilon^p_{ij}. \tag{10}$$

Assuming local linear elasticity the stress-strain relation (Hooke's law) is

$$\sigma_{ij} = L_{ijkl}\epsilon^e_{kl}, \tag{11}$$

where $L_{ijkl}$ is the elastic modulus tensor. With Eq. (11) Eq. (10) reads as

$$(\partial_j u_i + \partial_i u_j)/2 = L^{-1}_{ijkl}\sigma_{kl} + \epsilon^p_{ij}. \tag{12}$$

In order to proceed further we introduce the incompatibility operator. It acts on a tensor field $\hat{A}$ as

$$(\mathrm{Inc}\hat{A})_{ij} = -e_{ikm}e_{jln}\partial_k\partial_l A_{mn}. \tag{13}$$

Two important properties of the *Inc* operator are:

- For any vector field $\vec{u}(\vec{r})$ the Inc of the symmetric part of its derivative vanishes:

$$\mathrm{Inc}\left(Sim\left[\frac{d\vec{u}}{d\vec{r}}\right]\right)_{ij} = 0, \tag{14}$$

  where

$$\left(Sim\left[\frac{d\vec{u}}{d\vec{r}}\right]\right)_{ij} = (\partial_i u_j + \partial_j u_i)/2. \tag{15}$$

- For any tensor field $\hat{A}(\vec{r})$

$$\partial_i(\mathrm{Inc}\hat{A})_{ij} = 0. \tag{16}$$

By taking the incompatibility of Eq. (12) one arrives at

$$-e_{ikm}e_{jln}\partial_k\partial_l L^{-1}_{mnop}\sigma_{op} = \eta_{ij}, \tag{17}$$

where the incompatibility field $\eta_{ij}$ defined as

$$\eta_{ij} = -(\mathrm{Inc}\hat{\epsilon}^p)_{ij} = e_{ikm}e_{jln}\partial_k\partial_l\epsilon^p_{mn} \tag{18}$$

is introduced. From Eq. (10) one gets that

$$\eta_{ij} = (\text{Inc}\hat{\epsilon}^e)_{ij}. \tag{19}$$

One can find that the incompatibility field is related to the dislocation density tensor as

$$\eta_{ij} = -\frac{1}{2}\left(e_{iln}\partial_n\alpha_{jl} + e_{jln}\partial_n\alpha_{il}\right), \tag{20}$$

but the source of incompatibility is not necessary related to dislocations. Any other defect such as grain boundary, disclination, stacking fault, inclusion, etc. also can be the source of incompatibility. According to the above equation the primary source of internal stress is the incompatibility field. Since, however, as it was mentioned above, the Inc of a symmetric part of a derivative of a vector filed vanishes, Eq. (17) is not sufficient to determine the stress field generated by $\eta_{ij}$. It has to be supplemented with the equilibrium equation

$$\partial_i\sigma_{ij} = 0, \tag{21}$$

the symmetry condition $\sigma_{ij} = \sigma_{ji}$, and the surface traction on the boundary (for other boundary conditions see below).

## 2.2 *Second Order Stress Function Tensor*

Like in electrodynamics it is useful to reformulate Eqs. (17), (21) into a potential theory. Let us introduce a second order stress function tensor $\chi_{ij}$ defined with the relation

$$\sigma_{ij} = (\text{Inc}\hat{\chi})_{ij} = -e_{ikm}e_{jln}\partial_k\partial_l\chi_{mn}. \tag{22}$$

Due to the identity (16) the (22) form of $\sigma_{ij}$ guarantees that the equilibrium condition (21) is fulfilled. With the stress function tensor introduced above Eq. (17) reads as

$$\eta_{ij} = e_{ikm}e_{jln}e_{oqv}e_{puw}L^{-1}_{mnop}\partial_k\partial_l\partial_q\partial_u\chi_{vw}. \tag{23}$$

For an anisotropic medium the above equation is rather difficult to solve, but for isotropic materials, with shear modulus $\mu$ and Poisson's ratio $\nu$, a general solution can be obtained. It is expedient to introduce another tensor potential $\chi'_{ij}$ defined as

$$\chi'_{ij} = \frac{1}{2\mu}\left(\chi_{ij} - \frac{\nu}{1+2\nu}\chi_{kk}\delta_{ij}\right) \tag{24}$$

$$\chi_{ij} = 2\mu \left( \chi'_{ij} + \frac{\nu}{1-\nu} \chi'_{kk} \delta_{ij} \right). \tag{25}$$

By inserting Eq. (25) into Eq. (23) one can find, if $\chi'_{ij}$ fulfills the gauge condition

$$\partial_i \chi'_{ij} = 0, \tag{26}$$

Eq. (22) simplifies to the biharmonic equation

$$\nabla^4 \chi'_{ij} = \eta_{ij}. \tag{27}$$

A remarkable feature of this equation is that the different components of $\chi'_{ij}$ obey separate equations making the problem much easier to solve. For an infinite medium the general solution of Eq. (27) is

$$\chi'_{ij}(\vec{r}) = -\frac{1}{8\pi} \int\int\int |\vec{r} - \vec{r}'| \eta_{ij}(\vec{r}') dV' \tag{28}$$

## 2.3 2D Problems

In the next section the statistical properties of an ensemble of parallel edge dislocations are discussed. In this case the stress and the strain do not vary along the dislocation line direction $\vec{l}$. Taking $\vec{l}$ parallel to the $z$ axis (with $\vec{l} = (0, 0, -1)$) in the above expressions the derivatives with respect to $z$ vanish ($\partial_z \equiv 0$). One can find that Eq. (22) simplifies to (Kröner 1981):

$$\sigma_{11} = -\partial_y \partial_y \chi, \qquad \sigma_{22} = -\partial_x \partial_x \chi, \qquad \sigma_{12} = \partial_x \partial_y \chi, \qquad \chi \equiv \chi_{33} \tag{29}$$

$$\sigma_{23} = -\partial_x \phi, \qquad \sigma_{13} = \partial_y \phi, \qquad \phi = -\partial_x \chi_{23} + \partial_y \chi_{31}. \tag{30}$$

Furthermore, from Eqs. (20), (23) one obtains that the two scalar fields $\chi$ and $\phi$ introduced above obey the equations

$$\nabla^4 \chi = \frac{2\mu}{1-\nu} \left( b_1 \partial_y - b_2 \partial_x \right) \left( \rho_{d+} - \rho_{d-} \right) \tag{31}$$

$$\nabla^2 \phi = \mu b_3 (\rho_{d+} - \rho_{d-}), \tag{32}$$

where $b_1$, $b_2$, and $b_3$ are the $x$, $y$, and $z$ directional components of the Burgers vector, respectively. The notations $\rho_{d+}$ and $\rho_{d-}$ stand for the dislocation densities with positive and negative signs, respectively. They are the sum of $\delta(\vec{r} - \vec{r}_i)$ Dirac delta functions, where $\vec{r}_i$ denotes the position of a dislocation. Here, for the sake of simplicity we assumed that all dislocations belong to the same slip system (single slip), but the expressions can be easily generalized for multiple slip.

For an infinite medium the solutions of Eqs. (31), (32) read as

$$\chi(\vec{r}) = \frac{\mu}{2\pi(1-\nu)} \int \left(b_1 \partial_{y'} - b_2 \partial_{x'}\right) [\rho_{d+}(\vec{r}\,') - \rho_{d-}(\vec{r}\,')] R^2 \ln R \, d^2\vec{r}\,' \quad (33)$$

and

$$\phi(\vec{r}) = -\frac{\mu b_3}{2\pi} \int [\rho_{d+}(\vec{r}\,') - \rho_{d-}(\vec{r}\,')] \ln(R) \, d^2\vec{r}\,', \quad (34)$$

where $R = |\vec{r} - \vec{r}\,'|$.

## 2.4 Variational Approach I

For developing the statistical theory of dislocations it is useful to formulate the results explained above to a variational formalism. We are going to explain two possible approaches. In the first one the elastic deformation $\epsilon^e_{ij}$ is considered as the variational field, while in the second one the stress plays the role of independent variable.

Let us first consider the defect free situation. According to thermodynamics principles if the Helmholtz free energy as a functional of the deformation tensor $\epsilon_{ij}$ is given the elastic response of the material considered is determined. It should be stressed that we do not have to restrict our considerations to linear elasticity, nonlinearity and nonlocality (the free energy may depend on the derivatives of the deformation tensor) can be allowed. For simplicity here, however, we exclude further possible fields, like curvature, dependence of the free energy. For a general description see chapter "Generalized Continua and Phase-field Models: Application to Crystal Plasticity".

The stress is the functional derivative of the free energy:

$$\sigma_{ij} = \frac{\delta A}{\delta \epsilon_{ij}}. \quad (35)$$

Since $\epsilon_{ij} = (\partial_i u_j + \partial_j u_i)/2$, the minimum condition

$$-\frac{\delta A}{\delta u_i} = 0 \quad (36)$$

leads to the common equilibrium equation

$$-\frac{\delta A}{\delta u_i} = \partial_j \frac{\delta A}{\delta \epsilon_{ij}} = \partial_j \sigma_{ij} = 0, \tag{37}$$

where for simplicity body force is neglected and surface terms are not discussed.

One arrives at an equivalent result minimizing the free energy with respect to the elastic deformation with the additional condition that Inc of $\epsilon_{ij}$ vanishes, ensuring that $\epsilon_{ij}$ is the symmetric part of the derivative of the vector field $u_i$. The additional condition can be taken into account by minimizing the functional

$$Q[\epsilon_{ij}, \chi_{ij}] = A(\epsilon_{ij}) - \int \chi_{ij} (\mathrm{Inc}\hat{\epsilon})_{ij} dV \tag{38}$$

with respect to $\epsilon_{ij}$ and the Lagrangian multiplier $\chi_{ij}$ (Gröger et al. 2010), leading to

$$\frac{\delta Q}{\delta \epsilon_{ij}} = \frac{\delta A}{\delta \epsilon_{ij}} - (\mathrm{Inc}\chi)_{ij} = \sigma_{ij} - (\mathrm{Inc}\hat{\chi})_{ij} = 0, \tag{39}$$

and

$$\frac{\delta Q}{\delta \chi_{ij}} = (\mathrm{Inc}\hat{\epsilon})_{ij} = 0. \tag{40}$$

It can bee seen from Eq. (39) that the Lagrangian multiplier introduced is the second order stress function and the equilibrium condition $\partial_j \sigma_{ij} = 0$ is automatically fulfilled.

Generalizing the method for defected media is straightforward. If one simply considers the functional

$$Q[\epsilon_{ij}^e, \chi_{ij}] = A(\epsilon_{ij}^e) - \int \chi_{ij} \left[ (\mathrm{Inc}\hat{\epsilon}^e)_{ij} - \eta_{ij} \right] dV \tag{41}$$

the variations with respect to $\epsilon_{ij}^e$ and $\chi_{ij}$ lead to the bulk equations obtained above for dislocated media.

It should be mentioned, however, that in the above derivation surface terms appearing during the variations were not considered. So, the results obtained are valid only for infinite body with fields go to zero at infinity or for periodic boundary conditions. With other words, the variational method explained gives only the bulk equations, boundary conditions have to be handled separately.

## 2.5 Variational Approach II

In order to derive the other variational method (Groma et al. 2006, 2010) let us consider the Gibbs free energy $G$ that is a functional of the stress. Its functional derivative with respect to the stress is the negative elastic strain:

$$-\frac{\delta G}{\delta \sigma_{ij}} = \epsilon^e_{ij}.$$

To ensure the equilibrium condition let us use the second order stress function ($\sigma_{ij} = (\text{Inc}\hat{\chi})_{ij}$) and introduce the functional

$$P[\chi_{ij}] = G((\text{Inc}\hat{\chi})_{ij}) + \int \chi_{ij}\eta_{ij} dV \tag{42}$$

called as "plastic potential" hereafter.

The minimum condition

$$\frac{\delta P}{\delta \chi_{ij}} = 0 \tag{43}$$

leads to

$$\frac{\delta P}{\delta \chi_{ij}} = \left(\text{Inc}\frac{\delta G}{\delta \hat{\sigma}}\right)_{ij} + \eta_{ij} = -(\text{Inc}\hat{\epsilon}^e)_{ij} + \eta_{ij} = 0. \tag{44}$$

This means, in accordance with the definition of the incompatibility field $\eta_{ij}$, the plastic potential is minimized with respect to $\chi_{ij}$ so that the incompatibility is the Inc of the elastic deformation. Like for the other variational method explained above, surface terms are not taken into account. So, it is assumed that the different fields go to zero at infinity.

For the further considerations it is important to realize that the value of the plastic potential $P[\chi_{ij}]$ at its minimum $\chi^{eq}_{ij}$ is the Helmholtz free energy of the system. In order to see this, let us substitute the relation (19) into Eq. (42):

$$P[\chi^{eq}_{ij}] = G(\sigma^{eq}_{ij}) + \int \chi^{eq}_{ij}(\text{Inc}\epsilon^e)_{ij} dV. \tag{45}$$

After a double partial integration in the second term of the right hand side we arrive at

$$P[\chi^{eq}_{ij}] = G(\sigma^{eq}_{ij}) + \int \sigma^{eq}_{ij}\epsilon^e_{ij} dV. \tag{46}$$

From Eq. (42) one concludes

$$P[\chi_{ij}^{eq}] = G(\sigma_{ij}^{eq}) - \int \sigma_{ij}^{eq} \frac{\delta G}{\delta \sigma_{ij}^{eq}} dV, \tag{47}$$

i.e. $P[\chi_{ij}^{eq}]$ is the Legendre transform of the Gibbs free energy that is the Helmholtz free energy. So, the plastic potential at its minimum is the free energy of the system at the given incompatibility $\eta_{ij}$. It follows that the negative gradient of $P[\chi_{ij}^{eq}]$ with respect to the dislocation segment position $\vec{r}^{seg}$ is the Peach Koehler force acting on the dislocation segment

$$F_i = -\frac{dP}{dr_i^{seg}} = e_{ikl} t_k \sigma_{lm} b_m. \tag{48}$$

## 2.6 *Local Linear Medium*

We first demonstrate the variational principle on the elementary example of a local, linear material obeying Hooke's law (Groma et al. 2010). In this case the Gibbs free energy is a quadratic functional of the stress as

$$G_0[\sigma] := -\int \frac{1}{2} \sigma_{ij} L_{ijkl}^{-1} \sigma_{kl} \mathrm{d}V, \tag{49}$$

where $L_{ijkl}^{-1}$ is the elastic compliance tensor. Hence, the plastic potential given by Eq. (42) reads as

$$P[\chi, \eta] = \int \left[ -\frac{1}{2} e_{iop} e_{jqr} (\partial_o \partial_q \chi_{pr}) L_{ijkl}^{-1} e_{kst} e_{luz} \partial_s \partial_u \chi_{tz} + \chi_{ij} \eta_{ji} \right] \mathrm{d}V. \tag{50}$$

One can find that the plastic potential given above gets its minimum if $\chi_{ij}$ fulfills Eq. (23).

## 2.7 *Variation for Plane Problems*

In the statistical theory of dislocations, plane (2D) problems play an important role. In this section we discuss the variational method outline above for a systems of straight dislocations extending parallel to the $z$ direction (Groma et al. 2010).

After a long but straightforward calculation one can find that for edge dislocations the plastic potential functional reads as

$$P[\chi, \alpha] = \int \left[ -\frac{1-\nu}{4\mu} (\Delta\chi)^2 + \chi(\partial_2\alpha_{31} - \partial_1\alpha_{32}) \right] \mathrm{d}^2 r, \tag{51}$$

where $\chi := \chi_{33}$ is now a single component stress function (the other components of $\chi_{ij}$ vanish). The components of the stress tensor are given by Eq. (29). In $P[\chi, \alpha]$ the incompatibility tensor $\eta_{ij}$ has been expressed by the dislocation density tensor $\alpha_{ij}$, which now has only two nonvanishing components. The minimum condition $\delta P/\delta\chi = 0$ leads to the fourth order partial differential equation:

$$\frac{1-\nu}{2\mu} \Delta^2 \chi = \partial_2\alpha_{31} - \partial_1\alpha_{32}. \tag{52}$$

Although later in this chapter we do not consider screw dislocations, we summarize their case as well (this variational problem was first discussed by Berdichevsky (2005)). The plastic potential now is

$$P[\phi, \alpha] = \int \left[ -\frac{1}{2\mu} |\nabla\phi|^2 + \phi\alpha_{33} \right] \mathrm{d}^2 r \tag{53}$$

with

$$\phi := -\partial_1\chi_{23} + \partial_2\chi_{31}. \tag{54}$$

The relevant stress components are

$$\sigma_{23} = -\partial_1\phi, \quad \sigma_{13} = \partial_2\phi. \tag{55}$$

The corresponding minimum condition leads to the Poisson's equation

$$\frac{1}{\mu}\Delta\phi = -\alpha_{33}. \tag{56}$$

It has to be mentioned that Eqs. (52), (56) obtained by the variational approach are certainly equivalent with the ones derived earlier. In this section we just demonstrated how the variational method works for a classical local linear medium.

## 3 Dislocation Core Regularization

The significance of dislocation core regularization is widely known. It is not only necessary to account for core effects, but also to eliminate singularities in a physically well founded manner in numerical simulations. There are many different propositions for dislocation core regularization (Aifantis 1999; Gutkin and Aifantis 1999; Lazar 2003) but, as it is explained below, the variational approach offers a natural way to regularize the singular stress at the dislocation line.

It is common in phase field theories that surface or size effects are captured by introducing appropriate 'gradient terms' in the energy functional. The concept can be applied in dislocation theory too, but as we have recognized above, the physical properties of a material are determined by the functional form of the Gibbs free energy–stress relation. So, the gradient terms have to be introduced in the Gibbs free energy by adding terms that depend on the gradient of the stress. In first order linear approximation one can consider the 'nonlocal' Gibbs free energy

$$G_{nonlocal}[\sigma] = G_0 - b^2 \int N_{ijklmn}(\partial_i \sigma_{jk})\partial_l \sigma_{mn} \mathrm{d}V, \tag{57}$$

where $N_{ijklmn}$ is a constant tensor with inverse stress dimension and $b$ is the Burgers vector. ($b^2$ is separated from $N_{ijklmn}$ to indicate the relative order between $G_0$ and the gradient dependent term.) From $G_{nonlocal}[\sigma]$ the corresponding $P[\chi, \eta]$ has to be constructed as it is explained above.

It should be mentioned that nonlocality could be introduced on a much more general way by taking the Gibbs free energy in the form

$$G_{nonlocal}[\sigma] = -\int \frac{1}{2}\sigma_{ij}(\vec{r})S_{ijkl}(\vec{r}-\vec{r}\,')\sigma_{kl}(\vec{r}\,')\mathrm{d}V\mathrm{d}V', \tag{58}$$

where $S_{ijkl}(\vec{r}\,')$ is a function which goes to zero fast enough if $|\vec{r}| \to \infty$, but in a first order approximation, if its range is of the order of the lattice constant, it obviously gives the same as Eq. (57).

To demonstrate how the nonlocal term introduced above results in dislocation core regularization let us consider a single straight dislocation. From Eqs. (51) and (57) for a single edge dislocation at the origin

$$P[\chi] = \int \left\{ -\frac{1-\nu}{4\mu}\left[|\Delta\chi|^2 + a^2|\nabla\Delta\chi|^2\right] + \chi\partial_2\delta(\vec{r}) \right\} \mathrm{d}^2 r, \tag{59}$$

where $a$ is a parameter with length dimension that is in the order of the lattice constant. Here, for the sake of simplicity, we considered only the simplest possible isotropic gradient term from Eq. (57) but the general case can be treated in a similar way. The corresponding equilibrium equation has the form

$$\Delta^2\chi - a^2\Delta^3\chi = \frac{2b\mu}{1-\nu}\partial_2\delta(\vec{r}). \tag{60}$$

The above equation has analytical solution. By taking its Fourier transform one can find that

$$\chi^F(q_1, q_2) = \frac{2b\mu}{1-\nu} \frac{iq_y}{(q_x^2+q_y^2)^4 + a^2(q_x^2+q_y^2)^6} \tag{61}$$

from which, according to Eq. (29), the Fourier transform of the resolved shear stress reads as

$$\sigma_{12}^{\mathrm{r},F}(q_1, q_2) = -\frac{2b\mu}{1-\nu} \frac{iq_x q_y^2}{(q_x^2+q_y^2)^4 + a^2(q_x^2+q_y^2)^6}. \tag{62}$$

By inverse Fourier transformation we obtain

$$\chi = \frac{b\mu}{2\pi(1-\nu)} \frac{ay}{r} \left[ -2K_1\left(\frac{r}{a}\right) + 2\frac{a}{r} + \frac{r}{a} \ln\left(\frac{r}{a}\right) \right], \tag{63}$$

where $K_1(x)$ is the modified Bessel function of the second kind. To demonstrate the difference between the regularized and the "classical" solutions more explicitly the shear stress along the $x$ axis is plotted in Fig. 1. As it can be seen, for $x$ larger than about $10\,a$, $\sigma_{12}^{\mathrm{r}}$ is close to the classical stress field $\sigma_{12} \propto 1/r$. So as it is expected, the 'gradient term' introduced above influences only the central core region.

In a similar way, for screw dislocations from Eqs. (53) and (57)

$$P = \int \left\{ -\frac{1}{2\mu} \left[ |\nabla\phi|^2 + c^2(\Delta\phi)^2 \right] + b\phi\delta(\vec{r}) \right\} \mathrm{d}^2 r, \tag{64}$$

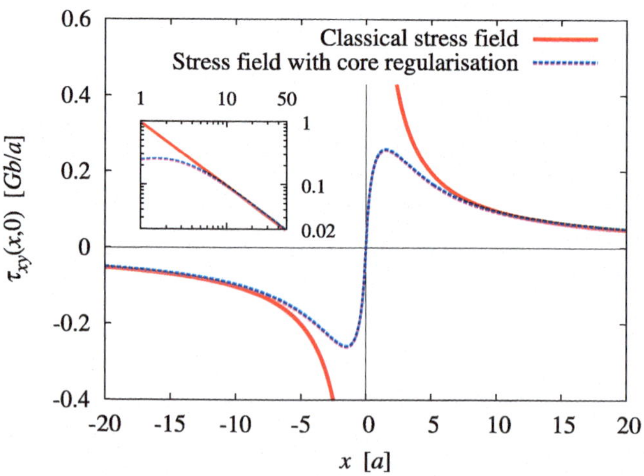


**Fig. 1** Shear stress along the $x$ axis obtained with and without core regularization ($G := \mu/[2\pi(1-\nu)]$)

where $c$ is a constant (see Berdichevsky (2005)). One can obtain that the minimum condition $\delta P/\delta\phi = 0$ is fulfilled if $\phi$ satisfies the equation

$$\Delta\phi - c^2\Delta^2\phi = -b\mu\delta(\vec{r}). \tag{65}$$

Like for edge dislocations, numerical solution of the above equation shows that the second term of the left hand side of the equation results in stress regularization in the vicinity of the dislocation line.

Comparing the present approach with other methods suggested by Aifantis (1999), Gutkin and Aifantis (1999), and Lazar (2003), although one can find some similarities, but the major difference is that in these works the core region is regularized by spreading out the dislocation density tensor $\alpha_{ij}$, while in our analysis the dislocation density tensor remains proportional to a Dirac delta, like in the classical non-regularized case. The core is regularized by the nonlocality in the Gibbs free energy.

## 3.1 *Dislocation-Solute Atom Interaction*

Solute atoms can strongly modify the collective properties of dislocations. Among other things they can lead to plastic instabilities (for a recent review see Ananthakrishna (2007)). In this section we show that the effect of solute atoms can be easily incorporated into the variational framework (Groma et al. 2007). We restrict our consideration for straight edge dislocations with Burgers vectors parallel to the $x$ axis, but it is straightforward to generalize the method for 3D.

The Gibbs free energy of a coupled system can be always given as the sum of the Gibbs free energy of the two uncoupled systems and a coupling term. Therefore, if we add to the plastic potential the Gibbs free energy contribution of the solute atoms we arrive at the 'plastic potential' of the dislocation-solute system. According to Eq. (51) the plastic potential of the parallel edge dislocation system considered is

$$P_{\mathrm{d}}[\chi,\kappa] = \int \left[ -\frac{1-\nu}{4\mu}(\Delta\chi)^2 + b\chi(\partial_2\kappa) \right] \mathrm{d}^2r, \tag{66}$$

where $\kappa$ is the signed dislocation density (geometrically necessary density, GND) defined as $\alpha_{31} = b\kappa$ that is the only nonvanishing component of the dislocation density tensor for the dislocation geometry considered.

For the solute atoms we assume that their concentration $c$ is close to the equilibrium concentration $c_\infty$. In this case the Gibbs free energy of the solute atoms can be given with the quadratic form

$$G_{\mathrm{c}}[c] = \int \alpha(c - c_\infty)^2 \mathrm{d}^2r, \tag{67}$$

where $\alpha$ is a constant (which may depend on $c_\infty$).

To determine the form of the coupling term we use the well-known fact that beside concentration gradient, the pressure gradient also causes solute atom diffusion. According to the principles of irreversible thermodynamics the solute atom current is proportional to the gradient of the chemical potential $\mu_c = \frac{\delta G}{\delta c}$. So, taking the coupling term in the form

$$G_{cp}[\chi, c] = \int \beta c p \, \mathrm{d}^2 r, \tag{68}$$

where $\beta$ is constant, the total plastic potential reads as

$$P[\chi, \kappa, c] = P_{\mathrm{d}}[\chi, \kappa] + G_{\mathrm{c}}[c] + G_{cp}[\chi, c] \tag{69}$$

$$= \int \left[ -\frac{1-\nu}{4\mu}(\Delta\chi)^2 + b\chi(\partial_2\kappa) + \alpha(c - c_\infty)^2 - \beta c \Delta\chi \right] \mathrm{d}^2 r, \tag{70}$$

and

$$\frac{\delta P}{\delta c} = \frac{\delta G_c + G_{cp}}{\delta c} = \mu_c. \tag{71}$$

It should be mentioned that Eq. (70) results in linear equations for the stress and the solute atom concentration. Nonlinearity can also be treated within the framework proposed, but it is out of the scope of the paper.

Since the purpose of the subsection is to demonstrate the way the variational approach works, we now restrict our analysis only to static problems. One can find from the equilibrium conditions $\frac{\delta P}{\delta \chi} = 0$ and $\frac{\delta P}{\delta c} = \mu_0$ that

$$\frac{1-\nu}{2\mu}\Delta^2\chi + \beta\Delta c = b\,\partial_2\kappa \tag{72}$$

and

$$\alpha(c - c_\infty) = \beta\Delta\chi + \mu_0, \tag{73}$$

where $\mu_0$ is the constant equilibrium chemical potential. By combining the two equations we get that

$$\left[\frac{1-\nu}{2\mu} + \frac{\beta^2}{\alpha}\right]\Delta^2\chi = b\partial_2\kappa. \tag{74}$$

A remarkable feature of the above equation is that, apart from a constant multiplier, the functional form of the stress function $\chi$ is not affected by the solute atoms. Moreover, the solute atom concentration is proportional to the pressure caused by the dislocations. Certainly the result obtained is not new, it is the well know Cottrell atmosphere of solute atoms around dislocation lines (Cottrell and Bilby 1949), but it illustrates very well the fact that the coupled system of dislocations and solute atoms can be treated with the variational framework suggested.

## 3.2 *Time Evolution of the Dislocation Density Tensor*

As it is explained above, if we know the dislocation density tensor (i.e. we know the dislocation line geometry) the internal stress field can be determined from Eq. (23). This is however, only a "static" description. In order to be able to describe the response of the dislocation system to external signals, the governing equations of the time evolution of the dislocation density tensor should be determined. Moreover, as it was mentioned earlier the dislocation density tensor $\alpha_{ij}$ does not determine completely the plastic distortion $\beta^p_{ij}$, however, the stress state is uniquely given by $\alpha_{ij}$ or more precisely by the incompatibility tensor $\eta_{ij}$. So, one have to put some additional physical input that determine the plastic distortion completely.

For this goal let us take the time derivative (denoted by "$\cdot$" ) of Eq. (7):

$$\dot{\alpha}_{ij} + e_{ikl}\partial_k j_{lj} = 0, \tag{75}$$

where

$$j_{ij} = -\dot{\beta}^p_{ij} \tag{76}$$

is called dislocation current density (Landau and Lifshitz 1986). The above equation is the "conservation law of the Burgers vector" in differential form. Indeed, if we integrate both sides of Eq. (75) for an arbitrary area contoured by the closed curve $L$, according to Eq. (8), we obtain that

$$\frac{db_j}{dt} = -\oint_L j_{ij} ds_i \tag{77}$$

It is obvious from this relation that $\hat{j}$ is the Burgers vector carried by the dislocations crossing a unit length part of the contour line $L$ per unit time.

For an individual dislocation one can find that

$$j_{ik} = e_{ilm} l_t v_m b_k \delta^{(2)}(\xi), \tag{78}$$

where $\vec{v}$ is the velocity of the dislocation line at a given point. It is important to note that if we added the gradient of an arbitrary vector field to $j_{ij}$ given above, this would also satisfy the conservation law (75). The problem is obviously related to the non-uniqueness of the plastic distortion discussed earlier. However, expression (78) is the only one which is physically meaningful. One expects that there is no plastic current anywhere else but at the dislocation line. Nevertheless, strictly speaking we have to postulate this.

With this postulation the plastic distortion is given by the time integration of $j_{ij}$ if $\beta^p_{ij}$ is known at a given moment. (One can often assume that the plastic distortion vanishes initially.)

For the better understanding of the problem related to the nonuniqueness of $\beta^p_{ij}$ let us consider a dislocation loop. One has to take into account that a defect in the materials is determined completely only if the cut of surface (where we have the jump in the displacement) is given. Taking another cut of surface ended on the same line corresponds to another defect. As it was explained above, however, the stress generated by the two defects is the same, it is determined only by the dislocation line loop. Nevertheless, there is a "natural" cut of surface that is generated during the expansion of the loop, i.e. the surface the dislocation line passed while the loop is formed.

The above results clearly show that $j_{ij}$ has to be considered as an independent quantity. In order to be able to describe the time evolution of the dislocation system we have to set up a constitutive relation giving how $j_{ij}$ depends on the dislocation density tensor and the external stress. Due to the long range nature of the dislocation-dislocation interaction, the constitutive relation is obviously non-local in $\alpha_{ij}$ and should also depend on the total amount of dislocation line per unit volume commonly called the statistically stored dislocation density. Beside this, the constitutive relation has to be able to account for several different "local" phenomena (self loop interaction, junction formation, annihilation, etc.) making even more difficult to determine its form.

One possible approach to handle this problem is to set up the constitutive relation from phenomenological considerations. During the past years several phenomenological expressions were proposed and successfully applied for modeling certain phenomena (Aifantis 1984, 1987, 1994; Fleck and Hutchinson 2001; Gurtin 2002; Svendsen 2002) but the problem is far not completely solved.

Another widely used approach to study the time evolution of dislocation systems is discrete dislocation dynamics (DDD) simulation in which the dislocation loops are considered individually. After setting up velocity laws for the dislocation segments the dislocation loop geometry is updated numerically. Describing the actual numerical techniques used in DDD simulations is out of the scope of this chapter. The details can be found in chapter "Multiscale Dislocation-Based Plasticity".

## 3.3 *Time Evolution of the Displacement Field*

In the previous part we have discussed how the stress field generated by the dislocations can be determined and what can be said in general about the time evolution of the dislocation density tensor. However, in many applications it is important to determine the displacement field $\vec{u}(\vec{r})$, too.

Let us go back to our starting Eq. (6), multiply it with the elastic modulus tensor $L_{ijkl}$, and take the div of the equation. With Eqs. (11), (21) one obtains

$$\partial_i L_{ijkl} \partial_l u_k = \partial_i L_{ijkl} \beta^p_{kl}. \tag{79}$$

This is formally equivalent with the common equilibrium equation of elasticity with body force density

$$f_j = -\partial_i L_{ijkl} \beta^p_{kl}. \tag{80}$$

Since, as it is explained earlier, the dislocation density tensor does not determine the plastic distortion uniquely, the above equation is not enough to determine the displacement field. Taking, however, the time derivative of Eq. (79), with the (76) definition of $j_{ij}$, one arrives at

$$\partial_i L_{ijkl} \partial_l \dot{u}_k = \partial_i L_{ijkl} j_{kl}. \tag{81}$$

As it is discussed above, based on physical arguments, $j_{ij}$ can be uniquely defined, so the deformation velocity field $\dot{u}_i$ can already be determined if $j_{ij}$ is known. Integrating it with respect to time gives the change of the displacement field that is the quantity one can really measure.

# 4 Statistical Continuum Theory of Dislocations

## 4.1 General Issues

In this section we analyze in details the statistical properties of a system of straight parallel edge dislocations in single slip, that is the simplest possible dislocation configuration one can envisage, but as it is demonstrated by DDD simulations (Kubin and Canova 1992; Rhee et al. 1998; Ghoniem and Sun 1999; Groma and Bakó 2000; Devincre et al. 2001; Gomez-Garcia et al. 2006; Ispánovity et al. 2010, 2014) this system can reproduce several key properties of the dislocation system. Moreover, the structure of the continuum evolution equations derived on a systematic manner can guide us to develop a more general statistical continuum theory of dislocations. Possible directions of the generalization are discussed in the last section.

Before we start to derive the statistical continuum theory we shortly summarize those key issues one faces developing a theory for the collective behavior of dislocations:

- The dislocation-dislocation interaction is long range. The force acting between two straight parallel dislocations is inversely proportional to their distance $F \propto 1/r$.

- The dislocation motion is strongly dissipative. It is commonly assumed that the dislocation motion is over-damped meaning that the inertial term ($ma$) is negligible beside the friction force. For small dislocation velocity one can assume that the friction force is proportional to the dislocation velocity. So, in our considerations the velocity is taken to be proportional to the Peach Koehler force, $v \propto F$ resulting that the equation of motion of a dislocation is a first order ordinary differential equation. Generalization is discussed in chapter "Multiscale Dislocation-Based Plasticity".
- At low enough temperature dislocation climb is negligible beside glide. In our considerations climb will be neglected. As a consequence dislocations cannot leave their slip plane. From statistical physics point of view this means that there is a quenched disorder in the system defined at the generation of the initial dislocation configuration. Since there is no any physical reason introducing a characteristic length scale but the average dislocation spacing the slip planes are assumed to be placed randomly.
- The last issue we have to investigate in more details is the role of thermal noise on the motion of dislocations.

Let us consider $N$ parallel edge dislocations with dislocation line direction parallel to the $z$ and Burger vector parallel to the $x$ axes, respectively (see Fig. 2).

With the over-damped dynamics the equations of motion of the dislocation system is

$$\frac{dx_i}{dt} = M_0 b s_i \left( \sum_{j=1}^{N} s_j \tau_{ind}(\vec{r}_i - \vec{r}_j) + \tau_{ext} \right), \tag{82}$$

where $\vec{r}_i$ is the position, $s_i = \pm 1$ is the sign of the $i$th dislocation, $M_0$ is the dislocation mobility, $b$ is the Burgers vector, $\tau_{ext}$ is the external shear stress, and $\tau_{ind}(\vec{r})$ is the shear stress generated by a dislocation at point $\vec{r}$ (Groma et al. 2003). The above equation is the one solved numerically in 2D discrete dislocation dynamics (DDD) simulations.

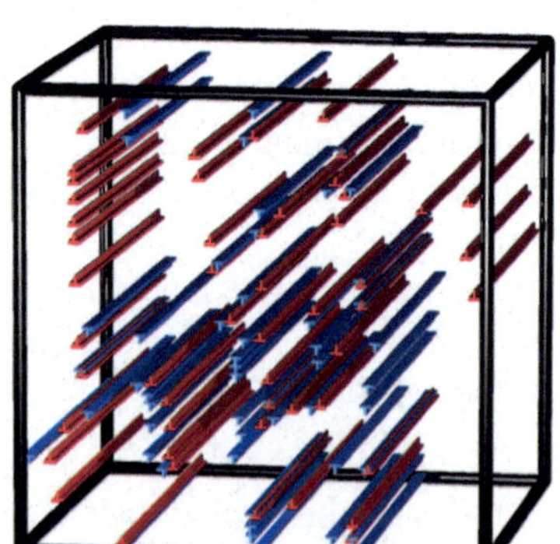

**Fig. 2** 2D dislocation configuration

According to Eq. (29) $\tau_{ind}(\vec{r} - \vec{r}_j)$ is related to the second order stress function $\chi_{ind}$ as

$$\tau_{ind}(\vec{r}) = \partial_x \partial_y \chi_{ind}(\vec{r}) \tag{83}$$

with $\chi_{ind}$ fulfilling the biharmonic equation

$$\nabla^4 \chi_{ind} = \frac{2\mu}{1-\nu} b \partial_y \delta(\vec{r}) \tag{84}$$

obtained from Eq. (31). Denoting

$$U_{ind}(\vec{r}) = -\partial_y \chi_{ind}(\vec{r}) \tag{85}$$

one gets

$$\tau_{ind}(\vec{r}) = -\partial_x U_{ind}(\vec{r}). \tag{86}$$

By introducing the dislocation-dislocation interaction energy

$$V(\vec{r}_1, \vec{r}_2, \ldots \vec{r}_N) = \frac{b}{2\sqrt{\rho}} \sum_{i \neq j} s_i s_j U_{ind}(\vec{r}_i - \vec{r}_j) \tag{87}$$

Eq. (82) can be given as

$$\frac{dx_i}{dt} = -M_\rho b \frac{\partial V}{\partial x_i}. \tag{88}$$

with $M_\rho = M_0\sqrt{\rho}$. It should be noted that since $bU_{ind}$ is energy per unit length, for the further analysis it is useful to multiply it with a characteristic length scale of the problem leading to a quantity with energy dimension. A natural characteristic length scale for the problem is the dislocation spacing $1/\sqrt{\rho}$. This is why $1/\sqrt{\rho}$ is introduced in $V$.

Thermal noise can be incorporated into the equation of motion of dislocation by adding a random force term to Eq. (88):

$$\frac{dx_i}{dt} = -M_\rho b \partial_{x_i} V + \sqrt{2M_\rho k_b T} \zeta_i(t), \tag{89}$$

where $\zeta_i(t)$ is a random noise with time correlation $< \zeta_i(t)\zeta_j(0) >= \delta_{ij}\delta(t)$, $T$ is the temperature, and $k_b$ is the Boltzmann constant.

For the probability density $p_N(t, \vec{r}_1, \vec{r}_2, \ldots \vec{r}_N)$ the Fokker-Planck equation corresponding to the stochastic differential equation (89) reads as

$$\frac{\partial p_N}{\partial t} = M_\rho \sum_{i=1}^{N} \partial_{x_i} \left( (\partial_{x_i} V) p_N + k_b T \partial_{x_i} p_N \right). \tag{90}$$

One can easily see that the steady state solution of the above equation is

$$p_N^\infty(\vec{r}_1, \vec{r}_2, \ldots \vec{r}_N) = \frac{1}{Z} e^{-\frac{V}{k_b T}}, \tag{91}$$

where $Z$ is a normalizing factor. This means, although a dislocation ensemble is not a Hamiltonian system, so the common methods of statistical physics cannot be applied, the probability of a given dislocation configuration is proportional to a Boltzmann factor. In many theories developed for describing the collective properties of dislocations Eq. (91) is used as starting point.

There is, however, an issue related to time scales one has to take into account. The right hand side of Eq. (90) contains two terms. Taking into account that energy of the dislocation system introduced above $V \propto \mu b^2/\sqrt{\rho}$ and the characteristic length scale of the problem $l \propto 1/\sqrt{\rho}$ one can introduce two characteristic time scales

$$t_V = \frac{1}{M_\rho \mu b^2 \sqrt{\rho}} \text{ and } t_T = \frac{1}{M_\rho k_b T \rho} \tag{92}$$

corresponding to the first and second term of the right hand side of Eq. (90), respectively. With typical values the ration

$$\frac{t_V}{t_T} = \frac{k_b T}{\mu b^2} \sqrt{\rho} \tag{93}$$

is less than $10^{-4}$, i.e. the characteristic time related to thermal noise is much longer than the one corresponding to elastic dislocation-dislocation interaction. This is why for most of the problems the thermal noise is negligible in the equation of motion of dislocations. So, by neglecting the noise term, from Eqs. (87), (90) the time evolution equation of the $N$ particle probability density $p_N(t, \vec{r}_1, \vec{r}_2, \ldots \vec{r}_N)$ reads as

$$\begin{aligned} \frac{\partial p_N}{\partial t} &= M_0 b \sum_{i \neq j}^{N} \partial_{x_i} \left[ s_i s_j \left( \partial_{x_i} U_{ind}(\vec{r}_i - \vec{r}_j) \right) p_N \right] \\ &= -M_0 b \sum_{i \neq j}^{N} \partial_{x_i} \left[ s_i s_j \tau_{ind}(\vec{r}_i - \vec{r}_j) p_N \right]. \end{aligned} \tag{94}$$

It should be noted, that the above equation is mathematically equivalent with the equation of motion of dislocations given by Eq. (82), its solution can be easily constructed from the solution of Eq. (82). Equation (94) acts, however, as the "starting" equation of the statistical physics based continuum theory of dislocations.

The result obtained that the thermal noise is often negligible beside the elastic interaction force has an important consequence for the time dependence of the elastic energy $E(t) = V(\vec{r}_1(t), \vec{r}_2(t), \dots \vec{r}_N(t))$ of the dislocation system. Let us calculate the rate of energy change:

$$\frac{dE}{dt} = \frac{d}{dt} V(\vec{r}_1(t), \vec{r}_2(t), \dots \vec{r}_N(t)) = \sum_{i=1}^{N} \frac{dx_i}{dt} \frac{\partial}{\partial x_i} V(\vec{r}_1, \vec{r}_2, \dots \vec{r}_N). \tag{95}$$

From Eq. (88) one obtains

$$\frac{dE}{dt} = -\frac{1}{M_\rho b} \sum_{i=1}^{N} \frac{dx_i}{dt} \frac{dx_i}{dt} \leqslant 0 \tag{96}$$

meaning that the elastic energy of the dislocation system cannot increase during the evolution of the system. So, the dislocation system stacks in the "closest" local energy minimum. Unlike a classical thermal system it cannot leave the local energy minimum by thermal fluctuation (or more precisely, the time needed to overcome an energy barrier is much longer than the elastic relaxation time).

## 4.2 *Coarse-Graining*

The dislocation density tensor introduced above is a highly singular quantity. It is infinite along the dislocation lines and vanishes elsewhere. More precisely, it is proportional to a delta function along the dislocation lines. The same holds for the dislocation current density. The conservation law (94) guarantees that during the evolution of the dislocation system this delta function does not "spread out", only the shape of the loops can change. This is certainly what we expect physically. This means, if we want to follow the evolution of the system we have to follow the track of each dislocation loop as it is done in DDD simulations.

We may hope, like for many other physical systems, to predict the macroscopic response of the dislocation system, we do not need this detailed knowledge of the evolution of the dislocation configuration. One should try to operate with locally averaged quantities. Locally averaged fields for the dislocation density tensor, stress, dislocation, current density, etc., can be obtained from the "singular" ones by convolving them with a window function. This is commonly called as "coarse-graining" or "homogenization".

One can immediately raise the question what is the appropriate function we should use for the shape of the window function, and what determines its half width. There is not a general recipe how to resolve these problems. Nevertheless, we can hope that within certain limits the result obtained by the coarse-graining is not sensitive to the actual window function shape and its width. If this is not the case, this clearly indicates that all the microscopical details are important. So, the coarse-graining procedure always requires extra care. It is important to stress that, before the equations obtained by coarse-graining are applied for a given problem, one always has to study the relevance of the homogenization.

In order to indicate the difficulties, as a simple example (see Groma et al. (2007)), let us consider again a set of parallel edge dislocations with $\pm\vec{b}$ Burgers vectors parallel to the $x$ axis. For this case Eq. (31) simplifies to

$$\nabla^4 \chi = \frac{2\mu b}{1-\nu} \partial_y \kappa_d, \tag{97}$$

where $\kappa_d = \rho_{d+} - \rho_{d-}$ is the signed dislocation density that is a sum of delta functions. If we take the convolution of Eq. (97) with a window function $w(\vec{r})$ we obtain that

$$\int w(\vec{r}-\vec{r}') \nabla'^4 \chi(\vec{r}') d^2 r' = \frac{2\mu b}{1-\nu} \int w(\vec{r}-\vec{r}') \partial y' \kappa_d(\vec{r}) d^2 r'. \tag{98}$$

After partial integrations we get that

$$\nabla^4 \int w(\vec{r}-\vec{r}') \chi(\vec{r}') d^2 r' = \frac{2\mu b}{1-\nu} \partial y \int w(\vec{r}-\vec{r}') \kappa_d(\vec{r}) d^2 r'. \tag{99}$$

As it can be seen, the coarse-grained fields denoted by

$$< \chi > = \int w(\vec{r}-\vec{r}') \chi(\vec{r}') d^2 r' \qquad < \kappa > = \int w(\vec{r}-\vec{r}') \kappa_d(\vec{r}') d^2 r' \tag{100}$$

are related to each other as

$$\nabla^4 < \chi > = \frac{\mu b}{1-\nu} \partial y < \kappa > \tag{101}$$

which is formally equivalent with Eq. (97). With a similar argument, from Eq. (29) one can find that

$$\begin{aligned} < \sigma >_{11} &= -\partial_y \partial_y < \chi >, \\ < \sigma >_{22} &= -\partial_x \partial_x < \chi >, \\ < \sigma >_{12} &= \partial_x \partial_y < \chi > . \end{aligned} \tag{102}$$

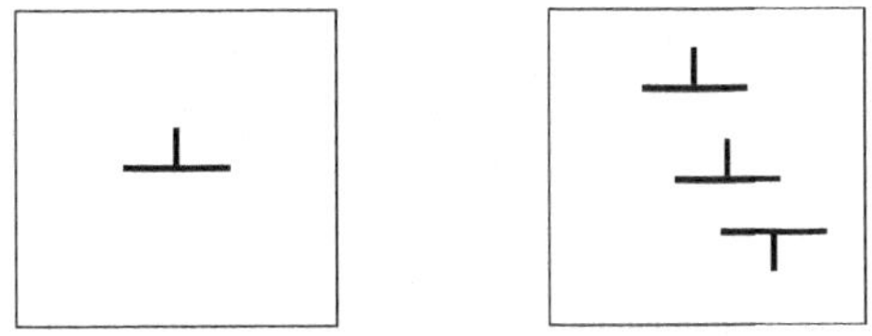

**Fig. 3** Two strongly different dislocation configurations giving the same $\kappa$ if they are coarse-grained for the areas indicated by the boxes

We can see, the coarse-grained fields are related to each other as the "discrete" ones.

We can see, however, important information is lost during coarse-graining. If we consider two dislocation configurations indicated in Fig. 3 and coarse-grain them for the square area indicated by the boxes, we get the same signed dislocation density value. On the other hand, it is obvious that the response of the two configurations are strongly different, if one applies an external shear. So, in a continuum theory of dislocations, in which we operate with smooth fields, the coarse-grained dislocation density tensor is not enough to characterize the state of the system. In the next subsection we discuss how a continuum theory can be derived from the equation of motion of straight parallel dislocations and what relevant quantities are needed to have an appropriate description of this simple dislocation system on the mesoscopic scale.

## 4.3 *Coarse-Graining of the Equations of Motion of Dislocations*

In order to derive the statistical physics based continuum theory of dislocation we have to link directly the microscale description of the evolution of the dislocation system to a mesoscopic scale where we operate with continuous dislocation density fields. This goal can be achieved by a systematic coarse-graining of the equations of motion of dislocation given by Eq. (82) (Groma 1997; Zaiser et al. 2001; Groma et al. 2003, 2006, 2007, 2010, 2015, 2016; Valdenaire et al. 2016).

For simplicity let us first assume that each dislocation has the same sign ($s_i = 1$) and the external load is zero (see Groma et al. 2007). In this case Eq. (82) reads as

$$\frac{dx_i}{dt} = M_0 \left( \sum_{j=1}^{N} F(\vec{r}_i - \vec{r}_j) \right) \tag{103}$$

with $F = b\tau_{ind}$.

As a first step let us multiply Eq. (103) with $\delta(\vec{r} - \vec{r}_i)$ and take its derivative with respect to $x$:

$$\partial_x \left\{ \frac{dx_i}{dt} \delta(\vec{r} - \vec{r}_i) \right\} = M_0 \partial_x \left\{ \left( \sum_{j \neq i}^{N} F(\vec{r}_i - \vec{r}_j) \right) \delta(\vec{r} - \vec{r}_i) \right\}, \tag{104}$$

where $\vec{r} = (x, y)$ is an arbitrary point. It is useful to introduce the "discrete" dislocation density

$$\rho_d(\vec{r}) = \sum_{i=1}^{N} \delta(\vec{r} - \vec{r}_i) \tag{105}$$

that is the same as $\rho_{d+}$ defined in Sect. 2.3, but since in the present analysis only one type of dislocation is considered, the subscript + is dropped. With this, the summation on the right hand side of Eq. (104) can be replaced by a weighted integral. Furthermore, taking into account that

$$\partial_x \left\{ \frac{dx_i}{dt} \delta(\vec{r} - \vec{r}_i) \right\} = -\frac{dx_i}{dt} \partial_{x_i} \delta(\vec{r} - \vec{r}_i) = -\frac{d}{dt} \delta(\vec{r} - \vec{r}_i), \tag{106}$$

from Eq. (104) we get that

$$\begin{aligned} &-\frac{d}{dt} \delta(\vec{r} - \vec{r}_i) \\ &= M_0 \partial_x \left\{ \left( \int F(\vec{r} - \vec{r}')[\rho_d(\vec{r}') - \delta(\vec{r} - \vec{r}')] d^2\vec{r}' \right) \delta(\vec{r} - \vec{r}_i) \right\}, \end{aligned} \tag{107}$$

where $\delta(\vec{r} - \vec{r}')$ beside $\rho_d(\vec{r}')$ is needed to avoid self dislocation interaction. By summing up with respect to $i$ we conclude

$$-\frac{d}{dt} \rho_d(\vec{r}) = M_0 \partial_x \left\{ \left( \int F(\vec{r} - \vec{r}')[\rho_d(\vec{r}') - \delta(\vec{r} - \vec{r}')] d^2\vec{r}' \right) \rho_d(\vec{r}) \right\}, \tag{108}$$

which is a nonlinear strongly non-local equation for the "discrete" dislocation density $\rho_d(\vec{r})$. Like it was done with the field equation (97), to get rid of the singular character of $\rho_d(\vec{r})$ we can coarse-grain Eq. (108). By introducing the coarse-grained quantities

$$\rho_1(\vec{r}) = < \rho_{disc}(\vec{r}) > \tag{109}$$

$$\rho_2(\vec{r}_1, \vec{r}_2) = < \rho_{disc}(\vec{r}_1) \rho_{disc}(\vec{r}_2) - \rho_{disc}(\vec{r}_1) \delta(\vec{r}_1 - \vec{r}_2) >, \tag{110}$$

we get

$$\frac{\partial \rho_1(\vec{r}_1, t)}{\partial t} + \int \partial_{x_1} \{ \rho_2(\vec{r}_1, \vec{r}_2, t) F(\vec{r}_1 - \vec{r}_2) \} d^2 r_2 = 0. \tag{111}$$

The procedure applied above clearly shows that the form of Eq. (111) does not depend on the actual form of the window function applied for the coarse-graining.

The densities $\rho_1(\vec{r})$ and $\rho_2(\vec{r}_1,\vec{r}_2)$, however, can certainly depend on $w(\vec{r})$. This is not a problem until we do not assume some relation between $\rho_1(\vec{r})$ and $\rho_2(\vec{r}_1,\vec{r}_2)$. We can say that Eq. (111) is exact but it is not enough to describe the time evolution of the dislocation density because the time derivative of the one particle density $\rho_1(\vec{r})$ depend on the two particle density $\rho_2(\vec{r}_1,\vec{r}_2)$. One can derive equation for $\rho_2(\vec{r}_1,\vec{r}_2)$ but it depends on the three particle density function. In general one can obtain a hierarchy of equations where the time derivative of the $k$ particle density depends on the $k+1$ particle density (see below).

In order to get a closed theory we need a closure approximation. Before we discuss how this can be obtained, the above results have to be generalized for the case where Burgers vector of the dislocations are not the same. The simplest generalization is if we allow that the Burgers vectors of the dislocations can differ in sign (Groma et al. 2007). This is still a strong simplification of a real dislocation ensemble but an important step forward. Without going into the details with a similar procedure explained above one can find that

$$\frac{\partial \rho_+(\vec{r}_1,t)}{\partial t} = \tag{112}$$
$$-M_0\partial_{x_1}\left[\rho_+(\vec{r}_1)b\tau_{ext} + \int \{\rho_{++}(\vec{r}_1,\vec{r}_2) - \rho_{+-}(\vec{r}_1,\vec{r}_2)\}\ F(\vec{r}_1-\vec{r}_2)d^2r_2\right]$$

$$\frac{\partial \rho_-(\vec{r}_1,t)}{\partial t} = \tag{113}$$
$$+M_0\partial_{x_1}\left[\rho_-(\vec{r}_1)b\tau_{ext} - \int \{\rho_{--}(\vec{r}_1,\vec{r}_2) - \rho_{-+}(\vec{r}_1,\vec{r}_2)\}\ F(\vec{r}_1-\vec{r}_2)d^2r_2\right],$$

where the subscripts "+" and "−" indicate the sign of the Burgers vector the different density functions are corresponding to. External load is also added. We mention here that the negative signs in front of $\rho_{+-}$ and $\rho_{-+}$ in Eqs. (112) and (113) come from the simple fact that the interaction force acting between dislocations with opposite signs is $-F$.

By adding and substituting the two equations one arrives at:

$$\frac{\partial \rho(\vec{r}_1,t)}{\partial t} + M_0 b\partial_{x_1}[\kappa(\vec{r}_1,t)\tau_{ext} + \int\{\rho_{++}(\vec{r}_1,\vec{r}_2,t) + \rho_{--}(\vec{r}_1,\vec{r}_2,t)$$
$$-\rho_{+-}(\vec{r}_1,\vec{r}_2,t) - \rho_{-+}(\vec{r}_1,\vec{r}_2,t)\}\tau_{ind}\vec{r}_1 - \vec{r}_2)d^2r_2] = 0, \tag{114}$$

$$\frac{\partial \kappa(\vec{r}_1,t)}{\partial t} + M_0 b\partial_{x_1}[\rho(\vec{r}_1,t)\tau_{ext} + \int\{\rho_{++}(\vec{r}_1,\vec{r}_2,t) - \rho_{--}(\vec{r}_1,\vec{r}_2,t)$$
$$-\rho_{+-}(\vec{r}_1,\vec{r}_2,t) + \rho_{-+}(\vec{r}_1,\vec{r}_2,t)\}\tau_{ind}(\vec{r}_1 - \vec{r}_2)d^2r_2] = 0, \tag{115}$$

where $\rho(\vec{r},t) = \rho_+(\vec{r},t) + \rho_-(\vec{r},t)$ is the total and $\kappa(\vec{r},t) = \rho_+(\vec{r},t) - \rho_-(\vec{r},t)$ is the signed dislocation density. ($\kappa$ is the same as $< \kappa >$ introduced in Eq. (100) but to have shorter equations the brackets $< .. >$ were omitted.)

## 4.4 Direct Averaging of $p_N$

Equation (111) derived by the direct coarse-graining of the equations of motion of dislocations can be obtained from the evolution equation of the $N$ particle probability density $p_N$ given by Eq. (94) too (Groma 1997). For one type of dislocations it reads as

$$\frac{\partial p_N}{\partial t} = -M_0 \sum_{i \neq j}^{N} \partial_{x_i} \left[ F(\vec{r}_i - \vec{r}_j) p_N \right]. \tag{116}$$

As it was mentioned earlier for many applications we do not need that detailed description represented by the $N$ particle probability density function. A less detailed description of the system is the $k$-th order probability density function defined as

$$p_k(\vec{r}_1, \vec{r}_2, .., \vec{r}_k) = \int \int \dots \int p_N(t, \vec{r}_1, \vec{r}_2 \dots \vec{r}_N) d^2\vec{r}_{k+1} d^2\vec{r}_{k+2} \dots d^2\vec{r}_N. \tag{117}$$

After integrating Eq. (116) with respect to the variables $\vec{r}_{k+1}, \vec{r}_{k+2}, \dots \vec{r}_N$, from the above definition of $p_k$ (117) we obtain that

$$\frac{\partial p_k}{\partial t} = -M_0 \sum_{i=1}^{N} \sum_{j=1, j \neq i}^{N} \int \partial_{x_i} \left\{ p_N F(\vec{r}_i - \vec{r}_j) \right\} d^2\vec{r}_{k+1} d^2\vec{r}_{k+2} \dots d^2 r_N. \tag{118}$$

After a long but straightforward calculation (for details see Groma 1997) we get that

$$\frac{\partial p_k}{\partial t} + M_0 \sum_{i=1}^{k} \sum_{j=1, j \neq i}^{k} \partial_{x_i} \left\{ p_k F(\vec{r}_i - \vec{r}_j) \right\} \tag{119}$$

$$+(N-k) \int \partial_{x_i} \left\{ p_{k+1} F(\vec{r}_i - \vec{r}_{k+1}) \right\} d^2\vec{r}_{k+1} = 0.$$

As it can be seen, equation for the $k$-th order probability distribution function depends on the $k+1$-th order one. So, the reduction procedure applied results in a hierarchy of the equations. In fluid dynamics and plasma physics this is called as BBGKY hierarchy.

For our further consideration the equations for $p_1$ and $p_2$ play an important role, so we give their explicit forms (Groma 1997):

$$\frac{\partial \rho_1(\vec{r}_1, t)}{\partial t} + M_0 \int \partial_{x_1} \{\rho_2(\vec{r}_1, \vec{r}_2, t) F(\vec{r}_1 - \vec{r}_2)\} d^2\vec{r}_2 = 0 \tag{120}$$

and

$$\frac{\partial \rho_2(\vec{r}_1, \vec{r}_2, t)}{\partial t} + \left(\partial_{x_1} - \partial_{x_2}\right) \rho_2(\vec{r}_1, \vec{r}_2, t) F(\vec{r}_1 - \vec{r}_2)$$
$$+\partial_{x_1} \int \rho_3(\vec{r}_1, \vec{r}_2, \vec{r}_3, t) F(\vec{r}_1 - \vec{r}_3) d^2\vec{r}_3 + 1 \leftrightarrow 2 = 0, \tag{121}$$

where the notations $\rho_1 = Np_1$, $\rho_2 = N(N-1)p_2$, $\rho_3 = N(N-1)(N-2)p_3$ were introduced. As it can be seen for $\rho_1$ Eq. (111) is recovered indicating that the course graining procedure and the averaging method explained here are equivalent.

## *4.5 Evolution of the Plastic Shear*

Before we discuss how a closed theory can be obtained for the evolution of $\rho$ and $\kappa$ it is useful to analyze the evolution of plastic shear. For the dislocation geometry considered the only non-vanishing component of the dislocation density tensor is

$$\alpha_{31} = b\kappa. \tag{122}$$

For the plane problem considered the only component of the plastic distortion contributing to $\alpha_{31}$ is $\beta^p_{21}$ and

$$b\kappa = -\partial_x \gamma, \tag{123}$$

where the notation $\gamma = \beta^p_{21}$ commonly used is introduced. Equation (123) means, to get spatially varying plastic shear one has to introduce dislocations. This is why $\kappa$ is often called geometrically necessary dislocation (GND) density.

Taking the time derivative of Eq. (123) we get

$$\frac{\partial \kappa}{\partial t} = -\partial_x \dot{\gamma}. \tag{124}$$

By comparing this with Eq. (115) we obtain an explicit expression for the plastic shear rate $\dot{\gamma}$:

$$\dot{\gamma} = \rho(\vec{r}_1)b^2\tau_{ext} \tag{125}$$
$$+\int\{\rho_{++}(\vec{r}_1,\vec{r}_2) - \rho_{--}(\vec{r}_1,\vec{r}_2) - \rho_{+-}(\vec{r}_1,\vec{r}_2)$$
$$+\rho_{-+}(\vec{r}_1,\vec{r}_2)\}b^2\tau_{ind}(\vec{r}_1 - \vec{r}_2)d\vec{r}_2.$$

## 4.6 Self-consistent Field Approximation

In order to have a closed continuum theory describing the evolution of the dislocation system, the (119) hierarchy of equations has to be cut at some order $k$. For this, from some considerations independent from the Eq. (119), we have to give how the $k+1$ order density function can be built from the lower order ones. The simplest possible assumption is that the two particle density functions are the products of the one particle density functions (Groma 1997; Groma and Balogh 1999), i.e.

$$\rho_{ss'}(\vec{r}_1,\vec{r}_2,t) = \rho_s(\vec{r}_1)\rho_{s'}(\vec{r}_2), \quad s,s' \in \{+,-\}. \tag{126}$$

This means, dislocation-dislocation correlations are neglected. As it is explained below this leads to a self-consistent field theory. Similar approximation is often used in other fields of physics.

By substituting Eq. (126) into Eqs. (114), (115) we arrive at

$$\frac{\partial\rho(\vec{r},t)}{\partial t} + M_0 b\partial_x[\kappa(\vec{r},t)\{\tau_{sc}(\vec{r},t) + \tau_{ext}\}] = 0, \tag{127}$$

$$\frac{\partial\kappa(\vec{r},t)}{\partial t} + M_0 b\partial_x[\rho(\vec{r},t)\{\tau_{sc}(\vec{r},t) + \tau_{ext}\}] = 0, \tag{128}$$

where

$$\tau_{sc}(\vec{r}) = \int \kappa(\vec{r}_1,t)\tau_{ind}(\vec{r} - \vec{r}_1)d^2r_1 \tag{129}$$

is the shear stress field generated by the coarse-grained signed dislocation density. This is why $\tau_{sc}$ is often called as self-consistent or mean stress field. The quantity $\tau_{sc}$, however, is not a "new" quantity. From Eqs. (82)–(86) one can see that $\tau_{sc}$ fulfill the field equations

$$\Delta^2\chi = \frac{2b\mu}{(1-\nu)}\partial_y\kappa(\vec{r}), \qquad \tau_{sc} = \partial_x\partial_y\chi \tag{130}$$

By comparing Eq. (130) with Eqs. (101), (102) we can see that $\tau_{sc}$ is nothing but the coarse-grained shear stress $<\sigma>_{12}$.

It is important to note that dislocation multiplication and annihilation can also be taken into account by adding a source term $f(\rho, \tau_{ext} + \tau_{sc}, \ldots)$ to the right hand side of Eq. (127):

$$\frac{\partial \rho(\vec{r}, t)}{\partial t} + M_0 b \partial_x [\kappa(\vec{r}, t) \{\tau_{sc}(\vec{r}, t) + \tau_{ext}\}] = f(\rho, \tau_{ext} + \tau_{sc}, \ldots). \quad (131)$$

Determining the actual form of the source term is a difficult issue, that is out of the scope of this paper.

## 4.7 Dislocation-Dislocation Correlation

The self-consistent field theory explained above was obtained by assuming that the two particle density functions are the product of the corresponding one particle densities. This means that the probability finding two dislocations at points $\vec{r}_1$ and $\vec{r}_2$ is simple the product of finding one in point $\vec{r}_1$ and another one in point $\vec{r}_2$, i.e. we neglect any effects related to dislocation-dislocation correlation. This is obviously a strong simplification leading to effects not observed experimentally. Just to mention one, the elastic energy of dislocations placed into a box randomly diverges logarithmically with the system size (Zaiser 2013) resulting that the energy of a dislocation system is not en extensive variable. There is not any experimental evidence indicating this. An appropriate form of dislocation-dislocation correlation, however, can resolve the problem.

Without restricting generality, the two particle density functions can be given in the form:

$$\rho_{ss'}(\vec{r}_1, \vec{r}_2, t) = \rho_s(\vec{r}_1)\rho_{s'}(\vec{r}_2)(1 + d_{ss'}(\vec{r}_1, \vec{r}_2)) \;\; s, s' \in \{+, -\}, \quad (132)$$

where $d_{ss'}$ is called dislocation-dislocation correlation function. In order to be able to say something about the correlation function as a first step it is useful to analyze the properties of dislocation-dislocation correlations in an originally homogeneous relaxed dislocation system (Zaiser et al. 2001; Groma et al. 2003). Although the BBGKY hierarchy (see Eq. (121)) explained earlier gives the possibility to investigate the properties of $d_{ss'}$ analytically (assuming something about the three particle density functions), but due to the complicated nonlinear character of the equations, apart from some simple general statements, it is rather difficult to say anything about $d_{ss'}$.

For initially homogeneous, relaxed dislocation systems $d_{ss'}$ can be determined by solving numerically Eq. (82). Details of the appropriate numerical methods can be found in Ispánovity et al. (2008). For determining the correlation functions we do not have to study extremely large systems (about 500 dislocations is already

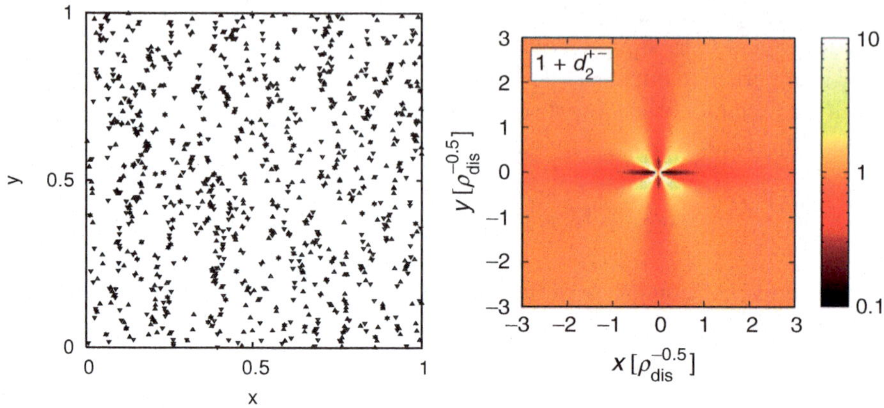


**Fig. 4** Typical relaxed 2D dislocation configuration (left box) and the $d_{+-}$ dislocation-dislocation correlation function (right box) determined numerically at single slip geometry

enough), but we need several (around 1000) relaxed configurations to have the necessary statistics. Knowing the relaxed positions of dislocations $d_{ss'}$ can be obtained by simply counting the number of dislocation pairs at different relative positions. In the simulations presented parallel straight edge dislocations were considered at single slip geometry ($\vec{b}$ is parallel to the $x$ axis). The number of dislocations was kept constant. Initially the dislocations were randomly distributed. By the numerical integration of Eq. (82) the relaxed dislocation configuration was determined at zero external stress. Figure 4 shows the correlation function $d_{+-}$ obtained numerically (Zaiser et al. 2001).

For our considerations the most important properties of the correlation function is that it decays to zero exponentially within a couple of average dislocation spacing. So, for originally random, relaxed dislocation configurations the dislocation-dislocation correlation is short-ranged. With other words, if the distance of two dislocations is larger than a couple of times the average dislocation spacing the correlation between them is negligible (Zaiser et al. 2001; Groma et al. 2003). We have to keep in mind, however, that this is valid only if the relaxed configuration is obtained from an initially random dislocation distribution. One can obviously set up initial configuration that relaxes to a strongly correlated state like for example a Taylor lattice. The problem is related to the constrained motion of dislocations. Since in the simulations only dislocation glide is allowed and dislocation multiplication is excluded, the number of dislocations in any narrow strip parallel to the Burgers vector cannot change during the relaxation of the system. It is determined by the initial configuration. The system does not "forget" fully the initial configuration. In reality, of course, the number of dislocations in a strip is determined by the dislocation multiplication. Since, however in 2D there is no "natural" law for dislocation multiplication one should investigate the correlation properties in 3D, but there is not any comprehensive result reported so far.

## 4.8 Local Density Approximation

Based on the results of the 2D simulations we can state that the dislocation-dislocation correlation functions are short range. So, it is plausible to assume that the correlation functions $d_{ss'}(\vec{r}_1, \vec{r}_2)$ defined by Eq. (132) can be approximated with the correlation function corresponding to a homogeneous system with dislocation density $\rho(\vec{r}_1)$. It follows that $d_{ss'}(\vec{r}_1, \vec{r}_2)$ practically depends only on $(\vec{r}_1 - \vec{r}_2)$, the direct $\vec{r}_1$ or $\vec{r}_2$ dependence is weak, it appears only through the spatial variation of the dislocation density, i.e.

$$\rho_{ss'}(\vec{r}_1, \vec{r}_2, t) = \rho_s(\vec{r}_1)\rho_{s'}(\vec{r}_2)(1 + d_{ss'}(\vec{r}_1 - \vec{r}_2, \rho(\vec{r}_1)) \quad s, s' \in \{+, -\}, \tag{133}$$

where we indicated that the correlation function certainly depends on the dislocation density. One can certainly raise the question if in $d_{ss'}(\vec{r}_1 - \vec{r}_2, \rho)$ one should take the dislocation density in point $\vec{r}_1$ or $\vec{r}_2$. Assuming, however, that the dislocation density varies slowly in the length scale of dislocation spacing, i.e. $|\nabla\rho| \ll \rho^{3/2}$, it does not make a difference which point is taken. Moreover, we also have to assume that the GND density is much smaller than the total one ($\kappa \ll \rho$) because in general the correlation function can also depend on the GND density too (Groma et al. 2003, 2016). Similar approximation is used successfully for many other systems like for example in first principle quantum mechanics calculations to estimate the exchange energy. It is called "local density approximation" (see also Zaiser (2015)).

With the two approximations mentioned above the $\rho$ dependence of the correlation function can be directly given from dimensionality argument. Namely, since $d_{ss'}$ cannot directly depend on $\vec{r} = \vec{r}_1 - \vec{r}_2$, a variable with length dimension, it should be normalized by a characteristic length scale of the problem. Since, if the two approximations mentioned above hold there is no other length scale in the problem but the dislocation spacing one concludes that

$$d_{ss'}(\vec{r}, \rho) = d_{ss'}(\sqrt{\rho}\vec{r}). \tag{134}$$

By substituting Eq. (132) into Eqs. (112), (113) one arrives at

$$\partial_t \rho_+(\vec{r}, t) + M_0 b \partial_x \{\rho_+ [\tau_{ext} + \tau_{sc} + \tau_+]\} = 0 \tag{135}$$

$$\partial_t \rho_-(\vec{r}, t) - M_0 b \partial_x \{\rho_- [\tau_{ext} + \tau_{sc} + \tau_-]\} = 0, \tag{136}$$

where $\tau_{sc}(\vec{r})$, given by Eq. (129), is the "self consistent fields stress" generated by the GND density,

$$\begin{aligned}\tau_+(\vec{r}) = \int &\left[\rho_+(\vec{r}')d_{++}(\vec{r} - \vec{r}')\right.\\ &\left. - \rho_-(\vec{r}')d_{+-}(\vec{r} - \vec{r}')\right] \tau_{ind}(\vec{r} - \vec{r}')d^2r',\end{aligned} \tag{137}$$

and

$$\tau_-(\vec{r}) = -\int \big[\rho_-(\vec{r}')d_{--}(\vec{r}-\vec{r}') - \rho_+(\vec{r}')d_{-+}(\vec{r}-\vec{r}')\big]\,\tau_{ind}(\vec{r}-\vec{r}')d^2r' \tag{138}$$

are stresses depending on dislocation-dislocation correlations. In these expressions, the first terms in the integrals express the stress contribution due to correlated arrangements of dislocations of the same sign in pile-ups or walls, whereas the second terms express the stress contribution due to the interaction of dislocations of opposite signs forming correlated, dipolar configurations.

For the further considerations let us introduce the quantities

$$\tau_v = \frac{\tau_+ + \tau_-}{2}, \tag{139}$$

$$\tau_a = \frac{\tau_+ - \tau_-}{2}. \tag{140}$$

With these quantities Eqs. (135), (136) read (Groma et al. 2016; Valdenaire et al. 2016)

$$\partial_t\rho_+(\vec{r},t) + M_0 b\partial_x\{\rho_+[\tau_{ext} + \tau_{sc} + \tau_v + \tau_a]\} = 0, \tag{141}$$

$$\partial_t\rho_-(\vec{r},t) - M_0 b\partial_x\{\rho_-[\tau_{ext} + \tau_{sc} + \tau_v - \tau_a]\} = 0. \tag{142}$$

In explicit form $\tau_v$ and $\tau_a$ are

$$\tau_v(\vec{r}) = \int \big[\rho(\vec{r}')d_a(\vec{r}-\vec{r}') + \kappa(\vec{r}')d_s(\vec{r}-\vec{r}')\big] * \tau_{ind}(\vec{r}-\vec{r}')d^2r', \tag{143}$$

$$\tau_a(\vec{r}) = \int \big[\rho(\vec{r}')d_p(\vec{r}-\vec{r}') + \kappa(\vec{r}')d_{a'}(\vec{r}-\vec{r}')\big] * \tau_{ind}(\vec{r}-\vec{r}')d^2r', \tag{144}$$

with

$$d_s = \tfrac{1}{2}(d_{++} + d_{--} + d_{+-} + d_{-+}), \tag{145}$$

$$d_p = \tfrac{1}{2}(d_{++} + d_{--} - d_{+-} - d_{-+}), \tag{146}$$

$$d_a = \tfrac{1}{2}(d_{++} - d_{--} - d_{+-} + d_{-+}), \tag{147}$$

$$d_{a'} = \tfrac{1}{2}(d_{++} - d_{--} + d_{+-} - d_{-+}). \tag{148}$$

It is important to summarize some symmetry properties of the pair correlation functions:

- the functions $d_{++}$ and $d_{--}$ must be invariant under a swap of the two dislocations resulting that they are even functions of $\vec{r}$.
- For dislocations with different signs one gets from the definition of correlation functions that $d_{+-}(\vec{r})=d_{-+}(-\vec{r})$.
- Hence $d_s(\vec{r})$ and $d_p(\vec{r})$ are even functions, while the difference $d_{+-} - d_{-+}$ appearing in $d_a$ and $d_{a'}$ is an odd function.

It is useful to introduce the notations

$$\tau_f(\vec{r}) = -\int \rho(\vec{r}')d_a(\vec{r}-\vec{r}')\tau_{ind}(\vec{r}-\vec{r}')d^2r' \tag{149}$$

referred to "friction stress" hereafter,

$$\tau_b(\vec{r}) = \int \kappa(\vec{r}')d_s(\vec{r}-\vec{r}')\tau_{ind}(\vec{r}-\vec{r}')d^2r' \tag{150}$$

commonly called "back stress",

$$\tilde{\tau}_b(\vec{r}) = \int \rho(\vec{r}')d_p(\vec{r}-\vec{r}')\tau_{ind}(\vec{r}-\vec{r}')d^2r' \tag{151}$$

called "diffusion stress", and

$$\tilde{\tau}_f(\vec{r}) = \int \kappa(\vec{r}')d_{a'}(\vec{r}-\vec{r}')\tau_{ind}(\vec{r}-\vec{r}')d^2r'. \tag{152}$$

The physical meaning and so the origin of the names of the stress like expressions introduced are discussed later.

Since $d_{++}$ and $d_{--}$ are even functions in Eqs. (149), (152) for nearly homogeneous systems the contribution of the difference $d_{++} - d_{--}$ to $\tau_f(\vec{r})$ and $\tilde{\tau}_f(\vec{r})$ can be neglected resulting in

$$\tilde{\tau}_f(\vec{r}) = \frac{\kappa(\vec{r})}{\rho(\vec{r})}\tau_f(\vec{r}). \tag{153}$$

From Eqs. (143), (144), (153) one obtains

$$\tau_v = -\tau_f + \tau_b \tag{154}$$

and

$$\tau_a = \frac{\kappa}{\rho}\tau_f + \tilde{\tau}_b. \tag{155}$$

After substituting Eqs. (154), (144) into Eqs. (141), (142) we conclude

$$\partial_t \rho_+(\vec{r}, t) =$$
$$-M_0 b \partial_x \left\{ \rho_+ \left[ \tau_{mf} + \tau_b - \left( 1 - \frac{\kappa}{\rho} \right) \tau_f + \tilde{\tau}_b \right] \right\} \quad (156)$$

$$\partial_t \rho_-(\vec{r}, t) =$$
$$+M_0 b \partial_x \left\{ \rho_- \left[ \tau_{mf} + \tau_b - \left( 1 + \frac{\kappa}{\rho} \right) \tau_f - \tilde{\tau}_b \right] \right\} \quad (157)$$

with $\tau_{mf} = \tau_{ext} + \tau_{sc}$.

By adding and subtracting the above equations one gets

$$\partial_t \rho(\vec{r}, t) =$$
$$-M_0 b \partial_x \left[ \kappa \tau_{mf} + \kappa \tau_b + \rho \tilde{\tau}_b \right] \quad (158)$$
$$\partial_t \kappa(\vec{r}, t) =$$
$$-M_0 b \partial_x \left[ \rho \tau_{mf} + \rho \tau_b - \rho \tau_f + \frac{\kappa^2}{\rho} \tau_f + \kappa \tilde{\tau}_b \right]. \quad (159)$$

As it was discussed above, according to numerical simulations the correlation functions decay to zero within a few dislocation spacing $1/\sqrt{\rho}$. So, in the above expressions for $\tau_v$ and $\tau_a$ the densities $\rho(\vec{r}')$ and $\kappa(\vec{r}')$ can be approximated by their Taylor expansion around the point $\vec{r}$. For slowly varying dislocation density field we can retain only the lowest-order nonvanishing terms. Since $\tau_{ind}(x, y) = -\tau_{ind}(-x, y)$ and $\tau_{ind}(x, y) = \tau_{ind}(x, -y)$, from the symmetry properties of the correlation functions mentioned above one concludes that up to second order

$$\tau_f(\vec{r}) = -\mu b C \sqrt{\rho(\vec{r})},$$
$$C(\tau_{mf}) = \int d_a^*(\vec{\tilde{r}}) \tau_{ind}^*(\vec{\tilde{r}}) d^2\tilde{r}, \quad (160)$$

$$\tau_b(\vec{r}) = -G b \frac{D}{\rho} \partial_x \kappa(\vec{r}),$$
$$D = \int \tilde{x} d_s^*(\vec{\tilde{r}}) \tau_{ind}^*(\vec{\tilde{r}}) d^2\tilde{r}, \quad (161)$$

and

$$\tilde{\tau}_b(\vec{r}) = -Gb\frac{A}{\rho}\partial_x\rho(\vec{r}),$$

$$A = \int \tilde{x}d_p^*(\vec{\tilde{r}})\tau_{ind}^*(\vec{\tilde{r}})d^2\tilde{r}, \tag{162}$$

where $\vec{\tilde{r}} = \sqrt{\rho}\vec{r}$, $\tilde{x} = \sqrt{\rho}x$, $d_a^* = \rho d_a$, $d_s^* = \rho d_s$, $d_p^* = \rho d_p$, and $\tau_{ind}^* = \tau_{ind}^*/(\mu b\sqrt{\rho})$ are dimensionless quantities, and $G = \mu/(2\pi(1-\nu))$.

It should be pointed out that in general the correlation functions are stress dependent. As a consequence, the parameters $C$, $D$, and $A$ introduced above can depend on the long-range stress $\tau_{mf}$. More precisely, due to dimensionality reasons, parameters may depend on the dimensionless parameter $\tau_{mf}/(\mu b\sqrt{\rho})$. From the symmetry properties of the correlation function we can see that $C$ is an odd, $D$, and $A$ are even functions of $\tau_{mf}$. As a consequence, at $\tau_{mf} = 0$, $C$ vanishes, while $D$, and $A$ have finite values and so they can be approximated up to second order in $\tau_{mf}$ by constants.

By substituting Eqs. (161), (162) into the evolution equations (158), (158) we arrive at (Groma et al. 2016)

$$\partial_t\rho = -M_0 b\partial_x\left\{\kappa\tau_{mf} - GbD\frac{\kappa}{\rho}\partial_x\kappa - GbA\partial_x\rho\right\},$$

$$\partial_t\kappa = -M_0 b\partial_x\left\{\rho\left[\tau_{mf} - \left(1-\frac{\kappa^2}{\rho^2}\right)\tau_f\right]\right.$$

$$\left. -GbD\partial_x\kappa - GbA\frac{\kappa}{\rho}\partial_x\rho\right\}. \tag{163}$$

To establish the stress dependence of the parameter $C$ we note that from Eqs. (125), (163) an explicit expression for the plastic shear rate in a homogeneous system is given by

$$\dot{\gamma} = \rho b M_0\left[\tau_{mf} - \left(1-\frac{\kappa^2}{\rho^2}\right)\tau_f\right]. \tag{164}$$

If we consider a system without excess dislocations ($\kappa = 0$), such a system exhibits a finite flow stress due to formation of dislocation dipoles or multipoles. For stresses below the flow stress, the strain rate is zero. It must therefore be that

$$C = \begin{cases} \frac{\alpha}{\mu b\sqrt{\rho}}\tau_{mf}, & |\tau_{mf}| < \tau_{flow} \\ \alpha, & |\tau_{mf}| \geq \tau_{flow}, \end{cases} \tag{165}$$

where $\tau_{flow} = \alpha\mu b\sqrt{\rho}$ is the flow stress. In a system where excess dislocations are present, the excess dislocations cannot be pinned by dipole/multipole formation but their effective mobility is strongly reduced. The above argument explains why $\tau_f$ is called "friction stress". Below a certain stress level it prevents plastic shear, while above it, $\tau_f$ is a stress independent constant with a value of the flow stress.

Concerning $\tau_b$, according to Eq. (161), it is proportional to the gradient of the GND density. Such a term is commonly introduced in phenomenological strain gradient plasticity (SGP) theories to account for size effects and it is termed as "back stress" (Aifantis 1984, 1987, 1994; Fleck and Hutchinson 2001; Gurtin 2002; Svendsen 2002). There is, however, a major difference between $\tau_b$ obtained here and the "back stress" introduced in SGP theories. In SGP theories a length scale considered as material parameter is always introduced to "compensate" the length dimension related to the derivation in the back stress. Here the length scale is the dislocation spacing $1/\sqrt{\rho}$ that is an evolving quantity, and so, it is not a material parameter.

The third stress like quantity related to dislocation-dislocation correlation is $\tilde{\tau}_b$. It acts a diffusion term because it is proportional to the gradient of the total dislocation density. For the first sight its rather unusual feature is that it moves both the positive and negative dislocations in the same direction (see the different signs in Eqs. (156) and (157)), but it simple means, it moves the dislocation dipoles.

## 5 Phase Field Approach

In this subsection we show that the evolution equations for the two densities of positive and negative dislocations can be cast into the framework of phase field theories (Groma et al. 2006, 2007, 2010, 2015; Zaiser 2015; Groma et al. 2016). It should be stressed that the phase field formalism introduced is established on a phenomenological ground, but it leads to the same evolution equations obtained by coarse-graining, so it is justified by that.

For a system of straight parallel edge dislocations with Burgers vectors parallel to the $x$ axis the evolution of dislocation densities $\rho_+$ and $\rho_-$ is described by the balance equations (Groma et al. 2007, 2015; Dogge et al. 2015)

$$\partial_t \rho_\pm + \partial_x [\rho_\pm v_\pm] = \pm f(\rho_+, \rho_-), \tag{166}$$

in which we consider only dislocation glide. Here $v\pm$ is the glide velocity of positive or negative signed dislocations, and $f(\rho_+, \rho_-)$ is a term accounting for dislocation multiplication or annihilation. Since multiplication terms cannot be derived for 2D systems (straight dislocations cannot multiply) they need to be introduced via ad-hoc assumptions, we assume that the number of dislocations is conserved, i.e., we consider the limit $f(\rho_+, \rho_-) = 0$. In the following we focus on the $\rho_\pm$ dependence of the velocities $v_+$ and $v_-$.

In the previous subsections over-damped dislocation motion was considered, i.e. the velocity of the dislocations are proportional to the stress. Keeping this in mind we can assume that $v_\pm$ is also proportional to the stress defined as the functional derivative of $P$ with respect to the elastic deformation

$$\tau = \frac{\delta P}{\delta \gamma^e}. \tag{167}$$

However, as it is discussed above, the dislocation configuration itself does not uniquely define the elastic and plastic distortions while the stress is determined if the dislocation density is given. It follows that $P$ should depend only on the dislocation density. Taking into account that $b\kappa = \partial_x \gamma^e$

$$P[\kappa] = P[\partial_x \gamma^e / b]. \tag{168}$$

From this one find that

$$\tau = \frac{\delta P}{\delta \gamma^e} = -\partial_x \frac{\delta P}{\delta \kappa} \frac{1}{b}. \tag{169}$$

From thermodynamical analogy $\nu_c = \delta P / \delta \kappa$ can be considered as the chemical potential of the dislocation system. So, assuming that $v_\pm \propto \tau$ one gets from Eq. (169)

$$v_\pm \propto -\partial_x \frac{\delta P}{\delta \kappa}, \tag{170}$$

i.e. it is proportional to the negative gradient of the chemical potential. It should be noted, however, that the analogy is somewhat formal because as it is seen above the thermal fluctuation does not play a role in the evolution of the dislocation system.

Since in the system considered there are two types of dislocations the above result has to be generalized (Groma et al. 2016):

$$v_+ = -M_0 \left\{ \partial_x \left[ \frac{1+\zeta}{2} \frac{\delta P}{\delta \rho_+} - \frac{1-\zeta}{2} \frac{\delta P}{\delta \rho_-} \right] \right\}, \tag{171}$$

$$v_- = -M_0 \left\{ \partial_x \left[ \frac{1+\zeta}{2} \frac{\delta P}{\delta \rho_-} - \frac{1-\zeta}{2} \frac{\delta P}{\delta \rho_+} \right] \right\}, \tag{172}$$

where $P[\rho_+, \rho_-]$ is the phase field functional and $\zeta$ is a parameter. It should be noted that the above form fulfill the symmetry properties that swapping the + and − indexes must not influence the equations. This is expected because it is up to us to define which type of dislocation is considered as positive or negative.

From Eqs. (166), (171), (172) the evolution equations for the dislocation densities derive as

$$\partial_t \rho_+ - \partial_x \left\{ \rho_+ M_0 \left[ \partial_x \frac{\delta P}{\delta \kappa} + \zeta \partial_x \frac{\delta P}{\delta \rho} \right] \right\} = 0, \tag{173}$$

$$\partial_t \rho_- + \partial_x \left\{ \rho_- M_0 \left[ \partial_x \frac{\delta P}{\delta \kappa} - \zeta \partial_x \frac{\delta P}{\delta \rho} \right] \right\} = 0. \tag{174}$$

Accordingly we find

$$\partial_t \rho = \partial_x \left\{ \kappa M_0 \partial_x \frac{\delta P}{\delta \kappa} + \zeta \rho M_0 \partial_x \frac{\delta P}{\delta \rho} \right\}, \tag{175}$$

$$\partial_t \kappa = \partial_x \left\{ \rho M_0 \partial_x \frac{\delta P}{\delta \kappa} + \zeta \kappa M_0 \partial_x \frac{\delta P}{\delta \rho} \right\}. \tag{176}$$

Concerning the actual form of $P[\rho_+, \rho_-]$ it is useful to split it into two parts, a "mean field" or "self consistent" part $P_{sc}$ and a "correlation" part $P_c$ defined below.

According to Eqs. (43), (51) the equation for the mean field stress $\tau_{mf}$ can be obtained from a variational principle. By taking $P_{sc}[\chi, \rho_+, \rho_-]$ in the form

$$P_{sc}[\chi, \rho_+, \rho_-] = \int \left[ -\frac{1-\nu}{4\mu} (\Delta \chi)^2 + b \chi \partial_y \kappa \right] d^2 r, \tag{177}$$

the minimum condition

$$\frac{\delta P_{sc}}{\delta \chi} = 0 \tag{178}$$

leads to the equation

$$\frac{1-\nu}{2\mu} \Delta^2 \chi = b \partial_y \kappa, \tag{179}$$

and $\tau_{mf} = \partial_x \partial_y \chi$. The general solution of Eq. (179) is $\tau_{mf}$ given by Eq. (129) plus the external stress.

Let us first see what one obtains by substituting Eq. (177) into (172) and (171). After a straightforward calculation one gets

$$\partial_t \rho_+(\vec{r}, t) + M_0 b \partial_x \left( \rho_+ \tau_{mf} \right) = 0, \tag{180}$$

$$\partial_t \rho_-(\vec{r}, t) - M_0 b \partial_x \left( \rho_- \tau_{mf} \right) = 0. \tag{181}$$

As it is seen $P_{sc}$ recovers the mean field equations (127), (128) but not the terms which are related to dislocation-dislocation correlations. It thus needs to be complemented by a "correlation" part that is in close analogy with the "energy error" introduced by Mesarovic et al. (2010) (see details in chapter "Physical Foundations of Mesoscale Continua").

From coarse-graining of the energy of the discrete system one can have some indication how the correlation part should look like (Zaiser 2015). but the one suggested here for the present dislocation system is rather an educated guess (Groma et al. 2007, 2016):

$$P_{corr} = \int \left[ Gb^2 A\rho \ln\left(\frac{\rho}{\rho_0}\right) + \frac{Gb^2 D}{2} \frac{\kappa^2}{\rho} \right] d^2 r. \tag{182}$$

It is justified by the evolution equations obtained from it.

Since we consider only weakly polarized dislocation arrangements, terms of higher than first order in $\kappa/\rho$ and $\partial_x\rho/\rho^{3/2}$ can be neglected. With these we find that

$$\partial_t \rho_+ = -\partial_x \left[ \rho_+ M_0 b \left( \tau_{mf} - Gb\frac{D}{\rho}\partial_x \kappa - Gb\zeta\frac{A}{\rho}\partial_x \rho \right) \right], \tag{183}$$

$$\partial_t \rho_- = +\partial_x \left[ \rho_- M_0 b \left( \tau_{mf} - Gb\frac{D}{\rho}\partial_x \kappa + Gb\zeta\frac{A}{\rho}\partial_x \rho \right) \right]. \tag{184}$$

From the above equations the evolution equations for $\kappa$ and $\rho$ read

$$\partial_t \rho = -M_0 b \partial_x \left\{ \kappa \tau_{mf} - GbD\frac{\kappa}{\rho}\partial_x \kappa - GbA\partial_x \rho \right\}, \tag{185}$$

$$\partial_t \kappa = -M_0 b \partial_x \left\{ \rho \tau_{mf} - GbD\partial_x \kappa - GbA\frac{\kappa}{\rho}\partial_x \rho \right\}. \tag{186}$$

With $\zeta = 1$, apart from the term containing the "friction" stress $\tau_f$, Eqs. (185), (186) are equivalent to Eqs. (163), (163). So, by applying the standard formalism of phase field theories, with the appropriate form of the correlation term in the phase field functional, we recover the evolution equations of the dislocation densities derived by ensemble averaging the equations of motion of individual dislocations. However, the friction stress $\tau_f$ playing a crucial role in the plastic deformation of any material can not be directly derived from the coarse-grained energy functional.

For resolving this issue we modify Eq. (176) by allowing a non-linear dependency on the driving force $\delta P/\delta\kappa$ (Groma et al. 2016). The modified equation is given by

$$\partial_t \kappa = \partial_x \left\{ \rho M \left( \partial_x \frac{\delta P}{\delta \kappa} \right) + \kappa M_0 \partial_x \frac{\delta P}{\delta \rho} \right\}, \tag{187}$$

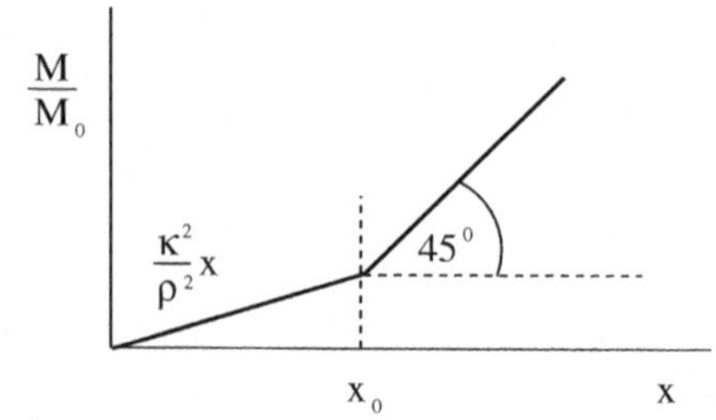


**Fig. 5** The $M(x)$ mobility function

were $M(x)$ is a nontrivial mobility function defined as

$$M(x) = M_0 \begin{cases} \frac{\kappa^2}{\rho^2} x & \text{if } |x| < x_0 \\ sgn(x) \left[ |x| - x_0 \left( 1 - \frac{\kappa^2}{\rho^2} \right) \right] & \text{if } |x| \geqq x_0 \end{cases} \tag{188}$$

with $x_0 = \alpha \mu b^2 \sqrt{\rho}$ (see Fig. 5). It is easy to see that this mobility function recovers Eq. (163).

To sum up, it is obtained that the continuum theory of dislocations can be cast into the general framework of a phase field theory. Nevertheless, the phase field theory is different from the ones commonly applied for other problems. Due to the nontrivial on/off type mobility function there are infinite different stationary dislocation states. As the system enters into a state where the total stress (including the back and diffusion stresses) is everywhere below the local flow stress, resulting that the dislocation mobility is zero everywhere, the system stops evolving.

The most important advantage of the phase field formalism explained is that it opens the possibility of a systematic generalization of the theory. Without going into the details we mention that extending the continuum theory into 2D multiple slip is quite straightforward. For the mean field part of the phase field functional one have to replace $P_{sc}$ given by Eq. (177) with the more general form of Eq. (51) with $\alpha_{ij}$ corresponding to the GNDs in the different slip systems. For the $P_{corr}$ correlation part one simple can take the sum of the ones corresponding to each slip systems. It can be generalized by adding appropriate cross terms too. Generalization to 3D is much more difficult. Although there are some promising attempts (see below), there is not a generally accepted one.

## 6 Dislocation Patterning

The continuum theory of dislocations presented can be validated by comparing its predictions with the results of 2D discrete dislocation dynamics simulations. Many direct comparisons indicate that the theory is able to reproduce the most important features of the collective evolution of the 2D dislocation system. For details the reader is refereed to the papers (Yefimov et al. 2004; Groma et al. 2006; Dogge

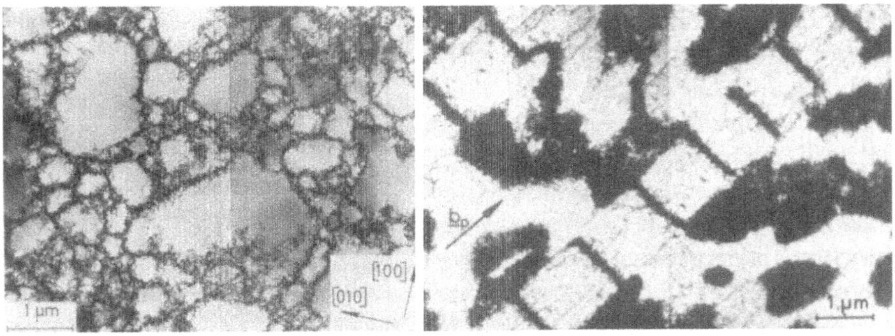


**Fig. 6** Fractal like cell structure obtained on Cu single crystal deformed by uniaxial tension (left box). The ladder structure developing in fatigued Cu single crystal

et al. 2015). In this paper we discuss a rather important issue only, namely if the continuum theory is able to predict dislocation patterning (Groma et al. 2016). It is known for a long time that the arrangement of dislocations in deformed crystals is practically never homogeneous. During the deformation dislocations arrange themselves into heterogeneous patterns. Two beautiful patterns can be seen on Fig. 6.

There are many different models proposed to predict the pattern formation. Most of them, however, are based on analogies with pattern formation in other physical systems. Thus, it has been suggested that dislocation patterns can be understood as minimizers of some kind of energy functional, because the dislocations try to minimize elastic energy (Hansen and Kuhlmann-Wilsdorf 1986). On the same line, in analogy with the Cahn Hilliard models of spinodal decomposition, Holt (1970) proposed a theory but some of the predictions of his model have never been observed experimentally. Another idea is that dislocations in a deforming crystal constitute a driven system far from equilibrium where patterns may form as dissipative structures. This has led to the formulation of nonlinear sets of partial differential equations for dislocation densities (Walgraef and Aifantis 1985; Pontes et al. 2006) giving rise to a variety of different dislocation patterns.

The most important shortcoming of these theories is that it is not clear how they are related to the properties of individual dislocations. So, they are completely phenomenological ones. In the following we show that although it is a rather simplified system the continuum theory derived for 2D single slip is able to predict pattern formation.

## *6.1 Theory of Pattern Formation*

In the following we discuss under what conditions the evolution equations derived above can lead to instability resulting in dislocation pattern formation (Groma et al. 2016). One can easily see that the trivial homogeneous solution $\rho = \rho_0$, $\kappa = 0$

and $\tau_{mf} = \tau_0$ satisfies Eqs. (163), (163), (179), where $\rho_0$ and $\tau_0$ are constants representing the initial dislocation density and the external shear stress, respectively. The stability of the trivial solution can be analyzed by applying the standard method of linear stability analysis. One can easily see that nontrivial behavior can happen only in the flowing regime i.e. if $|\tau_0| > \alpha\mu b\sqrt{\rho_0}$, so we consider only this case.

By adding small perturbations to the dislocation densities and the Airy stress function in the form

$$\begin{aligned} \rho(\vec{r},t) &= \rho + \delta\rho(\vec{r},t) \\ \kappa(\vec{r},t) &= \delta\kappa(\vec{r},t) \\ \chi(\vec{r},t) &= \tau_0 xy + \delta\chi(\vec{r},t) \end{aligned} \tag{189}$$

and keeping only the leading terms in the perturbations, Eqs. (163), (163, (179) become

$$\partial_t\delta\rho = M_0\partial_x\left[GbA\partial_x\delta\rho - \tau_0\delta\kappa\right], \tag{190}$$

$$\begin{aligned} \partial_t\delta\kappa = &-M_0\Theta_f\partial_x\left[\rho_0\partial_x\partial_y\delta\chi - GbD\partial_x\delta\kappa\right] \\ &-M_0\Theta_f\left[\tau^* - \alpha\mu b\frac{\sqrt{\rho_0}}{2}\right]\partial_x\delta\rho, \end{aligned} \tag{191}$$

$$\Delta^2\delta\chi = 4\pi Gb\partial_y\delta\kappa. \tag{192}$$

In these expressions, $\tau^* = \tau_0 - \alpha\mu b\sqrt{\rho_0}$, and the step function $\Theta_f = \Theta(\tau^*)$ is zero if the applied stress is below the flow stress in the homogeneous reference state, and 1 otherwise. To obtain the above equations it was taken into account that the first-order variation of the flow stress is given by

$$\delta\tau_f = \frac{\alpha\mu b\sqrt{\rho_0}}{2}\frac{\delta\rho}{\rho_0}. \tag{193}$$

The solution of Eqs. (192), (190), (191) can be found in the form

$$\begin{pmatrix} \delta\rho \\ \delta\kappa \\ \delta\chi \end{pmatrix} = \begin{pmatrix} \delta\rho_0 \\ \delta\kappa_0 \\ \delta\chi_0 \end{pmatrix} \exp\left(\frac{\lambda}{t_0}t + i\sqrt{\rho_0}\vec{k}\vec{r}\right), \tag{194}$$

were $\vec{k}$ is a dimensionless quantity. After substituting the above form into Eqs. (192), (190), (191) in the flowing regime ($\Theta_f = 1$) one obtains

$$\begin{pmatrix} \lambda + Ak_x^2, & i(\dot{\gamma}' + 2\alpha')k_x \\ i(\dot{\gamma}' - \alpha')k_x, & \lambda + Dk_x^2 + T(\vec{k}) \end{pmatrix} \begin{pmatrix} \delta\rho \\ \delta\kappa \end{pmatrix} = 0, \tag{195}$$

where the notations $t_0 = b^2 G\rho_0/B$, $T(\vec{k}) = 4\pi k_x^2 k_y^2/|\vec{k}|^4$, $\dot{\gamma}' = \tau_*/(Gb\sqrt{\rho_0})$, and $\alpha' = \pi(1-\nu)\alpha$ were introduced. Note that in the above equations each of the parameters are dimensionless and $\dot{\gamma}'$ is proportional to the average shear rate $\dot{\gamma} = M_0 b^2 \rho_0 \tau^*$.

Equation (195) has nontrivial solutions if

$$(\lambda + Ak_x^2)(\lambda + Dk_x^2 + T(\vec{k})) + k_x^2\beta = 0 \tag{196}$$

with $\beta = (\dot{\gamma}' + 2\alpha')(\dot{\gamma}' - \alpha')$. This leads to

$$\lambda_\pm = -\frac{(A+D)k_x^2 + T(\vec{k})}{2} \pm \frac{\sqrt{[(D+A)k_x^2 + T(\vec{k})]^2 - 4k_x^2[\beta + A(Dk_x^2 + T(\vec{k}))]}}{2}. \tag{197}$$

It follows that the condition for the existence of growing perturbations ($\lambda > 0$) is

$$[\beta + AT(\vec{k}) + ADk_x^2] < 0. \tag{198}$$

$T(\vec{k})$ cannot be negative and it vanishes if $\vec{k}$ is parallel to either the $x$ or to the $y$ axis. Thus, $\beta < 0$ is a necessary and sufficient condition for instability. This condition requires that

- the system is in the flowing phase,
- $\dot{\gamma}'$ must be smaller than $\alpha'$. In this case there exists a region in the $\vec{k}$ space in which perturbations grow.

Perturbations with wave vectors outside this region decay in time (see Figs. 7 and 8).

This results in a length scale selection corresponding to the fastest growing periodic perturbation $\vec{k}_{max}$ defined by the condition

$$\left.\frac{d\lambda_+(\vec{k})}{d\vec{k}}\right|_{\vec{k}_{max}} = 0. \tag{199}$$

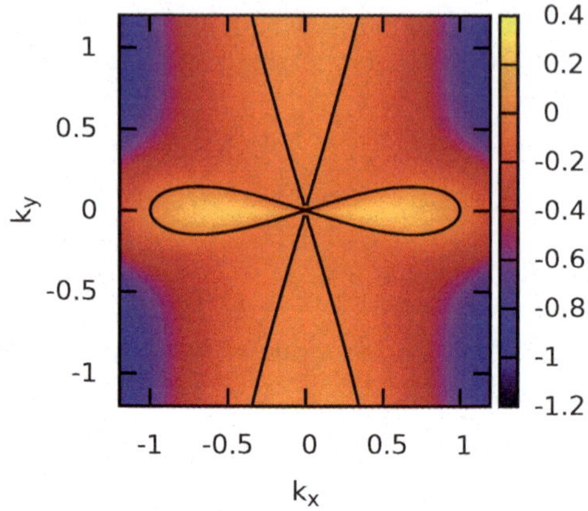


**Fig. 7** The $\lambda_+(k_x, k_y)$ function at $A = 1$, $D = 1$ and $\beta = -1$. The function is positive within the region marked by the contour line $\lambda_+(k_x, k_y) = 0$

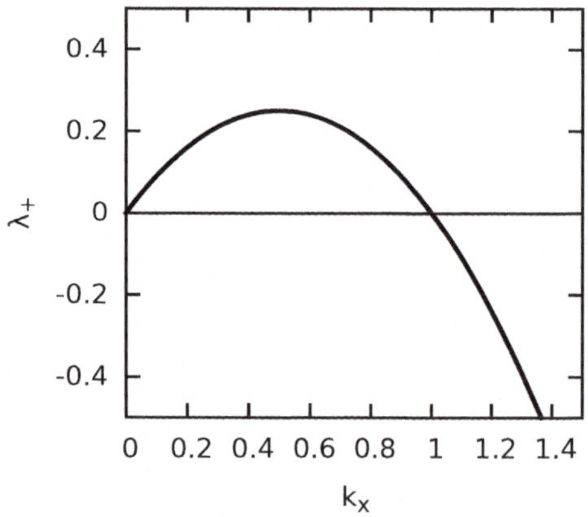


**Fig. 8** The $\lambda_+(k_x, 0)$ function at $A = 1$, $D = 1$ and $\beta = -1$

For negative $\beta$, the $\lambda_+(k_x, k_y)$ function has two equal maxima along the $x$ axis located at

$$k_x^2 = -2\beta \frac{-1 + \sqrt{1 + \frac{(A-D)^2}{4AD}}}{(A - D)^2}.$$

It should be stressed that according to Eq. (194) the actual wave vector of the fastest growing perturbation is $\sqrt{\rho_0}\vec{k}_{max}$. So, in agreement with the principle of similitude observed experimentally the characteristic pattern wavelength scales with the dislocation spacing $1/\sqrt{\rho_0}$. It is important to note at this point that both the back-stress $\tau_b$ and the diffusion-like $\tilde{\tau}_b$ term introduced here play a crucial role in characteristic wavelength selection. If either $A = 0$ or $D = 0$ perturbations of all wave vectors would grow and there would be no mode of maximum growth rate (see Eq. (162)). So we can say that the primary source of instability is the flow stress being proportional to $\sqrt{\rho}$, but for length scale selection observed experimentally we need both the back and the diffusion-like stresses. It should be stressed that in contrast to several earlier models which relate patterning to a particular elementary dislocation mechanism, like cross slip (Xia and El-Azab 2015), sweeping of dipoles with a moving curved dislocation (Kratochvil and Sedlacek 2003), etc., the continuum theory predict patterning under rather general conditions. The particular form of dislocation-dislocation correlations are not really important.

# 7 3D Continuum Theory

The generalization of the 2D continuum theory to 3D is far from straightforward. The key issue is that a dislocation loop is an extended object in space while a continuum theory operates with local fields. So, first we have to find the appropriate quantities the continuum theory should operate with.

A dislocation loop can be given by the parametric equation $\vec{r}(s)$, where $s$ is a scalar varying in a given interval. Since $\vec{r}(s)$ is extended in space, it is difficult to work with it in statistical sense. An equivalent description of the loop is the Taylor expansion of $\vec{r}(s)$ around a point $s_o$

$$\vec{r}(s) = \vec{r}(s_0) + \left.\frac{d\vec{r}}{ds}\right|_{s_0} (s - s_o) + \frac{1}{2}\left.\frac{d^2\vec{r}}{ds^2}\right|_{s_0} (s - s_o)^2 + \dots, \tag{200}$$

where derivatives are local quantities. Since, however they depend on the actual form of parametrization of the loop, one should use quantities related to the derivatives but independent of the parametrization. These are the tangent $\vec{l}$, the curvature $k$ (for simplicity we consider only planar loops), and quantities corresponding to higher derivatives (denoted by $\{\zeta_i\}$). With these, the $p$ probability density of finding the dislocation network in a give state can be given as a function of $\vec{r}$, $\vec{l}$, $k$, and $\{\zeta_i\}$. The probability $p(\vec{r}, \vec{l}, k, \{\zeta_i\})$ is a "local" function. In this section 3 different approaches operating with certain mean values of $p(\vec{r}, \vec{l}, k, \{\zeta_i\})$ are shortly summarized.

## 7.1 *Mean Field Theory*

Recently Xia and El-Azab (2015) have suggested a mean field approach for the time evolution of the the vector field $\vec{\rho}(\vec{r}) = \rho_a(\vec{r})\vec{l}_a(\vec{r})$ where

$$\rho_a(\vec{r}) = N \int p(\vec{r}, \vec{l}, k, \{\zeta_i\}) d\vec{l} dk d\zeta_i, \tag{201}$$

and

$$\vec{l}_a(\vec{r}) = \int \vec{l} p(\vec{r}, \vec{l}, k, \{\zeta_i\}) d\vec{l} dk d\zeta_i, \tag{202}$$

with $N$ denoting the number of dislocation loops.

According to Eq. (9) in this case the dislocation density tensor is

$$\alpha_{ij} = \rho_i b_j. \tag{203}$$

If we neglect the correlation between dislocation loops (mean field approximation) according to Eqs. (76), (78)

$$\dot{\alpha}_{ij} = -e_{ikl} \partial_k \dot{\beta}^P_{lj}, \tag{204}$$

and

$$\dot{\beta}^P_{ij} = -e_{ikl} u_k \alpha_{lj}, \tag{205}$$

where $\vec{u}(\vec{r})$ is the mean velocity of the dislocations at point $\vec{r}$. It has to be noted here that in general for a discrete dislocation system the mean value of $u_l \alpha_{ij}$ is not the product of the mean value of $u_l$ and $\alpha_{ij}$. In order to get Eq. (205) the dislocation-dislocation correlations have to be neglected. By combining Eqs. (203), (204, (205) one arrives at the evolution equation

$$\dot{\rho}_i b_j = e_{ipn} \partial_p (e_{nmk} u_m \rho_k b_j). \tag{206}$$

For getting a closed set of equation Xia and El-Azab (2015) assumed an over-damped dislocation motion with dislocation velocity proportional to the local shear stress generated by $\alpha_{ij}$.

By the numerical solution of the evolution equation given above Xia and El-Azab (2015) have obtained a modulation on the dislocation density field. Interestingly by allowing dislocation cross slip introduced by a probabilistic rule they have found a clear tendency for dislocation cell formation.

## 7.2 Hydrodynamics Approach

In the approach proposed by Kratochvil and Sedlacek (2003) the state of the material is also described by the density $\rho_a(\vec{r}, t)$, $\vec{l}_a$, and the velocity $\vec{u}(\vec{r}, t)$ fields, but $\rho_a(\vec{r}, t)$ and $\vec{l}_a$ are assumed to evolve separately. Since $\vec{l}_a$ is a unite vector the angle $\beta$ defined by the relation $\vec{l}_a = (\cos(\beta), \sin(\beta))$ is convenient to introduced.

For the sake of simplicity we assumed that all dislocation loops have the same Burgers vector and their slip planes are parallel.Taking the $z$ axis perpendicular to the slip plane of the loops, the two nonvanishing components of the dislocation density tensor are

$$\alpha_{11} = b\rho\cos(\beta), \qquad \alpha_{21} = b\rho\sin(\beta). \tag{207}$$

Since the dislocation density tensor is the curl of the plastic distortion, it has to be div free:

$$\frac{\partial \alpha_{ij}}{\partial r_i} = 0. \tag{208}$$

One can find from Eqs. (207), (208) that $\rho$ and $\beta$ have to satisfy the conservation law

$$\frac{\partial \rho\cos(\beta)}{\partial x} + \frac{\partial \rho\sin(\beta)}{\partial y} = 0. \tag{209}$$

On the other hand, from the general expression of the evolution of the dislocation density tensor given by Eq. (75),the following evolution equations can be deduced for $\alpha$ and $\beta$ fields(for details see Kratochvil and Sedlacek (2003))

$$\begin{aligned}\rho\dot{\beta} &= \cos(\beta)\frac{\partial \rho u}{\partial x} + \sin(\beta)\frac{\partial \rho u}{\partial y},\\ \dot{\rho} &= \sin(\beta)\frac{\partial \rho u}{\partial x} - \cos(\beta)\frac{\partial \rho u}{\partial y},\end{aligned} \tag{210}$$

where $u = |\vec{u}|$.

To have a closed theory a constitutive relation is needed between the three fields $\rho(\vec{r}, t)$, $\beta(\vec{r}, t)$ and $u(\vec{r}, t)$. Kratochvil and Sedlacek (2003) suggested the following constitutive relation for the velocity field:

$$Bu = \begin{cases} b\sigma_{13} + C\kappa_s - b\tau_0 - b\tau & if\ b\sigma_{13} + C\kappa_s > b\tau_0 + b\tau \\ 0 & if\ |b\sigma_{13} + C\kappa_s| < b\tau_0 + b\tau \\ b\sigma_{13} + C\kappa_s + b\tau_0 + b\tau & if\ b\sigma_{13} + C\kappa_s < -b\tau_0 - b\tau \end{cases} \tag{211}$$

where $b\sigma_{13}$ is the Peach-Koehler force due to the local shear stress, $C\kappa_s$ is the self-force, $b\tau_0$ is the friction force, and $b\tau$ represents the interaction between the gliding dislocations an the dislocation loops.

The self-force $C\kappa_s$ is considered in the line tension approximation, where $\kappa_s$ is the dislocation line tension. The curvature of a dislocation segment $C(\vec{r}, \beta, t) = -\mathrm{div}\vec{n}$, where $\vec{n}$ is the unit normal to the dislocation segment. As it is explained in details in Kratochvil and Sedlacek (2003) $C$ can be approximated by the expression

$$C = \cos(\beta)\frac{\partial \beta}{\partial x} + \sin(\beta)\frac{\partial \beta}{\partial y} \tag{212}$$

The most difficult problem is to set up an appropriate expression for $b\tau$. For this Kratochvil and Sedlacek (2003) suggested that

$$b\tau = Fc^{1/3} \tag{213}$$

were $c$ is the loop density, and $F$ is a constant.

According to detailed analytical and numerical investigations (Kratochvil and Sedlacek 2003) the model explained above is able to predict both dislocation patterning and size effect. Nevertheless, the justification of the assumptions used requires further investigation.

## 7.3 General Continuum Theory in 3D

Hochrainer et al. (2007, 2014) have taken a more generalized approach. They have considered the quantities

$$\rho_a(\vec{r}, \vec{l}) = N \int p(\vec{r}, \vec{l}, k, \{\zeta_i\}) dk d\zeta_i, \tag{214}$$

and mean value of the local curvature

$$k_a(\vec{r}, \vec{l}) = \int kp(\vec{r}, \vec{l}, k, \{\zeta_i\}) dk d\zeta_i, \tag{215}$$

depending on the position $\vec{r}$ and the local tangent $\vec{l}$. With these they have moved to a higher, 3+2 dimension $(\vec{r}, \varphi, \theta)$ where $\varphi$ and $\theta$ are the polar and azimuthal angles of $\vec{l}$. For plane problems (corresponding to $\theta = 0$) they have introduced the generalized nabla operator

$$\hat{\nabla} = \rho_a(\vec{r}, \varphi) + \partial_\varphi, \tag{216}$$

the 5D line direction

$$L(\vec{r}, \varphi) = (\cos(\varphi), \sin(\varphi), 0, k_a(\vec{r}, \varphi)) \tag{217}$$

and the 5D velocity vector

$$V(\vec{r}, \varphi) = (v_1, v_2, 0, -\nabla_L v) \tag{218}$$

with the 3D vector field perpendicular to the line direction

$$\vec{v} = (v_1, v_2, 0) = v(\vec{r}, \varphi)(\sin(\varphi), -\cos(\varphi), 0) \tag{219}$$

From the continuity of the dislocation line they derived the evolution equations for $\rho_a$ and $q = \rho_a k_a$

$$\partial_t \rho_a = -\hat{\nabla}(\rho_a V) + qv \tag{220}$$

$$\partial_t q = -\hat{\nabla}(q_a V) - \rho \hat{\nabla}_L \hat{\nabla}_L v \tag{221}$$

The two equations, given above, however, do not form a closed set of equations, because they depend on the unknown velocity $v$. They represent only the "kinematics" of the dislocation evolution. In spite of some promising attempts (Hochrainer 2016) at the moment it is not really developed how $v$ should depend on the $\rho_a$ and $q$ fields. The issue require further investigations.

In summary it can be stated that, in spite of the large amount of excellent works carried out on the problem, the 3D continuum theory of dislocations is not completely developed. It still remains a challenge to establish it.

**Acknowledgements** During the past years the author had the pleasure to work with Michael Zaiser, Erik van der Giessen, Géza Györgyi, Alphonse Finel, Botond Bakó, Ferenc Csikor, and Serge Yefimov. Most of the results presented in the paper were obtained together with them. Their outstanding scientific contributions and friendships are gratefully acknowledged.

Financial supports of the National Research, Development and Innovation Found of Hungary under contract number NKFIH-K-119561 is also acknowledged.

## References

E.C. Aifantis, On the microstructural origin of certain inelastic models. ASME J. Eng. Mater. Technol. **106**, 326–330 (1984)

E.C. Aifantis, The physics of plastic deformation. Int. J. Plast. **3**(3), 211–247 (1987)

E.C. Aifantis, Gradient effects at macro, micro, and nano scales. J. Mech. Behav. Mater. **5**(3), 355–375 (1994)

E.C. Aifantis, Gradient deformation models at nano, micro, and macro scales. J. Eng. Mater. Technol. **2**, 189–202 (1999)

G. Ananthakrishna, Current theoretical approaches to collective behavior of dislocations. Phys. Rep. **440**, 113–259 (2007)

V.L. Berdichevsky, Homogenization in micro-plasticity. J. Mech. Phys. Solids **53**, 2457–2469 (2005)

V. Bulatov, L. Hsiung, M. Tang, A. Arsenlis, M.C. Bartelt, W. Cai, J. Florando, M. Hiratani, M. Rhee, G. Hommes, T.G. Pierce, T. Diaz de la Rubia, Dislocation multi-junctions and strain hardening. Nature **440**, 1174–1178 (2006)

A.H. Cottrell, B. Bilby, Dislocation theory of yielding and strain ageing of iron. Proc. Phys. Soc. **62A**, 49–62 (1949)

B. Devincre, L.P. Kubin, C. Lemarchand, R. Madec, Mesoscopic simulations of plastic deformation. Mater. Sci. Eng. A **309**(SI), 211–219 (2001)

M.M.W. Dogge, R.H.J. Peerlings, M.G.D. Geers, Interface modeling in continuum dislocation transport. Mech. Mater. **88**, 30–43 (2015)

N.A. Fleck, J.W. Hutchinson, A reformulation of strain gradient plasticity. J. Mech. Phys. Solids **49**, 2245–2271 (2001)

N.M. Ghoniem, L.Z. Sun, Fast-sum method for the elastic field off three-dimensional dislocation ensembles. Phys. Rev. B **60**(1), 128–140 (1999)
D. Gomez-Garcia, B. Devincre, L.P. Kubin, Dislocation patterns and the similitude principle: 2.5D mesoscale simulations. Phys. Rev. Lett. **96**(12), 125503 (2006)
R. Gröger, T. Lookman, A. Saxena, Incompatibility of strains and its application to mesoscopic studies of plasticity. Phys. Rev. B **82**, 144104 (2010)
I. Groma, Link between the microscopic and mesoscopic length-scale description of the collective behavior of dislocations. Phys. Rev. B **56**(10), 5807–5813 (1997)
I. Groma, B. Bakó, Dislocation patterning: from micro-to mesoscale description. Phys. Rev. Lett. **84**(7), 1487 (2000)
I. Groma, P. Balogh, Investigation of dislocation pattern formation in a two-dimensional self-consistent field approximation. Acta Mater. **47**, 3647–3654 (1999)
I. Groma, F.F. Csikor, M. Zaiser, Spatial correlations and higher-order gradient terms in a continuum description of dislocation dynamics. Acta Mater. **51**(5), 1271–1281 (2003)
I. Groma, G. Gyorgyi, B. Kocsis, Debye screening of dislocations. Phys. Rev. Lett. **96**(16), 165503 (2006)
I. Groma, G. Gyorgyi, B. Kocsis, Dynamics of coarse grained dislocation densities from an effective free energy. Philos. Mag. **87**(8–9), 1185–1199 (2007)
I. Groma, G. Gyorgyi, P.D. Ispanovity, Variational approach in dislocation theory. Philos. Mag. **90**(27–28), 3679–3695 (2010)
I. Groma, Z. Vandrus, P.D. Ispanovity, Scale-free phase field theory of dislocations. Phys. Rev. Lett. **114**(1), 015503 (2015)
I. Groma, M. Zaiser, P.D. Ispánovity, Dislocation patterning in a two-dimensional continuum theory of dislocations. Phys. Rev. B **93**(21), 214110 (2016)
M.E. Gurtin, A strain gradient crystal plasticity analysis of grain size effects in polycrystals. J. Mech. Phys. Solids **50**, 313–324 (2002)
M.Y. Gutkin, E.C. Aifantis, Dislocations in the theory of gradient elasticity. Scripta Mater. **40**, 559–566 (1999)
N. Hansen, D. Kuhlmann-Wilsdorf, Low-energy dislocation-structures due to unidirectional deformation at low-temperatures. Mater. Sci. Eng. **81**(1–2), 141–161 (1986)
T. Hochrainer, Thermodynamically consistent continuum dislocation dynamics. J. Mech. Phys. Solids **88**, 12–22 (2016)
T. Hochrainer, M. Zaiser, P. Gumbsch, A three-dimensional continuum theory of dislocation systems: kinematics and mean-field formulation. Philos. Mag. **87**(8–9), 1261–1282 (2007)
T. Hochrainer, S. Sandfeld, M. Zaiser, P. Gumbsch, Continuum dislocation dynamics: towards a physical theory of crystal plasticity. J. Mech. Phys. Solids **63**, 167–178 (2014)
D.L. Holt, Dislocation cell formation in metals. J. Appl. Phys. **41**, 3197–3201 (1970)
P.D. Ispánovity, I. Groma, G. Györgyi, Evolution of the correlation functions in two-dimensional dislocation systems. Phys. Rev. B **78**(2), 024119 (2008)
P.D. Ispánovity, I. Groma, G. Györgyi, F.F. Csikor, D. Weygand, Submicron plasticity: yield stress, dislocation avalanches, and velocity distribution. Phys. Rev. Lett. **105**(8), 085503 (2010)
P.D. Ispánovity, L. Laurson, M. Zaiser, I. Groma, S. Zapperi, M.J. Alava, Avalanches in 2D dislocation systems: plastic yielding is not depinning. Phys. Rev. Lett. **112**(23), 1–5 (2014)
A.M. Kosevich, Crystal dislocations and the theory of elasticity, in *Dislocations in Solids*, vol. 1 (North-Holland, Amsterdam, 1979), pp. 33–142
L. Kovács, L. Zsoldos, *Dislocations and Plastic Deformation* (Pergamon Press, Oxford, 1973)
J. Kratochvil, R. Sedlacek, Pattern formation in the framework of the continuum theory of dislocations. Phys. Rev. B **67**(9), 094105 (2003)
E. Kröner, Continuum theory of defects, in *Physics of Defects*, ed. by R. Balian et al. (Elsevier, Amsterdam, 1981)
L.P. Kubin, G. Canova, The modeling of dislocation patterns. Scripta Metall. **27**(8), 957–962 (1992)
L.D. Landau, E.M. Lifshitz, *Theory of Elasticity*, Volume 7 of *Course in Theoretical Physics*, 3rd edn. (Pergamon, Oxford, 1986)

M. Lazar, A nonsingular solution of the edge dislocation in the gauge theory of dislocations. J. Phys. A **36**, 1415 (2003)

S.D. Mesarovic, R. Baskaran, A. Panchenko, Thermodynamic coarsening of dislocation mechanics and the size-dependent continuum crystal plasticity. J. Mech. Phys. Solids **58**(3), 311–329 (2010)

J. Pontes, D. Walgraef, E.C. Aifantis, On dislocation patterning: multiple slip effects in the rate equation approach. Int. J. Plast. **22**(8), 1486–1505 (2006)

M. Rhee, H.M. Zbib, J.P. Hirth, H. Huang, T. de la Rubia, Models for long-/short-range interactions and cross slip in 3D dislocation simulation of BCC single crystals. Modell. Simul. Mater. Sci. Eng. **6**(4), 467–492 (1998)

B. Svendsen, Continuum thermodynamic models for crystal plasticity including the effects of geometrically-necessary dislocations. J. Mech. Phys. Solids **50**, 1297–1329 (2002)

P. Valdenaire, Y. Le Bouar, B. Appolaire, A. Finel, Density-based crystal plasticity: from the discrete to the continuum. Phys. Rev. B **93**(21), 214111 (2016)

D. Walgraef, E.C. Aifantis, Dislocation patterning in fatigued metals as a result of dynamical instabilities. J. Appl. Phys. **58**(2), 688–691 (1985)

S. Xia, A. El-Azab, Computational modelling of mesoscale dislocation patterning and plastic deformation of single crystals. Model. Simul. Mater. Sci. Eng. **23**(5) (2015) (055009)

S. Yefimov, I. Groma, E. van der Giessen, A comparison of a statistical-mechanics based plasticity model with discrete dislocation plasticity calculations. J. Mech. Phys. Solids **52**(2), 279–300 (2004)

M. Zaiser, The energetics and interactions of random dislocation walls. Philos. Mag. Lett. **93**(7), 387–394 (2013)

M. Zaiser, Local density approximation for the energy functional of three-dimensional dislocation systems. Phys. Rev. B **92**, 174120 (2015)

M. Zaiser, M.C. Miguel, I. Groma, Statistical dynamics of dislocation systems: the influence of dislocation-dislocation correlations. Phys. Rev. B **64**(22), 224102 (2001)

L.A. Zepeda-Ruiz, A. Stukowski, T. Oppelstrup, V. Bulatov, Probing the ultimate limits of metal plasticity. Nature **550**, 492–495 (2016)

# Granular Materials: Micromechanical Approaches of Model Systems

**Jean-Noël Roux**

**Abstract** An overview is given of micromechanical approaches to the rheology of granular materials, from solidlike granular packs to large plastic strains and dense inertial flows, which essentially relies on the numerical simulation (by the "discrete element method" or DEM) of simple model systems. The main features of contact laws are presented, and then it is insisted on the importance of the geometry of disordered granular assemblies, such that some details of contact interactions are in fact often irrelevant. Some salient results, as obtained from DEM studies over the last decades, are presented about the variety of microstructures and internal states, depending on assembling processes; on elasticity and its (limited) role in quasistatic granular behavior; on plastic strains and the fundamental concept of critical states, and on its recent applications to the rheology of dense granular flows and suspensions.

## 1 Introduction

A wide variety of materials used in different engineering applications are granular assemblies (Andreotti et al. 2013), made of solid particles of different sizes: sands, or, more generally, soils (Mitchell and Soga 2005), powders used in pharmaceutical industries and food processing, building materials... Grain sizes range from micrometers for very fine powders (Castellanos 2005), verging on the colloidal realm, to meters for rockfill materials (Deluzarche and Cambou 2006), or even larger for some asteroids (Sánchez et al. 2017). As grains are brought into contact by confining external forces, possibly supplemented by mutual attraction, solid materials are formed, which may turn into liquidlike systems depending on the applied forces. Strongly agitated systems tend to form so-called granular gases (Jaeger et al. 1996).

---

J.-N. Roux (✉)
Laboratoire Navier, Université Paris-Est, Champs-sur-Marne, France
e-mail: jean-noel.roux@ifsttar.fr



S. Mesarovic et al. (eds.), *Mesoscale Models*, CISM International Centre for Mechanical Sciences 587, https://doi.org/10.1007/978-3-319-94186-8_4

At the continuum level the most sophisticated modeling attempts for the behavior of solid materials were proposed in the fields of geomechanics and geotechnique (Wood 1990; Mitchell and Soga 2005), in which quite elaborate experimental characterizations have also been exploited (Tatsuoka 2001; di Benedetto et al. 2003). The modeling of dense granular flows has known significant progress over the past 15 years (GDR MiDi 2004; Jop et al. 2006; Andreotti et al. 2013).

The micromechanics and micromorphology of granular assemblies have recently gained considerable interest (Radjaï et al. 2017), and attract nowadays a large research effort, for which the practice of the numerical simulation techniques known as "discrete element methods" (DEM), the granular analog of molecular dynamics for collections of molecules or atoms, is now an essential, widespread tool (Radjaï and Dubois 2011; O'Sullivan 2011).

The present contribution intends to supply a short, admittedly incomplete, review of some recent advances in the understanding of some essential features of granular mechanics, in connection with their microscopic origins, at the scale of solid grains and their interactions. It essentially relies on DEM results on model materials, with occasional illustrations in two-dimensional (2D) systems; most cited results, though, pertain to three-dimensional (3D) assemblies of spherical beads, with indications on the behavior of more general grain shapes, such as polyhedra.

Despite recent progress, our understanding of the rheophysics of granular materials is still considerably less advanced than in crystalline solids like metals and alloys. Describing the microscopic interactions, defining the material state, identifying such basic features of solid materials as an elastic range... all these essential operations carried out in the mechanical modeling of materials in relation to their elementary constituents are considerably more difficult with granular materials. Some of the most advanced attempts in micromechanics are not presented here. The aim of this chapter is, rather, to describe the landscape of current investigations, with some of its main landmarks, identifying constraints for future research and supplying possible guidelines.

It is organized as follows. First, Sect. 2 discusses the basis of mechanical modeling of granular materials, the contact interactions that derive from their geometric specificity, and some essential features of their macroscopic behavior that one wishes to relate to micromechanics. Section 3 introduces important tools needed to connect grain-level to macroscopic mechanics; among them statistical descriptors of material state, structural mechanics notions applied to contact networks. The following parts review some results on model systems that lead to a better understanding of the material state, as resulting from the assembling process (Sect. 4); of its properties as a solid material in quasistatic conditions (Sect. 5, dealing with the influence of the initial state, the role of elasticity, the "critical state" at large strains), and in dense flows (Sect. 6, about microstructure and constitutive modeling). Note that granular gases are not dealt with here. Section 7 is a brief conclusion.

# 2 Specificities of Granular Material Modeling

## *2.1 Contact Interactions and Discrete Degrees of Freedom*

One essential characteristic of granular materials, as dealt with in mechanical models, which enables their treatment by such numerical methods as DEM, involving a finite set of degrees of freedom, is that solid grains interact in contact regions which remain very small compared to their size. This is schematically illustrated in Fig. 1: solid objects only get deformed in regions close to the contacts. Outside those regions the material strains remain negligible. Those regions, shown as shaded areas in the figure, can safely be assumed to be very much smaller than the grains, and to be disjoint. The grains may thus be described as rigid, undeformable solid objects. They are merely sensitive to the total force and torque exerted on them by applied force fields (e.g., gravity) and other grains in contact with them. In Fig. 1, the force transmitted in the contact between grains 1 and 2 is only dependent on the motion of the same grains away from the contact region, where rigid body kinematics applies. It is, in particular, independent of the stress or strain fields within the region near the contact between grains 1 and 3. Denoting the grain diameter (or typical grain size) as $a$, contact regions, of typical (linear) size $l$, with $l \ll a$, may be dealt with as points. Interactions between a pair of contacting grains involve some surface force density over the contacting parts of their surfaces. The total force $\mathbf{F}$ and the total torque $\Gamma$ should be obtained on integrating this density. If the torque is evaluated at the centre of the contact region, then $||\Gamma||$ is of order $||\mathbf{F}||.l$, which may be neglected as the evaluation of global torques on grains involve terms of order $||\mathbf{F}||.a$. In most applications the local scale interaction law used in granular material modeling is a *contact law*, i.e. a relation between the motion of the pair of grains in contact and the *contact force*, regarded as a point force exerted on the surface of each grain.

One remarkable consequence of the scale separation between contact regions and grains is the validity of the *effective stress principle* (Mitchell and Soga 2005), for fluid-saturated granular materials. Considering (Fig. 2) a granular material with its pore space filled with a fluid at pressure $P_l$, the global stress tensor in such a

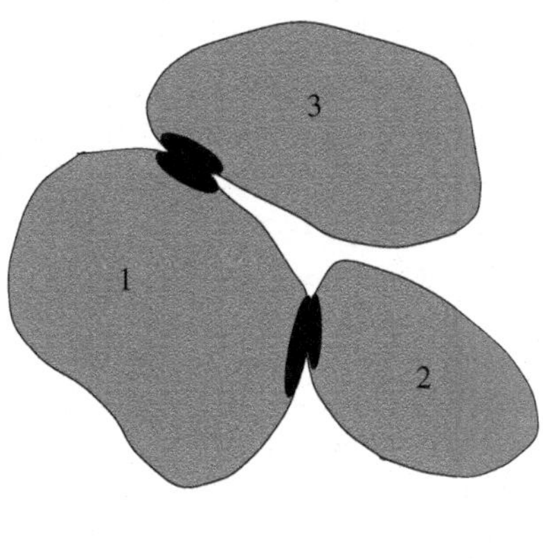


**Fig. 1** Three grains, with contacts between 1 and 2, and between 1 and 3. Material strains are only notable within darker contact regions (the size of which is exaggerated to ensure visibility on the picture)

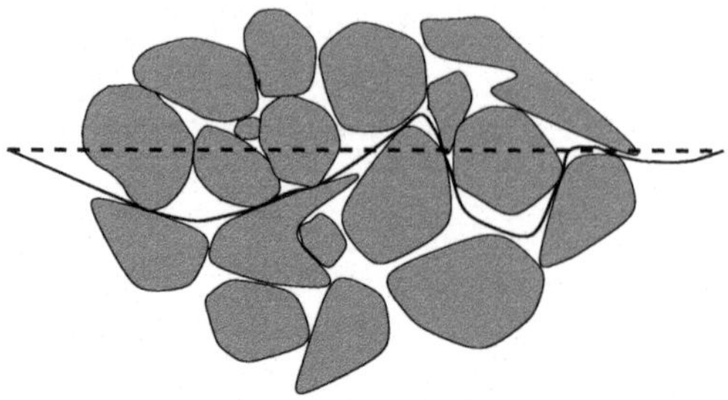

**Fig. 2** Two different surfaces through which stresses can be evaluated inside a saturated granular material, a flat one (dashed line) crossing through the grains, and another one (solid line), with hills and troughs, entirely comprised within pore space, except for very small contact regions

medium, assumed homogeneous and in equilibrium, may be deduced from the force density transmitted through a fictitious cutting surface. As illustrated by Fig. 2, this surface may be placed entirely within the interstitial fluid-filled space, except for very small intergranular contact regions. The stress tensor, $\underline{\underline{\sigma}}$, may thus be written as a sum:

$$\underline{\underline{\sigma}} = \underline{\underline{\sigma}}^{\text{cont}} + P_f \underline{\underline{\mathbf{1}}}, \tag{1}$$

in which $P_f$ is the fluid pressure and $\underline{\underline{\sigma}}^{\text{cont}}$ is the stress tensor associated with the contact forces (we adopt the soil mechanics convention, according to which compressive stresses are positive). As every grain is, in good approximation, entirely embedded within the fluid, the net force and torque due to the fluid pressure, which is uniform at equilibrium, vanish. Consequently, neglecting the compression of the solid grains caused by the surrounding pressure, the granular system with its contact network is in the same situation as in the absence of interstitial fluid, with contact forces balanced on each grain corresponding to stress $\underline{\underline{\sigma}}^{\text{cont}}$, as if this *effective stress* were applied to the dry material.

A counterexample of a system which should not, a priori, be dealt with as an ordinary granular material in the preceding sense, is illustrated in Fig. 3. Such objects with flat or conforming surfaces as shown in Fig. 3 may contact one another through a significant part of their periphery. Consequently the strains caused by contact stresses will extend through large domains inside the solid grains. Different contacts of the same grain may interfere with one another, as in the case (Fig. 3) of contact A-D, which should be affected, because of the Poisson effect, by the stresses in contacts A-B and B-C. Such systems should in principle require continuum mechanics boundary value problems to be dealt with in each solid grain, as opposed to granular systems in the sense of Fig. 1, for which the contact mechanics treatment of interactions usually regards each grain of a given pair as a (semi-infinite) half space. Obviously, the arguments used in connection with Fig. 2 justifying the effective stress principle in saturated granular materials do not necessarily apply in this case.

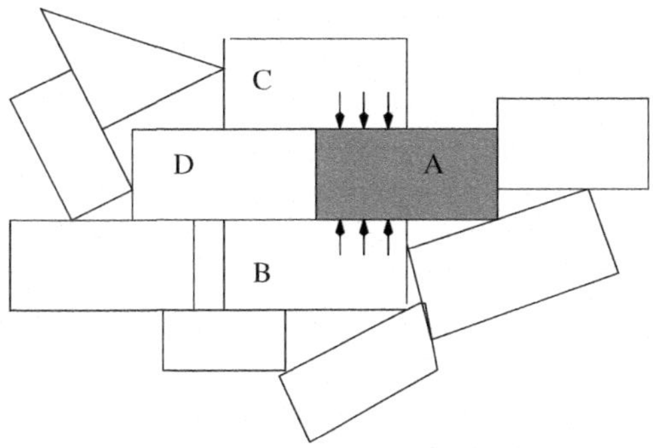


**Fig. 3** Brick-shaped grains may contact their neighbours through a notable part of their surface

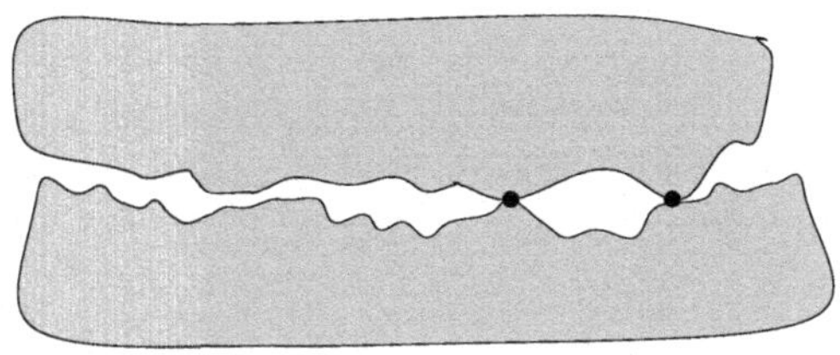

**Fig. 4** Schematic blown-up view of a contact between two nominally flat grain surfaces, showing their small scale roughness. Contact actually takes place via isolated points

However, apparently conforming surfaces, or flat ones, enabling large contact areas, may not actually exist in the presence of smaller scale roughness, as sketched in Fig. 4. In such a case, the granular modelling approach may be salvaged, because the strain will tend to be confined to a thin region, of a thickness similar to the size of the asperities. Note also that the effective stress principle might apply, since the saturating liquid will invade the interstitial region, save for the isolated asperities in contact. The practice of DEM modeling of polygonal grains (Azéma et al. 2012, 2013) may be justified thanks to this implicit assumption of roughness and contacts through isolated nearly punctual regions.

## 2.2 *Contact Laws*

From the separation of scales and the resulting possibility of a formulation of granular mechanics with the discrete degrees of freedom associated with a collection of rigid objects, as discussed in Sect. 2.1, the *contact law* is an essential input of a granular model at the microscopic scale. In general, contact behaviors differs significantly from bulk material behaviors, being influenced by the fine details of the surface geometry—such a basic property as the intergranular friction angle is not a property of the material which the grains are made of.

Contact mechanics is a complicated field, because local problems to be dealt with in the contact region between two grains involve boundary conditions on an a priori unknown part of the surfaces, where both objects contact each other. The treatise by Johnson (1985) deals with many aspects of contact mechanics, but some difficulties related to surface roughness are hardly addressed. The presentation given here evokes the most consequential (for granular mechanics) features of contact mechanics and mainly deals with frequently used simplified models, which hopefully contain the most important ingredients.

Fortunately, it is usually observed that the global, collective behavior of granular assemblies is not sensitively dependent on many features of the contact law. The relevance of contact models is often only evaluated a posteriori, on comparing numerical results to laboratory measurements on similar systems.

### 2.2.1 Friction

Coulomb friction is the most important mechanical property of intergranular contacts. While tribologists keep investigating realistic, complex, history-dependent models involving "third bodies" in the contact region (Richard et al. 2007), in the presence of a large number of grains and contacts as in a granular material sample, it is a common practice to stick to the simplest model for friction, the Coulomb model, with a constant friction coefficient $\mu$ in all contacts between similar bodies. The physical origin of friction is generally accepted to reside in the plasticity of small asperities through which the two bodies are contacting each other, because of their small scale roughness, despite the apparent smooth surfaces on the scale of the grain diameter or radius of curvature (Bowden and Tabor 1950). As the stress in such asperities coincides then with the plastic threshold $\sigma_c$ in normal indentation, the real contact area $A_c$ has to increase proportionally to the total normal force $F_N = A_c\sigma_c$ transmitted in the contact region. The plastic shear resistance $\tau_c$ then yields the total tangential force $F_T$ in sliding as $F_T = A_c\tau_c$, whence the Coulomb friction coefficient $\mu = \tau_c/\sigma_c$, independent of the normal load and of the apparent area of the contact. Many additional sophistications, involving aging and/or dynamic instabilities may affect the frictional behavior (Baumberger and Caroli 2006), but the simplest friction as defined by

$$\begin{aligned} ||\mathbf{F}_T|| &< \mu F_N \qquad \text{no sliding} \\ ||\mathbf{F}_T|| &= \mu F_N \qquad \text{sliding,} \end{aligned} \tag{2}$$

is most often deemed sufficient to provide a satisfactory description of granular material behavior. Friction is said to be fully mobilized in the contact in the second case (equality).

### 2.2.2 Elasticity

The elastic response of a contact region is of course strongly dependent on its shape, and differs for sharp edges, pointed corners or smooth surfaces with a well-defined curvature. In the latter case, for grains made of an elastic material with Young modulus $E$ and Poisson ratio $\nu$ the normal elastic force $F_N$ in the contact relates to the normal deflection, $h$, by the Hertz law, which reads for spherical beads of diameter $a$:

$$F_N = \frac{\tilde{E}}{3} a^{1/2} h^{3/2}, \tag{3}$$

in which notation $\tilde{E} = E/(1-\nu^2)$ is adopted. The non-linearity stems from the growth with $h$ of the contact region, a disk with radius $b$ proportional to $\sqrt{ah}$. Johnson (1985) gives a detailed derivation of (3) and other contact laws, based on the assumption, adequate for small enough deflections, that each contacting object might be locally dealt with as an infinite half space. Relation (3) is easily generalized (Johnson 1985) to beads of different diameters (just use $a = \frac{2a_1a_2}{a_1+a_2}$ for diameters $a_1, a_2$), or to objects with two different (positive) radii of curvature. Note that exponent $3/2$ may be obtained from a simple scaling argument: first, $b \propto \sqrt{ah}$ may be deduced from the contact geometry (the contact region radius being of the order of the "interpenetrated region" of non-deformed spherical balls); then a strain of order $h/b$ is assumed to be distributed over a volume of order $b^3$, whence an elastic energy scaling as $Ea^{1/2}h^{5/2}$ and a force scaling as $Ea^{1/2}h^{3/2}$. Other contact shapes could result in different forms of normal elastic forces, e.g. scaling as $h^2$ for a sharp angular edge (Johnson 1985). The Hertz law (3) does not apply for sands, but proves quite robust as model materials made of beads or smooth shapes are tested, for which elastic moduli might be measured (Jia et al. 1999; Kuwano and Jardine 2002). Anticipating that elastic contact deflections play a minor role in granular material mechanics, contact elasticity is also often assumed linear in DEM calculations: a constant stiffness $K_N = \frac{dF_N}{dh}$ (possibly dependent on particle radii) is assumed to relate force and deflection as $F_N = K_N h$, instead of the Hertz law implying

$$K_N = \frac{\tilde{E}}{2} a^{1/2} h^{1/2} = \frac{3^{1/3}}{2} a^{1/3} \tilde{E}^{2/3} F_N^{1/3}. \tag{4}$$

Such a simplification is adopted with the assumption that a certain limit of rigid grains, in which contact deflections are irrelevant, is approached.

The normal law relating $F_N$ and $h$ should be supplemented by a tangential contact law relating the variations of the tangential component of the contact force, $\mathbf{F}_T$, to the variation of tangential relative displacements. The modeling of such laws is quite complicated, because of the interplay of elasticity and friction. Friction is modeled locally, as a condition similar to (2) applying to the force density, i.e. to stress vector $\mathbf{T} = \underline{\underline{\sigma}} \cdot \mathbf{n}$ within the contact surface, $\mathbf{n}$ denoting the normal unit vector. The situation of two spherical objects pressed against each other by a *constant* normal force $F_N$, the contact being subjected to a *varying* relative tangential displacement $\delta U_T$, as investigated by Mindlin and Deresiewicz (1953) reveals a history-dependent distribution of sliding regions within the contact surface. As $\delta U_T$ increases from zero, sliding (relative tangential motion, where $||\mathbf{T}_T|| = \mu T_N$) takes place in an outer annulus of increasing width (Fig. 5), until all the contact may slide as the Coulomb equality applies everywhere to force density **T** and, consequently, to the global contact force. However, upon reversing the direction of the relative tangential motion, before this global sliding threshold is reached, a second annulus appears at the periphery of the contact region, within which sliding takes place in the opposite direction (Fig. 6). Accurate modeling of contact elasticity in a granular sample with many contacts thus seems to become almost hopelessly complicated, as records of all past changes in relative motion directions in the contacts should apparently be kept. This is not, however, the whole story: one should be able to predict the variations of contact forces for arbitrary relative motions of contacting

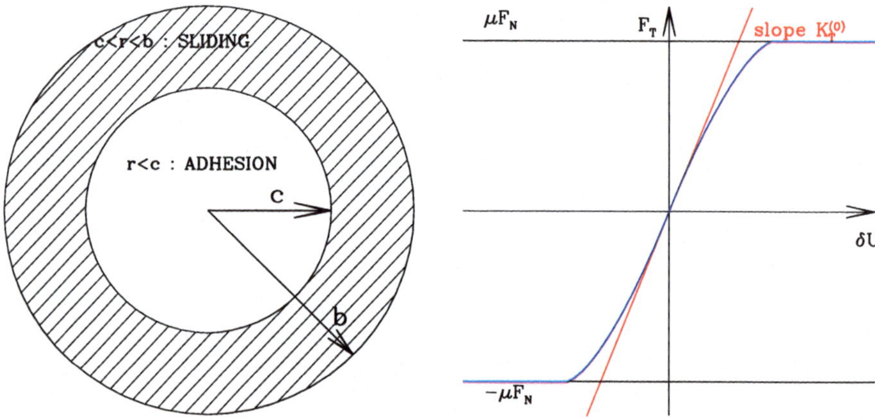


**Fig. 5** Left: initiation of a slip annulus at the periphery of the contact region as relative tangential displacement $\delta U_T$ is imposed. Right: tangential force $F_T$ versus $\delta U_T$

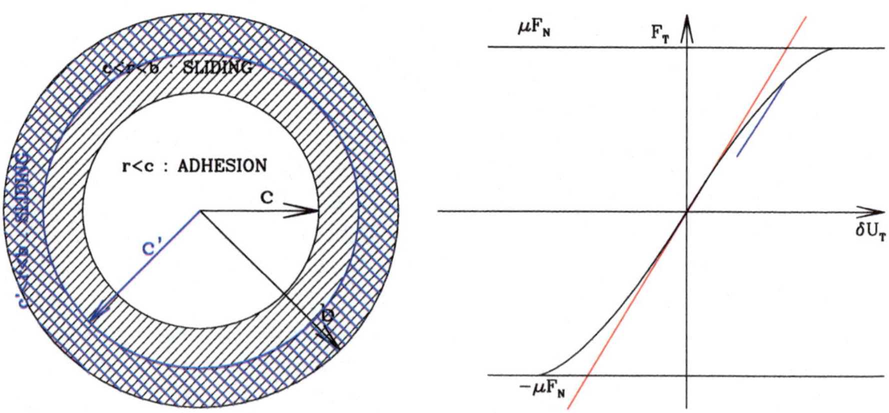


**Fig. 6** Left: appearance of a second slip annulus at the periphery as relative tangential displacement $\delta U_T$ starts to decrease. Right: tangential force $F_T$ versus $\delta U_T$ (note beginning of unloading path)

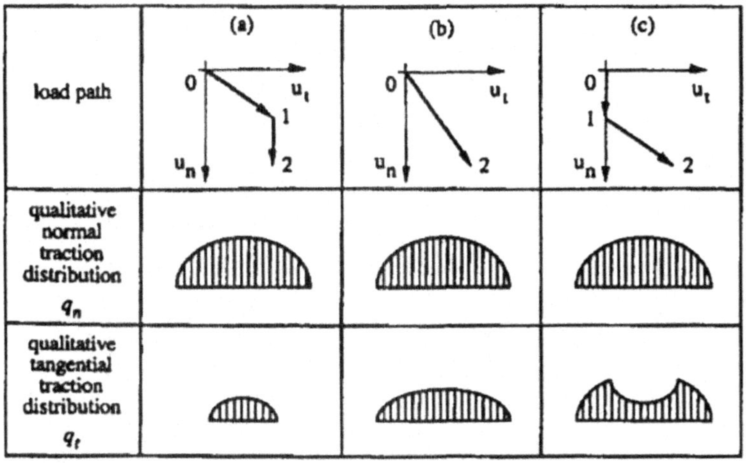


**Fig. 7** Different paths leading to the same final values of relative displacements (with notations $u_n$ for $h$ and $u_t$ for $\delta U_T$), normal ($q_n = T_N$) and tangential ($q_t = T_T$) force densities across disk-shaped contact region

grains. Elata and Berryman (1996) pointed out that, even if the Coulomb threshold is reached nowhere in the contact (as in the hypothetical situation of an infinite friction coefficient $\mu$) different stress and force density patterns in the contact region, and a different global tangential contact force $F_T$, may be obtained for the same values of deflection $h$ and relative tangential displacement $\delta U_T$, depending on the past history of those variables. This is visualized in Fig. 7, reproduced from this paper. The results of three different contact loading paths are shown, all ending at the same point in plane ($u_n = h, u_t = \delta U_T$): (a) increase $u_n$ and $u_t$ proportionally (1), along a straight line, so that $u_t$ reaches its final value, but not yet for $u_n$, then increase only $u_n$ (2); (b) increase both $u_n$ and $u_t$ proportionally to their final values; (c) proceed as along path (a), but reversing the roles of $u_n$ and $u_t$. Although the normal force density is the same, tangential force distributions are path-dependent. Elata and Berryman (1996) show this surprising conclusion to hold even in the absence of friction mobilization (i.e. in the limit of very large friction coefficient $\mu$). Strictly speaking, the contact never behaves elastically, even in the absence of friction effects.

Facing such difficulties, a widely accepted approximation consists in keeping a tangential stiffness $K_T = \dfrac{dF_T}{d(\delta U_T)}$ independent of $\delta U_T$, and coinciding with its value $K_T^{(0)}$ for $\delta U_T = 0$:

$$K_T = K_T^{(0)} = \frac{2-2\nu}{2-\nu} K_N = \alpha_T K_N = \frac{2-2\nu}{2-\nu} \frac{\tilde{E}}{2} a^{1/2} h^{1/2}, \tag{5}$$

for identical beads of diameter $a$—thus a function of deflection $h$ or normal elastic force $F_N$. Specific caution should however be exercised, requesting a suitable rescaling of $K_T$ when $h$ decreases—as suggested by Elata and Berryman (1996).

### 2.2.3 General Relative Motion, Objectivity Issues

Furthermore, it should be specified how the contact force evolves as the pair of grains move with arbitrary combinations of translations and rotations while maintaining the contact. While there appears to be no exact general solution, based on the detailed treatment of the problem of the contact between moving objects, in the available literature, adopted solutions should abide by the objectivity principle (Kuhn and Chang 2006), i.e., be such that if both contacting grains move as one non-deforming solid, then the contact force should follow this rigid body motion. Such a solution is described, e.g., by Agnolin and Roux (2007a).

### 2.2.4 Adhesion

All identical particles are attracted to one another by surface forces of different origins (Israelashvili 1991; Maugis 2000), among which van der Waals ones are the most universal type. Such attractive interactions introduce a characteristic force scale $F_0$ and a range, some length $D_0$ (on the nanometric scale for van der Waals interactions). This corresponds to some adhesion energy of order $F_0 D_0$. For ideal spherical particles of diameter $a$, $F_0$ is of order $\Gamma a$, $\Gamma$ denoting the interfacial energy of the grain surface. Comparing $F_0$ to other forces in a granular material (say, gravity, scaling as $a^3$ or applied pressure, resulting in contact forces scaling as $a^2$), it is usually observed that adhesive forces are quite negligible for grains sizes above the $10\,\mu\text{m}$ range—this is the reason why cohesion effects are notable in fine powders, such as flour, or fine soils like silts, but not in sands. Furthermore, adhesive forces, given their extremely small range, are most frequently dominated by roughness effects, and turn out to be of order $\Gamma d$, with $d \ll a$ corresponding to some asperity size (Castellanos 2005). Exact calculations combining adhesive forces and Hertz elasticity are available—in the so-called Johnson-Kendall-Roberts (JKR) and Deriaguin-Muller-Toporov (DMT) cases (Maugis 2000), corresponding to small or large values of the ratio (sometimes termed Tabor number, with definitions varying by factors of order 1) of elastic deflection under force $F_0$ to range $D_0$. Interpolation schemes (Castellanos 2005) are available in intermediate cases. In practice such ideal calculations do not quantitatively apply to experimental systems in which grains have irregular surfaces, and, in order to simulate large collections of grains, simplified models are adopted, introducing a force law with some maximum attraction $F_0$ and some range $D_0$.

One situation in which an accurate model might be used is that of wet grains, as liquid bridges form at contacts, or join pairs separated by a small distance, as sketched in Fig. 8. The contact law should then be replaced by an interaction law in which the grains are attracted to each other by a capillary force, which can be computed from Laplace's law (Lian et al. 1993; Pitois et al. 2000) (which states that the pressure within the liquid is lower than the external one, with difference $\gamma(\frac{1}{r_1} - \frac{1}{r_2})$, using the notations of Fig. 8, $\gamma$ being the surface tension). Assuming simple rules might be identified to identify the spatial distribution of the liquid, such interactions can be dealt with in DEM (Radjaï and Richefeu 2009a; Khamseh et al. 2015). The maximum attractive force $F_0$ is observed for small bridge volumes (filling angle $\phi \ll 1$ on the figure), at contact ($h = 0$), is then, for wetting angle $\theta$ and bead radius $R = a/2$:

$$F_0 = 2\pi R\gamma \cos\theta. \tag{6}$$

In the presence of attractive forces, which act at a small distance, the Coulomb condition characterizing the contact surface behavior applies to the repulsive elastic component of the normal force. Figure 9, from a DEM study of a model system by Gilabert et al. (2007), displays the Coulomb cone within which the point of coordinates $N_{ij}$, $T_{ij}$ (the normal and tangential force components in contact between grains $i$ and $j$) should remain. $N_{ij}$ is the sum of the adhesive contribution

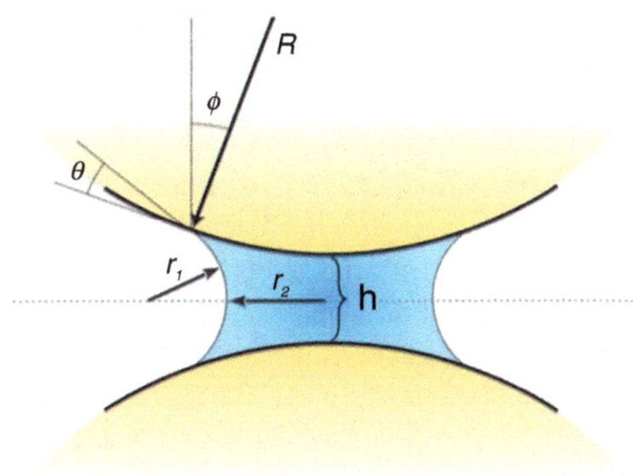


**Fig. 8** Liquid bridge joining two identical spherical beads

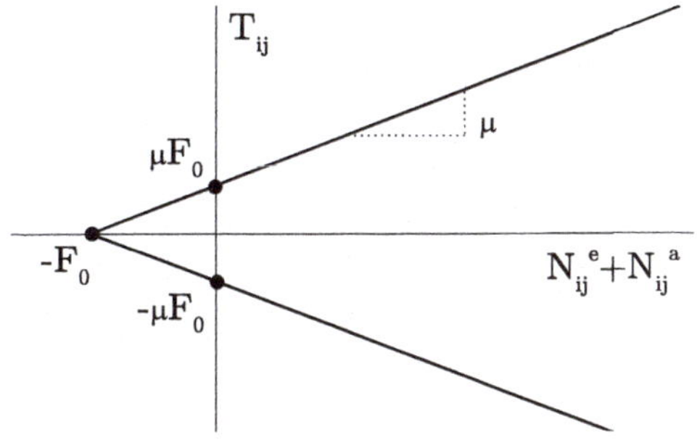


**Fig. 9** Coulomb condition in the presence of attractive force $-F_0$ in contact between grains $i$ and $j$

$N^a_{ij} = -F_0$ and the repulsive elastic one $N^e_{ij}$. The tip is no longer at the origin of coordinates. The tangential force may reach intensities as large as $\mu F_0$ when the total tangential force vanishes because the elastic repulsion compensates attraction $-F_0$.

### 2.2.5 Resistance to Rolling

It might be necessary in some cases to account for a finite lateral extension $l$ of the contact region (yet small compared to diameter $a$), which causes a local torque $\mathbf{\Gamma}$. Physical motivations of existing models for *rolling and pivoting resistance* should involve surface asperities such that grain pairs interact through several contact points separated by distance $l$. A *rolling friction* coefficient, $\mu_R$, and a *pivoting friction* coefficient $\mu_P$ are often introduced, such that inequalities analogous to (2) apply to normal and tangential components of $\mathbf{\Gamma}$, as follows.

$$\begin{aligned} |\Gamma_N| &\leq \mu_P F_N \qquad \text{(pivoting)} \\ ||\mathbf{\Gamma}_T|| &\leq \mu_R F_N \qquad \text{(rolling).} \end{aligned} \tag{7}$$

By definition, both coefficients $\mu_R$ and $\mu_P$ have the dimension of a length. They should be of order $l$. Below the threshold, torque components should vary more or less elastically with the relative rotation of both objects $i$ and $j$, i.e., $\mathbf{n}_{ij}$ denoting the unit normal vector, for small rotation vectors $\vec{\omega}_i$ and $\vec{\omega}_j$, $\Gamma_N$ relates to the pivoting angle, $\mathbf{n}_{ij} \cdot (\vec{\omega}_i - \vec{\omega}_j)$, while $\mathbf{\Gamma}_T$ is linked to the rolling motion, involving the tangential component $(\underline{\underline{\mathbf{1}}} - \mathbf{n}_{ij} \otimes \mathbf{n}_{ij}) \cdot (\vec{\omega}_i - \vec{\omega}_j)$. This model of rolling and pivoting friction is thus analogous to the classical model of sliding friction.

The implementation of rolling resistance with spherical or circular grains can be deemed analogous to the modeling of non-spherical shapes, especially angular ones, for which contacts may extend to parts of faces or edges on the grain periphery. Such a correspondence has been proposed in quantitative form for two-dimensional grains by Estrada et al. (2011).

In the presence of adhesive forces, analogously to the Coulomb condition for the tangential force, inequalities (7) apply with the sole elastic repulsive component of normal force $F_N$ in the right-hand side. Thus, in contacts where the total normal force vanishes, a finite torque $\mu_R F_0$ ($-F_0$ denoting the adhesive force as before), or $\mu_P F_0$ in the normal direction, might be transmitted. This has important consequences on the possible morphologies of assemblies of adhesive grains, as illustrated in Fig. 10, from Gilabert et al. (2007). These disk-shapes grains are assembled here under small external stress, and form rigid structures in equilibrium. Some resistance to rolling is necessary for the single particle strands to remain stable and rigidly transmit forces. In the absence of rolling friction, thin "arms" (rigid elongated structures) are made of at least two grains in the transverse direction.

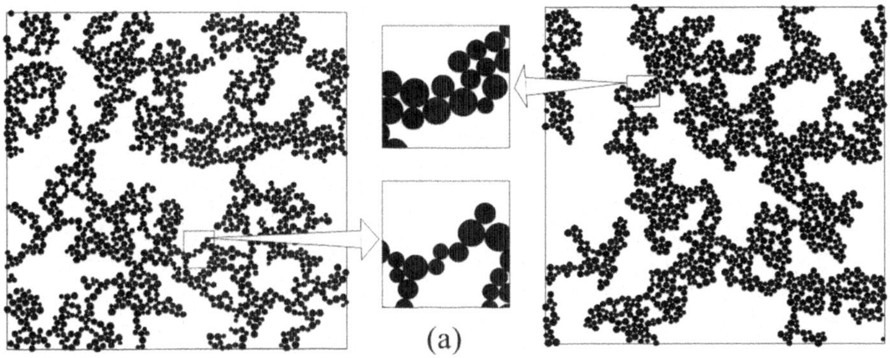


**Fig. 10** Aspect of cohesive grain clusters in a simple 2D model, with (left) and without (right) rolling friction. Blown-up details (**a**) show elongated parts joining denser regions (Gilabert et al. 2007)

### 2.2.6 Viscoelastic or Plastic Dissipation in Collisions

In addition to elastic and frictional forces (and, possibly, to adhesive ones), grains in contact exert onto one another viscous, dissipative force depending on their relative velocity. Typically, one writes normal and tangential forces proportional to corresponding relative velocities, and opposing the relative motion:

$$
\begin{aligned}
F_N^v &= \alpha_N \frac{dh}{dt} \\
\mathbf{F}_T^v &= \alpha_T \frac{d(\delta U_T)}{dt},
\end{aligned}
\tag{8}
$$

with damping coefficients $\alpha_N$, $\alpha_T$, possibly dependent on relative displacements $h$, $\delta U_T$ (or the elastic part of the latter). One physical origin of such viscous forces is the viscoelasticity of the grains (Ramirez et al. 1999; Brilliantov and Pöschel 2004), which always exists on small time scales. In practice, the choice of a damping model seldom relies on a true physical analysis of the microscopic origins of viscous dissipation. Damping coefficients are related to coefficients of restitution in binary collisions (Maw et al. 1976), which in general (and in particular in the case of viscoelastic effects in Hertzian contacts) depend on the initial relative velocity. The normal coefficient of restitution, $e_N$ is defined by the ratio, equal to $-e_N$, of the receding relative velocity in the normal direction after the collision to the approaching relative velocity in the normal direction before the collision. Its counterpart $e_T$ is defined analogously with tangential relative velocities. In the simplest case of linear unilateral elasticity, without adhesive forces, a constant coefficient $\alpha_N$ corresponds to a constant normal coefficient of restitution $e_N$, which is readily obtained on solving for the motion of a damped linear oscillator. For a pair of identical beads of mass $m$, the "critical" value of $\alpha_N$, separating the oscillating from the overdamped regimes is $\alpha_N^c = \sqrt{2mK_N}$ and $e_N$ depends on ratio $\zeta = \alpha_N/\alpha_N^c$, assumed below 1, as

$$e_N = \exp\left[\frac{\pi\zeta}{2\sqrt{1-\zeta^2}}\right]. \tag{9}$$

With Hertzian contact elasticity, viscous damping may be chosen in reference to the critical damping of a linear contact with stiffness $K_N$ equal to the instantaneous deflection-dependent value. Such a choice results in a velocity-independent coefficient of restitution.

## 2.3 Collective Behavior and Contact Behavior

The contact behavior may be represented as a combination of rheological elements, as shown in Fig. 11, (for a contact without adhesive forces, no resistance to rolling or pivoting, and no viscous force in the tangential direction). In an upscaling procedure, leading to macroscopic behavior, the characteristics of those elements, as well as the geometry of the grains and the contact network, are the microscopic inputs. We now recall some basic aspects of macroscopic granular mechanics and confront them to microscopic features, to get some insights on the feasibility of such an upscaling procedure and the relative importance of the different inputs.

### 2.3.1 Classical Quasistatic Stress-Strain Behavior: Triaxial Compression

The quasistatic stress-strain behavior of cohesionless granular materials is commonly measured in tests as triaxial compression, as schematized in Fig. 12. Starting from an equilibrated pack under isotropic stresses, one principal stress, $\sigma_1$ gradually increases, along with the conjugate strain $\epsilon_1$, while the lateral stresses, and thus the other two principal stresses, $\sigma_2 = \sigma_3$, are maintained fixed. Results are

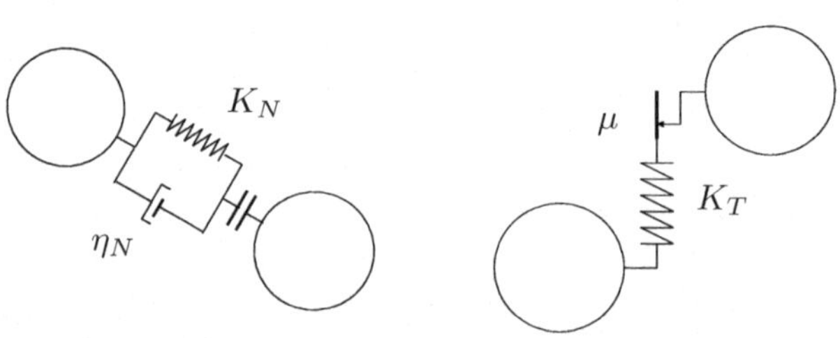


**Fig. 11** Rheological elements schematizing contact behavior: normal force (left) combines a spring (possibly nonlinear, with $K_N$ depending on elongation), a viscous dashpot, and a no-tension joint (no adhesion here). The tangential law (right) involves a plastic slider

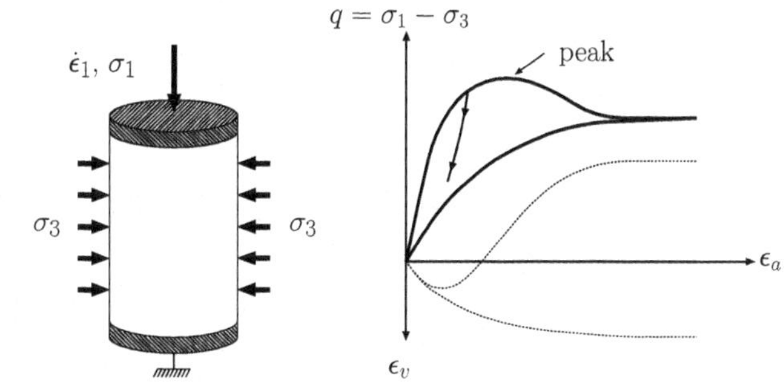


**Fig. 12** Left: schematic view of an axisymmetric triaxial compression test. Right: typical results in dense and loose systems. Deviator stress $q$ (solid lines) and volumetric strain $\epsilon_v$ (dotted lines) versus axial strain $\epsilon_1$. Note the deviator peak in the dense case

traditionally expressed in terms of deviator stress $q = \sigma_1 - \sigma_3$ and volumetric strain $\epsilon_v = 1 - (1 - \epsilon_1)(1 - \epsilon_3)(1 - \epsilon_3)$ plotted versus *axial strain* $\epsilon_a = \epsilon_1$. These curves differ according to the solid fraction $\Phi$ in the initial isotropically loaded state. $q$ steadily increases to an asymptotic value $q_c$ at large axial strain in the loose case, while it first passes through a maximum (the "deviator peak") in the dense case, before decreasing to the same large strain plateau value $q_c$, in initially dense systems. Meanwhile, loose systems contract, and dense systems, after some initial contraction, dilate, until some asymptotic value $\Phi_c$ of solid fraction is approached, which turns out to be the same whatever the initial density. A loose initial state is therefore defined by $\Phi < \Phi_c$, a dense one by $\Phi > \Phi_c$. This large strain state reached after the material has been monotonically deformed in the same direction is known as the *critical state* (Wood 1990)—we shall return, in Sects. 5 and 6, to the important role of the critical state in granular material rheology.

The maximum value of $q$ (either at the peak or at the final plateau) is associated to the internal friction coefficient $\varphi$ of the material as:

$$\frac{q}{\sigma_3} = \frac{2 \sin \varphi}{1 - \sin \varphi}. \tag{10}$$

A familiar notion, the internal friction angle should coincide with the maximum slope angle of a free surface of the material under gravity or angle of repose (Nedderman 1992), provided the plastic failure associated with the deviator peak (dense case) or the critical plateau (loose case) satisfies the Coulomb criterion—the yield criterion given as a function of principal stresses $\sigma_1 \geq \sigma_2 \geq \sigma_3 \geq 0$ by

$$f(\underline{\underline{\sigma}}) = \frac{\sigma_1 - \sigma_3}{2} - \frac{\sigma_1 + \sigma_3}{2} \sin \varphi \leq 0 \tag{11}$$

In practice, the yield properties of simple materials, such as sands or assemblies of spherical beads (Suiker and Fleck 2004; Peyneau and Roux 2008b), are better described by other forms of the yield criterion (Lade and Duncan 1975), and the apparent value of $\varphi$ as identified from (10) might be slightly different from the one applicable to a shear test (which is more directly the angle of repose).

### 2.3.2 (Over)Simplified Laws, Similarity with Contact Behavior

For simplicity (e.g., in engineering practice, for lack of very detailed information on a certain granular soil) the behavior as measured in a triaxial compression test of a cohesionless dense granular material might be described as shown in Fig. 13, assuming linear isotropic elasticity to apply up to the deviator peak, with a macroscopic Young modulus $E$ and Poisson ratio $\nu$, and then a constant slope (or *dilatancy*) $-\frac{d\varepsilon_v}{d\varepsilon_a}$, as the maximum deviator (confused here with the final plateau) is reached.

This is of course quite a gross simplification, unlikely to accurately describe a situation in which the softening after the peak (shown in Fig. 12, ignored in the simplified version of Fig. 13) plays an important role. Furthermore, the prepeak behavior is not elastic (as indicated in the sketch of an unloading curve in Fig. 12).

But, at first sight, the first graph of Fig. 13 is quite similar to the simplified version of the tangential contact law of Fig. 5, in which the slope is taken constant (equal to $K_T^{(0)}$ in Eq. (5)) and variations are assumed elastic before the Coulomb threshold $\mu F_N$ is reached. Both describe similar elastoplastic behaviors, and the analogy is strengthened by the use of a common vocabulary ("Coulomb threshold", "friction"...). Should one regard the increase of $q$ with $\varepsilon_a$ as a macroscopic consequence of the contact behavior and internal friction as a reflection of intergranular friction? As shown below, such a naïve view is however misleading.

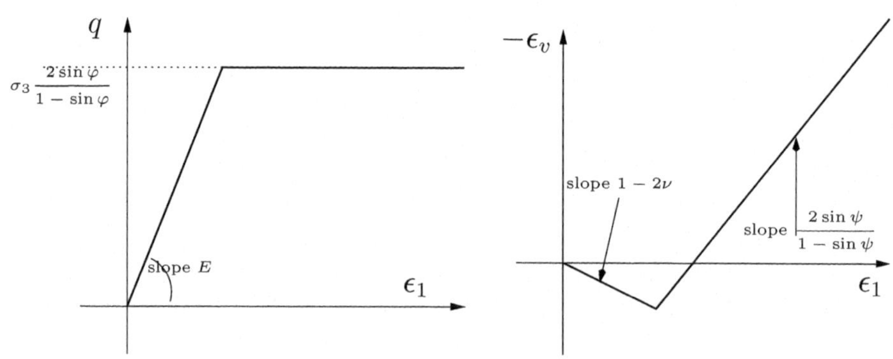


**Fig. 13** Simplified elastoplastic law for triaxial compression: deviator stress (left) and volumetric strain (right) versus axial strain

### 2.3.3 Dilatancy

One qualitative difference between microscopic and macroscopic behaviors is the existence of macroscopic dilatancy (or contractancy for loose systems). Dilatancy may be described as a flow rule, which for the volumetric strain behavior shown in Fig. 13 is associated with plastic potential $g(\underline{\underline{\sigma}})$ written below (i.e., the plastic strains are given by $\underline{\underline{\varepsilon}}^P = \lambda \frac{\partial g(\underline{\underline{\sigma}})}{\partial \underline{\underline{\sigma}}}$ with $\lambda > 0$):

$$g(\underline{\underline{\sigma}}) = \frac{\sigma_1 - \sigma_3}{2} - \frac{\sigma_1 + \sigma_3}{2} \sin \psi, \tag{12}$$

where a dilatancy angle, $\psi$, is introduced.

After Reynolds (1885), who coined the word "dilatancy", this property is usually interpreted as the result of deformation mechanisms on the scale of local grain arrangements (Goddard and Didwania 1998), in which grains or rows of grains slide onto one another, as depicted in an elementary 2D example in Fig. 14. Although rather simplistic, such an approach stresses the collective origins of dilatancy, as an effect of steric hindrance in relative grain motion. It will be questioned below in Sect. 5.

### 2.3.4 Granular Disorder: Forces and Displacements

Attempts at averaging the local behavior to obtain macroscopic laws are confronted with the characteristic disorder of granular materials. As shown in Fig. 15, force patterns comprise typical alignments of strongly loaded contacts (the *force chains*), while some regions (involving 5–10 grains in this case) carry little stress. Some grains (the *rattlers*) are not involved in the force-carrying contact network and are left free to move in the "cage" formed by their load-carrying neighbours. Many studies have been devoted to the statistical distribution of force values (Coppersmith et al. 1996; Radjaï 2015). The probability distribution function often decreases exponentially for large values, but forces, say, four times as large as the average represent a small, but notable fraction of the total number (say, of order $10^{-3}$). The role and the persistence of the stronger force chains, carrying stress anisotropy,

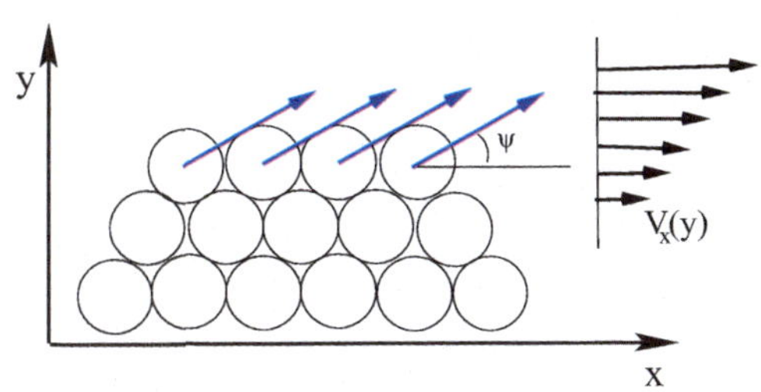


**Fig. 14** A simple example of alleged mechanism for dilatancy with 2D rows of disks sliding past one another in shear flow

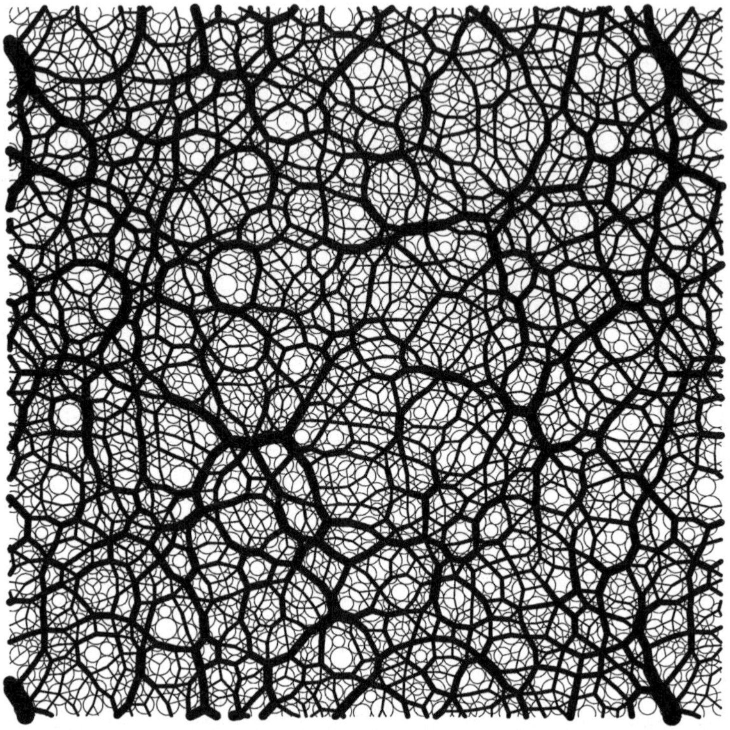

**Fig. 15** Equilibrium contact forces, balancing externally applied isotropic pressure, in 2D sample of disks. Stroke thickness is proportional to normal force intensity

while smaller forces prevent them from buckling, was also discussed (Radjaï et al. 1998). As to displacements, Fig. 16 evidences equally disordered patterns (Kuhn 1999; Roux and Combe 2002; Radjaï and Roux 2002), involving many vortex structures of large scale, comparable to the sample size. The figure actually shows the "non-affine part" of the displacements, i.e., for each grain $i$, the displacement $\mathbf{u}_i$ of its centre, positioned in $\mathbf{r}_i$, minus its value in a homogeneous continuum subjected to the same (small) strain $\underline{\underline{\epsilon}}$, i.e., for a certain choice of the origin and in the absence of macroscopic rotation (the sign being due to our convention that strains are positive for shrinking lengths):

$$\tilde{\mathbf{u}}_i = \mathbf{u}_i + \underline{\underline{\epsilon}} \cdot \mathbf{r}_i. \tag{13}$$

Figure 16 corresponds to $\varepsilon_1$, $\varepsilon_2$ of order $10^{-3}$, while $\epsilon_{12} = 0$. The importance of non-affine displacements might be assessed on evaluating, in a sample of $N$ grains

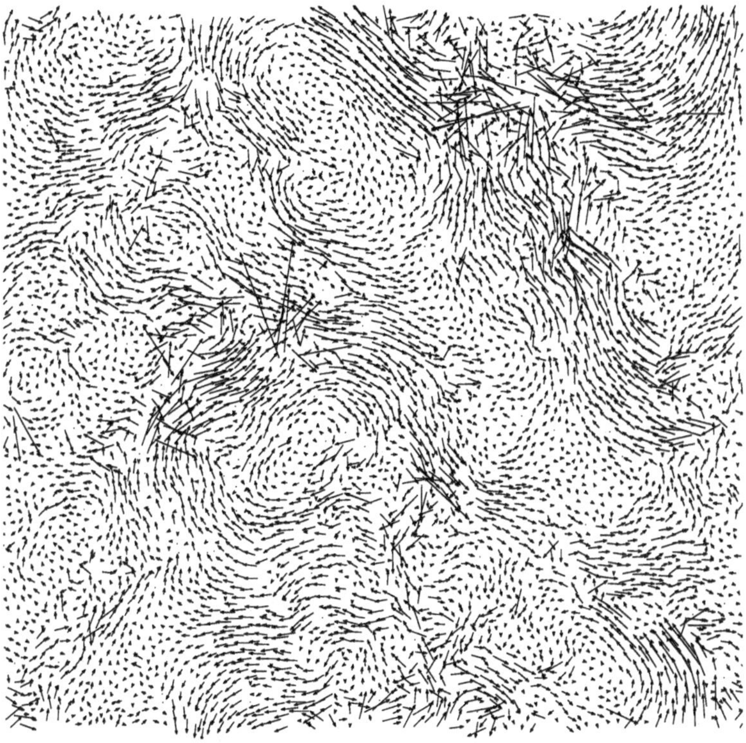

**Fig. 16** Non-affine part of grain displacements (arbitrary scale) corresponding to strain interval $10^{-3}$ in 2D disk assembly in biaxial compression

of average diameter $a$ the following ratio:

$$\Delta = \frac{1}{Na^2||\underline{\underline{\epsilon}}||^2} \sum_{i=1}^{N} ||\tilde{\mathbf{u}}_i||^2. \tag{14}$$

Values of $\Delta$ of order 10 or 100 are quite common.

### 2.3.5 The Role of Geometric Rearrangements of Contact Networks

Since the contact laws rule the mechanical properties at the grain scale, it may be tempting to expect that the macroscopic behavior could be retrieved on suitably averaging the contact behavior, as in homogenization approaches to the macroscopic properties of microscopically heterogeneous materials (Nemat-Nasser and Hori 1993). Can one regard the granular sample as a network of rheological elements as shown in Fig. 11? The following simple example (Roux 2000) shows that the deformation corresponding to certain changes in the applied load may not result

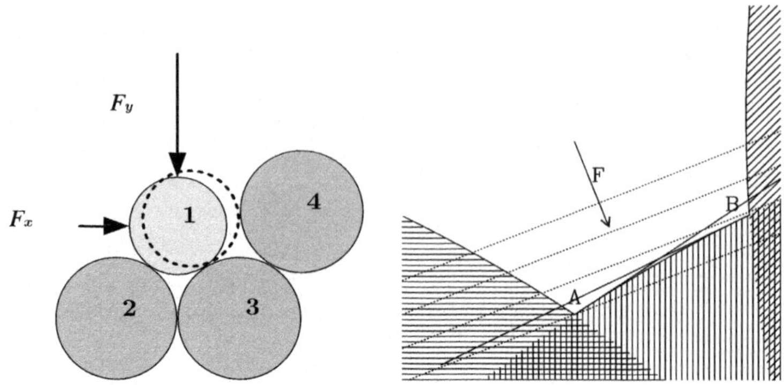


**Fig. 17** Left: simple model system. One mobile disk (1), in contact with fixed ones (2, 3 and 4), and subjected to external force with components $F_x$, $F_y$. All disks are rigid and frictionless. Two possible equilibrium positions: light grey disk, dashed contour. Right: equipotentials (dotted lines) and regions of plane forbidden to centre of disk 1 by steric exclusion (hashed zones). A and B mark equilibrium positions with two contacts

from contact mechanics averaged on a larger scale. Consider the set of four rigid frictionless disks shown in Fig. 17 (left graph), with only one mobile grain (disk 1), subjected to an external force. Depending on the orientation of **F**, disk 1 might find an equilibrium position (grey disk) with contacts with disks 2 and 3, or another one in contact with 3 and 4 (disk outlined with dashed perimeter). With no friction and no contact elasticity, the potential energy $W = -F_X.x - F_Y.y$, constant on lines orthogonal to vector **F**, is to be minimized at equilibrium, under the constraint that disks do not overlap.These are the two possible equilibrium positions marked A and B (right graph). As the direction of **F** gradually changes, the disk will move from one equilibrium position to the other, as soon as the steric exclusion constraints enable a motion with decreasing $W$. The concavity of the boundary of the accessible region for coordinates $x$, $y$ (the intersection of the exterior parts of circles) entails that the relation between $Q = F_X/F_Y$ and the equilibrium position $x$ of mobile disk 1 takes the form shown in Fig. 18. Position $x$ change from $x_A$ to $x_B$ by sudden jumps, and corresponding values of $Q$ are associated with destabilizations of equilibrium points A and B, as the equipotential line in the second graph of Fig. 18 becomes tangent to the excluded region (hashed zone).

One thus obtains a hysteretic relation, analogous to some effective friction law, between $x$—an analog of strain—and $Q$—an analog of stress ratio. But the "strain" is entirely geometric, corresponding to a change in the contact list, and has no relation to the contact law (here reduced to its bare minimum: the grains cannot interpenetrate).

It should thus be expected that the macroscopic features of granular material mechanics stem from the geometry of granular packings and of their rearranging contact networks, as much as from contact laws.

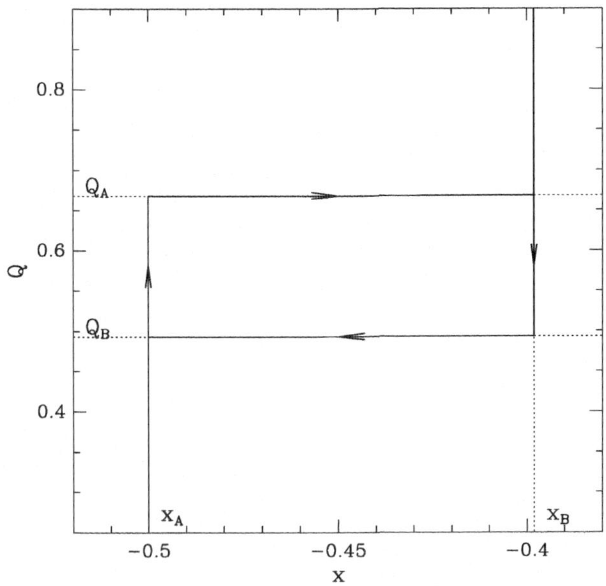


**Fig. 18** Hysteretic variations of equilibrium position $x$ of mobile disk 1 of Fig. 17 (with possible values $x_A$ and $x_B$ corresponding to points A and B) with force parameter $Q = F_x/F_y$

# 3 Collective Properties of Granular Assemblies

We now introduce important tools for the description of granular materials from a micromechanical point of view.

## *3.1 State Variables*

### 3.1.1 Solid Fraction

The first variable characterizing the state of a granular material is its density or the *solid fraction*, $\Phi$ defined as the ratio of the volume of the grains to the volume occupied by the material sample (geotechnical practice tends to favour the *void ratio*, defined as $e = -1 + 1/\Phi$). Everyday experience with sand, ground coffee or potatoes shows that, as already recalled in connection with the behavior under deviatoric load (see Fig. 12), solid objects can be assembled in stable, solidlike packs with different solid fractions. Identical disks in 2D achieve their densest arrangement if their centres are placed on the sites of a regular triangular lattice with spacing equal to their diameter, reaching area fraction $\Phi = \pi/(2\sqrt{3}) \simeq 0.91$. The densest possible structure of identical (3D) spherical balls are obtained on stacking such 2D lattices on top of one another, as shown in Fig. 19. Such a dense

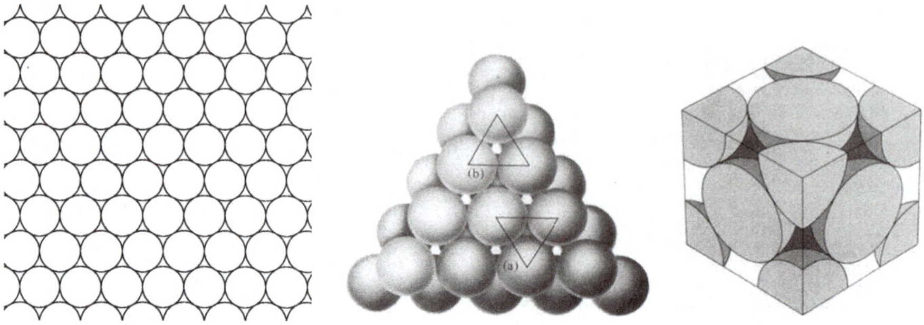


**Fig. 19** Regular lattices achieving maximum solid fraction. Left: triangular lattice for disks. Middle: building 3D maximum density lattices for beads on stacking such layers. Right: fcc lattice

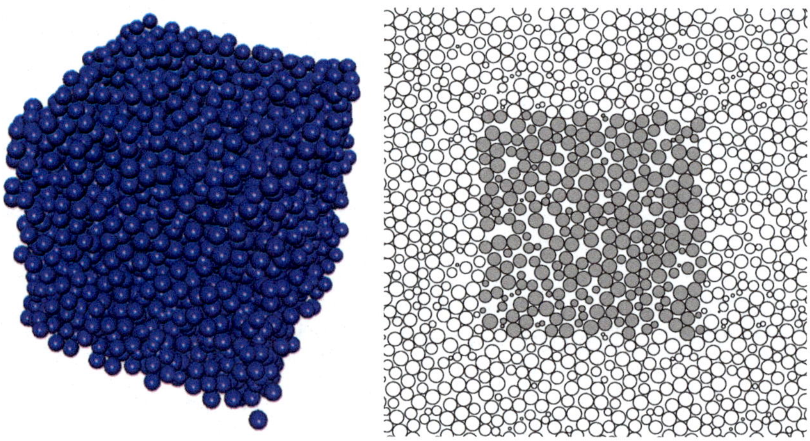

**Fig. 20** Spherical balls in RCP state: aspect of cubic sample (left), cut parallel to a face (right). Grey particles are within the periodic cell used in simulations

stacking should alternate between three possible horizontal placements, and may result in the face-centered cubic lattice (fcc), the hexagonal compact one, or hybrids thereof. The achieved solid fraction in all cases is $\Phi = \pi/(3\sqrt{2}) \simeq 0.74$. Although such lattice structures are important in solid-state physics (Ashcroft et al. 2016) and the search for maximum density structures is a time-honoured endeavour for which mathematical proofs were only recently obtained (Aste and Weaire 2000; Szpiro 2003), they are certainly not representative of the generic disorder in granular materials.

More relevant (see Fig. 20) is the classical concept of *random close packing* (RCP), referring to a disordered state of maximum solid fraction $\Phi^*$ with no ordering, no "germ" of incipient crystallisation. The corresponding solid fraction, as observed both in experiments and simulations, is $\Phi^* \simeq 0.64$ for spherical balls. The absence of crystalline nuclei can be checked with suitable order parameters (Volkov et al. 2002; Agnolin and Roux 2007a). 2D assemblies of monodisperse disks crystallize very easily, and thus should be avoided as a model material. 3D packs of equal-sized beads, on the other hand, are easily maintained in disordered states. DEM studies have revisited the RCP, which may be defined (Agnolin and Roux 2007a) as an equilibrium state of rigid frictionless grains under isotropic pressure. Such states in which confining force are balanced by steric repulsion between hard objects are often referred to as *jammed states* in the recent physics literature (O'Hern et al. 2003; Somfai et al. 2007). Minimizing the potential energy of an applied isotropic pressure amounts to maximizing density, and normal contact forces play the role of Lagrange multipliers corresponding to impenetrability constraints. Compacting procedures thus appear as strategies to avoid the effects of friction. Some studies have shown that the RCP state is not unique—some larger densities might be obtained, even in systems that cannot crystallize because of the diameter distribution (Chaudhuri et al. 2010). RCP states, though, may still be defined as the disordered "jammed states" forming in the limit of fast assembling procedures. By construction, such states are local solid fraction maxima in configuration space. Fastly assembled ones turn out to be statistically similar and share the same solid fraction.

### 3.1.2 Coordination Numbers

The coordination number, $z$, is defined as the average number of force-carrying contacts per grain. Thus, if $N_c$ is the number of contacts and $N$ the number of grains, one has $z = 2N_c/N$. As visible in Fig. 15, a proportion $x_0$ of "rattlers" carry no force. $x_0$ is about 1.5% in frictionless RCP states, and tends to vary with the friction coefficient. For $\mu$ of order 0.1–0.5, it may reach 10–15% in equilibrated frictional monosized sphere packs under uniform stress. Excluding those rattlers from the averaging of contact numbers, one may define a corrected coordination number for the force-carrying structure (sometimes called the *backbone*):

$$z^* = \frac{z}{1 - x_0}. \tag{15}$$

The coordination numbers of the lattices of Fig. 19, $z = 6$ for disks and $z = 12$ for spheres, are unrealistically high compared to generically disordered granular materials, for which, as shown in Sect. 3.3 below, $z^*$ is bounded for rigid, undeformable grains, to much lower values. Thus $z^* = 6$, obtained with rigid frictionless beads, is an upper bound in the presence of friction. As may be inferred from Fig. 20, a measurement of coordination numbers is difficult, even with sophisticated tomography techniques (Aste et al. 2005). Furthermore, the increase of pair correlation functions near contact (O'Hern et al. 2003; Somfai et al. 2007; Agnolin and Roux 2007a) enhances the difficulty to distinguish really contacting pairs.

### 3.1.3 Fabric and Force Anisotropy

The contact network will influence the material behavior by its density (number of contacts per unit volume), conveniently expressed as

$$\overline{n}_c = \frac{z\Phi}{2v_1}, \tag{16}$$

in which $v_1$ denotes the average grain volume. It is also characterized by the statistics of contact orientations, as expressed by the distribution $\mathcal{P}(\mathbf{n})$ of normal unit vectors on the unit sphere $\Sigma$, such that $\int_\Sigma \mathcal{P}(\mathbf{n})d^2\mathbf{n} = 1$, and $\mathcal{P}(-\mathbf{n}) = \mathcal{P}(\mathbf{n})$ since $\mathbf{n}$ and $-\mathbf{n}$ play the same role. $\mathcal{P}(\mathbf{n})$ may be expressed with spherical coordinates $\theta$, $\varphi$ and expanded on the basis of spherical harmonics. $\mathcal{P}(\theta,\varphi)$, its density with respect to the uniform measure $\frac{1}{4\pi}\sin\theta\, d\theta\, d\varphi$, reduces to a function of $\theta$ for axisymmetric states (no dependence on $\varphi$), which is in fact an even function of $\cos\theta$, expressible as a linear combination of Legendre polynomials of even order, as follows:

$$\begin{aligned}
p(\cos\theta) &= 1 + b_2\frac{3\cos^2\theta - 1}{2} + b_4\frac{35\cos^4\theta - 30\cos^2\theta + 3}{8} + \ldots, \quad \text{with} \\
b_2 &= \frac{15}{4}\left[\langle\cos^2\theta\rangle - \frac{1}{3}\right]; \\
b_4 &= \frac{9}{16}\left\{35\left[\langle\cos^4\theta\rangle - \frac{1}{5}\right] - 30\left[\langle\cos^2\theta\rangle - \frac{1}{3}\right]\right\}; \\
&\ldots\ldots
\end{aligned} \tag{17}$$

Such forms are adequate in systems assembled under gravity and/or subject to an axisymmetric compression process (e.g., triaxial, Fig. 12). Isotropic systems are such that $|\cos\theta|$ is uniformly distributed between 0 and 1, whence $\langle\cos^{2k}\theta\rangle = 1/(2k+1)$ for any $k \geq 1$, and all coefficients $b_{2k}$ in (17) vanish. The few first terms are often sufficient for a good representation of contact orientation distributions, as evidenced in Fig. 21: probability distribution function $P(|\cos\theta|)$ (twice $p(\cos\theta)$ of Eq. (17) because of normalization) is well described by the expansion to order 4 (i.e., truncated after the terms explicitly written in Eq. (17)).

The average normal force among all contact being denoted as $\langle F_N\rangle$, it is often useful to know how forces differ according to contact orientations. One thus defines $\mathcal{F}^N(\mathbf{n})$ as the average normal force carried by the contacts with normal direction $\mathbf{n}$, normalized by $\langle F_N\rangle$ (such that $\mathcal{F}^N(\mathbf{n})$ is uniformly equal to 1 in isotropic systems under isotropic pressure). $\mathcal{F}^N(\mathbf{n})$ may then be expanded in spherical harmonics or in Legendre polynomials in the axisymmetric case, just like $\mathcal{P}(\mathbf{n})$.

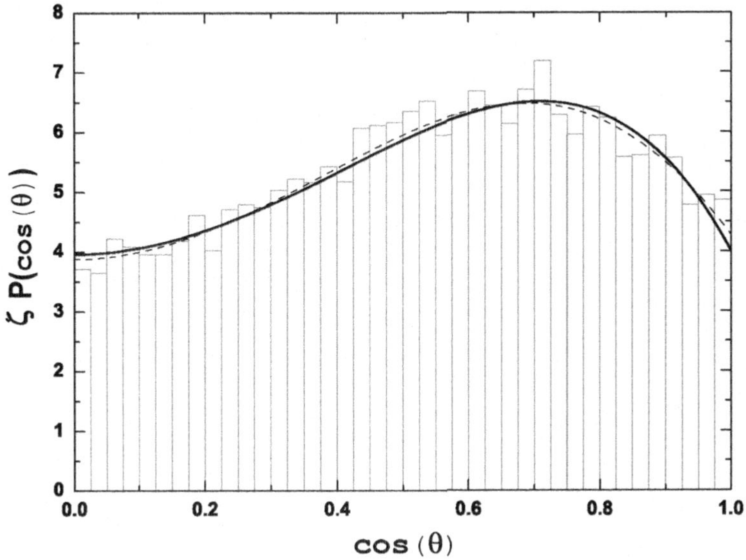


**Fig. 21** $\zeta P(|\cos\theta|)$ (coordination number $\zeta \simeq 5.2$ in this case) versus $|\cos\theta|$ in bead sample assembled under gravity. Histogram: numerical data. Expansion (17) to order 4 (solid line), and to order 6 (dotted line)

## *3.2 Contact Forces and Macroscopic Stress*

Consider a granular sample, of volume $\Omega$, made of grains labelled with indices $i$, $1 \leq i \leq N$, with masses $m_i$ and velocities $\mathbf{v}_i$. Let us define $\mathbf{F}_{ij}$ the contact force, exerted by grain $i$ onto its contacting neighbour $j$, and the *branch vector* $\mathbf{r}_{ij}$, pointing from the centre of $i$ to the centre of $j$. Then if the system is subjected to a uniform stress $\underline{\underline{\sigma}}$, one has the following relation, in which $\alpha$, $\beta$ are indices of coordinates, and the second sum runs over all contacts:

$$\sigma_{\alpha\beta} = \frac{1}{\Omega}\left[\sum_{i=1}^{N} m_i v_i^{\alpha} v_j^{\beta} + \sum_{i<j} F_{ij}^{\alpha} r_{ij}^{\beta}\right]. \tag{18}$$

This formula may be proved in various ways (Christoffersen et al. 1981; Iwashita and Oda 1999), e.g., averaging the stress field within the grains, or dealing with the momentum transmission through cutting surfaces inside the sample. The first term of (18) vanishes in equilibrium (evaluating velocities in the frame of the centre of mass). One may then write

$$\sigma_{\alpha\beta} = \overline{n}_c \langle F_{ij}^{\alpha} r_{ij}^{\beta} \rangle, \tag{19}$$

from which, using (16), a simple, convenient relation may be extracted for spherical grains of diameter $a$, between the average stress $P = \mathrm{tr}\underline{\underline{\sigma}}/3$ and the average normal force:

$$P = \frac{z\Phi}{\pi a^2}\langle F_N \rangle. \tag{20}$$

This formula gives a quantitative form to the estimation of an order of magnitude of typical contact forces as $a^2 P$.

Denoting as $\underline{\underline{\sigma}}^N$ the contribution of normal force components to the stress tensor, and assuming, in an equilibrated assembly of spherical grains, that the average branch vector length is equal to the average diameter $a$, and uncorrelated to the contact force, one has

$$\sigma^N_{\alpha\beta} = a\overline{n}_c \langle F_N \rangle \int_\Sigma \mathcal{F}^N(\mathbf{n})\mathcal{P}(\mathbf{n}) n_\alpha n_\beta d^2\mathbf{n} \tag{21}$$

In axisymmetric systems like those assembled under gravity and subjected to triaxial compression in the vertical direction, one may also define $\mathcal{F}^T$ as the force density, normalized by $\langle F_N \rangle$, in the tangential direction contained in the azimuthal plane—along vector $\mathbf{t}$ with non-negative coordinate along the axial direction from which angle $\theta$ is measured. In such cases, choosing this axis as axis of coordinate, one has for all diagonal components

$$\sigma_{\alpha\alpha} = a\overline{n}_c \langle F_N \rangle \int_\Sigma \mathcal{P}(\mathbf{n}) \left[ \mathcal{F}^N(\mathbf{n}) n_\alpha n_\alpha + \mathcal{F}^T(\mathbf{n}) t_\alpha n_\alpha \right] d^2\mathbf{n}. \tag{22}$$

Then, useful approximation formulae are obtained on keeping the dominant anisotropic terms in expansions of $\mathcal{P}$ (defining coefficient $b_2$ as in expansion (17)), of $\mathcal{F}^N$ (defining, analogously, coefficient $b_2^N$), and $\mathcal{F}^T$. In this latter case, one should pay attention to the different symmetry and write:

$$\mathcal{F}^T(\mathbf{n}) = b^T \sin\theta \cos\theta. \tag{23}$$

Keeping only the dominant anisotropic terms in the integral and neglecting their products, convenient formulae are obtained (Azéma et al. 2009, 2013), which often prove quite accurate, expressing axial, $\sigma_1$, and lateral, $\sigma_3$, stresses as

$$\begin{aligned} \sigma_1 &= a\overline{n}_c \langle F_N \rangle \left[ \frac{1}{3} + \frac{2}{15}\left(b_2 + b_2^N + b^T\right) \right] \\ \sigma_3 &= a\overline{n}_c \langle F_N \rangle \left[ \frac{1}{3} - \frac{1}{15}\left(b_2 + b_2^N + b^T\right) \right]. \end{aligned} \tag{24}$$

Similar formulae may be derived in systems under shear tests (Peyneau and Roux 2008b; Azéma and Radjaï 2014), and, for grains of arbitrary shapes and polydispersities, the anisotropy of branch vectors might be accounted for (Azéma et al. 2009, 2013).

## 3.3 Contact Networks and Rigidity Matrices

### 3.3.1 Definitions

The *rigidity matrix* is a central object in the relation between grain kinematics, contact behavior, and global properties of contact networks. It should not be confused with the *stiffness matrix*, expressing elastic or elastoplastic behavior. The rigidity matrix is a geometric object—its name originates in the theory of rigidity of structures assembled with cables, bars and joints (Thorpe and Duxbury 1998). For a pair of grains $i$, $j$ in contact or very nearly in contact (separated by a very small distance $h_{ij}$), as represented in Fig. 22, one defines the corresponding semi-branch vectors $\mathbf{R}_i^j$, pointing from the (arbitrarily chosen) centre of $i$ to the contact point or the nearest point to the surface of $j$, and, similarly, $\mathbf{R}_j^i$, and the unit vector $\mathbf{n}_{ij}$ normal to the contact plane (or the incipient one). We choose one of the grains, say $i$, as the first object. Then, the contact laws of Sect. 2.2 involve the *relative displacement* of the contact point, i.e., the difference in displacements, $\delta U_{ij}$, according to the (rigid body) motions of the first and the second object. Denoting as $\mathbf{u}_i$, $\mathbf{u}_j$, the displacements of grain centres and $\vec{\omega}_i$, $\vec{\omega}_j$ their small rotations (displacements are actually assumed infinitesimal, like velocities), one has

$$\delta U_{ij} = \mathbf{u}_i + \vec{\omega}_i \times \mathbf{R}_i^j - (\mathbf{u}_j + \vec{\omega}_j \times \mathbf{R}_j^i). \tag{25}$$

Contact laws are more conveniently implemented on distinguishing the normal (scalar $\delta U_{ij}^N = \mathbf{n}_{ij} \cdot \delta U_{ij}$) and tangential (vector $\delta U_{ij}^T \perp \mathbf{n}_{ij}$) components of relative displacements. If the contact law admits rolling and pivoting torques,

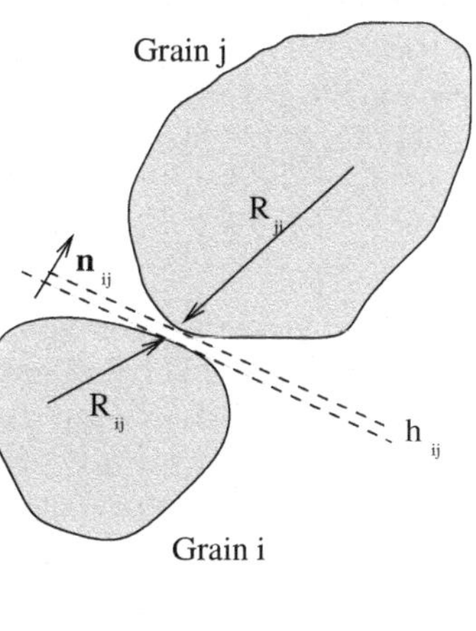


**Fig. 22** A pair of grains nearly in contact

then (25) should be supplemented by the definition of relative rotation vector $\vec{\omega}_i - \vec{\omega}_j$. Assembling all degrees of freedom, one gets an $N_f$-dimensional vector, $\mathbf{U}$, whose coordinates comprise all those of displacements and rotations of the $N$ grains. Depending on specific boundary conditions, involving walls, fixed objects, periodicity conditions... $N_f$ might slightly differ from $ND(D+1)/2$ in dimension $D$ (3 or 2). Likewise, let us define, with $N_c$ contacts, a $DN_c$-dimensional vector $\mathcal{U}$ containing all normal and tangential components of relative displacements $\delta U_{ij}$ in the contacts. Then (25) defines a linear operator or matrix we denote as $\underline{\underline{\mathbf{G}}}$, transforming the coordinates of $\mathbf{U}$ into those of $\mathcal{U}$:

$$\mathcal{U} = \underline{\underline{\mathbf{G}}} \cdot \mathbf{U}. \tag{26}$$

$\underline{\underline{\mathbf{G}}}$ is the *rigidity matrix*, with $N_f$ columns and $DN_c$ lines, attached to the structure and geometry of the contact network. Note that the elements of $\underline{\underline{\mathbf{G}}}$ contain normal unit vectors and semi-branch vectors, and pertain therefore to one specific set of grain positions and orientations. The *kernel* of the rigidity matrix, $\ker \underline{\underline{\mathbf{G}}}$, is the subspace of $\mathbb{R}^{N_f}$ containing the coordinates of *mechanism motions*, i.e., those displacements and rotations causing no relative displacement at contacts. It may include some (a small number in usual applications) trivial such motions, in which the whole grain assembly moves as one rigid body. Its dimension, which we denote as K, is the degree of displacement (or velocity) indeterminacy, or degree of hypostaticity. The *range* of $\underline{\underline{\mathbf{G}}}$, $\mathcal{R}\left(\underline{\underline{\mathbf{G}}}\right)$, is the subset of $\mathbb{R}^{N_c}$ containing the normal and tangential components of *compatible* relative displacements, i.e. those values which are actually achieved for some displacements and rotations of the grains. By the rank theorem, the dimension of $\mathcal{R}\left(\underline{\underline{\mathbf{G}}}\right)$—the rank of matrix $\underline{\underline{\mathbf{G}}}$—is $N_f - \mathrm{K}$.

Contact forces, defined in each contact as the forces exerted by the first grain onto the second one, may also be gathered in a $DN_c$-dimensional vector $\mathcal{F}$, with normal and tangential components in the same order as in vector $\mathcal{U}$. Forces and torques may be externally applied onto the grains, and their coordinates may be suitably gathered into an $N_f$-dimensional *load vector* $\mathbf{F}^{\text{ext}}$—such that its work in small displacement $\mathbf{U}$ is simply the scalar product $\mathbf{F}^{\text{ext}} \cdot \mathbf{U}$. The equilibrium condition, for forces and torques, requests that $\mathbf{F}^{\text{ext}}$ is balanced by the net internal forces and torques on each grain, the coordinates of which form vector $\mathbf{F}^{\text{int}}$. It is easy to check that this condition simply writes

$$\mathbf{F}^{\text{ext}} = -\mathbf{F}^{\text{int}} = {}^{\mathbf{T}}\underline{\underline{\mathbf{G}}} \cdot \mathcal{F} \tag{27}$$

involving the transposed rigidity matrix. The kernel of ${}^{\mathbf{T}}\underline{\underline{\mathbf{G}}}$ contains all those sets of contact forces in equilibrium without any applied load—self-balanced contact forces. Its dimension H is the degree of force indeterminacy or degree of hyperstaticity of the contact structure. The range of ${}^{\mathbf{T}}\underline{\underline{\mathbf{G}}}$ contains all load vectors which may be balanced by some set of contact forces: it is the set of supportable loads (defined here without any sign or inequality condition on forces). The statement that the matrices appearing in (26) and (27) are transposed to each other is some kind of generalized theorem of virtual work: work may be evaluated as $\mathbf{F}^{\text{ext}} \cdot \mathbf{U} = \mathcal{F} \cdot \mathcal{U}$ whatever the choice of those vectors provided $\mathcal{U} = \underline{\underline{\mathbf{G}}} \cdot \mathbf{U}$ and $\mathbf{F}^{\text{ext}} = {}^{\mathbf{T}}\underline{\underline{\mathbf{G}}} \cdot \mathcal{F}$. Since the range of a matrix is the orthogonal complementary subspace to the kernel of its transpose, this provides a condition for compatibility of $\mathcal{U}$ (orthogonality to all sets of self-balanced contact forces) and a condition for supportability of $\mathbf{F}^{\text{ext}}$ (orthogonality to all mechanism motions). Finally, combining the relations on subspace dimensions stemming from the rank theorem and from $\mathcal{R}\left(\underline{\underline{\mathbf{G}}}\right) = \left[\ker {}^{\mathbf{T}}\underline{\underline{\mathbf{G}}}\right]^{\perp}$, one finds the following relation between H, the degree of hyperstaticity (force indeterminacy) and K, the degree of hypostaticity (velocity indeterminacy):

$$N_f + \text{H} = DN_c + \text{K}. \tag{28}$$

Variants of relation (28) apply to frictionless grains, for which only normal relative displacements and contact forces are relevant. Restricting accordingly the definition of the rigidity matrix and the appropriate spatial dimensions, one finds

$$N_f + \text{H} = N_c + \text{K}. \tag{29}$$

In the presence of rolling and pivoting resistance in the contacts, relative displacements are to be supplemented with relative rotations, and contact forces are to be supplemented with contact torques, thereby increasing the number of lines of matrix $\underline{\underline{\mathbf{G}}}$ to $D(D+1)N_c/2$ (i.e. $3N_c$ in 2D and $6N_c$ in 3D). And (28) becomes

$$N_f + \text{H} = \frac{D(D+1)}{2} N_c + \text{K}. \tag{30}$$

Contact structures devoid both of hyperstaticity ($\text{H} = 0$) and of hypostaticity ($\text{K} = 0$) are *isostatic*. Equivalently, matrix $\underline{\underline{\mathbf{G}}}$ is square and invertible. It is also customary to regard as isostatic networks those that only possess trivial mechanism motions (global rigid body motions, or rotations for frictionless spheres) which are easily eliminated from the list of degrees of freedom. Note that isostatic systems have specific value of the coordination number, obtained on writing $N_c = zN/2$ and $N_f = D(D+1)N/2$, for $N$ grains, in (28).

### 3.3.2 Properties

One fundamental characteristic feature of granular materials is the scarcity of contacts, which in some particular cases reaches the limit that the contact network becomes devoid of hyperstaticity ($\text{H} = 0$). This is in particular, the generic case with frictionless grains in the rigid limit, i.e., when the confining stress is small enough and/or the stiffness of contacts large enough for the elastic deflections to be negligible—a condition that will be specified in more quantitative terms in the following. The absence of force indeterminacy for rigid frictionless objects, as discussed, e.g., by Roux (2000), is well established in generically disordered systems (Silbert et al. 2002; O'Hern et al. 2003; Agnolin and Roux 2007a; Donev et al. 2007) and stems from the impossibility (in the statistical sense of events of vanishing probability) of satisfying specific relations involving grain positions and sizes. A familiar example is that of a four-legged table, which is generically wobbly on a hard floor—thereby suppressing force indeterminacy. As a property originating in geometric genericity, it applies to the contact network of hard grains, with or without friction: generically, self-balanced sets of normal contact forces cannot be supported. This entails an upper bound to the number of contacts in assemblies of rigid grains. Applying relation (29) to large contact networks, and eliminating rattlers (which would contribute $6x_0N$ to both $N_f$ and $\text{K}$), one has $N_f = 6N$ (or $N_f = 3N$ in 2D), while $N_c$ is equal to $z^*N/2$. Upper bounds on $z^*$ follow: $z^* \leq 12$ ($z^* \leq 6$ in 2D) in general. If a lower bound is known to $\text{K}$, then the upper bound on $z^*$ will be more stringent. With frictionless spheres (or disks) all rotations are mechanism motions, whence $\text{K} \geq 3N$ and $z^* \leq 6$ (or $\text{K} \geq N$ and $z^* \leq 4$ in 2D). With objects of revolution, $\text{K} \geq N$ and $z^* \leq 10$. These bounds on (rattler-corrected) coordination number $z^*$ apply, as already mentioned, to frictional objects in general. The isostatic value of $z^*$ for frictionless objects coincides with this upper bound. Considering truly frictionless rigid objects, it is established (Roux 2000) that the force-carrying structure in sphere assemblies under a certain load is isostatic ($\text{K} = \text{H} = 0$), while a non-vanishing degree of hypostaticity ($\text{K} > 0$) may be retained in stable packs of frictionless grains of other shapes (which still satisfy $\text{H} = 0$), as, e.g., ellipsoids (Donev et al. 2007). In the presence of friction, values well below the upper bound might be obtained. From (28), the isostatic value is $z^* = 4$ ($z^* = 3$ in 2D), whatever the grain shape. In general, contact structures of frictional objects are observed to possess little or no hypostaticity [the small proportion of spheres maintained between two contacts (Roux 2000) is an exception], whence a lower bound: $z^* \geq 4$. The isostatic value is not very closely approached (one typically obtains $z^* = 4.5$ in poorly coordinated sphere assemblies), except on choosing unphysically large friction coefficients ($\mu \to \infty$) (Agnolin and Roux 2007a). However, with ratio $\text{H}/N_f$ of order 15%, it may be concluded that the paucity of contacts remains an important characteristic of granular materials. It explains the wide distribution of force values. While elasticity, which rules the distribution of forces in well connected networks, tends to share them evenly, geometrically determined force values, as modeled in some statistical approaches (Radjaï 2015) tend to spread force values over wider intervals. With contacts capable of resistance to rolling and pivoting, it is easy to show that any continuous contact network is devoid of non-trivial mechanism motion, any chain of contacting grains forms a solid structure (see Fig. 10). In agreement with (30), the isostatic coordination number is equal to 2, the value characterizing a tree or loopless structure. Any cycle in the contact network creates hyperstaticity. Upon assembling, with adhesive grains, equilibrium contact networks by some dynamical process, loops will spontaneously appear as preassembled structures break and come into contact with one another. But if the

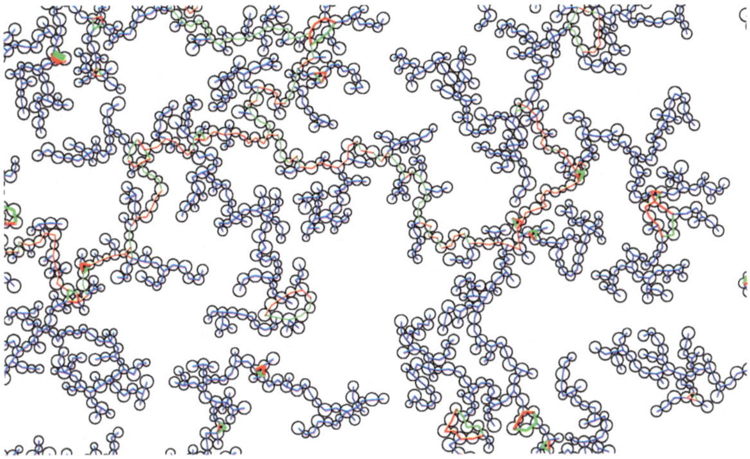

**Fig. 23** Detail of nearly loopless structure formed on slowly assembling cohesive grains with rolling resistance in the contacts (Gilabert et al. 2007)

assembling process is gentle enough, adhesive contacts will not break, and, as cycles only form in the unlikely event of two rigid structures coming in contact simultaneously by two different points, loopless, isostatic structures will tend to form, as shown in Fig. 23.

### 3.3.3 Rigidity Matrix and Stiffness Matrix

The *rigidity matrix* and its transpose are objects of constant use in numerical computations of granular micromechanics: while equations of motion are written for the coordinates of **U**, involving those of $\mathbf{F}^{\text{ext}}$ and $\mathbf{F}^{\text{int}}$, the contact behavior relates $\mathcal{F}$ to $\mathcal{U}$; matrix $\underline{\underline{\mathbf{G}}}$ and its transpose relate $\mathcal{U}$ to **U** and $\mathbf{F}^{\text{int}}$ to $\mathcal{F}$. *Stiffness matrices* express the response of the whole grain collection to small displacements. They are mostly used in quasistatic conditions (although they might also contain some viscous terms associated with wave damping). Their relation to rigidity matrices is explicitly shown here, for elastic contact behavior: small changes in relative displacements $\Delta\mathcal{U}$ are assumed to correspond to small contact force increments $\Delta\mathcal{F}$ as

$$\Delta\mathcal{F} = \underline{\underline{K}} \cdot \Delta\mathcal{U}, \tag{31}$$

where the *local stiffness matrix* $\underline{\underline{K}}$, a square matrix of order $DN_c$, expressing contact elasticity, contains zeros except on its principal diagonal, the elements of which are the normal and tangential stiffnesses $K_N$ and $K_T$ for all the contacts (values of $K_N$ and $K_T$ depend on the contact geometry and its deflection). Using the rigidity matrix, $\Delta\mathcal{U}$ results from small displacements and rotations, coordinates of $\Delta\mathbf{U}$;

and the increment in the net forces and torques on the grains, $\Delta \mathbf{F}^{\text{int}}$, results from $\Delta \mathcal{F}$:

$$\Delta \mathbf{F}^{\text{int}} = -\,{}^{\mathbf{T}}\underline{\underline{\mathbf{G}}} \cdot \Delta \mathcal{F} = -\,{}^{\mathbf{T}}\underline{\underline{\mathbf{G}}} \cdot \underline{\underline{K}} \cdot \Delta \mathcal{U} = -\,{}^{\mathbf{T}}\underline{\underline{\mathbf{G}}} \cdot \underline{\underline{K}} \cdot \underline{\underline{\mathbf{G}}} \cdot \Delta \mathbf{U} = -\underline{\underline{\mathbf{K}}} \cdot \Delta \mathbf{U}$$

This defines the *stiffness matrix* of the granular sample, a square matrix or order $N_f$, $\underline{\underline{\mathbf{K}}}$, as

$$\underline{\underline{\mathbf{K}}} = {}^{\mathbf{T}}\underline{\underline{\mathbf{G}}} \cdot \underline{\underline{K}} \cdot \underline{\underline{\mathbf{G}}}. \tag{32}$$

$\underline{\underline{\mathbf{K}}}$ is symmetric if $\underline{\underline{K}}$ is symmetric (which is insured by its diagonal form in the elastic case), and positive if $\underline{\underline{K}}$ is positive (as for elastic contacts with $K_N > 0$ and $K_T > 0$). If $\underline{\underline{K}}$ is diagonal with non-vanishing elements, the kernel of $\underline{\underline{\mathbf{K}}}$ coincides with the kernel of $\underline{\underline{\mathbf{G}}}$, i.e., contains the mechanism motions. Stiffness matrices are used to numerically measure elastic moduli and discuss theoretically their values and relations to microstructure (Agnolin and Roux 2007c; La Ragione and Jenkins 2007). In case the force in some of the contacts reaches the Coulomb threshold, causing potential sliding, one may still express the elastoplastic response through a local stiffness matrix coupling. $\underline{\underline{K}}$ remains block-diagonal, with $3 \times 3$ ($D \times D$) blocks corresponding to each contact, possibly coupling tangential forces to normal displacements for contacts with full friction mobilization—this matrix now depends on the direction of incremental displacements (McNamara and Herrmann 2006; Roux and Combe 2011). Such elastoplastic stiffness matrices are useful in the study of quasistatic contact network response and stability (Roux and Combe 2002; Welker and McNamara 2009; Roux and Combe 2011).

### *3.4 Dimensionless Control Parameters*

In view of the potential complexity of contact behavior, it has proved useful to extract the essential parameters governing material behavior in dimensionless form (Roux and Chevoir 2011). Such dimensionless groups combine parameters appearing in contact laws (such as material elastic properties) and those defining the mechanical test one wishes to model. Measurements of granular rheology typically involve some stress control, say some pressure level $P$, and some strain rate $\dot{\varepsilon}$—with plastic-like response for small enough rates, and different states away from mechanical equilibrium in flow.

Table 1 summarizes the dependence of material behavior in different regimes on control parameters. One relevant classification of the different regimes of mechanical behavior of granular materials is in terms of solidlike (statically resisting shear stress) or liquidlike (flowing) rheology, supplemented, by analogy with collections of molecules, with the case of "granular gases", in which the grains are strongly

**Table 1** Influence of initial state and dimensionless control parameters on mechanical behavior (Y=Yes, N=No)

| | Initial state | $I$ | $\kappa$ ($> 10^3$) | $e$ or $\zeta$ | $\mu$, $\mu_{R,P}/a$ | $P^*$ |
|---|---|---|---|---|---|---|
| Assembling | Y | Y | N | Y | Y | Y |
| Quasistatic (I) | Y | N | Y | N | Y | Y |
| Quasistatic (II) | Y | N | N | N | Y | Y |
| Critical state | N | N | N | N | Y | Y |
| Dense flow | N | Y | N | N | Y | Y |
| Collisional | N | Y | N | Y | Y | Y |

agitated and interact with one another in sequences of collisions. These categories correspond, in part, to the lines of Table 1. The columns are labelled according to the different relevant dimensionless parameters, and the intersection of, *e.g.*, column $\mu$ (together with rolling or pivoting friction coefficients in dimensionless form), contains "Y" on all lines for which $\mu$ ($\mu_R$ and $\mu_P$ as well) have some influence on the material behavior in the situation considered. Friction coefficients turn out to affect all rheological regimes.

The inertial number, $I$, quantifies the distance to equilibrium by multiplying $\dot{\varepsilon}$ by a characteristic time. Let us consider one grain, of diameter $a$ and mass $m$, in interaction with its neighbours by contact forces that may vary as the material deform and the contact network rearranges. The net force, accelerating it, is of order $Pa^2$. New contacts will form within some inertial time $\tau_i$, of the order of the time within which the grain, accelerated from rest, moves some distance proportional to $a$, i.e. $\tau_i = \sqrt{m/aP}$. One may thus define a dimensionless strain rate or *inertial number* as

$$I = \dot{\varepsilon}\sqrt{\frac{m}{aP}}. \tag{33}$$

(Some use variant $I = a\dot{\varepsilon}\sqrt{\rho/P}$, $\rho$ being the mass density within the grain).

The quasistatic limit, in which the material behaves like a solid, is the limit of $I \to 0$. A definition might depend on the accuracy with which inertial effects and kinetic energy may be neglected. In practice, for simple systems (e.g., frictional spherical beads with narrow diameter distribution), a solid regime is achieved below $I = 10^{-3}$ or $10^{-2}$. Frictionless grains require smaller values for the quasistatic behavior to be approached (Peyneau and Roux 2008a). This defines the "quasistatic" lines of the table. In such regimes, $I$, by definition, becomes irrelevant, and so are the viscous forces—whence the "N" in the column corresponding to viscous damping ($\zeta$) or restitution coefficients ($e$). Such ingredients of the contact model turn out to have little influence in dense granular flows too, i.e., in the regime of $I$ values up to order 0.1 (da Cruz et al. 2005). But more strongly agitated systems are sensitive to the level of dissipation in collisions ($e$ or $\zeta$).

Contact forces, under pressure $P$, or of order $Pa^2$ (with some effect of density and coordination, see relation (20)). With Hertzian contacts, this entails some deflection $h$ of order $[Pa^2/(a^{1/2}\tilde{E})]^{2/3}$, from (3). Comparing $h$ to diameter $a$ defines a stiffness number:

$$\kappa = \left(\frac{\tilde{E}}{P}\right)^{2/3}, \tag{34}$$

such that $(h/a) \propto \kappa^{-1}$, quantifying the geometric effect of contact deflection. The limit of rigid grains is defined as that of $\kappa \to \infty$. With usual materials (say, for sands or glass beads), $\kappa$ usually remains above $10^3$, unless grains are subjected to such high pressures that the stress concentration in contact regions should cause breakage (Agnolin and Roux 2007b). $h/a$ may be regarded as an order of magnitude for macroscopic strains, if their microscopic origin is the material strain in the contact regions. Anticipating on Sect. 5.1, such a strain regime is referred to as "type I" in Table 1, while strains associated to contact network rearrangements—as in the simple example of Fig. 17—are termed "type II strains". Type I strains are naturally of order $\kappa^{-1}$, while larger strain scales, within a solidlike quasistatic regime, are insensitive to $\kappa$ if $\kappa > 1000$. With linear contact elasticity, involving a constant stiffness parameter $K_N$, $\kappa$ is appropriately defined as $K_N/(aP)$.

Column $P^*$, in Table 1, concerns cohesive materials, for which force scale $F_0$ (the tensile strength, introduced in connection with Eq. (6) and Fig. 9) should be compared to confining forces of order $Pa^2$. This defines a reduced pressure, $P^*$ (last column in the table), as

$$P^* = \frac{a^2 P}{F_0}. \tag{35}$$

Cohesive effects dominate for $P^* \ll 1$ and become negligible for $P^* \gg 1$. Intergranular adhesion stabilizes open structures as shown in Figs. 10 and 23. For growing $P^*$ such structures collapse, and as $F_0$ becomes negligible, the dense force networks of Fig. 15 are retrieved.

The assembling process by which granular samples are prepared, before rheological tests are carried out, is both important and little known, because models of granular materials have been focussing on solid deformation or steady flow, whereas assembling processes are inherently hybrid situations, in which agitated or flowing materials come to rest in a short time. It tends to depend on all factors and parameters mentioned in the table, and is briefly discussed in the next section.

## 4 Assembling Process and Compression

The assembling processes condition the initial states of granular materials prior to mechanical testing. Most frequently such processes should include a compression stage, which brings the material to a certain controlled prestressed state. Unlike ordinary solids such as metals, granular materials have no natural state independently of the confining stress. The classical approaches to the mechanics of solidlike granular materials describe their response to *changes* or *increments* of external stress. Assembling methods escape such models as elastoplasticity applied to soil mechanics, and the knowledge of these processes is still largely empirical. For cohesionless materials, the method of *controlled pluviation* offers the advantage to produce homogeneous samples of varying densities. It consists in raining grains onto the free surface of a growing sample (Rad and Tumay 1987; Benahmed et al. 2004), with constant flow rate per surface area and constant vertical velocity of the falling grains. Simulations (Emam et al. 2005) enable homogeneity checks. Figure 24 shows the lack of homogeneity due to a varying height of free fall during the sample fabrication.

Many assembling processes have been tested by DEM (Combe and Roux 2011). They are often simply characterized by the values of the final state variables, rather than striving to mimic laboratory methods. The slow compression of loose configurations with initially non-contacting grains, until a contact network resists further compression, is a convenient method if carried out homogeneously within a periodic box of decreasing size. This results in a variety of microstructures, hopefully representative of the possible material states under low stresses. In the following, results will be reported on such ideal structures, initially isotropic. Specifically, we briefly report on the salient results obtained by Agnolin and

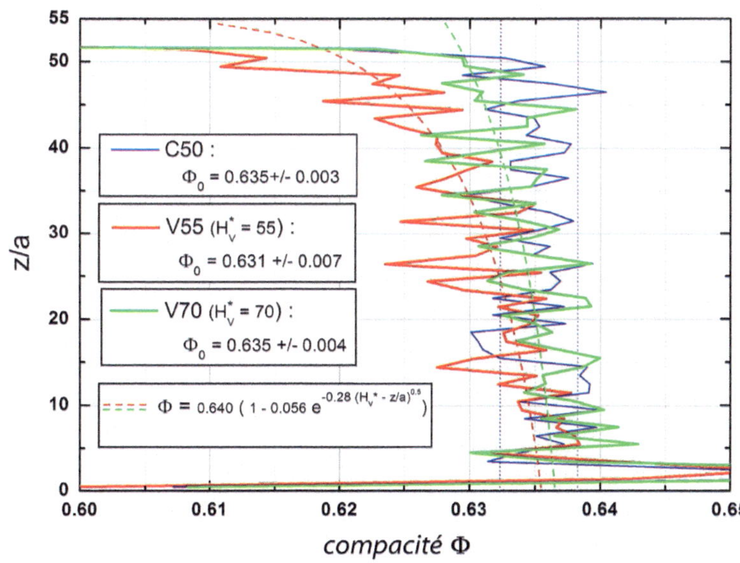


**Fig. 24** Vertical solid fraction profile $\Phi(z)$ in DEM-simulated samples assembled by pluviation. Blue line (C50): homogeneous system assembled with constant height of free fall $H$. Red (V55) and green (V70) lines: varying $\Phi$ obtained on dropping grains from constant elevation. Dotted lines: model of such variations from $H$-dependent controlled pluviation results

Roux (2007a,b,c) on various isotropic states: very dense, highly coordinated states (denoted as A), with $\Phi_A \simeq 0.64$ and $z^*_A = 6$ under low pressure (RCP state); very dense, poorly coordinated ones (denoted as C) with $\Phi_C \simeq \Phi_A$, but $z^*_C \simeq 4.6$; intermediate ones (B) such that $\Phi_B \simeq 0.628$ and $z_B \simeq 5.8$; and looser ones (D), with $\Phi_D \simeq 0.593$ and $z^*_D \simeq 4.55$. (These values are obtained under low pressure, corresponding to $\kappa \simeq 39,000$, or glass beads under 10 kPa). Among these systems those with low coordination initially have a proportion $x_0$ of rattlers above 10%. A result of DEM studies (unknown in the previous literature on dense bead packs) is thus the existence of very dense states with low coordination numbers, as $z^*_C$ is not larger than $z^*_D$, while $\Phi_C$ is very close to the maximum value of RCP. As to state B, it is looser than C, but better coordinated. Although the process leading to states A and B could be argued to be similar to lubricating the grains (ideally for A, assembled with $\mu = 0$, imperfectly for B) and the preparation of C bears some analogy to vibrating grains in a dense configuration, for lack of accurate measurements of coordination numbers in large grain assemblies it is not known in general which value of $z^*$ corresponds to experimental dense bead packs.

Figure 25 shows the evolution of density and coordination in such systems under an isotropic compression cycle, along with the shape of the force distribution, normalized by average $\langle F_N \rangle$ (or distribution of $f = F_N/\langle F_N \rangle$). The friction

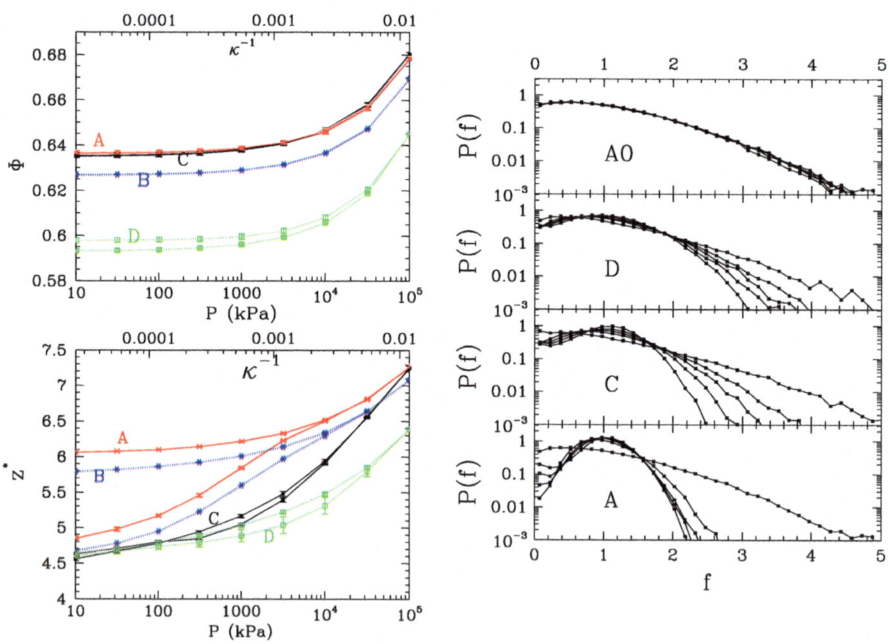


**Fig. 25** Left: variations of $\Phi$ and $z^*$. Right: distribution of normal forces, normalized by their averages (Agnolin and Roux 2007b), for five values of $P$, $\kappa$ decreasing from 39,000 to 84 by factors of $10^{2/3}$. A0-labelled results correspond to A systems kept frictionless during the compression

coefficient is equal to 0.3 in all cases. Changes in $\Phi$ and $z^*$ are moderate as long as $\kappa > 10^3$ (although more notable for $z^*$ when initially low). The growing force indeterminacy as $z^*$ increases well above the minimum value 4 leads to gradual narrowing of distribution $P(f)$—quite fast in system A, for which friction is "plugged in" at the lowest pressure, and slowly in frictionless case A0, which remains close to isostatic over a large pressure range. Remarkably, the evolution of $\Phi$ in the pressure cycle is almost reversible (with the small difference between A and C retrieved upon decompressing). Sands are supposed to behave plastically, with a significant density increase under isotropic compression. The different DEM observations are very likely due to the absence of damage in the contact regions in the numerical model. On the other hand, it is quite remarkable that the coordination number, if initially high, decreases by a large amount in the compression cycle (Fig. 25)—a phenomenon that occurs even for smaller pressure cycles. The reversibility is only apparent, and complex contact network evolutions take place. *Oedometric* compression (Khalili et al. 2017), in which $\sigma_1$ is increased with no lateral strain ($\varepsilon_2 = \varepsilon_3 = 0$) reveals a similar behavior to isotropic compression.

Cohesive systems, as announced in Sect. 3.4, if initially stabilized in very loose structures (Gilabert et al. 2007; Than et al. 2017), as apparent in Figs. 10 and 23, will irreversibly collapse to denser states under growing stress intensity. Their plasticity in isotropic (or oedometric) compression, unlike the subtle and hidden irreversible evolution of the model cohesionless systems of Fig. 25, is very conspicuous. Figure 26 shows the simulated behavior of wet glass beads, attracting one another at contacts through small liquid bridges as in Fig. 8. Most of the irreversible density increase occurs in $P^*$ range between 0.1 and a few units. Note that the numerical observations (right graph) are similar to experimental ones (Than 2017), despite a certain lack of reproducibility of the latter.

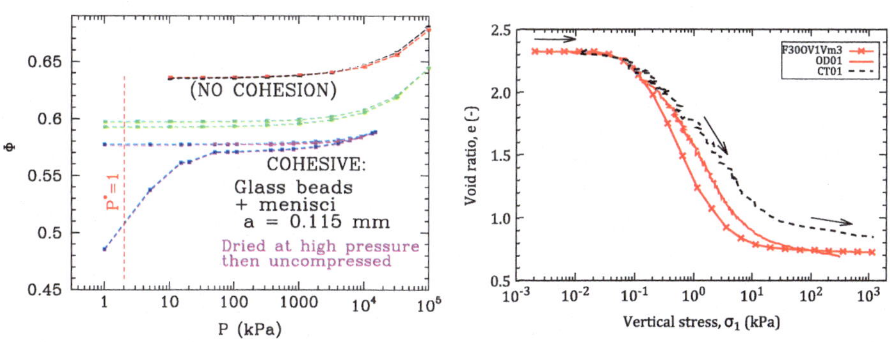


**Fig. 26** Left: $\Phi$ versus $P$ in model wet bead assemblies, comparing systems with and without capillary cohesion. Right: Void index $e = -1 + 1/\Phi$ versus pressure in experiments (red solid line and black dashed one for a different test) and simulations (red connected points) with wet beads (Than 2017) of diameter 0.1 mm

Loose cohesionless systems might be obtained on suppressing capillary cohesion (as if drying, or, equivalently for the solid, quasistatic behavior, saturating the system) under large $P^*$, and then decompressing to a low pressure (see first graph of Fig. 26). Still using $\mu = 0.3$, one thus gets samples denoted as L, with $\Phi = 0.571 \pm 0.005$ and $z^* \simeq 4.7$.

# 5 Quasistatic Behavior

We now turn to the response of solidlike granular systems in classical mechanical tests. Illustrations mainly focus on triaxial compressions of the cohesionless model material (assemblies of spherical beads, $\mu = 0.3$) the preparation of which in different states is presented in Sect. 4. What do we learn from the study of such materials in triaxial compression, with the full microscopic information accessible in numerical, grain-level modeling via DEM? The results reported here aim at clarifying (Sect. 5.1) the role of elasticity and the deformation mechanisms of granular materials, depending on their initial state, which should not only be classified according to its density. We also relate the approach of the critical state at large strain with internal material state evolutions (Sect. 5.2).

## 5.1 *Initial State, Small Strains, "Prepeak" Behavior*

### 5.1.1 Initial Elastic Behavior

Pressure dependent elastic moduli, in isotropic compression, might be obtained using the elastic stiffness matrix of Sect. 3.3. Figure 27 shows the values of the bulk ($B$) and shear ($G$) moduli of the isotropic systems of Fig. 25, versus pressure.

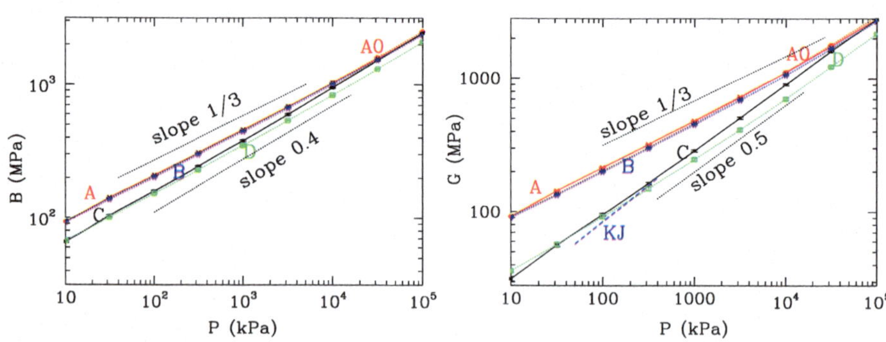


**Fig. 27** Elastic moduli in the different systems of Fig. 25 in isotropic compression [from Agnolin and Roux (2007c)]. Data set marked "KJ" correspond to experimental results on loose pack of glass beads (Kuwano and Jardine 2002). Left: $B$ versus $P$. Right: $G$ versus $P$

It should be remarked that the values of moduli are more sensitive to coordination number $z$ than to density: well-coordinated systems A and B are separated from poorly coordinated ones C and D (even though C is denser than B). Elastic moduli, which may be measured in the laboratory, may thus be regarded as indirect measurements of coordination number. It is expected from relations (4) and (20) that moduli grow as $P^{1/3}$. Best linear fits on the logarithmic graphs of Fig. 27 give however somewhat larger exponents. To some extent, this is due to the increase of the contact density in compression (variation of $z$, see Fig. 25). However, this does not explain the different behavior of shear moduli, which are anomalously small in poorly coordinated systems. This peculiar behavior is reflected by the performance of the Voigt approximation for the moduli, based as usual on the assumption that displacements associated to macroscopic strain $\underline{\underline{\varepsilon}}$ coincide with the values of the corresponding affine field at grain centres. This approximation predicts values $B^{\mathrm{V}}$, $G^{\mathrm{V}}$, involving the constant stiffness ratio $\alpha_T$ defined in (5) and $Z(1/3) = \frac{\langle F_N^{1/3}\rangle}{\langle F_N\rangle^{1/3}}$, characteristic of the force distribution:

$$B^{\mathrm{V}} = Z(1/3)B^e = \frac{1}{2}Z(1/3)P^{1/3}\left(\frac{z\Phi\tilde{E}}{3\pi}\right)^{2/3} ;\ G^{\mathrm{V}} = \frac{6+9\alpha_T}{10}B^{\mathrm{V}}. \tag{36}$$

The Voigt approximation grossly overestimates $G$ especially for small $z$. The explanation was first contributed by Wyart (2006), who argued that $G$ should be proportional to the degree of hyperstaticity. On the other hand, $B$, because it expresses the response to a stress increment proportional to the preexisting stress, possesses a Reuss-like lower bound (Agnolin and Roux 2007c), obtained from trial force increments proportional to the preexisting forces. This Reuss estimate is also proportional to $B^e$ defined in (36). Thus the bulk modulus is bracketed in some satisfactory estimation interval (Fig. 28). The second graph of Fig. 28, in

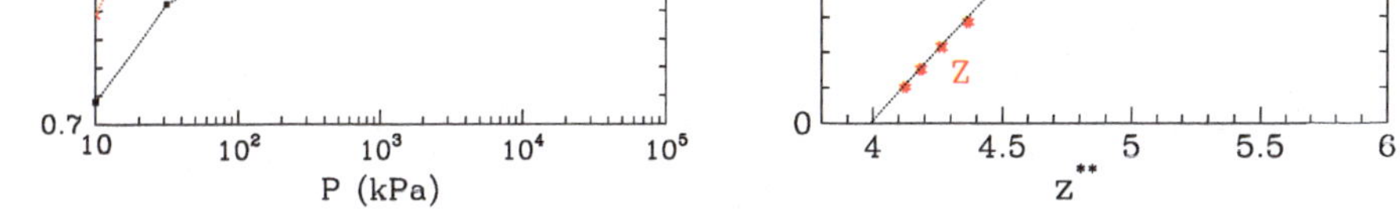


**Fig. 28** [From Agnolin and Roux (2007c)] Left: ratio $B/B^e$ (data points with error bars, joined by solid line), with $B^e$ defined in (36), bracketed by Voigt (crosses) and Reuss (square dots) estimates, in A (red) and C (black) states versus $P$. Right: $G$, normalized by average contact stiffness and density, versus corrected $z^*$ value. "Z" data points: samples with $\mu = \infty$. "LRJ" points: prediction of model by La Ragione and Jenkins (2007)

which shear moduli, divided by obvious factors of average stiffness and density, are plotted versus $z^{**}$ (a slightly corrected value of $z^*$ such that the degree of hyperstaticity H is proportional to $z^{**}-4$) shows that the shear modulus indeed tends to vanish proportionally to degree of hyperstaticity H (Somfai et al. 2007; Agnolin and Roux 2007c). To explore this tendency, additional samples with infinite friction coefficients were assembled by Agnolin and Roux (2007c), in which $z^{**} = 4$ is approached at low pressure. The model by La Ragione and Jenkins (2007), a sophisticated self-consistent approximation scheme, provides improved estimates of $G$ but fails to capture its tendency to vanish proportionally to H.

### 5.1.2 Prepeak Behavior and Initial Coordination Number

With initial isotropic state C, compressed to different pressures, the prepeak behavior, expressed with stress ratios and strains, as shown in Fig. 29, shows little dependence on initial pressure $P$. The peak deviator is reached for $\varepsilon_a \sim 3.10^{-2}$. But, of course, on much smaller scale, one should first observe, close to the initial state, the quasielastic regime. The corresponding strain interval, shown in the insets of the graphs of Fig. 29, is of order $10^{-6}$–$10^{-4}$, as in experiments on sands. The moduli, as deduced from the stiffness matrix approach, based on the assumption of a stable contact network in which all contacts behave elastically, describe the initial slope of stress-strain curves. This quasi-elastic regime is observed (Agnolin and Roux 2007c) to grow approximately as $P^{2/3}$, which, assuming moduli proportional to $P^{1/3}$, would correspond to constant relative stress increments $\delta\sigma/\sigma$.

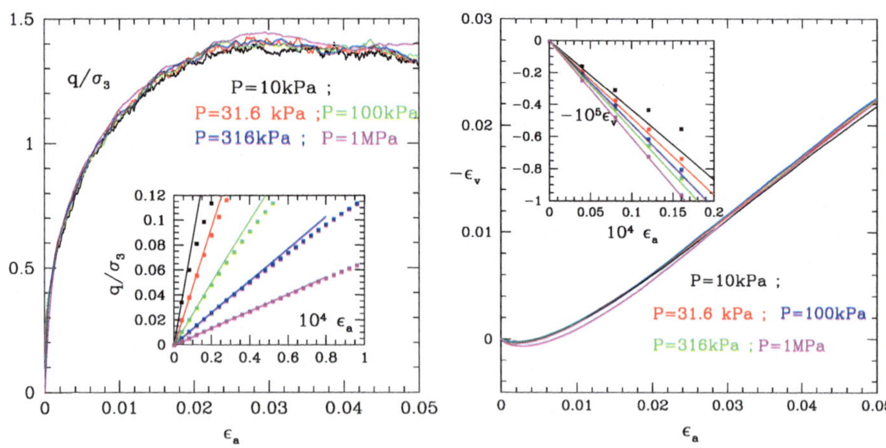


**Fig. 29** Deviator stress normalized by constant lateral stress $q/P$, in prepeak regime (left), and volumetric strain (right) versus axial strain, in dense initial states with small coordination number (C), for different values of $P$, and elastic properties of glass beads. Insets: initial quasielastic regime (note blown-up strain scale), straight line slope given by Young modulus (left) or $-(1-2\nu)$ (right)

Interestingly, elastic moduli may also be measured for different states along the curves of Fig. 29. The stress deviator increase along the curve reflects a non-elastic behavior, because of friction mobilization and network restructuring. On maintaining constant stresses, rather than imposing the axial strain rate, small creep strain intervals are observed, leading to well-equilibrated states (Fig. 30). Those states, in which contact networks are stabilized with force values without full friction mobilization, first respond elastically upon resuming the imposed strain rate test. Such elastic properties reflect the growing anisotropy of the systems in triaxial compression, as shown in Fig. 31. Thus the longitudinal moduli become larger in the axial direction ($C_{11}$), than in the transverse ones ($C_{22}$), because of the increasing fabric and force anisotropies. Contact normal directions are more numerous near the axial direction (with growing parameter $b_2$), and also tend to carry larger forces ($b_2^N$ increases).

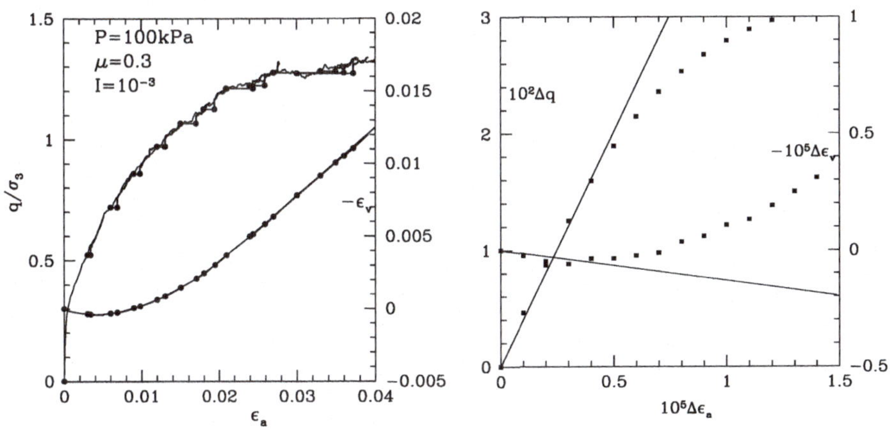


**Fig. 30** Left: creep to equilibrium (between dots) and resumed compression test along triaxial loading path. Right: resumed, strain rate controlled, triaxial compression (dots). Straight line slopes given by elastic moduli

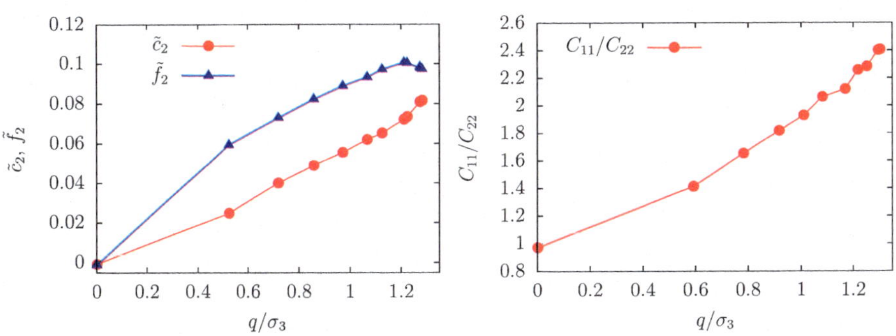


**Fig. 31** Growing anisotropy in triaxial test for dense (C-type) initial state: $\tilde{c}_2 = 4b_2/15$ and $\tilde{f}_2 = 4b_2^N/15$ (left); $C_{11}/C_{22}$ (right) versus $q/\sigma_3$

The dependence of the triaxial compression curve on the initial coordination number is visualized in Fig. 32, showing, on the same graph, the deviator and the volumetric strain curves obtained from initial states A and C, characterised by a very similar density ($\Phi$ near the maximum, RCP value) but different coordination numbers: $z^*$ near 6 for A, about 4.6 for C (and $z \simeq 4$). While the height of the peak deviator stress, given the level of statistical uncertainty on measurements on finite samples (see right graph in Fig. 32), should be regarded as identical, the shape of the stress-strain curve is notably different, with a much steeper initial rise of deviator stress in the well coordinated system. Figure 33, showing, analogously to Fig. 29, triaxial test results in A samples under varying initial pressure (and constant

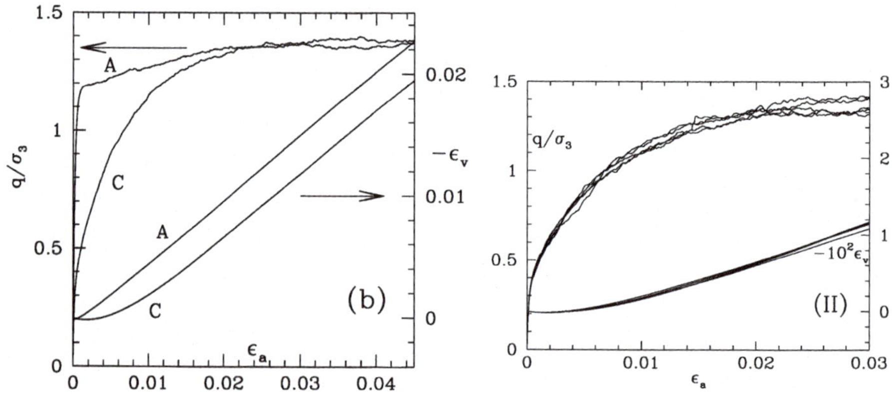


**Fig. 32** (Roux and Combe 2010). Left: triaxial compression response for initial states A and C ($\kappa \simeq 8000$). Right: response of different C-type samples

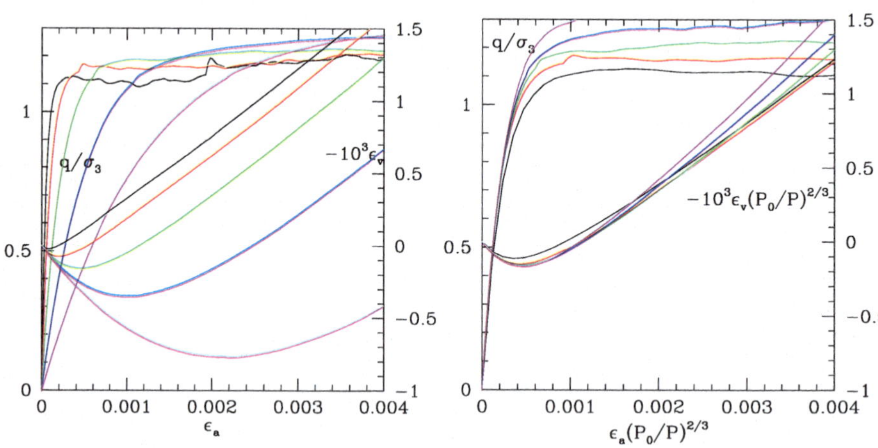


**Fig. 33** (Roux and Combe 2010). Triaxial tests in dense systems A with initially large coordination number. Left graph: $q/P$ and $\epsilon_v$ shown versus $\varepsilon_a$ for different $P$. Right graph: same data, shown versus rescaled strain $\varepsilon_a(P_0/P)^{2/3}$

lateral stress) $P$, reveals a striking difference: strain amplitudes strongly vary with $P$ for the part of the response corresponding to the steep increase of deviator stress. On rescaling strains, using the scale of elastic strains, proportional to $P^{2/3}$ (because moduli tend to grow like $P^{1/3}$), or like contact deflections $h/a \propto \kappa^{-1}$, thus using strain coordinates proportional to $\kappa\varepsilon$, then these different curves tend to superimpose in the initial range. This is characteristic of *type I strains*, which stem from strains in the contact regions: as long as $q/P$ stays below 0.8 or 0.9, A-type systems deform because contact regions deform. This contrasts with the type II strains exhibited by systems in initial state C, for which macroscopic strains are not sensitive to $\kappa$, as shown above with Fig. 29 and reported in Table 1. In a type I strain regime (Roux and Combe 2002, 2010; Roux 2015), the granular system actually behaves like a network of rheological elements as shown in Fig. 11. Although the strains are not elastic, due to intergranular friction (the plastic sliders in the network of rheological elements), elasticity (the springs) sets the strain scale. Within such type I regimes, the creation of new contacts plays a minor role: it is possible to observe the same behavior with simulations in which only the initial contact network is dealt with, as long as this network does not break. It is also possible to simulate this behavior with purely static methods, based on elastoplastic stiffness matrices (Welker and McNamara 2009; Roux and Combe 2011). Beyond regime I (say for $q/P > 1$ in systems A, and much sooner, for $q/P > 0.3$ in systems C) contact networks get repeatedly broken and repaired, in rearrangements involving instabilities at the microscopic scale. "Quasistatic" stress-strain curves are strictly continuous sets of equilibrium configurations in regime I. In regime II they have to be understood as discontinuous sets of equilibrium configurations involving jumps during which the system gets accelerated first, and then stabilizes with a new contact network. To obtain a macroscopic constitutive law from microscopic ingredients one should therefore follow different routes according to the type of strain. Regime II is more challenging, as the amplitude of strain does not originate in contact behavior, but in the complex geometry of rearranging grain packs.

Type I strains are observed for stable contact networks, and naturally extend to larger stress intervals for large coordination numbers (whence the behavior of A samples). They also occur over notable intervals on reversing the loading direction in tests such as the triaxial compression: load reversal, causing tangential forces to return inside the Coulomb cone, reduces the number of sliding contacts and may suppress the contact network instability due to friction mobilization. Systems with small hyperstaticity tend to possess less stable networks and typically exhibit type II response. In the extreme case of rigid, frictionless grains, forming contact networks with no force indeterminacy (isostatic for spheres or disks), it was explicitly shown (Combe and Roux 2000) that the range of stability of equilibrium contact structures, in terms of stress interval, vanishes in the limit of large samples. Any stress increment, however small, causes a rearrangement and strains in the macroscopic limit. Frictionless grain assemblies (Peyneau and Roux 2008a,b; Azéma et al. 2015), in which stress-strain curves are particularly elusive—non-existent, according to Combe and Roux (2000), if grains are perfectly rigid—exhibit finite macroscopic friction, but, at least for circular, spherical or polygonal shapes, are devoid of dilatancy, contradicting naive intuitions based on mechanisms like the one of Fig. 14. Instabilities, producing combined complex displacement fields as shown in Fig. 16, turn out to produce contraction as much as dilation (Combe and Roux 2000; Azéma et al. 2015).

## 5.2 Larger Strains and Approach to Critical State

Figure 34 shows how initially isotropic states L, D and A, made of frictional beads ($\mu = 0.3$) assembled with solid fraction ranging from $\simeq 0.57$ (L) to 0.64 (A), approach the same critical state at large axial strain $\varepsilon_a$. As expected, all three initial states evolve towards the same plastic plateau value ($q_c/\sigma_3 \simeq 0.9$) for deviator stress and the same density (solid fraction $\Phi_c \simeq 0.595$) for axial strains above 0.3. With, initially, $\Phi_c < \Phi_D < \Phi_A$, systems A and D classify as dense: deviator stress, increasing with $\varepsilon_a$, goes to a maximum, the larger the denser the initial state, and then decreases to the critical plateau value $q_c$. Meanwhile, solid fraction $\Phi$ decreases (after some contractant initial phase for D) towards the critical value $\Phi_c$. On the other hand, with $\Phi_L < \Phi_c$, initial state L classifies as loose: as axial strain $\varepsilon_a$ grows, its solid fraction $\Phi$, which decreases, and its deviator stress $q$, which increases, both vary monotonically and asymptotically approach their critical values, $\Phi_c$ and $q_c$, at large $\varepsilon_a$.

As apparent in Fig. 35, specific values for all internal state variables, which are common to different initial states, are approached for large strain in monotonic triaxial compression—this confirms the existence of a well-defined critical *state*, with a specific *plastic flow structure*. The second graph illustrates the success of approximation (24), relating stresses to leading fabric and force anisotropy

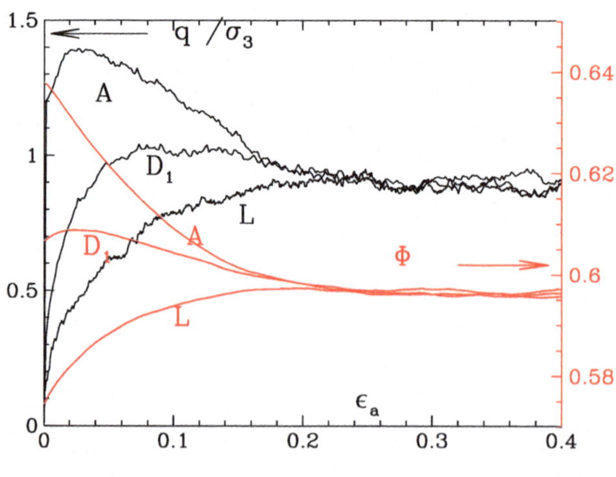


**Fig. 34** Deviator stress (left axis) and solid fraction (right axis) versus axial strain for dense (A), medium dense (D) and loose (L) initial state

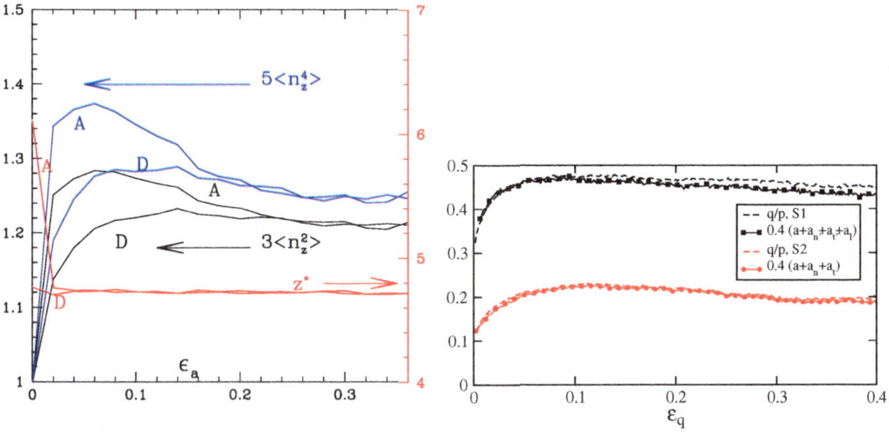


**Fig. 35** Left: Coordination number and fabric parameters versus axial in triaxial tests of Fig. 34 for initial states A and D. Right (Azéma et al. 2009): ratio $q/(3P + q)$ versus deviatoric strain $\varepsilon_q = \varepsilon_1 - \varepsilon_3$ (dashed lines) versus prediction of Eq. (24) (dots). Lower red curve: beads ($\mu = 0.5$); upper black curve: polyhedra ($\mu = 0.5$)

parameters—suitably supplemented with adequate branch vector anisotropy parameter in the results on polyhedra given here, from Azéma et al. (2009).

The critical state thus acts as an *attractor* to which all initial states converge, after large enough monotonically growing strains. It has therefore to be included in constitutive modelling, as a central concept, and state parameters are often specified in terms of the distance to the critical state. Dilatancy, which vanishes in the critical state, may thus be related to some stress combinations (Wood 1990). Plasticity models have been designed to account for specific fabric anisotropies, by which critical states are different, e.g., in simple shear or in triaxial compression (Manzani and Dafalias 1997). The statement that the critical state, for a given loading direction, only depends on $\mu$ for cohesionless materials (and possibly on rolling and pivoting friction, if present) as made here in Table 1, is somewhat at odds with the tradition of soil mechanics, according to which the critical state volume fraction $\Phi_c$ (or the void ratio $e_c = -1 + 1/\Phi_c$) varies with pressure, on the so-called "critical state line". This classical behavior of sands is believed however to be related to contact damage, which standard DEM approaches usually ignore—whence the essentially $\kappa$-independent behavior recorded in Fig. 29, as soon as $\varepsilon_a$ exceeds the small range of regime I strains. The critical state internal friction coefficient increases with $\mu$ [from its initial value near 0.1 for $\mu = 0$, Peyneau and Roux (2008a)] but usually reaches a plateau for $\mu > 0.25$. As to $\Phi_c$, the critical solid fraction, it coincides with the RCP value with frictionless spherical grains (which are devoid of dilatancy), and steadily decreases for growing $\mu$ (Lemaître et al. 2009). Due to its rheological importance (the material cannot be continuously sheared above $\Phi_c$) and independence on preparation, the critical state has often been characterized in DEM, for varying grain shapes (Azéma et al. 2013), or size distributions (Voivret et al. 2009). Contact elasticity being irrelevant for large strains, such studies may be carried out using models of rigid grains, as implemented in the "Contact Dynamics" simulation method (Radjaï and Richefeu 2009b; Radjaï and Dubois 2011).

# 6 Dense Granular Flow

Quasistatic critical states, in triaxial compression, or in simple shear, with controlled normal stress (as in Fig. 36), only depend on the friction coefficient. Steady dense flows, in which the material state departs notably from the quasistatic limit, may be regarded as generalizing critical states. Steady uniform flows are most easily described in simple shear, with a velocity field in direction 1, constant gradient in direction 2, and controlled stress $\sigma_{22}$. Using shear rate $\dot{\gamma} = \frac{\partial v_1}{\partial x_2}$ to define the inertial number as in (33) (suppress diameter $a$ in the denominator for an appropriate 2D definition), the state of the flowing material only depends on $I$, with the critical state in the quasistatic limit of $I \to 0$. The change in the state of the flowing material with $I$ is quite conspicuous in Fig. 36, as the "force chain" pattern of quasistatically deformed materials (similar to Fig. 15, but with the characteristic anisotropy of shear flow) becomes much more tenuous for $I$ in the 0.1 range, as the flowing systems dilates. The force network gradually evolves towards a set of binary interactions, isolated in space and time, in the collisional flow regime at larger $I$. Originated in soil mechanics, the idea to characterize flowing granular materials under controlled normal stress (da Cruz et al. 2005; Jop et al. 2006) proved much more convenient and efficient than more traditional approaches inspired by fluid mechanics (Campbell 2006). Figure 37 directly compares controlled normal stress to controlled volume measurements in a simulation of frictionless beads in slow steady shear flow with $I = 10^{-4}$, close to the quasistatic limit but with still notable differences in that case, as shown by Peyneau and Roux (2008a). For shear strain $\gamma < 1$, the normal stress is fixed ($\sigma_{22} = 1$ in the simulation units) while solid

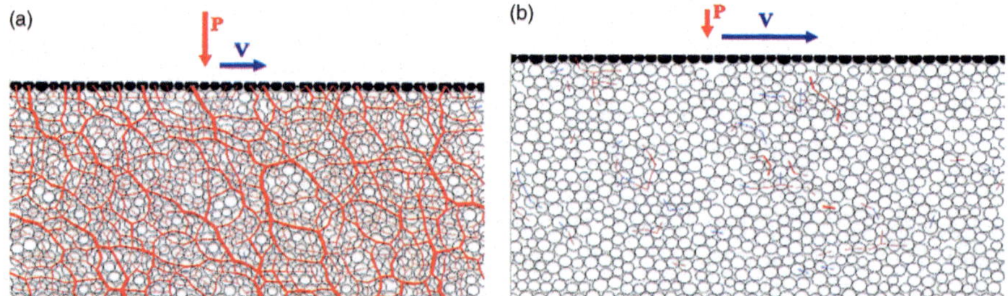


**Fig. 36** Sheared disk system, under controlled normal stress $\sigma_{22}$ (denoted as $P$ on the figure), with drawing of contact forces, with $I = 0.01$ (**a**) and $I = 0.2$ (**b**) (note volume change) (da Cruz et al. 2005)

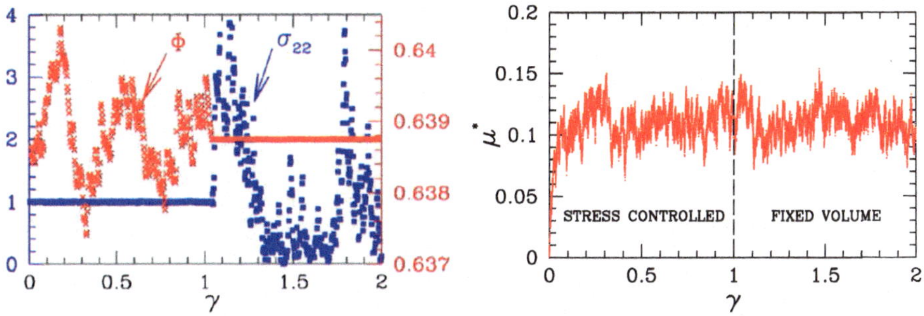


**Fig. 37** Left: normal stress $\sigma_{22}$ (left axis) and $\Phi$ (right axis) in simulated shear flow of frictionless beads, versus shear strain $\gamma$. For $\gamma \le 1$, $\sigma_{22} = 1$ is imposed, $\Phi$ fluctuates with average $\overline{\Phi} \simeq 0.6378$; for $\gamma > 1$, $\Phi$ is fixed at value $\overline{\Phi}$, $\sigma_{22}$ fluctuates. Right: ratio $\mu^* = |\sigma_{12}|/\sigma_{22}$ versus $\gamma$

fraction $\Phi$ fluctuates, between 0.637 and 0.640 (slightly below the critical value), with average $\overline{\Phi} \simeq 0.6378$. For $\gamma > 1$, as $\Phi$ is now fixed to $\overline{\Phi}$, $\sigma_{22}$ fluctuates between 0 and 4, so that measurement of an average stress is quite problematic. The second graph of Fig. 37 shows that the coefficient of internal friction, $\mu^* = |\sigma_{12}|/\sigma_{22}$, may be correctly identified in both situations.

Such an $I$-dependent coefficient of internal friction characterizes the material rheology, together with the $I$-dependent solid fraction. It should in principle be supplemented by measurements of normal stress differences ($\sigma_{22} - \sigma_{11}$ and $\sigma_{33} - \sigma_{22}$), which are usually quite small (Jop et al. 2006; Khamseh et al. 2015). This defines what is now known as the "$\mu^*(I)$ rheology" (Jop et al. 2006; Forterre and Pouliquen 2008; Andreotti et al. 2013) for dense granular flows.

Provided a steady state is achieved, with a uniform thickness, the flow, under gravity, of a granular layer down a plane inclined at angle $\phi$ with respect to a horizontal plane (with a rough surface, in order to avoid sliding), directly characterizes $\mu^*(I)$: the constant shear stress to normal stress (i.e., normal to the substrate or the free surface) ratio in the material fixes $\mu^*(I) = \tan\phi$. This has been applied experimentally (Jop et al. 2006) and numerically, e.g. in the study by Azéma et al. (2012), for a certain type of polygons known as pinacoids, as shown in Fig. 38. The resulting functions $\mu^*(I)$ and $\Phi(I)$ are shown in Fig. 39, and compared to the results for spherical beads. The linear variation of functions $\mu^*(I)$ (increasing) and $\Phi(I)$ (decreasing) is often observed in range $I \sim 0.1$ (da Cruz et al. 2005; Forterre and Pouliquen 2008), with a plateau of $\mu^*$ at larger values. For smaller $I$, as the quasistatic limit is approached the differences $\mu^*(I) - \mu_c^*$ and $\Phi_c - \Phi(I)$ with the quasistatic critical state values tend to vanish with different exponents (Hatano 2007; Peyneau and Roux 2008a). This is illustrated, for frictional ($\mu = 0.3$) bead assemblies, in Fig. 40—which also contains results obtained for different $P^*$ in systems endowed with capillary cohesion. The "$\mu^*(I)$ rheology" thus proves a robust constitutive approach. A particularly convincing illustration of its efficiency was supplied by Jop et al. (2006), who applied it to predict complex velocity profiles in

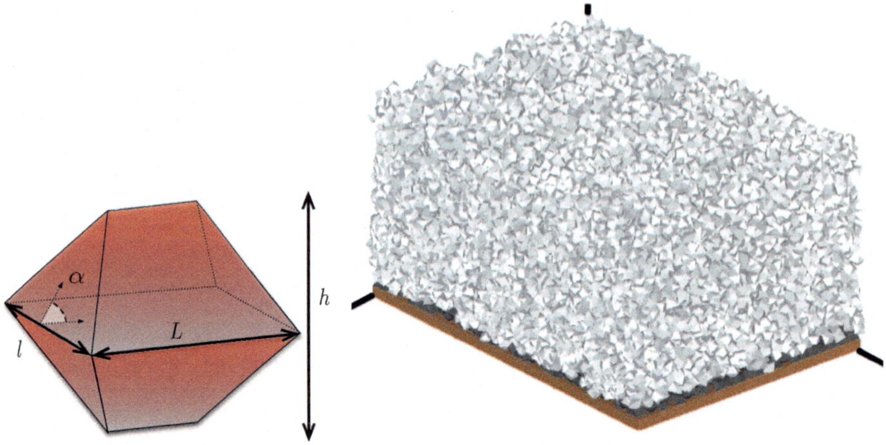


**Fig. 38** Left: one of the "pinacoids" used in study by Azéma et al. (2012). Right: sample (19,000 such polyhedra) used in inclined plane flow

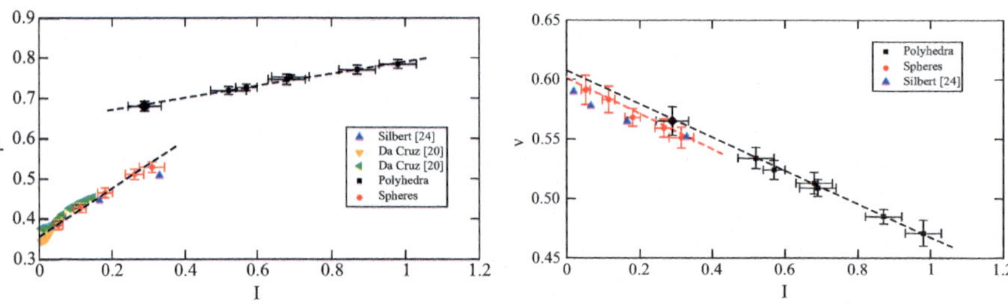


**Fig. 39** (Azéma et al. 2012) Left: $\mu^*(I)$ for polyhedra (upper curve) and spheres (bottom left, with comparisons with literature data). $\mu = 0.4$ for both shapes. Right: $\nu(I)$ [$= \Phi(I)$] for polyhedra and spheres

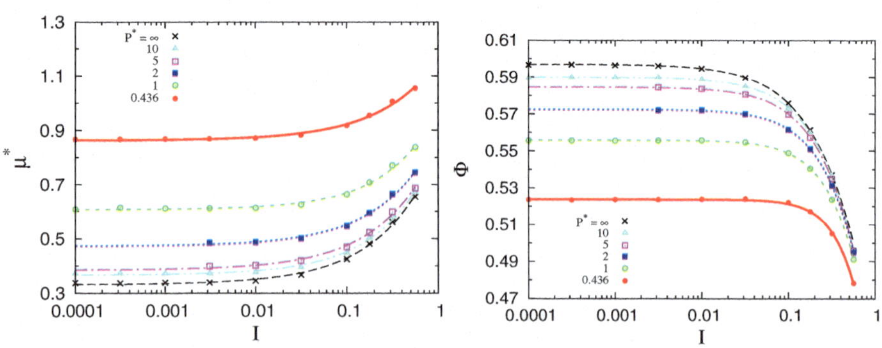


**Fig. 40** (Khamseh et al. 2015) $\mu^*(I)$ (left) and $\Phi(I)$ (right) for simulated dry ($P^*$ infinite) and wet beads, for different $P^*$, defined using Eq. (35) (with $\sigma_{22}$ instead of $P$) and (6). Departures from $I \to 0$ limit fitted (solid lines) as power laws, with exponents near 0.7 for $\mu^*$ (whatever $P^*$), between 1 and 1.7 (depending on $P^*$) for $\Phi$

surface flows of granular layers on top of static samples, between lateral walls. As the constitutive laws (suitably generalized to tensorial form) were initially measured in flow down inclined planes, this study involved no adjustable parameter.

An interesting development is the treatment of very dense suspensions with a similar approach, in which inertial number $I$ is replaced by a "viscous number", $Vi = \eta\dot{\gamma}/P_p$, involving the normal stress $P_p$ transmitted to the solid particles (not to the whole suspension) and the viscosity $\eta$ of the (Newtonian) suspending liquid (Cassar et al. 2005; Lemaître et al. 2009; Boyer et al. 2011). This approach emphasizes the importance of frictional solid contacts in the rheology of very dense suspensions. The relevance of the same critical state as for dry grains in the quasistatic limit of $Vi \to 0$ implies in particular that the solid fraction at which the effective viscosity of the suspension diverges coincides with the critical state solid fraction $\Phi_c$.

# 7 Concluding Remarks

This incomplete tour of current research on micromechanical approaches to granular mechanics reveals first our enduring inability to predict some basic features of material behavior from grain-level phenomena: a quantitative determination of the properties of critical states, for instance, from particle geometry and intergranular friction coefficient, is still unavailable. Many difficulties are related to the complexity of packing geometry, and to how their force-carrying networks responds and rearranges under varying load. No such well-defined objects as crystal dislocation has been identified at the origin of granular plasticity. Some inspiration could be gained from recent advances in the modeling of the rheology of amorphous materials [see, e.g., Maloney and Lemaître (2006)], although two major features set granular materials apart: intergranular friction (the network properties are no longer expressible in terms of potential energy landscape), and very stiff interactions (causing singular behaviors linked to reduced hyperstaticity). Some useful classification of mechanical regimes and material states however emerge, such as type I versus type II strains, or well coordinated versus poorly coordinated systems. The fundamental concept of critical states, as first identified in macroscopic soil mechanics, proves a robust tool in granular material modeling, including in the presence of inertial and/or viscous effects. Current research perspectives are being pursued in the treatment of systems of growing complexity (with different grain shapes, different grain interactions, as for hard cohesive colloidal particles), but basic issues regarding the nature and microscopic origins of granular plasticity should still be investigated.

# References


I. Agnolin, J.-N. Roux, Internal states of model isotropic granular packings. I. Assembling process, geometry, and contact networks. Phys. Rev. E **76**(6), 061302 (2007a)

I. Agnolin, J.-N. Roux, Internal states of model isotropic granular packings. II. Compression and pressure cycles. Phys. Rev. E **76**(6), 061303 (2007b)

I. Agnolin, J.-N. Roux, Internal states of model isotropic granular packings. III. Elastic properties. Phys. Rev. E **76**(6), 061304 (2007c)

B. Andreotti, Y. Forterre, O. Pouliquen, *Granular Media: Between Fluid and Solid* (Cambridge University Press, Cambridge, 2013)

N.W. Ashcroft, N.D. Mermin, D. Wei, *Solid State Physics* (Cengage Learning Asia, Singapore, 2016)

T. Aste, D. Weaire, *The Pursuit of Perfect Packing* (Institute of Physics Publishing, Bristol, 2000)

T. Aste, M. Saadatfar, T.J. Senden, The geometrical structure of disordered sphere packings. Phys. Rev. E **71**, 061302 (2005)

É. Azéma, F. Radjaï, Internal structure of inertial granular flows. Phys. Rev. Lett. **112**, 078001 (2014)

É. Azéma, F. Radjaï, G. Saussine, Quasistatic rheology, force transmission and fabric properties of a packing of irregular polyhedral particles. Mech. Mater. **41**, 729–741 (2009)

É. Azéma, Y. Descantes, N. Roquet, J.-N. Roux, F. Chevoir, Discrete simulation of dense flows of polyhedral grains down a rough inclined plane. Phys. Rev. E **86**, 031303 (2012)

É. Azéma, F. Radjaï, F. Dubois, Packings of irregular polyhedral particles: strength, structure, and effects of angularity. Phys. Rev. E **87**, 062223 (2013)

É. Azéma, F. Radjaï, J.-N. Roux, Internal friction and absence of dilatancy of packings of frictionless polygons. Phys. Rev. E **91**, 010202(R) (2015)

T. Baumberger, C. Caroli, Solid friction, from stick-slip down to pinning and aging. Adv. Phys. **55**, 279–348 (2006)

N. Benahmed, J. Canou, J.-C. Dupla, Structure initiale et propriétés de liquéfaction statique d'un sable. Comptes-Rendus Académie des Sciences, Mécanique **332**, 887–894 (2004)

F.P. Bowden, D. Tabor, *The Friction and Lubrication of Solids*, vol. I (Clarendon Press, Oxford, 1950)

F. Boyer, É. Guazzelli, O. Pouliquen, Unifying suspension and granular rheology. Phys. Rev. Lett. **107**, 188301:1–5 (2011)

N.V. Brilliantov, T. Pöschel, Collision of adhesive viscoelastic particles, in *The Physics of Granular Media*, ed. by H. Hinrichsen, D.E. Wolf (Wiley-VCH, Berlin, 2004), pp. 189–209

C.S. Campbell, Granular material flows – an overview. Powder Technol. **162**, 208–229 (2006)

C. Cassar, M. Nicolas, O. Pouliquen, Submarine granular flows down inclined plane. Phys. Fluids **17**, 103301 (2005)

A. Castellanos, The relationship between attractive interparticle forces and bulk behaviour in dry and uncharged fine powders. Adv. Phys. **54**, 263–376 (2005)

P. Chaudhuri, L. Berthier, S. Sastry, Jamming transitions in amorphous packings of frictionless spheres occur over a continuous range of volume fractions. Phys. Rev. Lett. **104**, 165701 (2010)

J. Christoffersen, M.M. Mehrabadi, S. Nemat-Nasser, A micromechanical description of granular material behavior. J. Appl. Mech. **48**, 339–344 (1981)

G. Combe, J.-N. Roux, Strain versus stress in a model granular material: a devil's staircase. Phys. Rev. Lett. **85**, 3628–3631 (2000)

G. Combe, J.-N. Roux, Construction of granular assemblies under satic loading, in *Discrete-Element Modeling of Granular Materials*, chap. 6 (Wiley-ISTE, London, 2011), pp. 153–180

S.N. Coppersmith, C.H. Liu, S. Majumdar, O. Narayan, T.A. Witten, Model for force fluctuations in bead packs. Phys. Rev. E **53**, 4673–4685 (1996)

F. da Cruz, S. Emam, M. Prochnow, J.-N. Roux, F. Chevoir, Rheophysics of dense granular materials: discrete simulation of plane shear flows. Phys. Rev. E **72**, 021309 (2005)

R. Deluzarche, B. Cambou, Discrete numerical modelling of rockfill dams. Int. J. Numer. Anal. Methods Geomech. **30**, 1075–1096 (2006)
H. di Benedetto, T. Doanh, H. Geoffroy, C. Sauzéat (eds.), *Deformation Characteristics of Geomaterials: Recent Investigations and Prospects* (Swets and Zeitlinger, Lisse, 2003)
A. Donev, R. Connelly, F.H. Stillinger, S. Torquato, Hypostatic jammed packings of nonspherical hard particles: ellipses and ellipsoids. Phys. Rev. E **75**, 051304 (2007)
D. Elata, J.G. Berryman, Contact force-displacement laws and the mechanical behavior of random packs of identical spheres. Mech. Mater. **24**, 229–240 (1996)
S. Emam, J.-N. Roux, J. Canou, A. Corfdir, J.-C. Dupla, Granular packings assembled by rain deposition: an experimental and numerical study, in *Powders and Grains 2005*, ed. by R. García Rojo, H.J. Herrmann, S. McNamara (Balkema, Leiden, 2005), pp. 49–52
N. Estrada, É. Azéma, F. Radjaï, A. Taboada, Identification of rolling resistance as a shape parameter in sheared granular media. Phys. Rev. E **84**, 011306 (2011)
Y. Forterre, O. Pouliquen, Flows of dense granular media. Annu. Rev. Fluid Mech. **40**, 1–24 (2008)
GDR MiDi, On dense granular flows. Eur. Phys. J. E **14**, 341–365 (2004)
F.A. Gilabert, J.-N. Roux, A. Castellanos, Computer simulation of model cohesive powders: Influence of assembling procedure and contact laws on low consolidation states. Phys. Rev. E **75**(1), 011303 (2007)
J.D. Goddard, A.K. Didwania, Computations of dilatancy and yield surfaces for assemblies of rigid frictional spheres. Q. J. Mech. Appl. Math. **51**, 15–43 (1998)
T. Hatano, Power-law friction in closely packed granular materials. Phys. Rev. E **75**, 060301(R) (2007)
J.N. Israelashvili, *Intermolecular and Surface Forces* (Academic, New York, 1991)
K. Iwashita, M. Oda, *Mechanics of Granular Materials: An Introduction* (Taylor & Francis, Leiden, 1999)
H.M. Jaeger, S.R. Nagel, R.P. Behringer, Granular solids, liquids, and gases. Rev. Mod. Phys. **68**(4), 1259–1273 (1996)
X. Jia, C. Caroli, B. Velický, Ultrasound propagation in externally stressed granular media. Phys. Rev. Lett. **82**, 1863–1866 (1999)
K.L. Johnson, *Contact Mechanics* (Cambridge University Press, Cambridge, 1985)
P. Jop, Y. Forterre, O. Pouliquen, A constitutive law for dense granular flow. Nature **441**, 727–730 (2006)
M.H. Khalili, J.-N. Roux, J.-M. Pereira, S. Brisard, M. Bornert, A numerical study of one-dimensional compression of granular materials: I. Stress-strain behavior, microstructure and irreversibility. Phys. Rev. E **95**, 032907 (2017)
S. Khamseh, J.-N. Roux, F. Chevoir, Flow of wet granular materials: a numerical study. Phys. Rev. E **92**, 022201 (2015)
M.R. Kuhn, Structured deformation in granular materials. Mech. Mater. **31**, 407–429 (1999)
M.R. Kuhn, C.S. Chang, Stability, bifurcation and softening in discrete systems: a conceptual approach for granular materials. Int. J. Solids Struct. **43**, 6026–6051 (2006)
R. Kuwano, R.J. Jardine, On the applicability of cross-anisotropic elasticity to granular materials at very small strains. Géotechnique **52**, 727–749 (2002)
L. La Ragione, J.T. Jenkins, The initial response of an idealized granular material. Proc. R. Soc. A **63**(2079), 735–758 (2007). ISSN 1364–5021
P.V. Lade, J.M. Duncan, Elastoplastic stress-strain theory for cohesionless soil. J. Geotech. Eng. Div. **101**, 1037–1053 (1975)
A. Lemaître, J.-N. Roux, F. Chevoir, What do dry granular flows tell us about dense non-brownian suspension rheology? Rheol. Acta **48**, 925–942 (2009)
G. Lian, C. Thornton, M.J. Adams, A theoretical study of the liquid bridge forces between two rigid spherical bodies. J. Colloid Interface Sci. **161**(1), 138–147 (1993)
C.E. Maloney, A. Lemaître, Amorphous systems in athermal, quasistatic shear. Phys. Rev. E **74**, 016118 (2006)
M.T. Manzani, Y.F. Dafalias, A critical state two-surface plasticity model for sands. Géotechnique **47**(2), 255–272 (1997)

D. Maugis, *Contact, Adhesion and Rupture of Elastic Solids* (Springer, Berlin, 2000)

N. Maw, J.R. Barber, J.N. Fawcett, Oblique impact of elastic spheres. Wear **38**(1), 101–114 (1976)

S. McNamara, H.J. Herrmann, Quasirigidity: some uniqueness issues. Phys. Rev. E **74**, 061303 (2006)

R.D. Mindlin, H. Deresiewicz, Elastic spheres in contact under varying oblique forces. ASME J. Appl. Mech. **20**, 327–340 (1953)

J.K. Mitchell, K. Soga, *Fundamentals of Soil Behavior* (Wiley, New York, 2005)

R.M. Nedderman, *Statics and Kinetics of Granular Materials* (Cambridge University Press, Cambridge, 1992)

S. Nemat-Nasser, M. Hori, *Micromechanics. Overall Properties of Heterogeneous Materials* (North-Holland, Amsterdam, 1993)

C. O'Hern, L.E. Silbert, A.J. Liu, S.R. Nagel, Jamming at zero temperature and zero applied stress: the epitome of disorder. Phys. Rev. E **68**(1), 011306 (2003)

C. O'Sullivan, *Particulate Discrete Element Modeling, A Geomechanics Perspective* (Spon Press, London, 2011)

P.-E. Peyneau, J.-N. Roux, Frictionless bead packs have macroscopic friction, but no dilatancy. Phys. Rev. E **78**, 011307 (2008a)

P.-E. Peyneau, J.-N. Roux, Solidlike behavior and anisotropy in rigid frictionless bead assemblies. Phys. Rev. E **78**, 041307 (2008b)

O. Pitois, P. Moucheront, X. Chateau, Liquid bridge between two moving spheres: an experimental study of viscosity effects. J. Colloid Interface Sci. **231**, 26–31 (2000)

N.S. Rad, M.T. Tumay, Factors affecting sand specimen preparation by raining. ASTM J. Geotech. Test. **10**, 31–37 (1987)

F. Radjaï, Modeling force transmission in granular materials. C. R. Phys. **16**(1), 3–9 (2015)

F. Radjaï, F. Dubois, *Discrete-Element Modeling of Granular Materials* (Wiley, New York, 2011)

F. Radjaï, V. Richefeu, Bond anisotropy and cohesion of wet granular materials. Phil. Trans. R. Soc. A **367**, 5123–5138 (2009a)

F. Radjaï, V. Richefeu, Contact dynamics as a nonsmooth discrete element method. Mech. Mater. **41**, 715–728 (2009b)

F. Radjaï, S. Roux, Turbulentlike fluctuations in quasistatic flow of granular media. Phys. Rev. Lett. **89**(6), 064302 (2002)

F. Radjaï, D.E. Wolf, M. Jean, J.-J. Moreau, Bimodal character of stress transmission in granular packings. Phys. Rev. Lett. **80**, 61–64 (1998)

F. Radjaï, S. Nezamabadi, S. Luding, J.-Y. Delenne (eds.), *Powders and Grains 2017*. EPJ Web of Conferences, vol. 140 (EDP Sciences, Les Ulis, 2017)

R. Ramirez, T. Pöschel, N.V. Brilliantov, T. Schwager, Coefficient of restitution of colliding viscoelastic spheres. Phys. Rev. E **60**, 4465–4472 (1999)

O. Reynolds, On the dilatancy of media composed of rigid particles in contact. Philos. Mag. (5th Series) **20**, 469–481 (1885)

D. Richard, I. Iordanoff, Y. Berthier, M. Renouf, N. Fillot, Friction coefficient as a macroscopic view of local dissipation. J. Tribol. Trans. ASME **129**, 829–835 (2007)

J.-N. Roux, Geometric origin of mechanical properties of granular materials. Phys. Rev. E **61**, 6802–6836 (2000)

J.-N. Roux, Pre-peak deformation of model granular materials: A DEM study, in *Geomechanics from Micro to Macro*, ed. by K. Soga, K. Kumar, G. Biscontin, M. Kuo. (CRC, Taylor & Francis, Boca Raton, London, 2015)

J.-N. Roux, F. Chevoir, Dimensional analysis and control parameters, in *Discrete-Element Modeling of Granular Materials*, chap. 8 (Wiley-ISTE, London, 2011), pp. 199–232

J.-N. Roux, G. Combe, Quasistatic rheology and the origins of strain. C. R. Phys. **3**, 131–140 (2002)

J.-N. Roux, G. Combe, How granular materials deform in quasistatic conditions, in *IUTAM-ISIMM Symposium on Mathematical Modeling and Physical Instances of Granular Flow*, ed. by J.D. Goddard, J.T. Jenkins, P. Giovine. *AIP Conference Proceedings*, vol. 1227 (2010), p. 260

J.-N. Roux, G. Combe, Quasi-static methods, in *Discrete-Element Modeling of Granular Materials*, chap. 3 (Wiley-ISTE, London, 2011), pp. 67–101

P. Sánchez, D. Scheeres, M. Hirabayashi, S. Tardivel, Looking into the evolution of granular asteroids in the Solar System, in *Powders and Grains 2017*, ed. by F. Radjaï, S. Nezamabadi, S. Luding, J.-Y. Delenne. EPJ Web of Conferences, vol. 140 (EDP Sciences, Les Ulis, 2017), p. 14004

L.E. Silbert, D. Ertaş, G.S. Grest, T.C. Halsey, D. Levine, Geometry of frictionless and frictional sphere packings. Phys. Rev. E **65**(3), 031304 (2002)

S. Somfai, M. van Hecke, W.G. Ellenbroek, K. Shundyak, W. van Saarloos, Critical and noncritical jamming of frictional grains. Phys. Rev. E **75**(2), 020301 (2007)

A.S.J. Suiker, N.A. Fleck, Frictional collapse of granular assemblies. ASME J. Appl. Mech. **71**, 350–358 (2004)

G. Szpiro, *Kepler's Conjecture* (Wiley, New York, 2003)

F. Tatsuoka, Impacts on geotechnical engineering of several recent findings from laboratory stress-strain tests on geomaterials, in *Geotechnics for Roads, Rail Tracks and Earth Structures*, ed. by G. Correia, H. Brandle (Balkema, Lisse, 2001), pp. 69–140

V.-D. Than, Compression behavior of loose wet granular materials: experiments and discrete numerical simulations. PhD thesis, Université Paris Est (2017)

V.-D. Than, S. Khamseh, A.-M. Tang, J.-M. Pereira, F. Chevoir, J.-N. Roux, Basic mechanical properties of wet granular materials: a DEM study. ASCE J. Eng. Mech. **143**(SI1), C4016001 (2017)

M.F. Thorpe, P.M. Duxbury (eds.), *Rigidity Theory and Applications*. Fundamental Materials Research (Kluwer Academic, New York, 1998)

C. Voivret, F. Radjaï, J.-Y. Delenne, M.S. El Youssoufi, Multiscale force networks in highly polydisperse granular media. Phys. Rev. Lett. **102**, 178001 (2009)

I. Volkov, M. Cieplak, J. Koplik, J.R. Banavar, Molecular dynamics simulations of crystallization of hard spheres. Phys. Rev. E **66**(6), 061401 (2002)

P.R. Welker, S.C. McNamara, What triggers failure in frictional granular assemblies? Phys. Rev. E **79**, 061305 (2009)

D.M. Wood, *Soil Behaviour and Critical State Soil Mechanics* (Cambridge University Press, Cambridge, 1990)

M. Wyart, On the rigidity of amorphous solids. Ann. Phys. Fr. **30**, 1–96 (2006)

# Multiscale Modeling of Interfaces, Dislocations, and Dislocation Field Plasticity

**David L. McDowell**

**Abstract** The various scales of hierarchical structures of metallic materials range from nm to mm. The notion of crystalline plasticity modeling is generalized to a set of model constructs that address phenomena associated with evolution of dislocations in crystals across a range of corresponding length scales, with time scales ranging from fs to years. These model constructs include coarse-grained atomistic modeling (atomistics), microscopic phase field models, dislocation field models, discrete dislocation dynamics, statistical continuum dislocation models, and mesoscale generalized continuum models of gradient, micropolar or micromorphic type, as well as local continuum crystal plasticity that can be applied to polycrystals. We identify key phenomena of lattice dislocations and discuss how these are mapped onto the capabilities of various scale-specific model constructs. Concurrent and hierarchical multiscale model transitions in space and time are discussed, distinguishing between coarse-graining and spatial domain decomposition approaches for lower scale models. We focus on model order reduction methods for mesoscale to macroscale constructs. In terms of bridging scales, the practical importance of two-scale transitions between models of differing fidelity and/or resolution is emphasized, whether of concurrent or hierarchical nature. Various scale-specific models for crystalline plasticity are considered, along with examples. The chapter closes by summarizing some of the long-standing gaps in modeling dislocation plasticity in crystals and polycrystals.

## 1 Introduction: Hierarchy of Material Structure and Models

Engineering materials typically exhibit hierarchy of material structure, otherwise known as microstructure. Structure at various levels of hierarchy can be selectively tailored to achieve multiple property targets required for performance. Generation,

---

D. L. McDowell (✉)
Woodruff School of Mechanical Engineering, School of Materials Science and Engineering, Georgia Institute of Technology, Atlanta, GA, USA
e-mail: david.mcdowell@me.gatech.edu



S. Mesarovic et al. (eds.), *Mesoscale Models*, CISM International Centre for Mechanical Sciences 587, https://doi.org/10.1007/978-3-319-94186-8_5

motion, and interaction of dislocations control mechanical responses of metals to applied thermomechanical loading. These dislocations constitute non-equilibrium, metastable evolution of microstructure. Figure 1 depicts five levels of structure hierarchy, ranging from the atomic structure of lattices and interfaces (atomistics), to migration and interaction of dislocation line segments (discrete dislocations), to collective pattern formation of dislocations (dislocation substructure), to heterogeneous plastic flow within sets of grains/phases (polycrystal plasticity), and finally up to the scale of engineering applications where underlying structure is "smeared" by considering an equivalent macroscopic set of properties or responses. The minimum length scale typical of each of these levels is also shown in Fig. 1, and ranges from interatomic spacing to mean free path for dislocations, to grain size, and on up to characteristic dimensions of components or structures.

Time scales of interest for processes at each level of structure hierarchy is implicit in the dynamic to thermodynamic transition indicated from left to right in Fig. 1. Dynamic atomistic simulations typically range from fs to ns, while large scale discrete dislocation dynamics simulations can extend to the order of s. At far right in Fig. 1, polycrystal and macroscale plasticity models can address time scales relevant to that of large scale laboratory specimens or structural applications, ranging from ms to years.

Multiscale models of dislocation plasticity are necessary to simulate various properties/responses of interest over a wide range of length scales. Goals may differ according to the scales addressed. For example, scales that govern corrosion resistance or crack tip ductile fracture processes are often near the atomic structure scale,

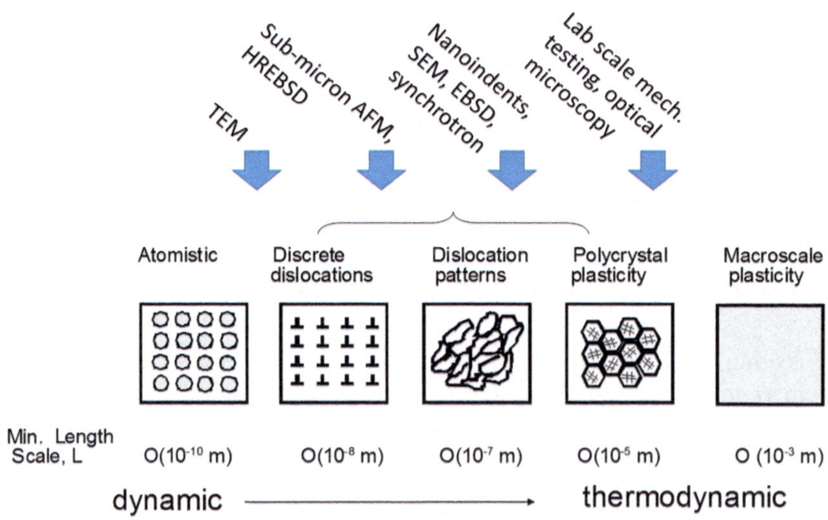


**Fig. 1** Hierarchy of length scales in metal plasticity ranging from atomic (resolution of dislocation cores), to configurations of dislocations, patterning at the mesoscale, and up to multiple grains/phases at the scale of structural applications

while those that control elastic stiffness or yield point of polycrystals involve many grains. Reduced order models are employed to address phenomena corresponding to scales to the right in Fig. 1 and reflect cooperative thermodynamics and kinetics of dislocated crystals. On the other hand, discrete models that apply to scales at the left in Fig. 1 are of fully dynamic character or employ some kind of overdamped dynamical scheme, tracking locations of individual atoms or defect segments. Degrees-of-freedom (DOF) necessary to characterize the structure of a fixed volume of material decrease from left to right in Fig. 1; in so doing, information necessary to characterize the dynamical state is discarded in favor of achieving a reduced order continuum thermodynamic description. Stochastic response of ensembles of dislocations is recognized as a hallmark of these intermediate scales or "mesoscales", demanding the use of statistical methods to represent material behavior.

The term "model construct" is considered to apply both to simulations and to methods for interpretation of associated experimental information. Model constructs that relate structure to responses or properties at various levels of the length scale hierarchy in Fig. 1 differ considerably, both in terms of their numbers of degrees of freedom and their fundamental character. As shown along the top in Fig. 1, experimental techniques accumulate information that is resolved or averaged over different length and time scales at various levels to measure evolving structure, either in synchronous or asynchronous manner with deformation. In fact, it is difficult to entirely separate the issue of the form or structure of a model from the type of experiments that are used to support and calibrate it, since interpretation of experimental information requires some form of model. Figure 2 illustrates the fundamental limits on time and space associated with capturing experimental signals to image displacements. Certain standard models serve as a basis for interpretation of such measurements of attributes (e.g., time-averaged atomic positions, explicit

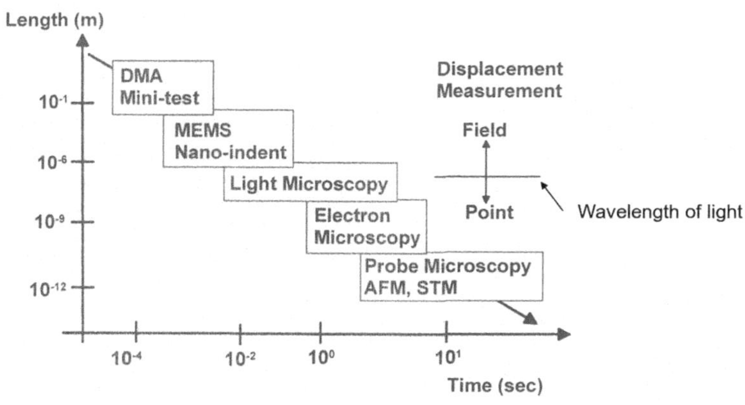


**Fig. 2** Field versus point displacement measurements using different experimental imaging techniques, showing that experiments to the left in Fig. 1 require accumulation of signal over time to build images (Gates and Hinkley 2003).

dislocation configurations, dislocation densities and lattice curvature, crystal deformations, and relative orientations of grains or phases).

Modeling evolution of dislocations in a lattice is challenging and highly complex. Dislocations evolve well away from equilibrium through sequences of metastable states characterized by a large spectrum of relaxation times, driven by short and long range internal stresses, and spatial heterogeneity of structure. Moreover, short range dislocation core effects can persist to large scales and affect dislocation-dislocation and dislocation-obstacle interactions. Mean field spatial statistical measures of evolving dislocation structure are often insufficient to describe statistical interactions, necessitating consideration of higher order spatial statistics. The conventional decomposition of immobile and mobile populations of dislocations is a concession to the dual requirements of modeling plastic strain while tracking dislocation substructure, and typically involves heuristics and subjectivity. In spite of more than a half century of research in metal plasticity, there is still no single model construct that seamlessly bridges amongst all scales.

Bottom-up predictive models such as atomistic or discrete dislocation dynamics can be used to inform both the form and parameters of reduced order model constructs that are framed at higher levels of hierarchy. Most models (even interatomic potentials) are commonly calibrated to information from laboratory measurements obtained at higher length and time scales. Experimental methods at lower length scales in Fig. 1 are typically based on signals acquired over substantial passage of time to interpret relevant physical phenomena, whereas optical methods at right can acquire images rapidly but at lower resolution (see Fig. 2). Models that address dislocation patterning and polycrystal plasticity are often calibrated and/or validated by using experimental information from top-down experiments that address length scales well above those of individual cells (patterns) or grains (polycrystal plasticity); this can complicate the calibration process by giving rise to non-uniqueness of parameter sets for these many-body experiments.

Quantifying uncertainty of experimental information and model constructs (both model form and parameters) is of great practical importance. Beyond the computational limits of space and time that can be accessed by each model construct, uncertainty effectively limits applicability of models to those specific scales for which the model form is best suited. For this reason, practical multiscale modeling of plasticity explores a range of methods to transition between various adjacent scales of hierarchy rather than seeking a unified standard approach across many scales. Some examples attempting the latter objective do exist (cf. Groh et al. 2009) but they have not typically been subjected to formal uncertainty quantification. The error involved at intermediate levels of hierarchy is unclear. A notional, provisional schematic of various sources of uncertainty associated with models and experiments at various length scales shown in Fig. 1 is presented in Fig. 3. While details vary among models and experimental methods, the key point is that uncertainty of models increases with length and time scale (predictive character diminishes with loss of information), while experimental information is least certain at short length and time scales (i.e., small sets of defects at limited time scales and spatial fields of view) in terms of capturing cooperative dislocation phenomena of many-body character. This

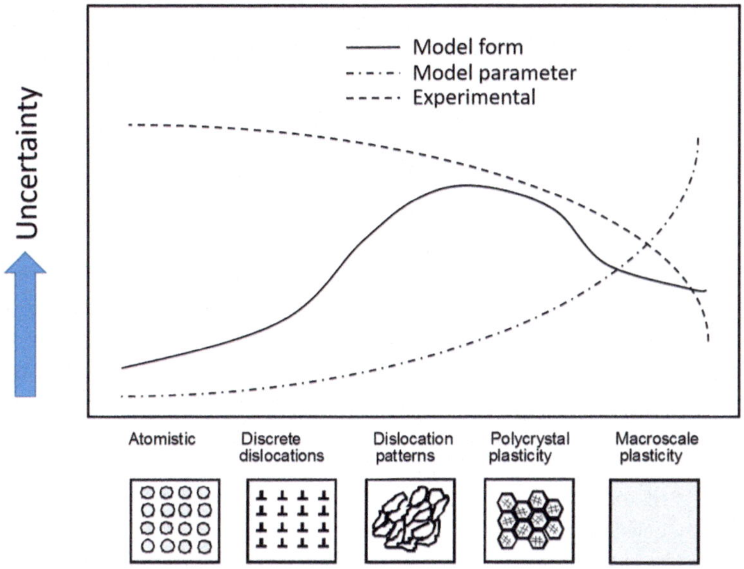


**Fig. 3** Notional representation of uncertainties relevant to models of dislocation plasticity at various length scales

is clearly a statistical sampling issue, in addition to other sources of uncertainty including operative mechanisms. We postulate that model form uncertainty has a peak at the mesoscale of dislocation patterning, as neither bottom-up simulations nor top-down calibration based on experiments can resolve model form uncertainty in this range as clearly as at lower or higher length scales, respectively.

Models pertaining to different levels of structure hierarchy in Fig. 1 are typically related in one of two ways, either via hierarchical (one-way, bottom up) or concurrent (two-way) multiscale schemes. *Concurrent multiscale modeling* schemes exercise simultaneous simulations for models with different fidelity or spatial resolution over the same timeframe, necessitated in cases where time scales are not separable for phenomena occurring at several length scales of interest. These models can either apply to the same spatial domain with different spatial resolution and degrees of freedom or to models with different fidelity employed in adjacent domains; the latter requires schemes for communication of model responses between these regions, and are typically referred to as domain decomposition methods. *Hierarchical multiscale modeling schemes* typically pass results from modeling conducted at each successive length/time scale to the next higher scale(s), with the intent to instruct model form and/or parameters of the latter. In some cases, they can pass information to models pertinent to much higher scales. For example, elastic constants and diffusion coefficients computed using density functional theory (DFT) or atomistic simulations can be employed in crystal plasticity models or macroscale plasticity models. They may be hierarchical in length and/or time, adding additional flexibility to the framing of the multiscale modeling problem. More details and examples will be given later in this chapter.

## 2 Key Phenomena in Generalized Crystalline Plasticity Models

Metal plasticity is fundamentally associated with processes of nucleation, generation, migration, interaction, trapping, and annihilation of dislocations in crystals and polycrystals (McDowell 2008, 2010). The term 'crystal plasticity' has been typically used to describe the classical continuum model (Asaro 1983a, b) that assumes a continuously distributed population of dislocations moving on slip systems within a crystalline lattice. From a more general perspective, it is necessary to generalize to a much larger range of model constructs that are pertinent to 'crystalline plasticity' across levels of material structure hierarchy. Typical physical phenomena that are addressed by crystalline plasticity models are listed in Table 1, ranging from those that are most commonly addressed (early in the list) to those less commonly treated (later in the list). This is rather comprehensive set of phenomena for any single model to consider (no single model construct addresses this entire set), with length scales ranging from nm to hundreds of μm and time scales ranging from ~ps to s. A range of model constructs that address subsets of these phenomena for dislocation plasticity in metals at various levels of spatial and temporal resolution appears in Fig. 4, superimposed on the scales of hierarchy in Fig. 1. This generalization of crystalline plasticity models expresses the inevitable compromises between consideration of scale-specific phenomena, computational feasibility, uncertainty, and the utility and purpose of the simulations.

Reduced order models for polycrystal plasticity, including those that may incorporate dislocation substructure effects on material yield and work hardening, employ continuous field assumptions and thermodynamic descriptions that rely heavily on forms of equilibrium statistical mechanics. High fidelity discrete models, whether based on particle descriptions (e.g., atomistics) or discrete dislocation representations, offer a more direct means to make contact with details of dislocation structure and interactions.

Atomistic approaches can be framed either in terms of 0 K energy minimization methods for the potential energy landscape (e.g., Molecular Statics or MS) or as fully dynamic (e.g., Molecular Dynamics or MD). In either case, the challenge of modeling extended line defects and their interactions in crystals via atomistics is substantial due to the length and time scales involved. Moreover, the utility of information derived may not warrant the computational costs of massive simulations for volumes with dimensions of hundreds of nm owing to uncertainty of the underlying interatomic potentials. This motivates study via discretized modeling of dislocation lines using discrete dislocation dynamics (DDD). DDD embeds dislocations within

**Table 1** Phenomena addressed by crystalline plasticity model constructs

| |
|---|
| 1. Anisotropy of yielding and plastic flow via slip |
| 2. Dislocation multiplication and recovery (implicit or explicit) |
| 3. Thermally activated flow (rate dependence) and temperature dependence |
| 4. Slip system interactions |
| 5. Polycrystalline and/or polyphase |
| 6. Lattice rotation driven by the skew symmetric part of the plastic velocity gradient |
| 7. Elastic anisotropy |
| 8. Long range elastic dislocation interactions |
| 9. Dislocation junction formation and short range interactions |
| 10. Behavior of screw and edge components of mixed character 3D dislocations (implicit or explicit) |
| 11. Effects of net Burgers vector for size effects, work hardening, and dislocation substructure formation |
| 12. Dislocation core effects in dislocation interactions and initial yielding |
| 13. Crystal connection for elastic and plastic incompatibilities for single dislocations and populations of dislocations |
| 14. Explicit obstacle bypass via thermal activation |
| 15. Explicit dislocation multiplication and recovery relations |
| 16. Dislocation mobility relations over a wide range of stress |
| 17. Correlated dislocation bypass of extended obstacles at intermediate to long time scales via thermal activation |
| 18. Dislocation cross slip |
| 19. Coupling with point defects, dislocation climb |
| 20. Distinct treatment of homogeneous and heterogeneous nucleation |
| 21. Explicit dislocation substructure effects on yield and work hardening |
| 22. Interface slip transfer phenomena |

a continuum representation of the lattice, and employs continuum elastic field solutions for long range interactions between dislocations, along with mobility relations and algorithms to represent curved dislocations and dislocation junctions. DDD models can be cast as having quasi-static, overdamped, or fully dynamic character. In practice, DDD applications are rarely truly dynamic in nature, but instead opt for an overdamped prescription via the mobility relations to evolve dislocation fields in time. Dislocation patterning processes can be modeled explicitly using DDD approaches, although progress in this regard has been relatively slow owing to computational challenges and uncertainty of short range interaction mechanisms and model forms for mobility of screw and edge segments, cross-slip, etc.

Certain continuum models have the capability to bridge to small length scales while retaining access to longer time scales relevant to thermally activated processes of dislocation bypass of obstacles. Examples include microscopic phase field (MPF) models and field dislocation mechanics (FDM) descriptions. These FDM descriptions share incompatibility mechanics based on the net dislocation density tensor with statistical continuum dislocation (SCD) models, differing mainly in terms of how distributions of dislocations are tracked. Only atomistic and coarse-grained

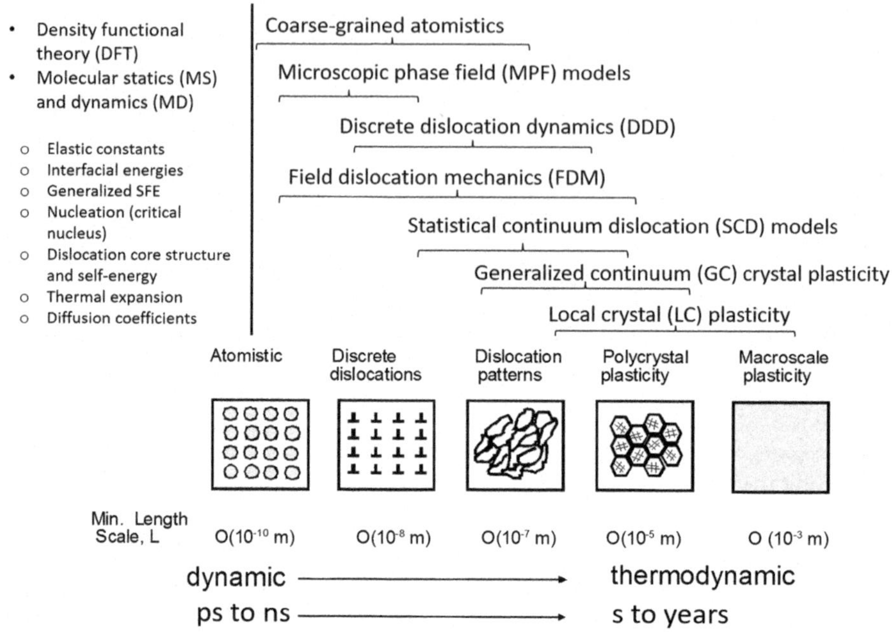


**Fig. 4** Various crystalline plasticity model constructs superimposed on range of associated material structure hierarchy. Certain properties/responses computed using DFT or atomistics shown at far upper left can directly to inform higher length scale models at right via hierarchical modeling approaches

atomistic approaches commonly address dislocations as naturally emergent line defects in the lattice. With the exception of atomistics (MS or MD), coarse-grained atomistics, or the application of MPF or FDM models with near interatomic spatial resolution, all other models introduce primitive representations of dislocations, e.g., their associated lack of crystal closure (Burgers vector). While all these models can be cast in either quasi-static or dynamic form, there are limitations in terms of representation of defect inertia stemming from velocity dependent core effects and long range interactions with regard to strongly dynamic loading conditions, as will be discussed later. Moreover, we will also discuss the challenges in addressing a broad range of thermally activated dislocation processes using coarse-grained atomistics or DDD.

As another general comment, models to the far left in Fig. 4 tend to address a greater fraction of the phenomena listed in Table 1, while intermediate (mesoscopic) DDD, SCD, and FDM models tend to be more selective in this regard, retaining only elements that are deemed to suffice to address certain key phenomena considered in the interest of efficiency and tractability. While relaxation times of lattice defects of short range nature are typically on the order of tens of ps, time scales required for relaxation of extended dislocation lines over complex energy landscapes can be much longer (by many orders of magnitude), depending on temperature. It is emphasized that atomistic or coarse-grained atomistic models cannot fully address phenomena at long time scales associated with thermal activation in cases involving large activation volumes and substantial configurational (entropic) resistance, such as that related to extended cross slip or cooperative obstacle bypass processes. Moreover, non-equilibrium excess point defect concentrations and point defect-dislocation interactions are challenging to address in continuum models, including DDD, and have been considered in more heuristic terms; they may be incorporated in atomistic models and highly resolved regions of coarse-grained atomistic models but with limitations that necessitate assumptions regarding separation of time scales.

Generalized continua (GC) crystal plasticity models share the attribute of nonlocality (albeit limited compared to DDD), but typically opt for a mean field evolution description of the dislocation distribution. SCD models vary in terms of detailed assumptions, but essentially incorporate more detailed statistical information regarding dislocation density and its evolution (cf. Groma 1997; El-Azab 2000, 2006; Groma et al. 2016). They have capability to predict emergent dislocation pattern formation that provides stress shielding, owing to the incorporation of long range elastic dislocation-dislocation interactions. Various forms of models differ with regard to how these interactions are incorporated, and how the dislocation segment density distributions are defined and tracked. The distinction between GC and SCD models may become somewhat blurred if distributions of dislocation segments are considered in the former, along with incompatibility associated with net Burgers vector and an associated slip system back stress approximation. Early approaches to model dislocation patterning were focused on constitutive relations and governing equations with solutions intended to mimic self-organization of dislocations into substructures, e.g., nonlocal reaction-diffusion and reaction-convection models (Bammann and Aifantis 1982, 1987; Walgraef and Aifantis 1985). The state-of-the-art in generalized crystalline plasticity models and statistical continuum models has advanced substantially over the past few decades.

Sufficient experimental understanding has been established using x-ray diffraction and EBSD methods to identify some key characteristics of dislocation substructure evolution (Kuhlmann-Wilsdorf 1989). Scaling laws have been deduced from experiments that express the observed similitude relations involving self-organization of dislocations into substructures with low and high angle boundaries (Hughes and Hansen 1995), for example the work of Hughes et al. (1997) regarding disorientation and spacing of dense dislocation walls as a function of applied effective strain. Consistent with these scaling relations, as a first step Horstemeyer and McDowell (1998) and Butler and McDowell (1998) modified the plastic spin of the lattice to promote more diffuse texture evolution in local continuum (LC) crystal plasticity.

DDD and SCD studies over the past 15–20 years have shed light on dislocation substructure formation in single crystals. One vein of research (Ortiz and Repetto 1999; Ortiz et al. 2000; Carstensen et al. 2002; Koslowski et al. 2004; Kochmann and Hackl 2011; Hackl et al. 2012) has appealed to the notion of bifurcation (self-organization) into laminated microstructures governed by energy minimization over a non-quasiconvex energy landscape. As a caveat, the metastability of the collective energy landscape depends on the non-equilibrium history of dislocation evolution; it is not at the potential energy ground state, unlike potential energy minimizers of individual dislocation segments or isolated, dilute dislocation-obstacle interactions. The slowly evolving non-equilibrium configuration of the dislocation network well away from global minimum energy is supported by sufficiently strong dislocation interactions that cannot be relaxed via conventional fast relaxation dynamics. The character of extended dislocation configurations and associated energy at finite temperature is complicated by constraints of large activation volumes (coordinated thermal fluctuation) of collaborative processes necessary to access lower energy states, sampling a wide spectrum of relaxation times. This necessitates consideration of the configurational entropy of the dislocation network, which gives rise to certain multiscale complexities (cf. Berdichevsky 2006). We will later discuss the related but distinct issue of the entropic barrier contribution to the activation free energy for collective modes of dislocation depinning or bypass of barriers, which complicates the issue of separating slow and fast processes associated with dislocation migration through a system of barriers. Suffice it to say that the entropic contribution to activation energy in the flow rule is significant, in general.

The non-equilibrium character of dislocation structures is fully acknowledged by continuum internal state variable theory, wherein the theoretical constructs of thermodynamic forms based on equilibrium statistical mechanics are adopted but with the caveat of admitting non-zero thermodynamic force at zero conjugate flux, reflecting an accompanying process of non-equilibrium character. This notion of a sequence of constrained equilibrium states (cf. Kestin and Rice 1970; Rice 1971; Muschik 1993; McDowell 2008), while perhaps subtle, is key to modeling non-equilibrium evolution of dislocation substructure.

One stream of approaches to modeling dislocation substructure has addressed nucleation and growth of sub-grain boundaries as partial disclination dipoles/quadropole or blocky dislocation substructures or microbands (Nazarov et al. 1996; Seefeldt et al. 2001a, b; Seefeldt 2001; Seefeldt and Klimanek 2002; Kratochvil and Sedlacek 2003; Sedlacek and Kratochvil 2005; Pantleon 2002; Kratochvil et al. 2007; Viatkina et al. 2007). The recent work of Klusemann and Kochmann (2014) provides a sort of combined strategy of incremental plasticity with laminate approaches to dislocation pattern formation. These continuum approaches naturally involve certain approximations in constructing the free energy and idealizing fast and slow relaxation processes for the dislocation network. Of course, DDD and particularly SCD models (cf. Xia and El-Azab 2015) offer a direct means of modeling dislocation substructure formation and evolution that accounts for the history dependence, including consideration of long range dislocation interactions, dislocation mobilities, and thermally activated generation and bypass of junctions and short range obstacles; they are limited in ability to model cooperative thermally activated dislocation processes at longer time scales across a complex energy landscape, requiring adoption of heuristics for this purpose.

# 3 Model Fidelity, Resolution, and Multiscale Concurrency

It is useful to distinguish between model fidelity and resolution. Fidelity relates to conceptual representation of the physical problem as embodied in the components of a model intended to address specific phenomena in Table 1. A high fidelity model can faithfully and explicitly represent key mechanisms, and can predict outputs from inputs in a way that compares with corresponding physical measurements. Resolution is the degree of refinement in space and/or time associated with the numerical solution procedure. Clearly, a multiscale model may be of high fidelity and high resolution (e.g., atomistics or coarse-grained atomistics), high fidelity and comparatively lower resolution (e.g., DDD), or low fidelity and high resolution (e.g., highly refined meshes using GC or LC). In many respects, existing schema for multiscale modeling can be classified accordingly. The notion of fidelity is often comparative. For example, high fidelity models in a given multiscale model chain that addresses dislocation substructure might be atomistic models, or in some cases DDD models. If the goal is to relate crystallographic texture of a polycrystal to macroscopic responses, the higher fidelity model might address individual grains in a polycrystal (e.g., LC models).

Another distinction should be made between coarse-graining and model reduction approaches to multiscaling at multi-resolution. Here we adopt the definition most commonly encountered in the discrete particle/molecular modeling communities: coarse-graining reduces the number of model degrees of freedom while retaining the underlying nature of the original description, averaging over space and/or time. For example, a model might replace a large number of atoms in a molecular system using a representation with coarse-graining error while retaining key information related to bonding. On the other hand, model reduction (sometimes referred to as model order reduction) seeks to reduce model complexity and fidelity by reducing the model's associated state space dimension and degrees of freedom, offering a simplified form of the original model, i.e., a metamodel. The key difference is that model form often changes in model reduction, whereas model form is essentially invariant in coarse-graining. As we move to the right in Fig. 4, successive constructs typically involve model reduction rather than coarse-graining; this greatly complicates pursuit of concurrent multiscale modeling approaches, as discussed later in this section.

We focus here on modeling evolution of dislocations in crystalline materials. Different goals may exist. From an applied physics perspective, multiscale modeling might mean statistical coarse-graining of a high fidelity model in a manner that preserves its essential structure and retains accuracy in comparison to the averaged fields from the fully resolved simulations. The aim is then to validate similar predictive accuracy of the resulting lower fidelity description. From a systems engineering perspective, multiscale modeling often pertains to introduction of a reduced order model that exhibits similar trends within ranges of interest of independent variables as the high fidelity model, but incurs model approximation error that can be quantified. To a data science expert, coarse-graining or model reduction amounts to a loss of information content which must be quantified. For the present purposes, we take the more general view that incorporates a broad range of objectives and perspectives: *multiscale modeling is the practice of building and exercising models for a set of physical processes that operate over a range of length and/or time scales, regardless of the form of linkage between models expressed at various levels of fidelity or resolution.* Accordingly, we can quite generally discuss coarse-graining of discrete atomistic models, or we can consider the transition from discrete models to continuous field descriptions, or reduction from high fidelity continuum models that explicitly account for microstructure heterogeneity to reduced order continuum models characterized by reduced heterogeneity contrast or even homogeneity. Instead of focusing on how accurately a reduced order model compares with the response of a more highly refined companion model for a given process, we may also consider the goal of applying models at different scales to extract statistical distributions of responses or to answer questions that pertain to the influence of microstructure (e.g., sensitivity) at various scales on some response function of interest. Responses measured at length and time scales using scale-specific experimental techniques, as shown in Fig. 1, might then serve to provide information to calibration or validate the model results or trends at these higher scales rather than pursuing purely bottom-up, deductionist strategy. One may view multiscale modeling as focusing more on utility of information gained by exercising models across a hierarchy of scales than on predictive accuracy of reduced order descriptions relative to a bottom-up coarse-graining or model reduction approximations (cf. McDowell et al. 2009). With this in mind, it is evident that reduced order continuum models (FDM, SCD) that reside more in the middle of the spatial hierarchy in Fig. 4 that address dislocation patterning have potentially higher uncertainty in terms of both model form and parameters (cf. Fig. 3) than lower scale constructs, and perhaps higher than those at higher scales such as GC or LC.

As mentioned in the opening discussion on material structure hierarchy, fully concurrent models seek to simultaneously conduct solutions for different spatial resolution and corresponding model fidelity. The solutions carried out at different scales are coupled and interact in two-way fashion. They are by nature temporally synchronized, in that the time/strain step is limited by the most severe resolution and/or convergence requirement of any domain, typically that of highest fidelity and/or resolution. Weakly concurrent multiscale models might employ a coupling of the coarse-grained or reduced order companion model(s) in a time-averaged sense based on the response of the underlying high temporal resolution domain; this sort of approach is hierarchical in time and concurrent in space, tending more towards one-way hierarchical coupling.

Concurrent multiscale modeling has at least two further classifications: (1) domain decomposition and (2) multi-resolution and/or multi-fidelity models exercised within or over the same spatial domain. In domain decomposition, a coarse-grained or reduced order model representation is employed over part of a overall spatial domain, with an abutting high fidelity representation used in the remainder. For example, we may require high resolution and fidelity near an interface or a crack tip, with low fidelity and/or resolution elsewhere. In such cases, the transition regions over which fine and coarse-grained/reduced models are bridged are an important part of the overall modeling strategy (e.g., overlapping "pad" region), as they can impede or mediate communication of information between the two descriptions.

For purposes of modeling dislocations, we will consider that fully concurrent modeling schemes require exchange/transport of dislocations between the two domains; weaker forms of concurrency might consider dislocation migration in only one domain. In multi-resolution or multi-fidelity (or both) modeling within a given spatial domain, solutions are pursued simultaneously with appropriate initial and boundary conditions, with the ability to zoom in at fine scale or to zoom out at coarse scale at any stage of the evolution process. The overall spatial domain of interest remains invariant. A good example is a concurrent multiscale analysis of a notch root region or 'hot spot' in an aircraft component, using the reduced order far field model representation to link to the length and time scales of a global structural analysis and the companion fine scale solution to track details of deformation and damage at relevant material structure scales (Ghosh et al. 2001, 2007; Kouznetsova et al. 2004). In such as case, it is typically necessary to employ the same incremental variational principle in the numerical implementation to solve the governing equations; the need to properly track the working rate of internal and external forces is crucial, including the dissipation associated with different model fidelity representations (thermodynamic force-flux pairing). Care must be taken with regard to boundary conditions in such approaches, for example those of higher order GC models (cf. Kouznetsova et al. 2002, 2004; Geers et al. 2003, 2010; Luscher et al. 2010, 2012; Forest and Trinh 2011), based on the generalized principle of virtual work. In adopting such an approach, it is non-trivial to achieve equivalence in accounting for the uncertainty associated with representation of dissipation for models having different order—most investigators simply state the equivalence in deterministic terms without reference to distributions associated with different model constructs or representations. While it is tempting to assign the same model form at each level of spatial hierarchy to circumvent these issues and to simplify the problem to one of multi-resolution coarse-graining, is it typically not the case in continuum models that many-body interactions at fine scale can be addressed by an equivalent coarse scale heterogeneous description with the same model form. Model order reduction is typically necessary in this case, for example DDD to GC or LC. Coarse-graining can only be pursued if similitude of model form can be maintained across scales in the mathematical description, as in atomistic coarse-graining methods reviewed by Chen et al. (2011) and those pertaining to dislocations to be discussed later in this chapter.

We further note that for coarse-grained or reduced order models exercised over the same spatial domain as their higher fidelity/resolution counterparts, separation of time scales often cannot be reconciled within the practical limitations on computation for a given evolving process. For example, while we can directly compute dislocation nucleation, multiplication and interaction via very large scale MD on supercomputers to scales on the order of hundreds of nm, it is understood that these processes, as modeled, are strongly overdriven and therefore do not adhere to the sequence of constrained equilibrium states corresponding to dislocation processes occurring under thermally activated conditions at much longer times and lower strain rates. This is not a problem that can be addressed simply by introduction of strategies that introduce local random noise/fluctuation near dislocation cores such as Langevin damping (cf. Marian et al. 2010) or local thermostatting; the statistics of coordinated bypass of obstacles by dislocations via thermal energy/fluctuation requires addressing longer ranged spatio-temporal correlations in a manner consistent with dislocation line vibration modes within the network at finite temperature (cf. Granato et al. 1964; Sobie et al. 2017a). These challenges were outlined by McDowell (1997, 1999, 2010) as pertaining to the need for advanced strategies to pursue "activation volume averaging" in hierarchical multiscale modeling, an extension of statistical mechanics to address a sequence of constrained equilibrium states for evolution of defect ensembles over finite volumes; it is quite distinct from Hill-type averaging in composite micromechanics or other self-consistent approaches, and requires assignment of chained, interdependent event probabilities in combination with kinematic constraints. For example, if we perform domain decomposition of a crack tip problem, with the near tip domain employing atomistics and the remote region using DDD, there is a lack of compatible strategies to address temporal evolution of both domains in the regime of thermally activated flow. This is typically the regime relevant to most applications.

Hierarchical multiscale models are of one-way character, typically informing parameters of models at the next successive length scale of modeling within the hierarchy shown in Fig. 4. As shown to the far left in Fig. 4, there are some exceptions. Certain properties computed using DFT or atomistics can directly inform models at various length scales, even up the to the grain scale. Examples include elastic lattice constants, interface energies, generalized stacking fault energies, and effects of the dislocation core structure and constriction on stress-state dependent yielding that persists through all scales (Racherla and Bassani 2007). Hierarchical coarse-graining involves informing parameters of the upscaled, lower fidelity representation; an example is that of MPF models which are exercised at the nanoscale but seek to explore very long time scale behaviors. Most commonly, hierarchical multiscale modeling pertains to transfer of information (often referred to as 'handshaking') obtained by executing a succession of reduced order models, e.g., atomistics to DDD, DDD to GC, GC to LC. Optimization schemes are often employed to transfer information from higher fidelity/resolution models to reduced order models, whether in terms of informing model parameters or instructing the form of reduced order models. Application of data science strategies for reduced order modeling and metamodeling (Kouchmeshky and Zabaras 2010; Ma and Zabaras 2011; Schöberl et al. 2017) is currently at the frontier of hierarchical multiscale modeling informatics, holding promise for model form adaptation to conform to the input of fine scale information in a statistically meaningful sense.

## 4 Challenges in Modeling Length and Time Scale Transitions

Multiscale modeling of dislocation field problems is particularly demanding, owing to the complexity of the physics. Progress has been slow and elusive on concurrent models in this regard, and no satisfactory concurrent multiscale models exist that bridge all length scales from dislocation cores (order of nm) to dislocation substructures (hundreds of nm) to those of typical laboratory specimens (μm to mm). A few hierarchical multiscale modeling schemes have been reported that span from atomistics to the level of structures; an example is the work of Groh et al. (2009) for dislocation plasticity in Al, moving from the atomic scale to laboratory scale specimen tensile test behavior through a cascade of atomistics, DDD, and LC models. Such an ambitious goal has not commonly pursued in the literature. It is generally agreed that hierarchical bridging across many scales carries with it the prospect of considerable uncertainty that must be quantified, and validation in terms of comparing results at the final scale of interest (e.g., laboratory scale specimens) may not sufficiently strong and/or robust with regard to the various intermediate model forms and handoffs. Transferability of these hierarchical schemes and their value in terms of scientific contribution remains unknown until uncertainty quantification and propagation through hierarchical model chains receives full attention.

Challenges to length scaling include:

- Dislocation core effects, strength of dislocation junctions, obstacle and short range interactions, and velocity and stress state dependent mobility functions of screw and edge dislocation segments, as well as mixed character segments.
- The long range nature of dislocation stress field interactions in a lattice that give rise to strong sensitivity of many body dislocation arrangements and interactions to the size, configuration and boundary conditions of the overall system. This is a major source of size effects in micron-scale crystalline plasticity.
- The short range nature of dislocation patterning and associated correlations of net signed Burgers vector that serve to reduce the energy of long range interactions through shielding.

We have already alluded to challenges in time scaling. Short range lattice relaxation of defected lattices is observed to typically occur on the order of tens of ps in a MD simulation. Approaches have been introduced to improve time scaling of both direct type (Kim et al. 2014; Binder et al. 2015) or statistical coarse-graining of short range interactions in cross slip of dislocations (cf. Deng and El-Azab 2010). However, the time scales required for a dislocation to overcome a single obstacle of several nm in scale through thermally activated processes may range from the ns regime at high temperature to ms at room temperature. At low applied stress levels, thermally activated bypass of larger obstacles or a field of obstacles with high activation volume requires coordinated thermal fluctuations having low probability of successful attempts. For these reasons, time scaling in reduced order models for thermally activated dislocation processes has almost universally focused on the idealized notion of a single "rate-limiting" mechanism or process that controls behavior. This may be consistent with a model that is both spatially and temporally coarse-grained (e.g., continuum crystal plasticity, either GC or LC), but not so in a model that attempts to resolve action of thermally activated processes of individual dislocation segments within a large population of such segments at the mesoscale, as in DDD or SCD models.

Another important and little explored issue is the activation entropy associated with dislocation nucleation and obstacle bypass phenomena. Recent works (Nguyen et al. 2011; Ryu et al. 2011; Saroukhani et al. 2016) have considered the limitations of harmonic transition state theory (hTST), since it does not represent anharmonic entropic effects at finite temperature, including temperature-dependence of the elastic shear modulus (thermal softening), thermal expansion, and other surface energies (e.g., stacking fault energy) that relate to dislocation bypass. The activation energy at finite temperature is the free energy $E_a(\sigma, T) = \Delta H(\sigma) - T\,\Delta S(\sigma)$, where $\Delta H(\sigma)$ is the activation enthalpy (energy difference between the saddle point and initial equilibrium well position on the 0 K potential energy surface), and $\Delta S\,(\sigma)$ is the activation entropy (Zhu et al. 2013). The rate of a thermally-assisted process then follows the relation $\nu_1 \exp\,(\Delta S(\sigma)/k_b) \exp\,(-\Delta H(\sigma)/k_b T)$, making clear the role of the entropic pre-factor that multiplies the enthalpic contribution in the second exponential function. Here, $\nu_1$ is the uncorrected attempt frequency. The empirical Meyer–Neldel (1937) compensation law can be used as an approximation for the activation entropy, i.e., $\Delta S(\sigma) \approx \Delta H(\sigma)/T_{MN}$, where $T_{MN}$ is the melt or disordering temperature (Zhu et al. 2013); however, the accuracy of its extension to more complex, cooperative reaction pathways involving relatively large activation volumes is unclear. For example, Ryu et al. (2011) have discussed the strong entropic barrier associated with dislocation nucleation, and have shown that the Meyer–Neldel compensation law, accounting primarily for anharmonic effects of thermal expansion and softening, is useful for heterogeneous dislocation nucleation but not homogeneous nucleation, likely owing to large nonlinear elastic strain in the latter process. Sobie et al. (2017a) used DDD to compute the entropic pre-factors in the expression for frequency of dislocation bypass of an array of self-interstitial atom loops based on an analysis of modal frequencies for pinned edge or screw dislocation segments, substantially correcting the attempt frequency. Analogous to entropically-stabilized quasicrystals and equilibrium vacancy concentration in a lattice at finite temperature, Kim and Tadmor (2014) have emphasized the importance of the free energy, and have pointed out that entropically stabilized dislocations can exist without occupying a corresponding potential energy well, with 3D ensembles of dislocations having more accessible microstates; this reinforces the need to be careful in using temporal scaling methods such as hTST based on analysis of the potential energy surface, whether using atomistic approaches or higher level models. Certainly, the activation energy barrier strength at finite temperature is affected. It also points to the interpretation of configurational entropy of dislocations as stabilizing networks away from minimum energy configurations in the PEL, effectively "locking" metastable substructures in place as a result of high activation energy to reconstruct via assistance of thermal fluctuations. Certainly, construction of reduced order forms for this configurational entropy is extremely challenging and should depend not only on dislocation density, but also on line character. Depinning kinetics for dislocation-obstacle interactions should depend on the statistical distributions both of obstacles and dislocation character (cf. Sobie et al. 2017c).

While energy storage and dissipation, in addition to the second law inequality with regard to dissipation, have long been key elements of thermodynamically consistent dislocation viscoplasticity (cf. Lemaitre and Chaboche 1990), the distinction of a non-equilibrium temperature has been made principally under conditions of fast dislocation microstructure evolution under overdriven dynamic conditions. Following up on the premise that the configurational entropy of microstructure, in this case dislocations in a network (cf. Berdichevsky 2006, 2008), can play an important and largely overlooked role in a physically-consistent reduced order continuum framework, Langer et al. (2010) and Langer (2015, 2016, 2017) have introduced a statistical framework (Langer-Bouchbinder-Lookman or LBL theory). In LBL theory, configurational (evolving with dislocation depinning from strong barriers) and fast kinetic-vibrational subsystems are decoupled and associated with independent measures of temperature, the latter sub-system with absolute temperature and the former with an effective temperature. To date this theory has been applied to viscoplasticity of metals (mainly pure metals) with promising results (cf. Langer et al. 2010; Le and Tran 2017; Le et al. 2017), effectively incorporating slow evolving configurational effects into the isotropic hardening response associated with dislocation density. As a reduced order model (GC or LC), this framework offers additional flexibility to capture effects of dislocation configuration/patterning and influence on yield, work hardening and thermal softening in the thermally activated regime at the scale of laboratory specimens. LBL theory appears to pose the barrier strength to depinning in the flow rule (Orowan equation) as purely enthalpic, in spite of the strong contribution of entropic resistance associated with collective vibrational modes of dislocations at finite temperature; as in other reduced order theories, statistically averaged high energy barrier (e.g., athermal) barriers in that case are subsumed into network isotropic hardening. Of course, more predictive lower scale discrete models such as atomistics or DDD explicitly address unit dislocation processes within an ensemble and naturally give rise to distributions of characteristic time scales for depinning; coarse-grained phase field modeling and SCD models, depending on the scale at which they are framed and implied nature of statistical averaging (cf. Berdichevsky 2006), may be amenable to a statistical expression of this decomposition in terms of probability distributions. In view of the fact that entropic effects on dislocation depinning for distributed obstacles may give rise to a spectrum of crossing frequencies distributed through the network, it may not be feasible to separate the time scales of processes into fast and slow, instead opting for a statistical distribution of associated slow processes (compared to very short sub-ns lattice relaxation times) and accompanying non-equilibrium effective temperatures. SCD would seem to be a logical setting to embed this kind of approach, rather than reduced order GC or LC models in which the primal description of configurational entropy may be clouded.

The discussion to this point has focused on limitations in modeling thermally activated dislocation processes at long time scales. The challenge of dynamic shock loading lies at the other end of the strain rate spectrum. Naturally, representation of strongly driven dynamics is challenging for continuum models. Under shock loading, the principle of causality dictates that dislocations ahead of the shock front should not interact with dislocations in the wake of the shock. However, typical DDD schemes consider long range interactions of all dislocations in the system. True discrete dislocation dynamics schemes must therefore be modified to convolve the time history of the shock front with the interactions, and to avoid spurious dislocation nucleation or source activation ahead of the shock front (e.g., Markenscoff and Clifton 1981; Gurrutxaga-Lerma et al. 2013). This is highly involved and computationally intensive. Further, the dislocation mobility and inertia depends upon velocity via phonon-dislocation core interactions (cf. Xiong et al. 2016) as the dislocation approaches the shear wave speed or enters the transonic regime. This is difficult for DDD or SCD approaches to address. On the other hand, dynamic atomistic and coarse-grained atomistic approaches directly address these issues but cannot access the same range of system length scales.

The foregoing discussion has largely focused on approximation error of coarse-grained or reduced order models. We close the discussion of challenges offered by dislocation field problems with an additional commentary on approximation error associated with multiscale model transitions—the handoff of information among scales. Any single model form or concurrent multiscale modeling strategy will struggle with rectifying the large spatial range of phenomena shown in Fig. 4. Given the fundamental shift in Fig. 1 from discrete (dynamic) to continuum (thermodynamic) models, it should be clear that upscaling error due to coarse-graining or model order reduction is unavoidable. Only a few phenomena or physical systems are amenable to coarse-graining or model reduction that serves as a fully predictive (rather than correlative) representation of the larger scale, collective response of the system. Inescapably, non-equilibrium evolution of dislocation microstructure is not such a phenomenon. Systematic coarse-graining or model reduction error is the rule. This source of error complicates prospects for concurrent multiscale models, specifically in terms of the relative contributions of model form and parameter uncertainties (cf. Fig. 3). In general, the scientific pathway to quantify uncertainty in hierarchical multiscale model transitions is perhaps more direct (cf. Kouchmeshky and Zabaras 2010), since it involves assessment of coarse-graining or model order reduction error. As a result, the combined consideration of multiscale model transitions and linking schemes with associated uncertainty quantification, traditionally somewhat loosely connected and rarely pursued, is in fact of central scientific importance.

Scale transitions of hierarchical, bottom-up character are typically used to convey information to instruct model form and/or to estimate model parameters in a higher scale model. Scales that are bridged need not always be adjacent, although transitions between two adjacent scales are most common. Such two-scale transitions can be of concurrent, two-way nature. When transitions involve bridging more than two scales, they are almost always of hierarchical, bottom-up nature. Figure 5 emphasizes such two-scale transitions (primarily hierarchical) of models at various length and time scales, linking methods to connect these models, and the primary sources of uncertainty associated with these transitions (Panchal et al. 2013). As an example, Fan et al. (2010) have introduced an adaptive two-scale hierarchical nonlinear homogenization methodology that includes error estimates.

It is noted that the sources of uncertainties addressed in the right column in Fig. 5 have rarely been explicitly quantified in multiscale modeling. Beyond the details of model form and parameter uncertainty at a given scale, uncertainty quantification and propagation in multiscale model chains is relevant to the value of information provided by the modeling exercise. This is one of the reasons why the utility of information (cf. McDowell et al. 2009) of mesoscale modeling from DDD, SCM, and GC is still an open issue, reflecting the substantial uncertainty associated with the mesoscopic gap in modeling cooperative behavior of populations of dislocations.

In the following sections, we start with the motivation and essential elements of atomistic modeling, with emphasis on understanding the mechanical behavior of nanostructured metals and nanoscale dislocation processes, followed by discussion of concurrent multiscale modeling methods that bridge from atomistic to continuum descriptions, as well as challenges to time scaling in discrete model constructs. Then, we briefly discuss various classes of crystalline plasticity model constructs listed in Fig. 4, along with their applicability to different levels of structure hierarchy elaborated in Fig. 5. Examples are discussed for each of these model constructs.

## 5 Status of Model Constructs from Atomistics Upward

Nanostructured materials have single crystalline domains that extend only over the order of a few to perhaps tens of nm. As shown in Fig. 6, nanostructured materials have a wide range of morphologies ranging from multi-layers, to rod-shaped and equiaxed crystallites. There is a corresponding range of compositions available as well within these heteromaterials (from left to right in Fig. 6).

It is apparent that such nanostructured systems have two key characteristics. First, the mean free path along with dislocations can travel is limited to the size/spacing of crystallites, and only a small number of dislocations can be accommodated within the crystalline domains between interfaces. Second, interfaces have a very high density and strongly influence elastic properties and mediate dislocation absorption, desorption, direct transmission and reflect (slip transfer reactions). We also comment that interfaces themselves may contain a range of initial point and line defect densities that affect mechanical response.

We may distinguish regimes of behavior according to the characteristic length scale for mediation of dislocations in equiaxed polycrystals, closely associated with the grain size distribution and associated statistical mean free path for dislocations. For characteristic lengths below about 10 nm, response is controlled by nucleation of shear transition zones in interfaces, analogous to the response of amorphous solids, since the activity of bulk dislocations would require higher stress levels to overcome

| Length Scale | Time Scale | Scale-Specific Models | Examples of Two-Scale Transition Approaches | Primary Sources of Uncertainty |
|---|---|---|---|---|
| 2 nm | NA | DFT | | Assumptions in DFT method, placement of atoms, degree of relaxation |
| | | | Quantum-MD | Interatomic potential, updating |
| 200 nm | 10 ns | MD | | Interatomic potential, cutoff, thermostat and ensemble |
| | | | Domain decomposition, QC, CADD, CAC | Attenuation due to abrupt interfaces of models, passing defects, coarse-graining error for dislocations |
| 2 μm | s | DDD and SCD | | Discretization of dislocation lines, cores, reactions and junctions, mobilities, interfaces |
| | | | Multiscale crystal plasticity | Averaging methods for defect kinetics and lattice rotation |
| 20 μm | 1000 s | LC and GC methods | | Kinetics, slip system hardening (self and latent) relations, increasing # adjustable parameters |
| | | | RVE simulations, polycrystal/composite homogenization, direct numerical simulation | RVE size, initial and boundary conditions, eigenstrain fields, self-consistency |
| 200 μm | Days | Heterogeneous FE simulations with GC or LC | | Mesh refinement, convergence, model reduction |
| | | | Substructuring, variable fidelity, adaptive remeshing, multigrid | Loss of information, remeshing error, boundary conditions |
| >2 mm | Years | Structural FE | | Model reduction error, meshing, geometric modeling |

**Fig. 5** Characteristic length and time scales pertinent to some of the crystalline plasticity modeling approaches shown in Fig. 4, along with methods for two-scale hierarchical transitions and associated primary sources of uncertainty in linking. Adapted with permission from Table 2 of Panchal et al. (2013)

interactions and image forces. It is also the regime of so-called inverse Hall-Petch behavior that gives rise to a decrease of strength with further decrease of grain size. Above grain size of 10 nm, dislocations begin to play an important role in governing deformation behavior.

Nanocrystalline structures are typically regarded as having characteristic structural length scales below 100 nm. For scales of 10–15 nm, up to 50 nm or so, the behavior of partial dislocations is dominant, in view of typical stacking fault widths. Remnant stacking faults are commonly produced in deformed grains as a result of passage of only the leading partial dislocation. Activation of dislocation sources is more limited in scope, with grain boundary and triple junctions playing a more important role than in conventional polycrystals; non-Schmid effects are manifested in dislocation nucleation (Spearot et al. 2007a, b; Tschopp et al. 2008a; McDowell 2008). Above this scale regime, we may further distinguish ultrafine grain response (sub-micron grain size) from that of tradition microcrystalline metals, with the former much more strongly dependent upon slip transfer characteristics of the high density grain boundary network, source activation and slip transfer reactions at grain boundaries. As discussed by Misra et al. (2005) and Wang and Misra (2011), the response of multi-layered nanostructured materials in the top row in Fig. 6 depends strongly on the response of individual dislocations that thread through layers with thicknesses up to tens of nm; for higher thicknesses, the transfer of dislocations through interfaces is assisted by dislocation pile-ups and cooperative slip transfer mechanisms.

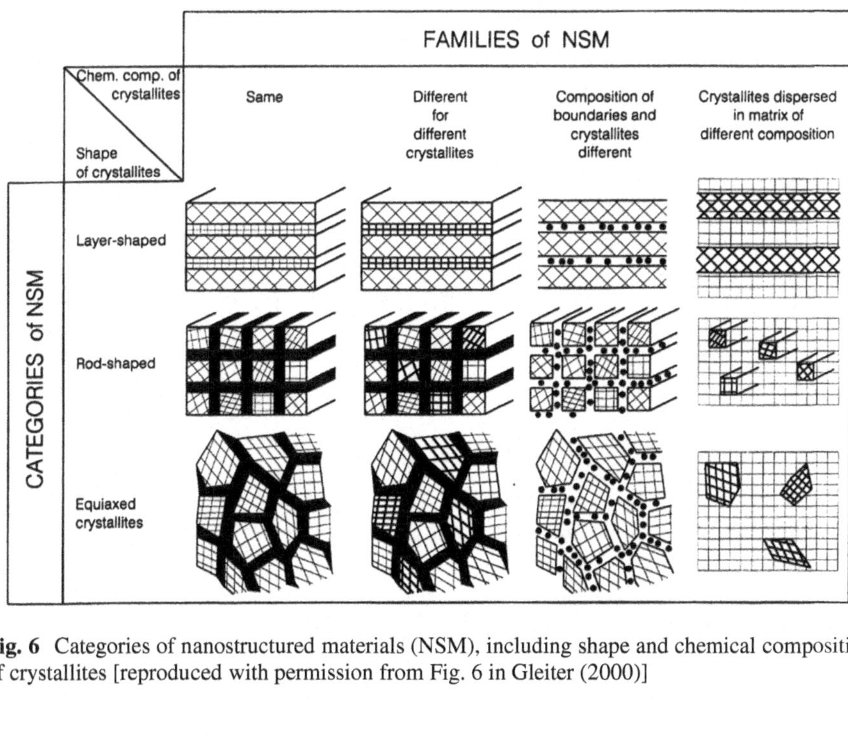


**Fig. 6** Categories of nanostructured materials (NSM), including shape and chemical composition of crystallites [reproduced with permission from Fig. 6 in Gleiter (2000)]

From a practical perspective, atomistic simulations can be pursued to explore response of nanostructured materials up to length scales on the order of 50 nm, with several caveats: (a) periodic BCs limit consideration of random variation of structures and limit size, (b) accessible time scales are limited, and (c) initial structures are typically formed using energy minimization and finite temperature relaxation. Unfortunately, nanostructured materials can be processed in different ways, some of which follow trajectories near equilibrium and some far from equilibrium. For example, ball milling or other methods for achieving severe plastic deformation (SPD) such as equi-angular channel extrusion result in high non-equilibrium dislocation densities, internal stresses, and remnant lattice distortion. Such processes are attractive because they can be performed rapidly and produce significant volumes of (bulk) nanocrystalline material. On the other hand, electrodeposition is slow and occurs near equilibrium, giving rise to significantly lower initial defect densities and internal stresses; however, only small volumes of material can be processed. High resolution TEM images of grain boundaries in electrodeposited nanocrystalline Ni and Cu produced by gas-phase condensation are shown in Fig. 7, exhibiting distinct, well-defined interfaces (Kumar et al. 2003).

The advantage of nanocrystalline form is evident in the high yield strengths in a shear punch test for electrodeposited nc Cu shown in Fig. 8 (Guduru et al. 2007). The strong strain rate sensitivity of nc Cu is also noted in this plot, and is characteristic of the reduced activation volume for dislocation mediated deformation processes in ultra-high strength nanocrystalline metals, as discussed by Zhu and Li (2010) and depicted in Fig. 9.

It is worth noting that not all nanostructured systems with the same characteristic scales are equivalent. Dislocation mediation also depends intimately on details of the interface structure. For example, the work of Lu et al. (2004) in Fig. 10 clearly shows that one can achieve ultra-high strength of nanostructured Cu processed using pulsed elecrodeposition (as opposed to direct current) if a high density of low energy coherent $\Sigma 3$ twins can be imparted during processing; note a level of ductility (prior to diffuse necking) in the nanotwinned Cu relative to conventional nanocrystalline Cu processed via SPD, which has nil ductility. The efficiency of slip transfer reactions to facilitiate dislocation transport in the former gives rise to this remarkable difference.

## *5.1 Principles of Atomistic Modeling*

**Interatomic Potentials** Since direct atomistic modeling or coarse-grained methods that employ an empirical interatomic potential offer the most direct means to model dislocations in nanostructured materials, we briefly consider the relevant foundational principles of this class of models. This leads to an understanding of the simplifications involved and the limitations in terms of modeling details of electronic charge distribution and local bonding characteristics/environments.

Since the latter limitations can limit efficacy of modeling defects in crystals, it is particularly important to understand what kinds of information must be emphasized or given preference in fitting these interatomic potentials. The form of the empirical potential is regarded as a modeling approximation, carrying with it so-called model form uncertainty that can be quantified by comparing performance of multiple potentials for model problems of interest (cf. Xu et al. 2017b).

The basis for modeling with empirical potentials using classical models of atomic interactions is the so-called Born-Oppenheimer approximation (BOA), employing two fundamental assumptions:

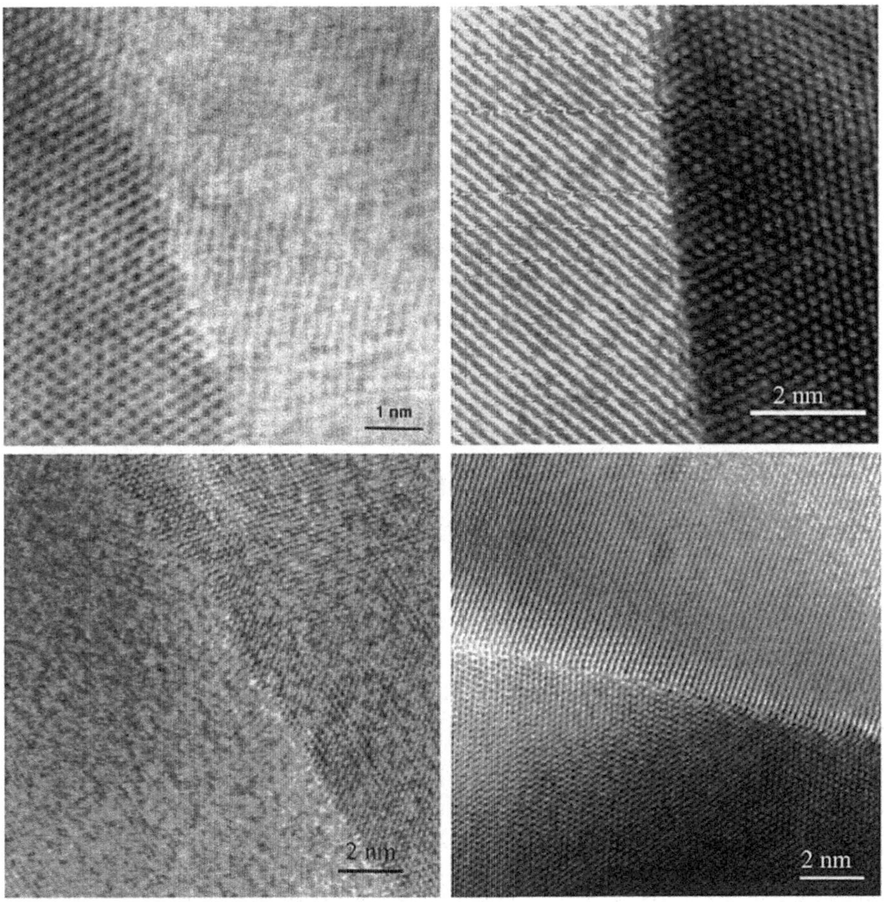


**Fig. 7** High resolution TEM images of grain boundaries in electrodeposited nanocrystalline Ni (top row) and (bottom row) nanocrystalline Cu produced by gas-phase condensation [reproduced with permission from Fig. 3 of Kumar et al. (2003)]

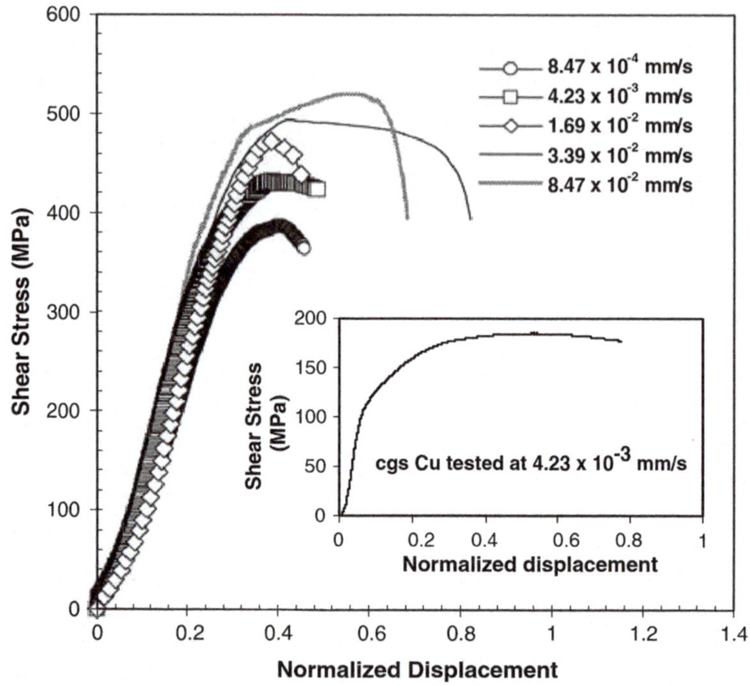


**Fig. 8** Stress-strain curves measured from a shear punch test for electrodeposited nanocrystalline (enc) Cu (30–80 nm grain size range) and conventional as-received (cgs) cold rolled Cu [reproduced with permission from Fig. 4 of Guduru et al. (2007)]

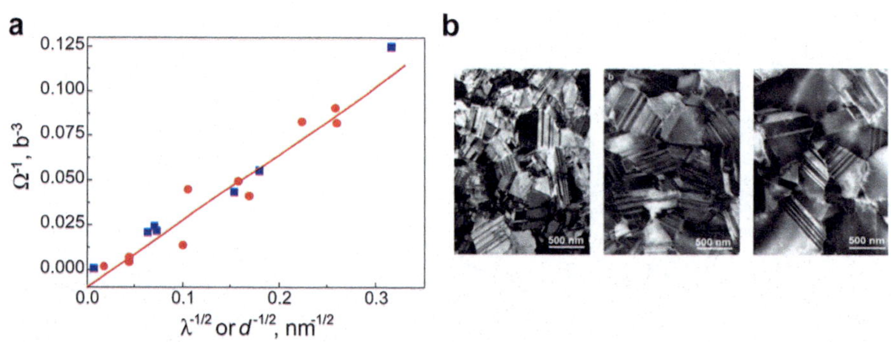


**Fig. 9** Size dependent activation volume versus inverse grain size (**a**) for nanotwinned and nanocrystalline Cu, and (**b**) TEM images showing the associated decrease of nanotwin density (from left to right) with the same average grain size of 500 nm [reproduced with permission from Fig. 9 of Zhu and Li (2010)]

- Electrons move "fast" compared to the nuclei. The BOA asserts that the electrons find the ground state configuration by responding instantaneously to the slow evolution of the nuclei.

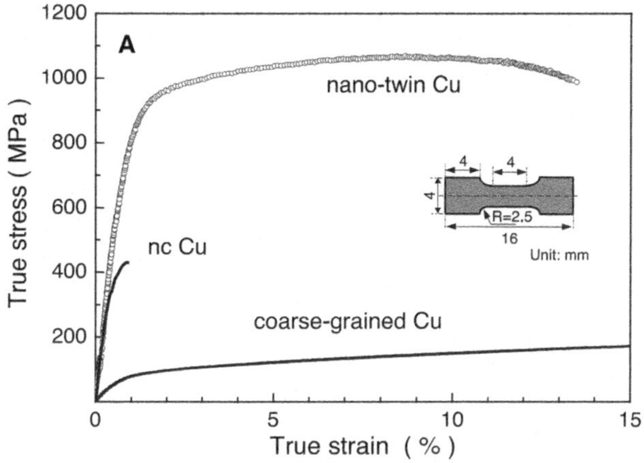


**Fig. 10** Tensile stress-strain curve for pulsed elecro-deposited nanocrystalline Cu sample with nano-twins compared to the responses of a coarse-grained polycrystalline Cu sample (with an average grain size larger than 100 μm) and a nanocrystalline Cu sample processed via conventional SPD (mean grain size ~ 30 nm) [reproduced with permission from Fig. 2 of Lu et al. (2004)]

- The response of the interaction of nuclei is modeled as that of rigid particles that move with fully nonlocal forces established by bond strengths at a distance that factor in the electron interactions and valence states.

The BOA is not valid for cases where the electron configurations do not instantaneously change with position of nuclei, for example dynamic fracture of covalently bonded insulators. From a quantum-mechanical perspective, the BOA effectively implies multiplicative separability of the wave function solution to the time dependent Schrodinger equation into (1) a wave function of electrons that depends only on the positions of nuclei, and (2) a time dependent wave function for nuclei. This removes explicit dependence of the electrons in the solution of the time dependent Schrodinger equation for the nuclei. Accordingly, nuclei are treated as classical particles that interact with nonlocal forces (action at a distance) affected by associated electronic states. If the de Broglie thermal wavelength $\Lambda = h/\sqrt{2\pi m k_B T}$ is much less than the interatomic spacing, then waves are spatially localized and atoms can be treated as classical particles. Here, $m$ is the particle mass and $h$ is the Planck constant. This approximation is valid to very low absolute temperatures for many atomic systems. Of course, quantum mechanics is still required if we desire positions of electrons, valence states, and/or details of changes of bonding with changes in relative position of nuclei.

The interatomic potential incorporates these interactions chiefly through description of the internal potential energy, which has contributions from Coulomb interactions between atoms as well both the potential and kinetic energy of electrons bound to their ground state via the BOA. For a finite system of atoms, we define the potential energy as $U = U^{\text{int}} + U^{ext}$, where $U^{\text{int}}$ is the interatomic, short range potential energy, and $U^{ext}$ is the potential energy due to interaction of system atoms with external fields (e.g., gravity, EM radiation) and interactions with atoms external to the system (mainly near external boundaries). The effects of interactions across external boundaries become less and less important as the system size increases.

Atomic interactions are fully nonlocal. As in quantum mechanics, forces on atoms ideally depend on all other atoms in the system. It turns out that most of the internal PE due to bonding is affected largely by only a few neighboring atoms—hence its qualification as short range. We will later introduce the notion of a "cutoff radius" for bonding in applying interatomic potentials to model atomistic systems for purposes of efficiency without sacrificing essential accuracy in describing interactions.

There are some nuances. Because the interatomic interactions are nonlinear and the energy is not quadratic, we must in general expect the system of atoms to behave in a nonlinear elastic manner for anything other than infinitesimal perturbations away from an equilibrium (energy minimized) configuration. Furthermore, the interatomic potentials much satisfy the requirements of continuum mechanics in terms of material frame indifference, rotation invariance, and material symmetry (cf. Tadmor and Miller 2011). These requirements impose powerful constraints that assist enormously in reducing the potentially acceptable forms of interatomic potentials.

For any system of interatomic interactions, the total internal force on atom a due to interaction with all other atoms in the system is given by

$$\underline{f}^{\mathrm{int},\alpha} = -\frac{\partial}{\partial \underline{r}^{a}}\left[U^{\mathrm{int}}\right] = \sum_{\substack{\beta \\ \beta \neq \alpha}} \underline{f}^{\alpha\beta} \tag{1}$$

For pairwise interactions, the simple pair potential $\varphi_{\alpha\beta}$ depends on the center-to-center distance between atoms and internal potential energy is defined as

$$U^{\mathrm{int}} = \frac{1}{2}\sum_{\substack{\alpha,\beta \\ \beta \neq \alpha}} \widehat{\varphi}_{\alpha\beta}\left(\underline{r}^{\alpha}, \underline{r}^{\beta}\right) = \sum_{\substack{\alpha,\beta \\ \beta \neq \alpha}} \varphi_{\alpha\beta}\left(r^{\alpha\beta}\right) \tag{2}$$

where $r^{\alpha\beta} = \left\|\underline{r}^{\beta} - \underline{r}^{\alpha}\right\|$. This leads to the relation for the central force between atoms $\alpha$ and $\beta$ as

$$\underline{f}^{\alpha\beta} = \varphi'_{\alpha\beta}\left(r^{\alpha\beta}\right)\frac{\underline{r}^{\beta} - \underline{r}^{\alpha}}{r^{\alpha\beta}}, \quad \varphi'_{\alpha\beta} = \frac{\partial U^{\mathrm{int}}}{\partial r^{\alpha\beta}} \tag{3}$$

The general form of the total potential for an N-atom system can extend well beyond pair potentials to incorporate dependence on three-body interactions and higher. Effectively, this incorporates much information regarding the configurational environment of atoms, including bond length distributions, bond angles, etc. It can be shown that Eq. (1) holds for any potential, including those with multi-body interactions; therefore, central force interactions are manifested even for multi-body potentials, not just pair potentials (cf. Tadmor and Miller 2011). Bond stiffness, closely related to linearized elastic constants, is given in the general case by

$$\kappa^{\alpha\beta\delta\gamma} = \frac{\partial^2 U^{\text{int}}\left(r^{\xi\eta}\right)}{\partial r^{\alpha\beta}\partial r^{\delta\gamma}} \tag{4}$$

It is understood that pair potentials offer an incomplete description of the bond stiffness.

Recognizing that the majority of central force interactions occur over a very limited distance, we introduce the notion of cutoff radius to facilitate implementation and code efficiency. The interatomic potential is effectively modified by this assumption and therefore it is an integral part of the framework, i.e.,

$$\varphi_{\alpha\beta} = \begin{cases} \psi_{\alpha\beta} & \text{if } r < r_{cut} \\ 0 & \text{otherwise} \end{cases} \tag{5}$$

where $r_{cut}$ is effectively a model parameter. Since forces and stiffnesses depend on derivatives of $\varphi_{\alpha\beta}$, the potential $\psi_{\alpha\beta}$ should approach 0 at $r = r_{cut}$ and beyond. Note that line defects in a lattice can have much longer ranged elastic interactions; hence, we distinguish between structure morphology and its interactions at a distance through a lattice and interatomic interactions. The quality of fit of the potential depends on the cutoff radius. A smoothing polynomial can be applied to potential in order to satisfy these requirements without introducing undesirable discontinuity. Also, we must track only the neighbor list of each atom, at a distance just beyond the cutoff, e.g., $1.2r_{cut}$.

The Lennard-Jones potential is an example of a simple pair potential, defined as by a competition of repulsive and attractive interatomic forces as (Tadmor and Miller 2011)

$$\sum_{\substack{\alpha,\beta \\ \beta\neq\alpha}} \varphi_{\alpha\beta}\left(r^{\alpha\beta}\right) = 4\varepsilon\left[\left(\frac{\sigma}{r^{\alpha\beta}}\right)^{12} - \left(\frac{\sigma}{r^{\alpha\beta}}\right)^{6}\right] \tag{6}$$

where $\varepsilon$ is the depth of the potential energy well and $\sigma$ is the value of interatomic distance at which the potential energy vanishes. The equilibrium distance is given by the condition $\underline{f}^{\alpha\beta} = \underline{0}$.

While pair potentials are simple and easy to use, they are limited in describing complexity of interatomic interactions in lattices, and they don't address the change of coordination of bonding very well. For example, Buehler (2008) has shown that fundamental differences in bonding and local atomic environment between bulk and surface atoms cannot be properly described using pair potentials. Fortunately, there is a hierarchy of interatomic potentials available to address these issues:

- Pair Potential—e.g., Lennard-Jones (https://en.wikipedia.org/wiki/Lennard-Jones_potential), Morse (https://en.wikipedia.org/wiki/Morse_potential)
- Pair Functional—EAM (Daw and Baskes 1984) (glue potential, effective medium) (https://en.wikipedia.org/wiki/Embedded_atom_model)
- Cluster Potential—CHARMM (https://en.wikipedia.org/wiki/CHARMM), AMBER (https://en.wikipedia.org/wiki/AMBER)
- Cluster Functional—ReaxFF (https://en.wikipedia.org/wiki/ReaxFF), MEAM (Baskes 1992, https://www.potfit.net/wiki/doku.php?id=meam), Tersoff and other bond order potentials (https://en.wikipedia.org/wiki/Bond_order_potential)

As we move from pair to cluster potentials, we incorporate many body interactions. As we move from potentials to functionals, we incorporate greater nonlinearity and sensitivity to the local atomic environment. Tadmor and Miller (2011) discuss how the maximum practical system size that can be modeled using 32–64 CPUs depends on the type of empirical or semi-empirical potential, with cluster potentials and functionals typically having more severe limitations (e.g., CHARMM/AMBER, ReaxFF).

The Embedded Atom Method (EAM) potential (Daw and Baskes 1984) is often preferred as a good compromise for purposes of modeling various aspects relating to structure and properties of dislocations in crystals, including the generalized stacking fault energy or $\gamma$-surface. Both the unstable and stable stacking fault energies play a key role in describing the initial separation into partial dislocations and stable width of stacking faults in crystals. It is widely employed for cubic crystals, and augments the pair potential with a background embedding energy term that is effective for nondirectional metallic bonding, i.e.,

$$U_i = \underbrace{G\left(\overline{\rho}_i\right)}_{\substack{\text{Embedding}\\ \text{Function}}} + \underbrace{\frac{1}{2}\sum_{j(\neq i)} \phi\left(R_{ij}\right)}_{\text{Pair Potential}}\ ,\quad \overline{\rho}_i = \sum_{j(\neq i)} \rho\left(R_{ij}\right) \tag{7}$$

The Modified Embedded Atom Method (MEAM) was introduced later by Baskes (1992) to consider additional effects of bond angles (e.g., three body terms) on response of crystals with more directional bonding (e.g., pure Al).

The stacking fault energy in FCC and HCP crystals pertains to the change of energy associated with the formation of a change of stacking sequence (stacking fault) corresponding to the passage of a leading Shockley partial dislocation, terminated at the trailing partial dislocation. In FCC crystals, the stacking fault sequence shifts the close-packed planes from ABCABC... (the parent lattice) to ABCACABCA... , with the fault adopting a local HCP stacking sequence (ABABAB...), and an attendant increase of enthalpy. Essentially, the $\gamma$-surface reflects the reaction pathway of dislocation dissociation into Shockley leading and trailing partial dislocations within the potential energy landscape (Hull and Bacon 2011).

In fitting an interatomic potential suitable for application to modeling dislocations, we may consider a number of quantities to consider, whether informed via experiments or bottom-up DFT computations. These include:

- Cohesive energy of an infinite crystal
- Vacancy formation energy $E_{vac}$, based on the relaxed energy of the simulation cell containing a single vacancy relative to the cohesive energy of a single crystal
- Surface energy (free surfaces, interfaces, etc.)
- Interface (e.g., grain boundary) structure (to be discussed later)
- The $\gamma$-surface, used to characterize the energy increase associated with transition pathways associated with partial dislocations that give rise to stacking faults (especially FCC, HCP)
- Elastic constants of a crystal

The fitting process is highly nontrivial, and there are only a handful of seasoned experts who are regarded as most effective in this regard. Breaking down the process more systematically (perhaps with an eye towards increased automation, for example in connection with machine learning), we may consider that fitting interatomic potentials requires both fitting and testing phases, and more iteration is required to develop potentials for complex phenomena. Moreover, fitting potentials for multicomponent alloys is quite involved (e.g., Moore et al. 2017). We expect much progress in such fitting protocols in the coming decade with enhanced dedication to the confluence of simulation, experiments, and data science.

Returning to the $\gamma$-surface and its key importance in modeling dislocations in crystals via atomistics, we consider the differences that are observed among various distinct EAM potential fits for pure Al and Cu. The huge variation of the GSFE curve among various fits to the EAM potential in Fig. 11 is striking. A few of these fits are more appropriate in terms of unstable and stable stacking fault energies based on DFT simulations. Some of these fits would be more appropriate in modeling of dislocations than others (e.g., Mishin et al. potentials 1999, 2001; Xu et al. 2017b). Of course, there is always an issue of the uncertainty of various types of DFT exchange correlation functional approximations, as well as experiments.

**Molecular Statics and Dynamics Methodologies** There are two types of atomistic simulations, molecular statics (MS) and molecular dynamics (MD). Both MS and MD rely on interatomic potentials to govern atomic interactions. As the names imply, MS is concerned with seeking atomic ensemble arrangements that minimize the global potential energy, i.e., static equilibrium atomic positions. The resulting arrangement approximates atomic structure at 0 K. Energy minimization is commonly pursued using conjugate gradient algorithms, resulting in increasing cost with the size of the ensemble. MD algorithms integrate the classical equations of motion simultaneously for all atoms at finite temperature, i.e.,

$$m^{\alpha}\ddot{\underline{r}}^{\alpha} = \underline{f}^{\alpha} = \underline{f}^{\alpha,\text{int}} + \underline{f}^{\alpha,ext} = -\frac{\partial U}{\partial \underline{r}^{\alpha}}. \tag{8}$$

with the correction on the equations of motion depending on the type of ensemble chosen (e.g., NPT, NVT, NVE, etc.). Steps in MD including providing input parameters (e.g., T, N, and time step size, typically on the order of fs), initializing the position and velocity of all atoms, computing forces, and integrating the equations of motion. The velocity-Verlet algorithm is most commonly employed to move the atoms at each time step (Buehler 2008; Tadmor and Miller 2011).

Some commonly accessible MD codes include:

- LAMMPS—http://lammps.sandia.gov/
- DL Poly—http://www.csar.cfs.ac.uk/user_information/software/chemistry/dl_poly.shtml
- Amber—http://ambermd.org/
- Carr-Parrinello MD—http://www.nnin.org/nnin_carrparrinello.html
- NAMD—http://charm.cs.uiuc.edu/research/moldyn/

MD has certain limitations. First, time scales are necessarily short since we must resolve fs scale atomic vibration in a lattice. Beyond this, the equation of motion in Eq. (8) neglects:

- Conductive heat transfer among atoms (considers only phonons), including conduction associated with free electrons
- Chemical reactions of successive bond breaking that redistributes the electron charge (unless ReaxFF (Buehler et al. 2006) or tight binding (Negre et al. 2017) potentials are used to reflect the evolving environment of charge density)

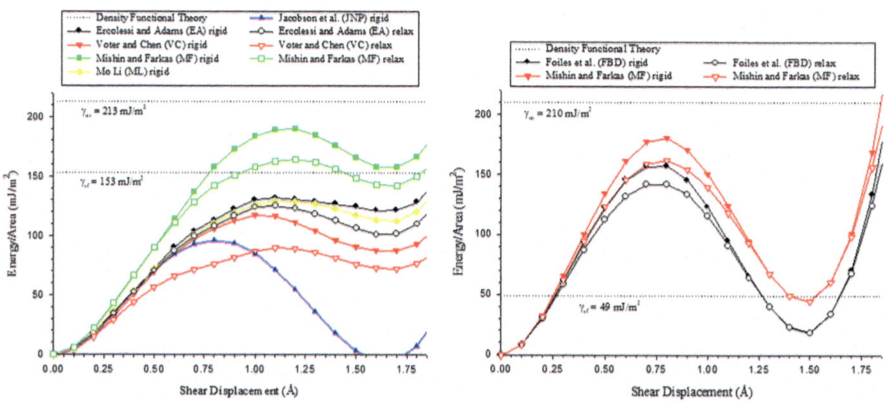


**Fig. 11** <112> Generalized stacking fault energy (GSFE) curves based on various fits to the Embedded-Atom Method (EAM) potential for Al (left) and Cu (right) (data from Zimmerman et al. 2000 and Spearot 2005)

Energy minimization approaches tend to focus on 0 K conditions, e.g., MS. Temperature of an atomic ensemble is defined based on atomic velocities relative to center of mass (Tadmor and Miller 2011), i.e.,

$$T_{inst} = \frac{2}{3k_B N} \sum_{\alpha=1}^{N} \frac{1}{2} m^{\alpha} \left\| v_{rel}^{\alpha} \right\|^2 \tag{9}$$

A thermostat modifies the thermal energy and is useful if temperature gradients are not too large within the ensemble/cell of interest. There are several common types of MD ensembles that stem from statistical mechanics (https://en.wikipedia.org/wiki/Molecular_dynamics), each having different characteristics according to the aims of the simulation:

- An adiabatic process with no heat exchange is termed as the *microcanonical* (NVE) ensemble. In this case, the system of atoms of interest is isolated from changes in number of atoms (N), volume (V), and energy (E).
- In the *canonical* (NVT) ensemble (NVT), N, V and temperature T are all conserved. Thermal energy of the atomic system is exchanged with surroundings by using a thermostat.
- The *isothermal–isobaric* (NPT) ensemble, N, pressure (P) and T are all conserved. A barostate is introduced along with a thermostat achieve the desired pressure by allowing the simulation "box" size to adjust until the desired pressure/stress is achieved.

Control of NPT or NVT ensemble temperature focuses on a range of techniques. They include velocity scaling, Langevin' 'white' noise (random forces are incorporated into the equations of motion), Berendsen friction (global heat bath), and the Nosé-Hoover algorithm (formal scaling). The interested reader can consult (Tadmor and Miller 2011) for more details, including chains of thermostats with regard to sensitivity of temperature to defect avalanching (Tiwari 2014). One of the issues with use of a thermostat in modeling evolution of defects such as dislocations or cracks is that velocity rescaling is typically performed over the entire ensemble, whereas temperature can locally rise to higher levels in conjunction with dissipative processes. Strategies that allow for local modification of velocity, such as Langevin dynamics, can be employed to this end. Care must be taken to adhere to proper heat conduction behavior in the equilibrium limit.

Thermodynamic interpretations of these ensembles appeal to the foundations of equilibrium statistical mechanics/thermodynamics, which is valid for processes occurring not too far from equilibrium. Of course, there is always a question in this regard for high rate processes commonly modeled using MD, but there is much precedence in the literature for modeling nonequilibrium dislocation evolution in a complex potential energy landscape based on forms consistent with equilibrium statistical mechanics (e.g., internal state variable theory, cf. Lemaitre and Chaboche 1990; Muschik 1993; McDowell 2005). The Helmholtz free energy for a canonical NVT ensemble (Tadmor and Miller 2011) is defined by $\psi = U - TS = \langle H \rangle - TS$, where $S = -\partial\Psi/\partial T$ is the entropy. The thermodynamic internal energy $U$ is based on the ensemble average of the Hamiltonian defined by

$$H\left(\underline{q}, \underline{p}\right) = \sum_{\alpha} \frac{\underline{p}^{\alpha} \cdot \underline{p}^{\alpha}}{2m^{\alpha}} + U^{\text{int}}\left(\underline{q}\right) \tag{10}$$

Here, $p_i$ and $q_j$ represent the momenta and positions of atoms in the ensemble in phase space, which is considered equilibrium and therefore Boltzmann distributed in terms of the partition of internal energy; in this case, the free energy function is based on the partition function $Z$ (probability of states) for a canonical NVT ensemble, i.e.,

$$\Psi = -k_B T \ln Z, \quad Z = \frac{1}{N!h^{3N}} \int_{\Gamma(\underline{F})} e^{-\mathrm{H}(q,p)/k_B T} d\underline{q}\, d\underline{p} \tag{11}$$

where $k_B$ is Boltzmann's constant and $h$ is Planck's constant. At finite temperature, when change of volume is allowed for the ensemble to accommodate a desired pressure (e.g., NPT), the Gibbs free energy is given by $G = \psi + PV$. The appropriate current configuration stress tensor measure (analogous to Cauchy stress) is the virial stress for the ensemble, i.e.,

$$\sigma_{ij} = -\frac{1}{V} \sum_{\alpha} \left\langle \frac{p_i^{\alpha} p_j^{\alpha}}{m^{\alpha}} + f_i^{\alpha,\text{int}} r_j^{\alpha} \right\rangle \quad , \quad V = V_o \ \det\left(\underline{F}\right) \tag{12}$$

The first term of the virial stress is the kinetic part, associated with velocity fluctuations relative to the mean velocity of the ensemble. The second term derives from bond forces, i.e., the potential energy function; accordingly, at 0 K only the second term is manifested. This stress measure is not useful for an individual atom or a small set of bonds per se at finite temperature, since the mean of the fluctuation velocities is not statistically representative. On the other hand, the second term has meaning on a per atom basis. While some have criticized the virial stress on these grounds, it is simply a reflection of the ensemble behavior in the context of statistical mechanics. There are limits to ensemble behavior assumptions when modeling small numbers of atoms.

We note that invoking the machinery of equilibrium statistical mechanics is not essential to interpret MD results (nor is it or even strictly correct). Irreversible microstructure rearrangement processes are inherently non-equilibrium when exercised at time scales below characteristic relaxation times associated with atomic structure, which is generally the case for extended defects such as dislocations. We discuss this issue later in the context of interpreting dislocation evolution as a sequence of constrained equilibrium states, for which equilibrium statistical mechanics measures of free energy are used as heuristic forms, with the equilibrium requirement of zero thermodynamic force at zero thermodynamic flux relaxed to admit description of evolving thresholds on ensemble average driving forces for evolution.

**Boundary Conditions in Atomistic Modeling** Owing to size limitations in atomistic simulations and the desire to employ a thermostat and/or barostat over some ensemble of atoms, the assumption of periodic boundary conditions (PBCs) is often imposed on a user-defined unit cell. This cell may or may not be statistically representative of the material structure of interest; it typically is not large enough to consider modeling set sof dislocations with net Burgers vector in many MS or MD simulations. Since the interatomic potential is translationally invariant and has permutation symmetry of indices (owing to material frame indifference), by repeating a unit cell in space, i.e., $\underline{r}^{\alpha}{}_{per}\,(n_1, n_2, n_3) = \underline{r}^{\alpha} + n_i\ \underline{L}_i$, where $n_i$ are integers and $\underline{L}_i$ are nonorthogonal basis vectors defining a "unit cell". Forces within each "image" are identical ("periodic copies") as shown in Fig. 12 in 2D. As a consequence, $U = U^{\text{int}}$ and the unit cells repeat to infinity (i.e., no external forces). PBCs are often used for perfect lattices (e.g., "bulk" behavior"), but we must be careful when using them for nonuniform structures and defects, as these images are replicated in each cell and can interact via image forces that are artifacts of the simulation. Dislocation simulations are often notoriously susceptible to influence of such image forces, since defects are also periodic and dislocations have long range stress fields. Moreover, long wavelength phenomena are challenging to model using PBCs since they require very large scale unit cells, e.g. phonons with wavelengths longer than the cell size or 3D dislocation networks. Some atomic scale phenomena can be modeled effectively using PBCs as long as the structure is well-represented and image forces are minimized, including elastic constants of a defect free crystal, thermal conductivity [e.g., using the Green-Kubo formalism in MD (Green 1954; Kubo et al. 1957)] and dislocation core structures and self energies).

In addition to commonly employed PBCs, one can apply external forces on atoms or one can specify the position of atoms. We typically impose such conditions using a "pad" region that includes some interior atoms, not just boundary atoms. The imposition of such boundary conditions must be consistent with the type of ensemble employed; for example, a generalized NPT ensemble that admits imposition of desired surface tractions (cf. Spearot 2005; Tschopp et al. 2008a) is useful for problems involving stress-control. For displacement-controlled situations, one can also impose pad region displacements. Imposition of mixed force-displacement boundary conditions are also possible. Constraints can further be applied on atomic motion within the domain, e.g., velocities in certain directions, for example to avoid introduction of shock waves under dynamic shear loading conditions (cf. Horstemeyer et al. 2003; Tucker et al. 2010).

To model isolated finite regions (rather than periodic domains) such as crack tip fields or dislocation core fields for an individual dislocation, we can sometimes impose the atomic displacements to conform to a linear elastic solution suitably far

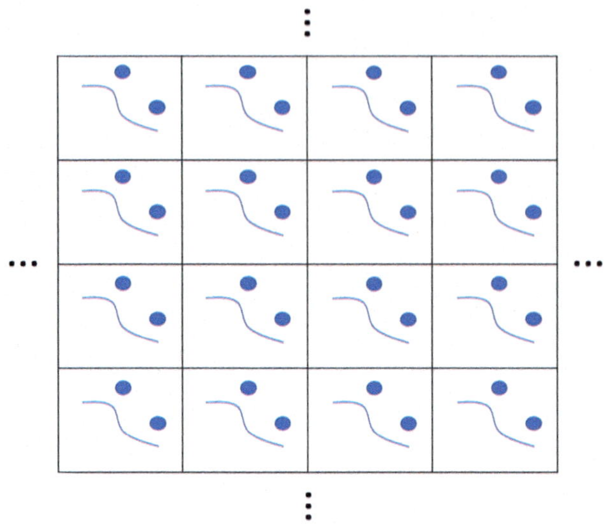

**Fig. 12** 2D rendering of periodic unit cell concept

from the region of interest. This will be discussed in a later section along with other domain decomposition methods.

**Potential Energy Landscape** Given the central importance of the potential energy of bonding in atomistic solutions, the concept of the Potential Energy Landscape (PEL) is key to understanding equilibrium configurations of atoms as a basis for estimating the probability of occurrence of nonequilibrium structure transitions with assist of temperature and external driving force(s). The configuration space is established by the positions of all atoms in the system, $\underline{r} = \left(\underline{r}^1, \underline{r}^2, \ldots, \underline{r}^N\right)$. The function $U = U^{\text{int}} + U^{ext} = U\left(\underline{r}\right)$ defines the potential energy landscape (PEL). Here, $U^{\text{int}}$ accounts for interactions between atoms comprising the system, while $U^{ext}$ accounts for both the influence of external fields (gravity, EM radiation) and bond interactions with atoms outside the system (but still near the boundary, lying within the cutoff radius). We note that in general the relative level of $U^{\text{int}}$ rank orders according to perfect crystal < defects < amorphous.

The severity of challenges in modeling structural transitions involving defect evolution is closely related to the complexity of the PEL. Figure 13 shows a complex PEL for protein folding, which is qualitatively similar to a relief map of a mountain range. At atomic fidelity, the PEL is well-defined for a defect-free, periodic lattice, and is less well-defined in the presence of defects, especially line and surface defects, depending on the defect type and density. Point defect migration in a period lattice is readily characterized with reference to the parent lattice PEL. However, dislocations are extended defects that can modify the PEL substantially, especially at high density and/or with strong core interactions. The PEL is biased by the internal stress fields of dislocations in addition to the applied stress. A dislocation network can manifest complexity and heterogeneity of the PEL in a manner analogous to the complexity of long chain polymer networks which have topological complexity and combinations of weak and strong bonds.

**Molecular Statics Energy Minimization** Minimization of PE at 0K based on statics ($\ddot{\underline{r}}^{\alpha} = \underline{0}$) is pursued to determine energetically favorable equilibrium configurations. The function $U\left(\underline{r}\right)$ is multidimensional and non-convex. Figure 14 shows a complex PEL, which is 3N dimensional and has numerous local minima or "basins" separated by dividing surfaces (Tadmor and Miller 2011).

In static equilibrium, the net force on each atom vanishes since its acceleration vanishes. This contrasts with MD, even if time averaged. Local minima that are reached depend on the starting point in the search; if a global minimum is desired, then many starting points must be considered. The goal is to achieve the relaxed, energy minimized configuration/structure. We may generalize the potential energy function to depend on a set of admissible configurations of atoms, $\underline{u}$. Since

$$\underline{f}\left(\underline{u}\right) = -\frac{\partial}{\partial \underline{u}} U\left(\underline{u}\right) = -\nabla_u U\left(\underline{u}\right), \tag{13}$$

the minimization process focuses on the solution of $\underline{u}_{\min} = \arg\min U\left(\underline{u}\right)$ corresponding to $\underline{f}\left(\underline{u}\right) = \underline{0}$. We may consider both local and global techniques to pursue this iterative minimization procedure:

Local (in the vicinity of initial guess for $\underline{u}_{\min}$):

- Method of Steepest descent (SD)—search direction along forces at each step
- Conjugate Gradient (CG) ➔ most widely used—search direction of steps are orthogonal
- Newton-Raphson (NR)—employs a Taylor Series expansion

Global methods:

- Simulated annealing (based on Monte Carlo sampling)

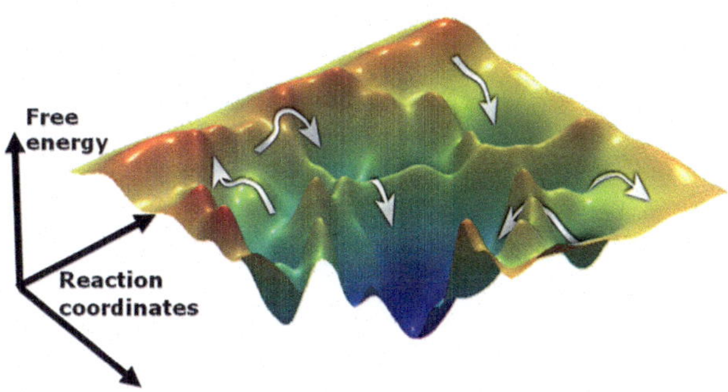


**Fig. 13** Schematic of a complex PEL for protein folding (https://www.weizmann.ac.il/chemphys/cfharan/sites/chemphys.cfharan/files/uploads/free_energy_map.png)

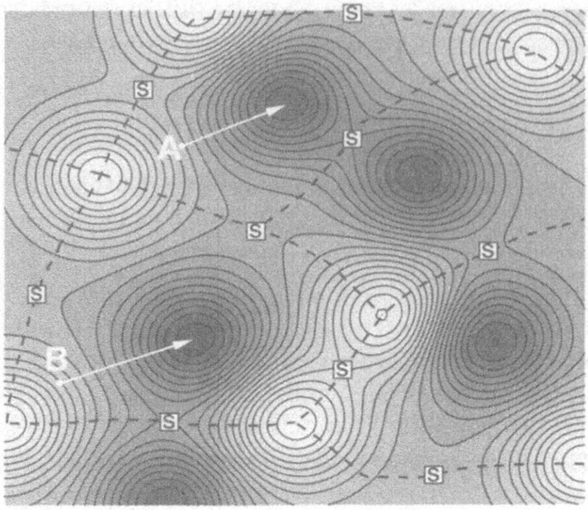


**Fig. 14** Projected schematic of a 3D PEL with local minima at A and B, and saddle points marked by "s". Solid lines are iso-energy contours [reproduced with permission from Fig. 6.1 of Tadmor and Miller (2011)]

- Genetic algorithms

Figure 15 presents a fairly typical example of how the total potential energy in an atomic/molecular ensemble is reduced with number of iteration steps towards a minimum. A nuance of energy minimization for systems with large numbers of degrees of freedom is the tendency to find a large number or local minima, depending on the starting configuration of the system. For this reason, global searching is necessary to assure proper sampling.

A simple but illustrative example of energy minimization is the process of finding energy minimized interfaces in bicrystals. The macroscopic grain boundary geometry is defined using five degrees of freedom (DOF) that fully describe the crystallographic orientation of one grain relative to the other (3 DOF) and the orientation of the boundary relative to one of the grains, i.e., the grain boundary plane (2 DOF). Such interfaces are commonly characterized by a rotation angle and a misorientation axis, reflecting five macroscopic degrees of freedom. Coincidence site lattice (CSL) notation is used to represent certain interfaces (including some of the most common symmetric tilt boundaries). For example, a $\Sigma 5$ (310) interface has the (310) boundary plane (Miller indices) and the interface contains one shared atom for both lattices every 5 sites along the interface.

On a microscopic level, the translation between the two adjoining crystal lattices requires three additional DOF. Atomistic simulations are used to explore how the grain boundary DOF affect the structure and properties of particular grain boundaries (Tschopp et al. 2008a, 2015). Owing to different translations of the grains with respect to each other and different densities of atoms at the interface, a

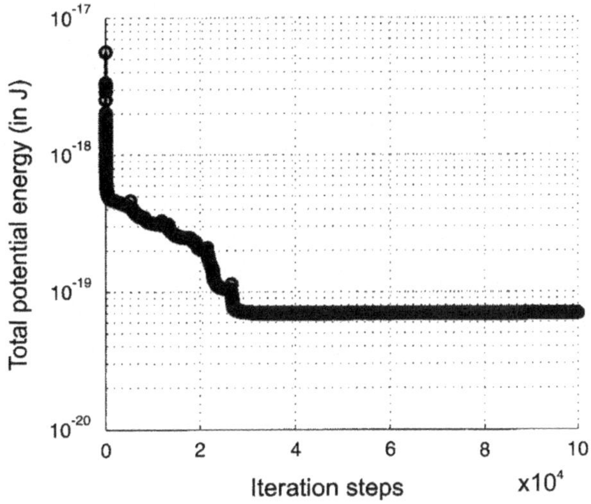


**Fig. 15** Example of reduction of total potential energy of an ensemble of atoms comprising molecular structure of a solvated protein [reproduced with permission from Fig. 2.6 of Buehler (2008)]

single set of macroscopic DOF can produce a multitude of different grain boundary structures and energies. Accordingly, there are multiple steps for obtaining the minimum energy structure: grain boundary initialization, rigid body translation, atom deletion criterion, and conjugate gradient energy minimization (Tschopp and McDowell 2007a). The (excess) interface energy is defined for each atom within some small region of the interface as $e_i^{excess} = e_i - e^{bulk}$, where $e^{bulk}$ is the potential energy of the bulk lattice. Then, the interface energy is averaged over the interface region according to $E^{\text{int}} = (1/A_{\text{int}}) \sum_{N^{\text{int}}} e_i^{excess}$. A wide range of near minimum energies can be considered, or a criterion of choice can be exercised by comparing with available TEM results (Tschopp et al. 2008a). Figure 16 presents interface energies as a function of misorientation angle for STGBs in Al and Cu (Tschopp et al. 2015).

Using NPT ensemble calculations (Spearot 2005; Tschopp et al. 2008a), various interface behaviors were explored in terms of nucleation of dislocations under applied stress, with interfaces exhibiting strong non-Schmid effects (Spearot et al. 2007b; Tschopp et al. 2008a) and tension-compression asymmetry (Tschopp and McDowell 2007b; Tschopp et al. 2008b). Good examples of application of atomistic modeling to responses of nanocrystalline ensembles can be found in Tucker et al. (2012), Tiwari et al. (2013), and Tiwari (2014).

## 5.2 *Rendering and Visualization of Atomistic Ensembles*

Any atomistic modeler can attest to the challenge of rendering the evolution of various kinds of point, line and surface defects for large scale simulations. To this end, several descriptors are commonly used. The bulk coordination number (CN) of a given atom in the interior of a crystal lattice is the number of atoms touching the given atom. Perfect HCP and FCC lattices have CN = 12, the highest among various Bravais lattices. BCC lattices have CN = 8. Atoms located at surfaces have different coordination than atoms lying in the bulk, in general. Since point, line and surface defects reflect, change of coordination. CN can be useful to distinguish among local environments to assist in visualization and interpretations.

The centrosymmetry parameter (Kelchner et al. 1998) is a commonly used descriptor for initially centrosymmetric microstructures. For FCC crystals, for example, with coordination of 12, the centrosymmetry parameter for the ith atom is defined by

$$c_i = \sum_{j=1}^{6} \left\{ \left| \sum_{k=1}^{3} r_{k,j} + r_{k,j+6} \right|^2 \right\} \tag{14}$$

where $r_k$ is the kth component of the bond vector. For a perfect FCC lattice, the ratio $c_i/a_o^2 = 0$, where $a_o$ is the equilibrium lattice parameter. Values of $c_i/a_o^2$ for a partial dislocation, stacking fault and surface atom in a FCC lattice are respectively given by 0.1423, 0.4966, and 0.16881 (Buehler 2008).

The slip vector is useful for determining the Burgers vector of a dislocation (Zimmerman et al. 2001). It is given by

$$s_i^{\alpha} = \frac{1}{n_s} \sum_{\alpha \neq \beta}^{n_n} \left( x_i^{\alpha\beta} - X_i^{\alpha\beta} \right) \tag{15}$$

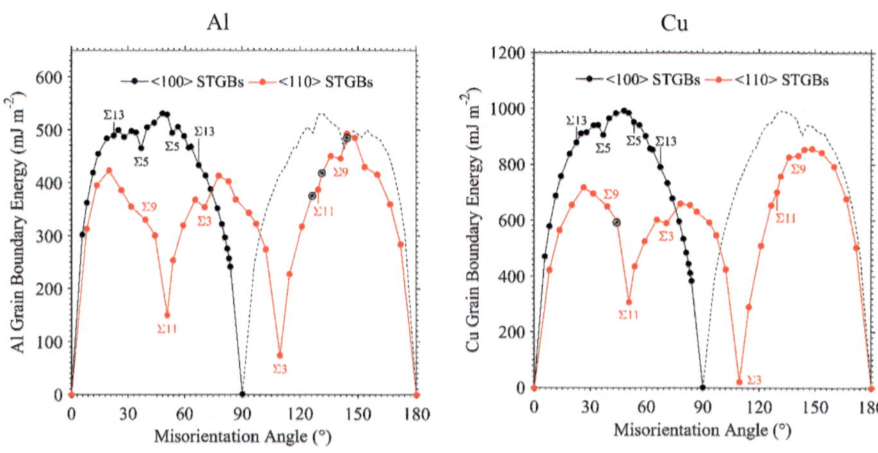


**Fig. 16** Grain boundary energy as a function of misorientation angle for <100> and <110> symmetric tilt grain boundaries (STGBs) in (left) Al and (right) Cu. The low-Σ boundaries are identified in each tilt system [reproduced with permission from Fig. 3 of Tschopp et al. (2015)]

where $n_n$ is the number of nearest neighbor (β) atoms, $n_s$ is the number of slipped neighbors, $x_i^{\alpha\beta}$ is the vectorial distance of atom α and neighbor β in the current (deformed) configuration, and $X_i^{\alpha\beta}$ is the vectorial distance of atom α and neighbor β in the reference (initial) configuration.

These measures (and others) are or can be employed in several widely used open source visualization tools such as OVITO (https://ovito.org/ Stokowski, Darmstadt) and AtomEye (http://li.mit.edu/A/Graphics/A/). We also mention some further advances in descriptors with an eye towards informing micromechanical and GC crystal plasticity models that often require information of individual "phases" or local lattice rotation. Tucker et al. (2010, 2011) have laid out schemes for estimating the local deformation gradient via optimization using neighboring atoms, performing right polar decomposition, and then defining the microrotation vector based on the skew symmetric part of the right polar rotation, i.e., $\phi_k = -(1/2)\varepsilon_{ijk}(R_{skew})_{ij}$. This magnitude and direction of this vector can be used to classify characteristic values for each structure of interest. For example, the evolution of parent FCC phase, stacking faults (HCP), and other (e.g., grain boundaries) in nanocrystalline Cu is shown in Fig. 17 (Tucker et al. 2012). From these results, it is clear that the peaks in the stress-strain behavior occurs at the circled inflection points where the density of stacking faults saturate in the grain interiors; from that point on, the grain boundaries accommodate the majority of the deformation, with a continuing role for (heterogeneous) rotation within the FCC lattice of the grain interior.

Once the local deformation gradient is estimated by optimally overfitting relative motion of neighboring atoms, the Green strain estimate for each atom is accessible, i.e.,

$$\mathbf{E} = \frac{1}{2}\left(\mathbf{F}^{\mathrm{T}} \cdot \mathbf{F} - \mathbf{I}\right) \tag{16}$$

The Green strain can be partitioned between phases (FCC parent lattice, grain boundaries and HCP structured-dislocation stacking faults. Tucker et al. (2012) show the partitioning of Green strain among phases evolves during deformation of a nanocrystalline ensemble of grains. They found that nanocrystalline Cu with larger mean grain size (15 nm) is more strongly influenced by dislocation activity, whereas grain boundary deformation modes account for a higher fraction of tensile Green strain in smaller grain structures (e.g., 5 nm). These computable metrics at the atomic scale offer a means by which otherwise qualitative arguments can be quantified. It is apparent that continuum micromechanical models of nanocrystalline structures can be assessed via atomistics using these descriptors or metrics.

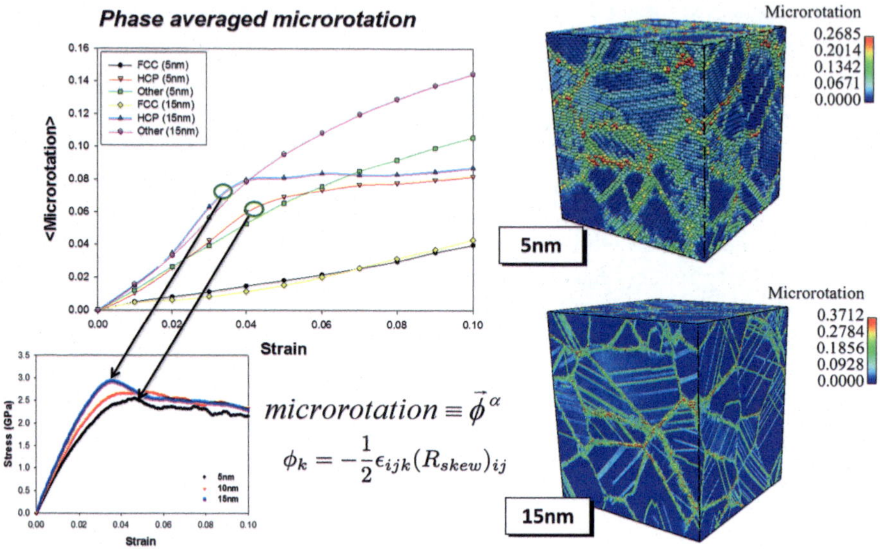


**Fig. 17** Uniaxial tension stress–strain behavior (lower left) from MD simulations of Cu nanocrystals at 10 K for mean grain sizes of 5, 10, and 15 nm at a strain rate of $10^9$ $s^{-1}$. The evolution of the simulation cell-averaged microrotation for the FCC (parent lattice), HCP (stacking faults), and other (grain boundary) atomic groups is shown at upper left, showing that the peak stress is associated with saturation of stacking faults in the grain interiors, after which grain boundary deformation dominates. Compiled from Tucker et al. (2012)

## 6 Overview: Multiscale Modeling from Atomistics to Continuum

Large scale direct MD simulations of dislocations are computationally highly intensive (Buehler et al. 2005) and are limited to short times and high strain rates. Historically, concurrent schemes for bridging atomistic and continuum methods in problems involving dislocation activity have employed spatial domain decomposition to address domains in which fully resolved atomistic models are desired in a localized region near a crack tip or a grain boundary interface, for example, with coarse-grained or reduced order descriptions employed elsewhere. In this way, the predictive power of atomistic models can be exploited for dislocation reactions at fine scale, while long range fields are also rigorously addressed.

Many variants of spatial domain decomposition have been introduced, with the notion of an overlapping pad region that patches the atomistic domain to a far field continuum model to alleviate intrinsic incompatibilities between the two formulations (Kohlhoff et al. 1991; Gumbsch 1995; Rudd and Broughton 1998, 2000; Rafii-Tabar et al. 1998; Weinan and Huang 2001; Qu et al. 2005; Zhang et al. 2017), as shown in Fig. 18. Incompatibilities include for example impedance mismatch in high frequency lattice vibrations, as well as ghost forces that arise in

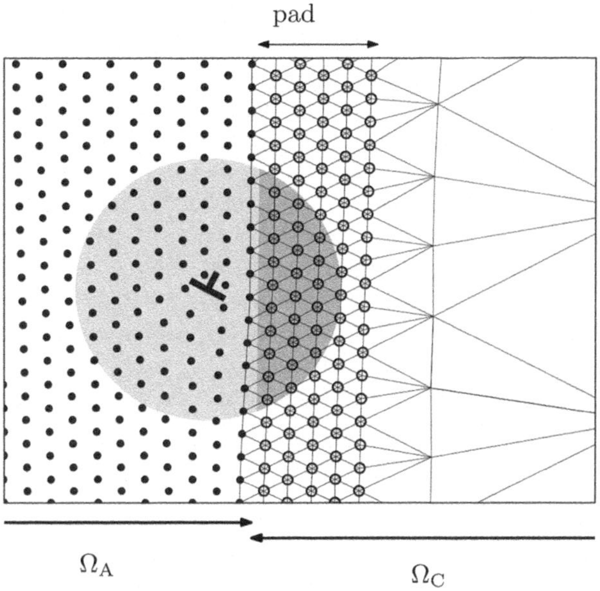


**Fig. 18** Notion of an overlapping pad region between atomistic ($\Omega_A$) and continuum ($\Omega_C$) domains, where atoms and finite element nodes may have one-to-one correspondence to connect the two domains in terms of displacements and communication of fields at the interface [reproduced with permission from Fig. 1 of Cho et al. (2015)]

the interface region owing to differences in the underlying hyperelastic potential, representation of defects, and solution methodology for the governing equations. The choice of the pad region size and its characteristics are key distinguishing characteristics among various methods. A number of these methods have been compared by Tadmor and Miller (2011). Approaches may be characterized as force or energy matching, with the latter dominating most recent developments. In general, such methods have been more commonly applied to static energy minimization applications rather than dynamics of defects at finite temperature, since the continuum description of waves in the lattice differs to a great extent from the atomistic domain, with strong implications for temperature and thermal transport as well.

We also mention the method of a bridging scale decomposition in coupling atomistic and continuum simulations proposed by Wagner and Liu (2003) and Liu et al. (2006), inspired by a variational multiscale method, that allows for much longer time scales in the continuum solution; this concurrent strategy applies to the same spatial domain for atomistic and continuum solutions, and can be used in spatial decomposition schemes. To our knowledge it has not been applied to dislocation field problems, with more of a focus on crack growth problems.

Spatial coarse-graining methods have been developed over the years in parallel with domain decomposition methods. Coarse-graining methods do not employ an interface between disparate atomistic and continuum descriptions of the underlying lattice hyperelasticity, but rather seek to reduce the model degrees of freedom while retaining the same overall lattice description, faithful to atomistics.

We focus here on concurrent methods that admit moving dislocations in both fully resolved and coarse-grained or continuum regions, as well as two-way exchange of dislocations between the two regions. This severely limits available methods. Clearly, the value of methods that bridge from atomistic to continuum descriptions is their potential for strongly predictive character (limited by the uncertainty in the interatomic potential and scale linking algorithms) compared to mesoscale continuum modeling strategies. This predictive quality offers more quantitative decision support for mechanistic modeling and materials design in terms of candidate nanostructure mechanisms at the unit process level (e.g., single dislocation reactions), and even into the mesoscale. These capabilities are advancing rapidly, limited mainly by the challenge of available and accurate interatomic potentials, particularly for multicomponent systems (cf. the online Knowledgebase of Interatomic Potentials (KIM), OpenKim 2016). Coarse-grained and spatial domain decomposition atomistic-continuum approaches can model the structure of interfaces and dislocations-interface reactions, as well as longer range interactions, as they are multi-resolution and concurrent. In contrast, mesoscale DDD must appeal to some form of hierarchical transfer of information to address effects of short range dislocation core interactions.

## *6.1 Concurrent Atomistic-Continuum Models*

We start the discussion by considering a spatial coarse-graining method. One of earliest and most widely documented atomistic-continuum multiscale modeling approaches is the Quasicontinuum (QC) method (Tadmor et al. 1996a, b; Shenoy et al. 1998, 1999; Miller et al. 1998a, b; Knap and Ortiz 2001; Tadmor and Miller 2011). QC represents a significant step forward in supporting predictive modeling of material response by extending predictive capabilities of atomistic modeling to higher length scales. QC accomplishes atomistic coarse-graining by linking lattice site positions to a reduced set of representative atoms, introducing summation rules that allow reduction of cost of ensemble sampling for force or energy, and schemes for adaptation that facilitate fully resolved atomistic simulation in regions of high local deformation (e.g., interfaces, crack tips, dislocations), while coarse-graining away from these evolving regions to minimize the degrees of freedom. In its earliest and perhaps most commonly used form, termed 'local' QC, the coarse-grained continuum region makes use of a Cauchy-Born assumption of an affine deformation within each region, interpolating between representative atoms that define elements. Accordingly, it offers a precise description of the underlying interatomic hyperelastic potential of the lattice even in continuum regions, subject to coarse-graining error in representing inhomogeneous deformation. In recent work, Γ-convergence of the QC method has been analyzed (Espanol et al. 2013). As commonly applied

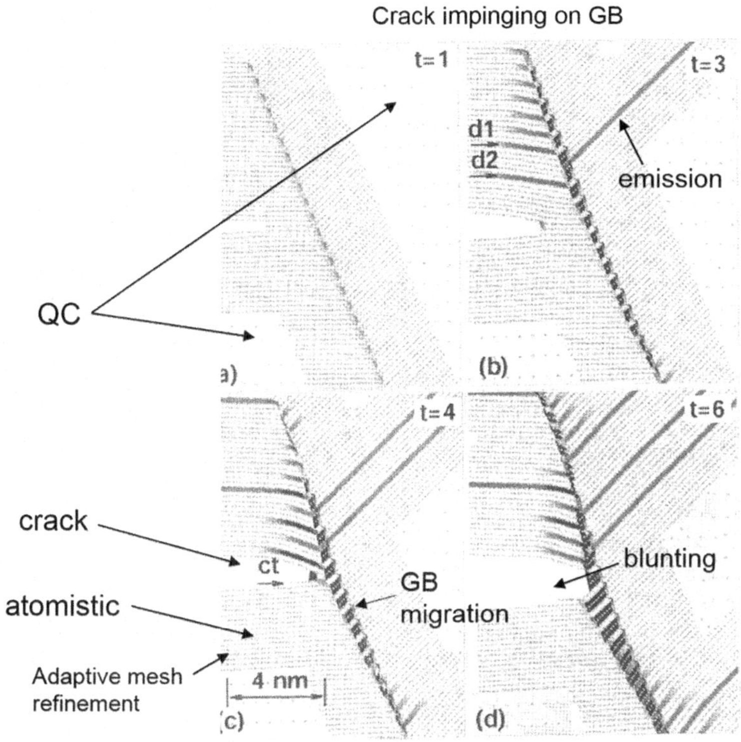


**Fig. 19** Application of quasi-static QC to a crack impinging on a Σ21(421) symmetric tilt grain boundary in Ni, delineating regions remeshed to full atomistic resolution and coarse-grained QC regions (white regions) [adapted with permission from Fig. 7 of Miller et al. (1998a)]. The Cauchy-Born assumption for underlying lattice cannot deal with the strong heterogeneity of deformation near dislocation cores

to dislocation field problems, QC may be regarded as an atomistic coarse-graining approach. However, it is actually more of an adaptive concurrent multiscaling method because dislocations cannot pass through coarse-grained regions that are subject to the Cauchy-Born assumption of affine deformation. QC requires either a priori assignment of fully atomistic and coarse-grained regions for dislocation migration and interactions, or adaptation/reassignment of associated of atoms and nodes to achieve a fully atomistic description near dislocations (Shenoy et al. 1999) or interfaces (Miller et al. 1998a). This is shown in Fig. 19 for quasi-static QC simulation of a crack impinging on a symmetric tilt grain boundary in Ni, with adaptive remeshing of regions around dislocations conducted as necessary in the vicinity of evolving dislocations.

Most applications to dislocation problems have employed quasi-static QC methods based on energy minimization at each step. A fully nonlocal or cluster QC version was developed (Knap and Ortiz 2001) to minimize ghost forces in coarse-graining, with extensions to energy-based rather than force-based cluster sampling

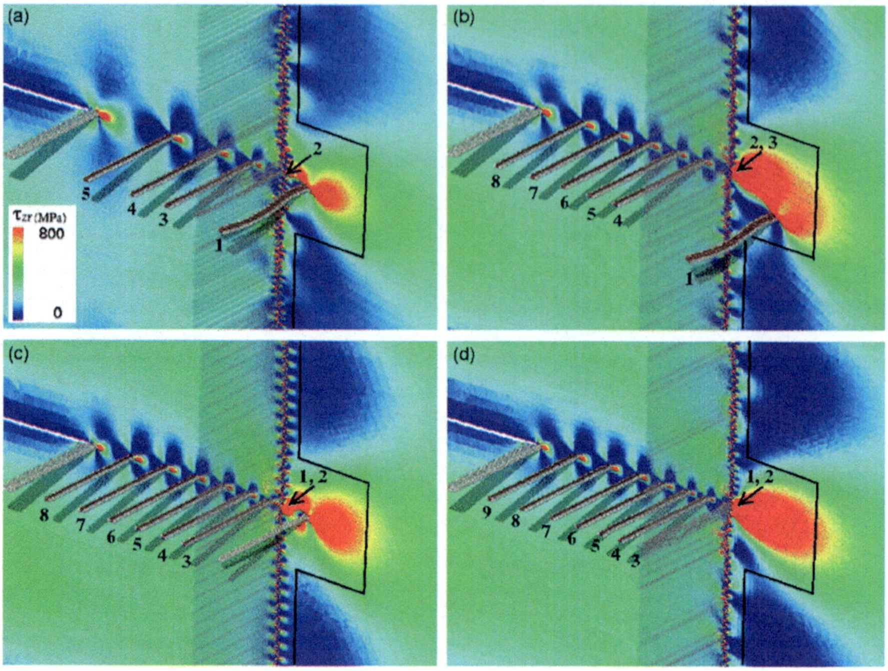


**Fig. 20** Use of the QC method to explore interactions between a lattice edge dislocation pile up emanating from a crack tip and a <112> asymmetric tilt grain boundary under shear deformation for four different misorientation angles of (a) 13°, (b) 29.5°, (c) 63°, and (d) 89.3° [reproduced with permission from Fig. 5 of Shimokawa et al. (2007b)]. The crack tip is approximately 40 nm from the grain boundary

schemes (Eidel and Stukowski 2009). Adaptive remeshing for evolving dislocations and for high dislocation densities is challenging. Although full atomistic resolution is still required in the vicinity of dislocations, the use of nonlocal or cluster QC facilitates reduction in coarse-graining error away from these regions. Kwon et al. (2009) have introduced higher order adaptive QC schemes to improve description of large scale inhomogeneous deformation as might occur in coarse-grained regions away from dislocations. QC has been applied to quantify lattice dislocation interactions with grain boundaries (Shimokawa et al. 2007a; Yu and Wang 2012, 2014), as shown in Fig. 20; this takes advantage of the capability to resolve short range interactions using full atomistic simulation near the interfaces, while also accounting for long range dislocation interactions in remote regions.

In practice, QC becomes quite computationally intensive for more general problems involving mixed character dislocations moving along arbitrary trajectories, when nucleating or generating new dislocations, dislocation interactions with other dislocations and interfaces, etc. Recent advances have sought to increase efficiency while handling such arbitrary fields of dislocations (Amelang et al. 2013, 2015). A fully nonlocal, energy-based QC method developed recently by Amelang et al.

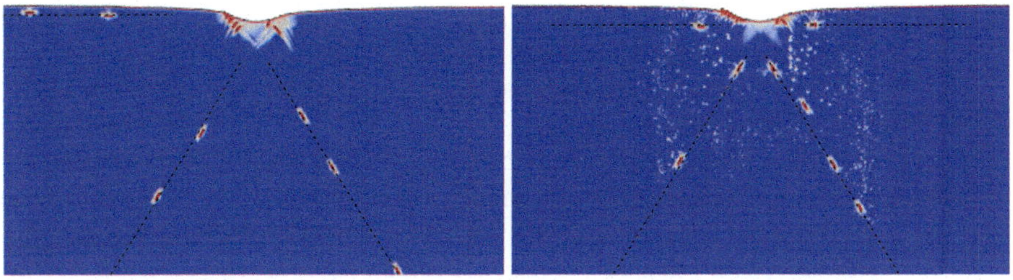

(a) fully atomistic (b) adaptive nonlocal QC

**Fig. 21** Emission of dislocations from 2D surface quasi-static nanoindentation of Cu using (**a**) molecular statics (**b**) a fully nonlinear energy-based QC method with second-order summation rules and automatic mesh adaptation. Dislocations are color-coded using the centrosymmetry parameter on a log-scale, with red being the highest level. Reproduced with permission from Fig. 28 of Amelang et al. (2015)

(2015) addresses seamless bridging that produces no residual or spurious force artifacts in centrosymmetric crystals under affine deformation in 2D, with marginal forces in 3D. This approach has been extended recently to consider automatically adaptive modeling of fully nonlocal QC to model dislocations moving through a lattice (Tembhekar et al. 2017).

A meshless QC method based on local maximum-entropy interpolation has been developed (Kochmann and Venturini 2014) that improves efficiency of model adaptation near regions of fully atomistic resolution to capture dislocations. A comparison of fully nonlocal QC with automatic mesh adaptation to a full static atomistic simulation for nanoindentation of Cu using an extended Finnis-Sinclair potential is shown in Fig. 21 (Amelang et al. 2015).

The finite-temperature quasicontinuum hot-QC method (Dupuy et al. 2005) and coarse-grained molecular dynamics (CGMD) methods (Rudd and Broughton 1998, 2000; Fish 2009) have facilitated coarse-grained multiscale MD simulations at finite temperature based on a local harmonic approximation to estimate entropic effects of the atoms not explicitly represented in the model. They are still limited to typical time scales of MD simulations. Kulkarni et al. (2008) extended the QC approach to model non-equilibrium conditions at finite temperature. Tadmor et al. (2013) distinguished between quasi-static (free energy minimization at each time step) and dynamic (fully dynamic evolution) forms of hot-QC that can address near equilibrium and highly non-equilibrium dislocation phenomena, respectively. One way to extend these methods to longer time scales is by employing hyperdynamics as in the Hyper-QC version (Kim et al. 2014; Binder et al. 2015). Hyper-QC involves a separation of time scales between short atomic vibration periods and long waiting times associated with switching of metastable states. Acceleration is achieved by modifying the hot-QC potential energy to reduce the energy barriers between metastable states to promote reaction paths during the simulation via transition state theory (TST). There are issues if there is more than one barrier type or strength in the system, requiring localization of the algorithm. These finite temperature methods are in early stages in terms of modeling dislocation field evolution.

The Coupled Atomistic Discrete Dislocation (CADD) method, introduced originally by Shilkrot et al. (2002a, b, 2004) with further advances and applications (Shiari et al. 2005; Dewald and Curtin 2007a, b, 2011), employs domain decomposition with a continuum discrete dislocation representation away from the grain boundary, admitting two-way exchange of dislocations between this remote continuum region and a full atomistically-resolved domain near the grain boundary. This kind of formulation is intended for studies of crack tip and interface plasticity. It is fully concurrent since dislocations can move from the atomistic to continuum regions and vice versa. With a focus on capturing the interaction stress fields of dislocations at long range, Dewald and Curtin (2007a) employed CADD to analyze dislocation-boundary reactions of edge dislocation pile-ups impinging on a Σ11 (113) tilt boundary in Al. Dewald and Curtin (2007a, b) suggested extensions of classical dislocation transmission criteria based on CADD studies that include information on stress state at the interface, step size of residual GB dislocations, and resolved shear stress acting on the incoming pile-up plane. It is quite likely that somewhat different additional slip transfer criteria apply for various kinds of boundaries. Progress has been achieved on passing general 3D mixed character dislocations between atomistic and continuum regions (Pavia and Curtin 2015), which was an early limitation of the CADD approach owing to the complexity of transferring curved segments through the interface between atomistic and continuum regions (Cho et al. 2015).

Variations of the CADD approach have been used by Curtin and collaborators to address problems such as fatigue crack growth (Curtin et al. 2010). A study of crack tip decohesion and shielding due to dislocations near the crack tip (Song et al. 2010) demonstrated that the commonly adopted assumption in micromechanics of a local singular field (and corresponding $K_{local}$) modified by dislocation shielding is not valid, which necessitates an approach such as CADD to estimate the near tip fields. Unlike QC, CADD uses a more conventional linear elastic elastic energy function in the remote regions—these regions are subjected to reduced order continuum models rather than atomistic coarse-graining. Accordingly, in addition to model reduction error, there are issues regarding incompatibility of the elasticity and the interpretation of temperature in the two regions. Moreover, there is impedance to wave propagation at the atomistic-continuum interface owing to the substantial jump from a discrete particle system to a continuum linear elastic domain with defects. These can be addressed by introducing damping heuristics in the interface region (Qu et al. 2005). It is noted that statistical coarse-graining also introduces impedance effects, albeit to a lesser degree. The CADD approach is likely most robust for modeling quasi-static dislocation interactions for specific, well-defined regions of atomistic resolution. In such cases, it should scale exceptionally well with increasing size of the remote domain compared to coarse-grained atomistics, whether or not of adaptive character. It is unclear how CADD might be used to provide enrichment of solutions for adaptive, evolving dislocation fields (e.g., nanoindentation with expanding regions of dislocation density). Initial implementations of CADD focused on 2D problems by virtue of complexities in addressing dislocation transfer across interfaces. Pavia and Curtin (2015) have advanced parallel algorithms for 3D CADD that enhance scalability at high accuracy for appropriate problems with long range fields, absent seamless passing of dislocations from atomistic to continuum domains (it is limited to dislocations within the atomistic domain and therefore not yet fully concurrent), including Langevin graded stadium damping to eliminate spurious wave reflections and interface forces even for non-equilibrium dynamics of dislocation processes within the atomistic domain.

The concurrent atomistic-continuum (CAC) method is a spatial coarse-graining strategy for concurrent multiscale modeling that was developed as an application of atomistic field theory (Chen 2009), with a local density function used to connect this continuum theory to atomistics via multiscale balance equations. The CAC method employs an integral finite element approach for coarse-grained atomistics that admits description of dislocation nucleation and migration with or without adaptive coarse-graining (Xiong et al. 2011, 2012a, b, 2015). Ghost forces arising from a change of the underlying continuum formulation and energy summation rules in many other approaches based on domain decomposition or coarse-graining are not an issue since the underlying integral formulation and constitutive framework are invariant with respect to coarse-graining. Dislocations can be modeled throughout the entire domain, whether at full atomistic resolution or coarse-grained. This sets it apart from fully concurrent methods that require full atomistic resolution at the dislocation core, such as QC. In contrast to the QC method, which has the objective of seeking convergence of the solution to that of the fully atomistic case for various field problems, CAC can have multiple purposes. On one hand, it can coarse-grain in regions away from atomistic domains of interest and capture long range fields of dislocations, as in CADD, while preserving high accuracy in fully resolved atomistic regions of interest. But it can also model dislocations across a range of length scales to access trends and provide support for mechanistic understanding of coarse scale behavior for fields of dislocations, smearing individual cores but preserving net Burgers vector and associated interactions. While QC typically seeks the most accurate and efficient solution to dislocation plasticity via adaptive remeshing of the domain near dislocations to full atomistic resolution, CAC can resolve full atomistics if necessary near interfaces or crack tips, but allows dislocations to nucleate, multiply, migrate, and interact even in the coarse-grained regions along interfaces between elements, introducing the option to coarse-grain dislocation fields over larger scales. Like QC, CAC employs the same interatomic potential in all regions. Such coarse-graining in CAC introduces systematic error, for example in representation of dislocation core structures and short range interactions; this error can be quantified and balanced with the high computational demands of remeshing, according to the purposes of the mesoscale modeling. For example, if trends of behavior or collective mechanisms are to be considered as a function of microstructure or stress state, as is often the case in DDD, then CAC may offer a means to support such parametric studies to inform DDD. Coarse-graining error in CAC can be minimized via adaptive remeshing (Xu et al. 2016b), based on the level of the nodal displacement between elements; this is necessary for general field problems to allow dislocation migration along arbitrary extended slip planes. However, remeshing need not be carried out to full atomistic representation, but can involve simply splitting larger elements into smaller ones. CAC does not employ the Cauchy Born assumption throughout the coarse-grain domain and admits dislocations (displacement discontinuities) to pass between elements by virtue of its integral formulation. As such, it admits gradual coarse-graining from full atomistic resolution upward.

More specifically, CAC employs a 3D integral formulation of the governing field equations. The quasi-static implementation enforces quenched dynamics at each time step, combined with periodic energy minimization applied every 50–100 time steps (Xu et al. 2015, 2016a). Use of quenched dynamics enhances efficiency for problems with defects by reducing the load on periodic energy minimization. The dynamic implementation (including full inertial effects that can address phonon interactions with wavelengths above the element size, cf. Xiong et al. 2014a, b, 2015; Pluchino et al. 2016), is based on an extension of the lattice statistical mechanics approaches of Irving and Kirkwood (Kirkwood 1946; Irving and Kirkwood 1950). Applications to date of the quasi-static implementation have been limited to monoatomic systems, while multicomponent systems have been considered in dynamic CAC applications. The quasi-static implementation is considered useful for modeling physically representative reaction pathways for thermally activated dislocation processes since it is not overdriven, as typical of MD simulations. Using CAC, energy minimization may be regarded to accord with the concept of a sequence of constrained equilibrium states in internal state variable theory (e.g., Muschik 1993) in the thermally activated flow regime, accessing local minima of the energy landscape based on the history of dislocation evolution that at each stage (even with no dislocation flux) corresponds to a non-zero thermodynamic (Peach-Koehler) force due to elastic interactions. Rice's seminal paper on internal variable theory in metal plasticity (1971) provided a clear exposition of this concept of a sequence of constrained equilibrium states.

Each element in CAC contains a number of atoms, with element boundaries corresponding to close-packed planes (i.e., slip planes), as shown in Fig. 22. For FCC crystals, rhombohedral isoparametric elements with (1) all surfaces that correspond to slip planes and (2) trilinear shape function are employed. Within each element, the atoms are interpolated from the nodes by a trilinear interpolation function, i.e., $R_k = \Phi_{k\xi} r_\xi$, where re are nodal displacements. Gaussian integration is employed. Since the positions of atoms within elements are subject to interpolation, interelement fields can capture only weak gradients of deformation, whereas displacement discontinuities such as dislocations can propagate along interelement boundaries. Figure 23 presents a simplified 2D schematic which illustrates the passage of an edge dislocation between coarse-grained and atomistic domains.

Figure 24 provides more insight into options that have been developed to date for the finite element representations in CAC. In order to better represent reconstruction of atomic environment due to defects between elements (e.g., dislocation cores or cracks), the outer layer (1NN elements) or two outer layers (2NN elements) of atoms within each element are subjected to different interpolation functions from those used in the element interior, increasing the number of integration points accordingly. These higher order elements avoid the limitations of affine deformation within elements in representing, for example, atomic level stacking sequence changes associated with the propagation of partial dislocations (stacking faults) between elements. Figure 25 compares the performance of 1NN and 2NN elements in terms of accuracy of representing fully resolved atomistic results for FCC Cu. Figure 26 shows how the stacking fault width (distance between leading and trailing partial dislocations) is captured in both coarse-grained and fully resolved atomistic domains for Cu using 2NN elements; note that dislocations can propagate between atomistic and coarse-grained domains, in either direction, without appreciable effect on the separation of partial dislocations for the degree of coarse-graining employed.

As for QC and CADD methods, one obvious application of CAC is the simulation of dislocation pile-ups at interfaces and progression of slip transfer reactions as multiple dislocations react at the interface (transmission, reflection, absorption, desorption) sequentially. This class of problem defies treatment using fully resolved atomistics at length scales large enough to capture pile-ups. Just as important is the requirement to incorporate realistic boundary conditions (e.g., free surface and specimen sizes) to make contact with *in situ* experiments, for example, regarding dislocation-interface reactions conducted using TEM thin foils. Figure 27 shows a configuration of a quasi-static CAC simulation used to simulate dislocation pile-ups with leading screw character against both (a) a $\Sigma 3$ coherent twin boundary and (b) a $\Sigma 11$ STGB in Ni (Xu et al. 2017b), with the interface represented with a thin "ribbon" of material on each side at full atomistic resolution, and coarse-grained elements used elsewhere. The size of the simulation domain approaches TEM foil dimensions reported in the literature, with traction free boundary conditions applied on the top and bottom surfaces, in order to realistically model dislocation migration and reactions from experiments. Coarse-grained elements with 2197 atoms per element are employed everywhere in this simulation domain (140 nm $\times$ 65 nm $\times$ 53 nm) except for a 7 nm wide fully atomistic region along the interface. Shear stress $\tau_{xz}$ is effectively applied to drive the screw dislocations into the interface.

Owing to improvement of efficiency of such large scale simulations (by 1–2 orders of magnitude relative to full atomistic simulation), CAC has been used

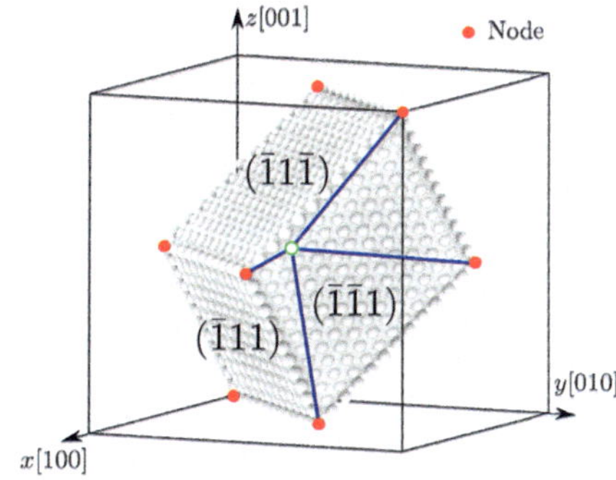


**Fig. 22** Geometric definition of rhombohedral finite element boundaries for a FCC crystal representation in CAC with faces on the (111) slip planes [adapted with permission from Fig. 2c of Xu et al. (2017a)]

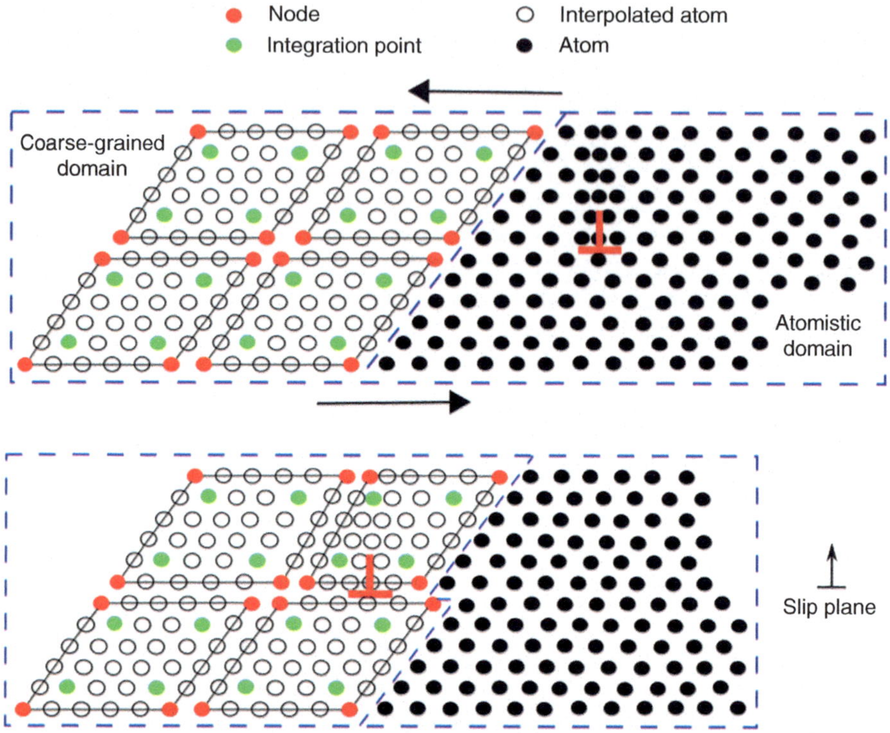


**Fig. 23** 2D schematic of fully resolved (right) and coarse-grained atomistic (left) domains in CAC, with discontinuous element boundaries shown at left, along with different representations of an edge dislocation core [reproduced with permission from Fig. 1 of Xu et al. (2015)]. The Burgers vector spreads out between discontinuous elements in the coarse-grained region

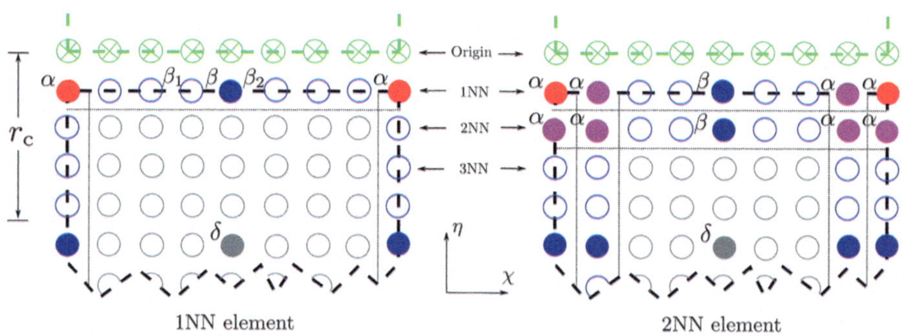


**Fig. 24** 2D Illustration of 1NN and 2NN elements in natural coordinates in CAC simulations, showing integration points (solid blue) and slave nodes (open blue), with $r_c$ as the cutoff distance of the interatomic potential. 1NN and 2NN elements have 27 and 125 integration points, respectively. Reproduced with permission from Fig. B.26 of Xu et al. (2015)

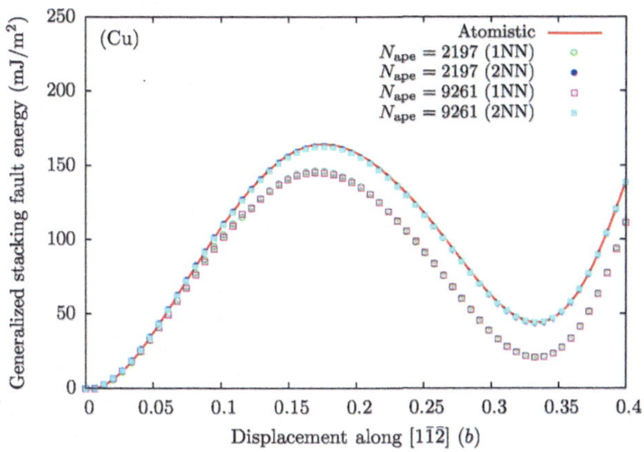


**Fig. 25** Comparison of generalized stack fault energy curve along the reaction pathway for partial dislocation dissociation for pure Cu using the Mishin et al. (2001) EAM potential, showing the 2NN elements with 2197 and 9261 atoms per element both exactly capture the results of fully resolved atomistic simulations. Reproduced with permission from Fig. 4a of Xu et al. (2015)

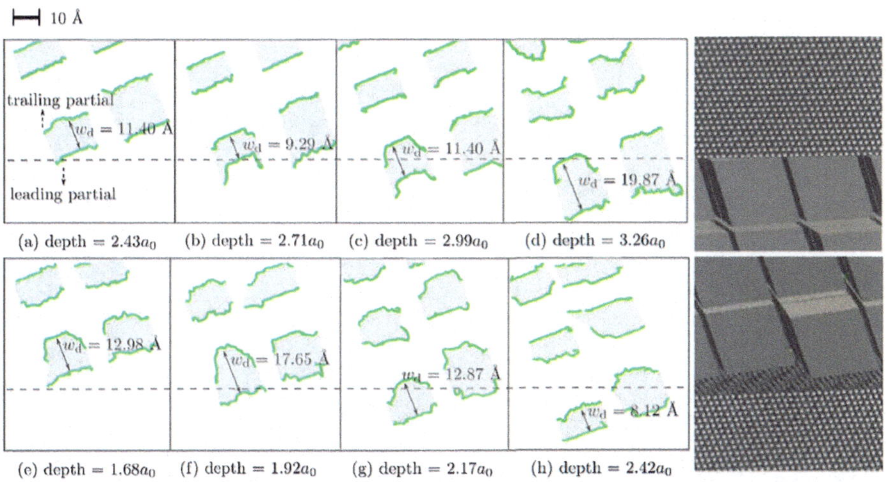


**Fig. 26** Snapshots of dislocation migration from the atomistic to coarse-grained domain (**a–d**), as well as in the opposite direction (**e–h**) in Al. The green curves represent leading and trailing partial dislocation lines, and the gray ribbons are the intrinsic stacking faults [reproduced with permission from Fig. 18 of Xu et al. (2015)]

to support parametric studies to compare the relative performance of various interatomic potentials, with an eye towards uncertainty quantification. For example, Fig. 28 shows the configuration facilitating a parametric study of the predictions of five EAM interatomic potentials in modeling successive pile-up dislocation interactions with Σ3 interfaces in Ni for thin foil specimens (Xu et al. 2017b).

Finite temperature formulations for quasi-static (near equilibrium) and non-equilibrium dynamic conditions, as well as heat transport via modes other than phonons remain to be fully developed for CAC. Present finite temperature applications have assumed isothermal conditions or dynamic behavior without applying a thermostat.

In closing this section, we note that other concurrent multiscale modeling approaches have been recently developed. Gracie and Belytschko (2011) introduced an adaptive concurrent multiscale method to simulate moving dislocation cores using highly refined atomistic regions, building on the bridging domain method (Xia and Belytschko 2004) combined with the Extended Finite Element Method (XFEM). In this case, adaptivity is based on a heuristic criterion of the distance of a dislocation from the bridging domain. In the adaptive XFEM-BDM framework, the displacement discontinuities are specified through a step function across the active slip planes in continuum region by means of XFEM enrichment; the mesh is refined and coarsened based on either of these two criteria: (1) the broken interatomic bonds and (2) the errors in atomic displacements associated with introducing a continuum field. Xu et al. (2016b) provide an overview of mesh refinement schemes for concurrent atomistic-continuum modeling of dislocations.

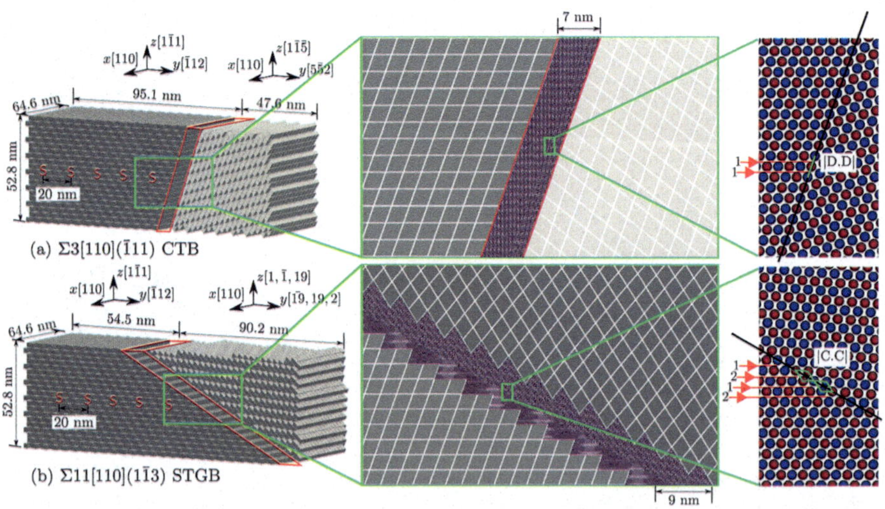


**Fig. 27** Bicrystal simulation cells used to study sequential slip transfer of five $(a_0/2)\,[110]\,\left(1\overline{1}1\right)$ dislocations across (**a**) a $\Sigma3\left(\overline{1}11\right)$ CTB and (**b**) a $\Sigma11\ \left(1\overline{1}3\right)$ STGB in Ni. An atomistic domain is meshed in the vicinity of both GBs; the jagged interstices at the cell boundaries are also filled in with atoms, which are not shown here. Away from the GBs and cell boundaries are coarse-grained finite elements, each containing 2197 atoms. All cell boundaries are assumed traction free to allow a full 3D description. Exploded views of the GB region are given at far right, where atoms in different (110) atomic layers have different colors. The far right exploded views show how the details of interface reactions of screw dislocations on incoming slip planes are affected by local interface structure for Σ3 and Σ11 STGBs. Reproduced with permission from Fig. 1 of Xu et al. (2017b)

Other gradual coarse-graining methods also exist, for example generalized particle dynamics (GPD) applied to atomistic systems (Fan 2011) or peridynamics (Silling et al. 2007). GPD would require remapping to full atomistic resolution to address dislocations, similar to QC. Peridynamics has an issue of model identifiability since it is more of a mesoscale strategy that does not employ the same underlying interatomic potential of the atomistic system. Still, there are certain parallels between CAC and peridynamics since they share characteristics of an integral formulation that admits displacement discontinuities in the coarse-

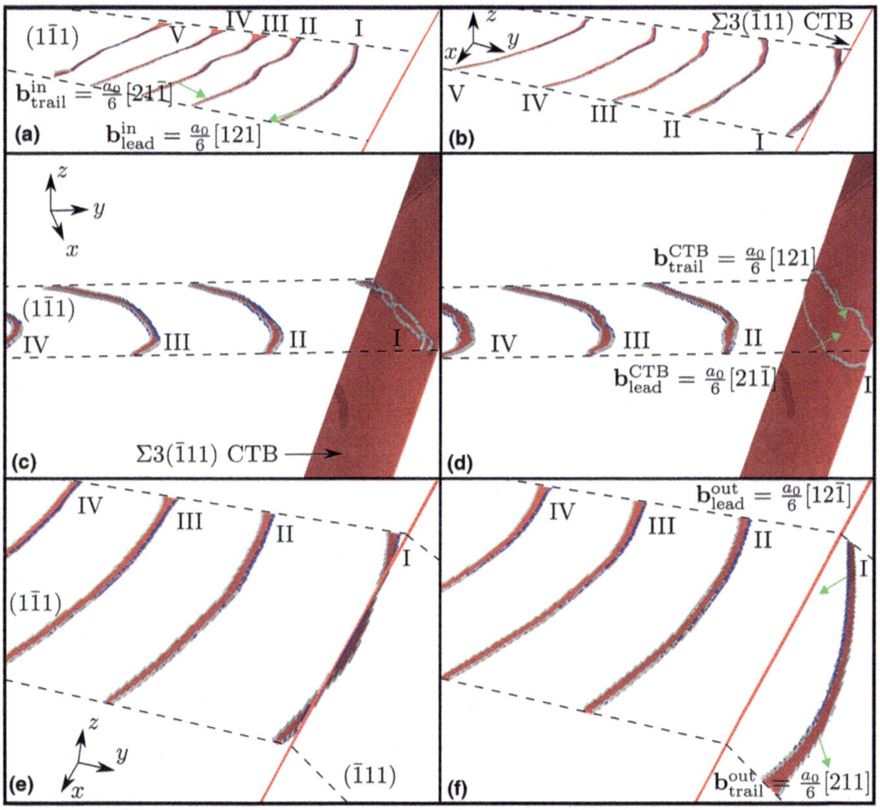


**Fig. 28** Successive encounter of a/2[110] dislocations with dominant leading screw character impinging on a Σ3{111} coherent twin boundary (CTB) in Ni, showing constriction of leading and trailing partial dislocations in (**b**). Atoms are colored by adaptive common neighbor analysis; red are of HCP local crystal structure, blue are not coordinated as either FCC or HCP, and all FCC atoms are deleted. In (**c**) and (**d**), three of the EAM potentials predict incorporation into the interface with two partial dislocations gliding on the twin plan in opposite directions. In (**e**)–(**f**), the other two EAM potentials considered predict that the dislocation cross slips into the outgoing twinned grain via redissociation into two partials. Hence, the need to conduct parametric studies of phenomena across a range of empirical potentials is established. Reproduced with permission from Fig. 2 of Xu et al. (2017b)

grained region. In the context of modeling dislocations at fine scale, it seems that peridynamics does not offer a clear path forward to address dislocations at the high fidelity necessary to be considered as an atomistic coarse-graining method, similar to coarse-grained phase field models discussed in the next section. However, peridynamics has recently been cast in terms of a nonlocal GC model for crystal plasticity by Sun and Sundararaghavan (2014).

## 6.2 Time Scaling of Dislocation Evolution in Discrete Atomistic and DDD Models

The intimate association of fully atomistic or CG-atomistic models with the underlying particle dynamics serves to limit accessible time scales of these models, compared to certain other model classes to be discussed later. MD is appropriate for strongly dynamic conditions. For longer time scales in the thermally assisted regime, time scaling based on Transition State Theory (TST) can be applied to unit processes amenable to atomistic simulation. However, larger scale, cooperative processes involving dislocation fields are much more challenging to consider in this regard. This has already been mentioned earlier in Sect. 4 on challenges in length and time scaling.

One of the challenges of classical MD modeling is that the accessible time scales, on the order of ps to perhaps tens of ns, are well below the temporal resolution of imaging methods with spatial resolution below tens of nm, such as TEM or probe microscopy techniques (Gates and Hinkley 2003), as shown in Fig. 2. These imaging methods require acquisition of signal over time. In contrast, full field imaging techniques with high temporal resolution (e.g., light microscopy) are effective to achieve spatial resolution at or above the wavelength of the light employed, typically on the order of μm and above.

MD can therefore address the kinetics of only a limited range of phenomena, for example, point defect migration in a lattice with a relatively simple PEL. Overcoming the time scaling issue is challenge for MD to consider kinetics of thermally activated processes, since the strain rates are on the order of $10^8$–$10^9$ $s^{-1}$. While massive parallelization of MD codes addresses spatial scaling (e.g., via domain decomposition methods), it does not address the issue of time scaling. Moreover, numerous relevant phenomena occur on a much larger time scale: chemical reactions, diffusion, phase transition, nucleation, thermally assisted dislocation bypass of obstacles, for example. The response of dislocations as complex, extended line defects is among the most problematic in this regard for accelerated MD strategies to address.

Let us first consider point defect (e.g., vacancy or interstitial) diffusion in a lattice at room temperature based on elementary harmonic Transition State Theory (hTST), which assumes an harmonic energy well (Zhu and McDowell 2015). In particular, we consider that the potential energy variation along the reaction pathway between adjacent low energy lattice sties/wells can be described using the harmonic form

$$E(\xi) = E_{\min} + \frac{K}{2}(\xi - \xi_{\min})^2 \tag{17}$$

where $\xi$ is the reaction coordinate along the PEL, and the activation energy is given by the difference in energy between the saddle point energy, $E_{\text{saddle point}}$, along this trajectory and that of the initial well, $E_{\min}$. At room temperature, the thermal energy $k_B T$ is approximately 0.025 eV. A typical activation energy barrier would be $E_{act} = E_{\max} - E_{\min} = E_{\text{saddle point}} - E_{\min} \sim 0.6\,\text{eV}$, which is about $25 k_B T$. Clearly, the barrier is much higher than available thermal energy at room temperature. Based on the harmonic oscillator model, the attempt frequency is given by

$$\nu = \frac{1}{2\pi}\sqrt{\frac{K}{m}} \sim 10^{13} s^{-1} \tag{18}$$

The barrier crossing rate is then given by multiplying the attempt rate by the probability of successful transition (as per statistical mechanics for processes occurring near equilibrium), i.e.,

$$\Lambda = \nu \exp\left(\frac{-E_{act}}{k_B T}\right) \approx 1 \;\; per \;\; ms \tag{19}$$

This crossing rate is 12 orders of magnitude longer than the typical time step of 1 fs used in MD simulations! Clearly, the kinetics of even such a diffusion process with simple reaction pathway are inaccessible to direct modeling via MD.

A number of techniques have been introduced to address time scaling limitations. The parallel replica method (Voter 1998) replicates the dynamics of an entire system as a parallel set of ensembles with the same initial conditions to explore transition states; the parallel replicates are then reset to the new state once an infrequent transition event is observed and another set of replicates are simulated in parallel. This avoids the need to conduct a single simulation to long times, and is best applied to consider limited configurations (e.g., unit processes). There is no requirement that the system adheres to TST in any specific form. Another strategy for a single simulation is Hyperdynamics (Voter 1997), in which a non-negative bias potential is introduced to raise the floor of the PEL without affecting the peaks; this approach assumes applicability of TST and essentially increases the probability of occurrence of thermally assisted events by decreasing the activation energy barrier. Hyperdynamics is also limited in applicability to situations for which the PEL is relatively simple and does not have a broad spectrum of peaks and corresponding range of competing mechanisms for structure evolution; modeling dislocation networks would be challenging using this strategy. Yet another strategy is Temperature-Accelerated Dynamics (TAD) (Sørensen and Voter 2000), in which temperature is increased to speed up the dynamics, again assuming applicability of TST. It is necessary in TAD to filter out transitions that should not have occurred at the original temperature of interest; this is challenging and important since the dynamics will otherwise steer towards entropically favored higher energy barrier transitions simply due to the artificially high temperature.

Other schemes have also been introduced, including on-the-fly Monte Carlo (Henkelman and Jonsson 2001), which involves a state-specific rate catalog of transition event probabilities constructed from atomistic modeling of the evolving PEL. Kinetic Monte Carlo methods often employ a pre-defined event table which may limit their ability to model an evolving PEL. It should be noted that there are various TST approaches, beyond hTST (local quadratic assumption of reaction pathway) (cf. Nguyen et al. 2011). Li et al. (2007) introduced the baby-bathwater scheme for coarse-graining in time.

Saddle points and minima in the PEL confirm to the stationary conditions $\delta U = 0$ , while minima additionally correspond to $\delta^2 U > 0$. The determination of the activation energy $E_{act}$ requires finding the saddle point energy relative to the minimum energy basin. Saddle points can be found by searching directions along eigenvectors of the stiffness or Hessian matrix with lowest eigenvalues, but this is expensive, based on the local quadratic (harmonic) approximation of hTST theory. Knowledge of the saddle point facilitates determination of the transition path and activation energy using the method of steepest descent. An alternative, more efficient technique rooted in hTST theory is to employ the so-called Nudged Elastic Band (NEB) method to assess the reaction pathway through the saddle point in the higher dimensional PEL, from one minimum energy basin to the next, with the attempt frequency and energy wells based assumed to follow the harmonic approximation (Henkelman et al. 2000). Essentially, starting with some assumed path, an "elastic band" is iteratively relaxed to conform to the lowest energy (most probable) pathway in the PEL between basins. For more details, refer to Tadmor and Miller (2011).

As shown in Fig. 29, applied stress σ has the effect of biasing or "tilting" the PEL in a manner that decreases the effective activation energy. The stress

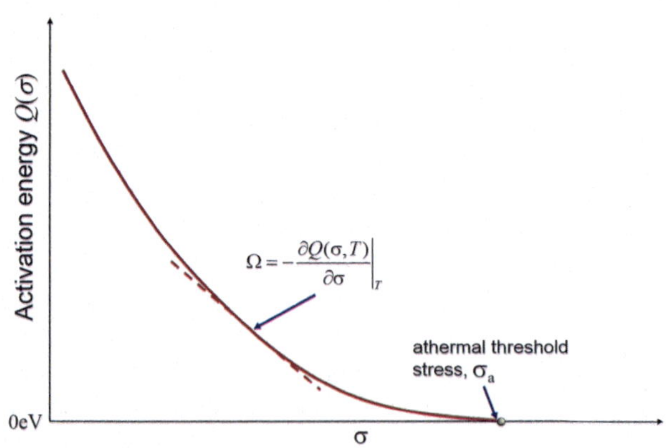


**Fig. 29** The influence of applied stress in reducing the activation energy for a thermally-assisted dislocation reaction/bypass to occur

systematically decreases the activation energy relative to the case without stress, i.e., $Q = Q(\sigma)$. This means that the same level of thermal energy/fluctuation is increasingly effective to affect state transition as stress increases. Moreover, at some point the activation energy vanishes due to tilting, giving rise to spontaneous transition at the athermal threshold (no need for thermal activation at all). The influence of biasing by applied stress is approximated to first order via the correction for effective activation energy given by $Q = E_{act} - \sigma\,\Omega$, where $\Omega$ is the activation volume, interpreted as the material volume associated with rearrangement processes during the state transition. The activation volume is the negative slope of the curve shown in Fig. 29. Zhu et al. (2007) report activation volumes of $0.1b^3$ for diffusion controlled processes, 10–20$b^3$ for nanoindentation, and up to 1000$b^3$ for bulk forest hardening in FCC crystals, where b is the magnitude of the Burgers vector. The strain rate sensitivity diminishes with increasing $\Omega$. Narayanan et al. (2014) have demonstrated application of free-end NEB to analyze the energetics of coordinated kink pair formation on screw dislocations in BCC Fe, using these results to inform the parameters of a Kocks-Argon-Ashby flow rule of crystal plasticity (Kocks et al. 1975; Kocks and Mecking 2003) by employing the hTST approximation.

Simulations of dislocation network interactions with a field of obstacles are particularly challenging with regard to time scaling. Chen et al. (2010) used level set DDD methods to simulate 2D dislocation bypass of random obstacle fields. Picu et al. (2009) showed that rate sensitivity and bypass kinetics for a distributed random field of obstacles can be governed by correlated motion of dislocations rather than the unit processes (see Fig. 30). Their simplified analyses based on the line tension model neglected internal stresses, and assumed that the obstacle size was small compared to distance between obstacles. For both closely nm scale obstacles near peak strengthening and for larger, more widely spaced obstacles interacting with multiple dislocations, these simulations provided qualitative insight into dislocation-obstacle field interactions at different levels of applied stress. At low stress levels, the dislocation bypassed by unzipping along the obstacle front, with no correlation of multiple pinned segments; at higher stress levels, jerky bypass was observed for which multiple segment bypass events were correlated in time. They found that long-range interactions, instead of the short-range interactions between an elastic line and immobile obstacles, give rise to the observed self-organized, stress independent critical plasticity.

Comprehensive DDD simulations of cooperative thermally assisted obstacle bypass by dislocations were undertaken by Hiratani et al. (2003), considering long range interactions along with wait time approximations at obstacles (cf. Benzerga 2008, 2009), and assuming that the attempt frequency is proportional to the inverse pinning distance of segments. They found very significant differences between periodic and random arrays of obstacles, with the latter leading to much lower applied stresses required to drive bypass due to correlated bypass events. Moreover, temperature dependence decreased substantially with applied stress/loading rate. The form of the wait time approximations are critical to the efficacy of this approach, and these forms are somewhat phenomenological in nature, lacking consideration of collective vibrational entropy effects in the thermally activated regime. Recent

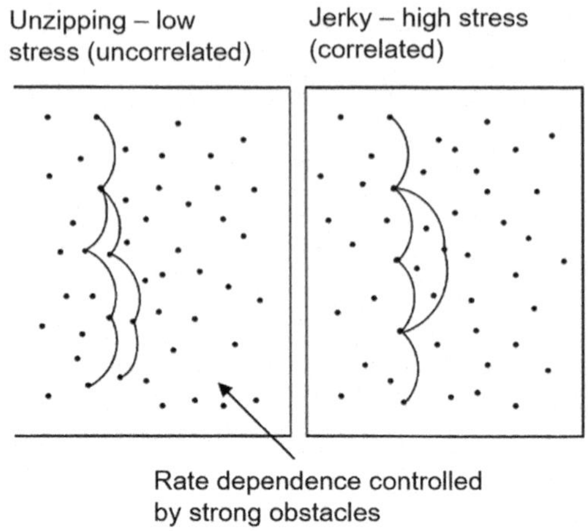


**Fig. 30** Schematic representation of the smooth ("unzipping") and jerky dislocation motion modes. In the unzipping mode the dislocation advances in average by the mean distance between obstacles after leaving an equilibrium configuration. Multiple obstacles are bypassed in jerky motion upon each activation event. Adapted with permission from Fig. 1 of Picu et al. (2009)

work of Baker and Curtin (2016) has considered energetics of thermally assisted dislocation climb limited by vacancy diffusion using a multi-time scaling strategy.

Picu et al. (2009) point out that the "trough model" initially proposed by Fisher (1955) appears to provide better agreement with experiments in terms of temperature and rate dependence of bypass than is offered by direct DDD simulations. In the trough model, the dislocation searches over an energy surface defined everywhere in the glide plane along a path corresponding to the lowest barriers. Hence, motion is controlled by the simultaneous interaction with many obstacles and long-range spatial correlations develop. In the discrete obstacle model, the dislocation bypasses obstacles in an apparently uncorrelated fashion (as a consequence, the activation energy for dislocation motion is identical to that for bypassing a single, isolated obstacle). We may consider that NEB sampling methods at the mesoscale of dislocation-obstacle interactions may more closely express the basic assumptions of the trough model. The NEB method has typically been used in the context of atomistic simulations to quantify the activation barriers for a given reaction. Recent works of Sobie et al. (2017a, b, c) have explored the use of NEB for DDD descriptions of evolving microstructure, since NEB can be applied to any coarse-grained continuum description with the following essential features:

- Dominant kinematic degrees of freedom for evolving attributes of microstructure are included.
- Various self- and interaction (potential) energies of evolving microstructure attributes are included, as necessary to coarse-grain the energy landscape, including short and long range interactions.
- A continuous representation of the state of the system is available from the continuum representation as microstructure evolves during state transitions.

Sobie et al. (2017b, c) generalized the NEB method to coarse-grain continuum representations of evolving dislocation-obstacle interactions using DDD. Activation energies were computed for a glide dislocation bypassing a [001] self-interstitial atom loop of size in the range of 4–10 nm with a spacing larger than 150 nm in $\alpha$-iron for a range of applied stresses and interaction geometries; results were compared in selected cases with full atomistic simulations. Sobie et al. (2017c) showed that reduced order continuum calculations can facilitate more extended mappings of dislocation-barrier interactions than are accessible to typical atomistic simulations, including cooperative/correlated bypass. Moreover, it was shown that the mean of the probability distribution of activation energy for correlated bypass events was far below the activation energy for unit bypass, and substantially above Friedel's simple approximation (1962). Extension of these simulations based on the Meyer-Neldel compensation rule for activation entropy effects was discussed, since temperature dependent sources of anharmonicity can be included in these particular simulations. This issue was considered in greater detail by Sobie et al. (2017a), which explored the attempt frequency for a dislocation bypassing a SIA loop using the continuum theory of defects. To pursue this, the fundamental modes of a finite dislocation segment with pinned ends were characterized. Then, the attempt frequency for a finite dislocation segment (length L) bowed against an SIA loop was determined and compared to theoretical estimates where possible. A more realistic configuration of an infinite dislocation bowed against an array of SIA loops was then considered and the attempt frequency was determined. The study concluded with discussion of the sensitivity of the numerical calculation of attempt frequency and the possible implications on predictions of the reaction rate, including the entropic pre-factor. Significant findings were as follows:

- The fundamental frequency of a dislocation scales inversely with the segment length L, as predicted from line tension modes.
- Higher modes do not scale linearly but with a power law, contrary to line tension predictions.
- The entropic pre-factor for SIA loop bypass is independent of dislocation length, resulting in an attempt frequency that scales with 1/L.
- The entropic pre-factor is on the order of 5 to 10, resulting in an attempt frequency that is much less than the Debye frequency for dislocation segments with lengths much greater than the lattice spacing.

## 6.3 *Microscopic Phase Field (MPF) Models*

Microscopic phase field (MPF) models (Wang et al. 2001; Shen and Wang 2003; Wang and Li 2010; Shen et al. 2014; Mianroodi and Svendsen 2015) hold great promise to bridge from first principles and atomistic modeling to describe partial dislocations and deal with time dependent deformation phenomena in continuum theory in a manner that is not limited to the time step size of atomistic simulations. It can be regarded as an application of phase field theory at or near atomic scale of refinement, including diffuse interface theory and associated gradient terms specific to that scale. While higher scale DDD employs the Volterra model of dislocations (Hirth 1985), MPF has spatial resolution at the dislocation core scale, similar to the Peierls model (Peierls 1940), and admits conveyance of information regarding core size, energy and interactions (Chen et al. 2014). Therefore, MPF is framed at a length scale that can resolve individual dislocation cores (including dissociated partial dislocations) via the generalized stacking fault energy function, representing dislocations using eigenstrain methods. It can address limited numbers of dislocations with near atomic scale fidelity. It can directly admit inputs from ab initio and atomistic modeling regarding physical properties, surface and generalized stacking fault energies, and elastic constants, and has the ability to predict fundamental properties of individual defects or small sets of defects, including formation energy, size, structure, saddle point configuration of the reaction pathway, and activation energy of defect nuclei, and the micromechanisms involved, including spatial interactions (Wang and Li 2010; Mianroodi and Svendsen 2015).

A significant advantage of MPF models compared to DDD models is their direct and unambiguous connectivity to underlying quantum and atomistic field information as input. When used in conjunction with transition state theory or other appropriate kinetics for time scale transitions for thermally activated processes, including diffusive phenomena, continuum phase field theory offers the potential to bridge to long time scales. This is quite useful as a means for predictive computational materials science to support design of interfaces, microstructures, etc. MPF serves to bridge from atomistic modeling of generalized stacking fault energy to describe partial dislocations in a higher scale continuum theory (Wang and Li 2010). It can also be regarded as a higher length scale analogue of the phase field crystal (PFC) model (Elder et al. 2002; Elder and Grant 2004), which is a representation for the continuum density functional theory of electronic structure that pertains to sub-angstrom length scales (Wang and Li 2010). MPF models combine the Cahn-Hilliard equation for chemical inhomogeneity combined with a Peierls description of the dislocation core and associated Burgers vector (Wang and Li 2010). This lends high levels of predictive character to MPF at the level of individual extended defects, including for example partial dislocations, and dislocation interactions with precipitates and interfaces, making it a useful tool to support decision-making in materials design that is sensitive to details of composition, segregation, and other complexities common in engineering alloys. Figure 31 shows an example of MPF applied to simulation of leading and trailing partial dislocations in a $\gamma - \gamma'$ Ni base superalloy interacting with small $\gamma'$ precipitates. This would require extremely large atomistic simulations that would be severely limited in time scale, and these processes simply cannot be addressed using DDD approaches. The only other viable route would be coarse-grained atomistics, and the interatomic potentials for various phases would require sufficient accuracy.

MPF models are much more efficient than fully resolved atomistic models at modeling complex systems at the nanoscale and don't rely on the availability or accuracy of an interatomic potential; there is likely some overlap of capabilities to model extended nanoscale systems between MPF and coarse-grained atomistic models for certain problems. A recent example application in modeling prismatic

HCP/FCC interfaces that may relate to martensitic transformations in TRIP steels, for example, is presented by Louchez et al. (2017).

MPF models differ substantially from coarse-grained phase field (CGPF) models (Wang and Chen 2000; Chen 2002; Wang et al. 2005), which we simply refer to as "Phase Field Models." CGPF models are often applied to nucleation, growth and coarsening of larger scale microstructures (e.g., grains, phases, interface networks) by employing linearized force-flux relations, defect energies, and mobilities as input from a combination of experiments, calibrations, and atomistic modeling results. CGPF models make use of the Cahn-Hilliard and the time-dependent Ginsburg-Landau (for non-conserved order parameters) equations, using diffuse interface theory with gradient energy to address various aspects of microstructure evolution involving composition and structure/order gradients. They are not as focused on nanoscale processes than MPF models, and do not have the same level of predictive character. Although not receiving significant attention here, there are recent advances in CGPF models that address 3D dislocation plasticity (Hunter et al. 2010; Chen et al. 2014) in crystals and polycrystals in manner that may complement standard crystal plasticity, with somewhat less detail and refinement than that required by DDD models. Such CGPF dislocation models are still in very early stages.

## 6.4 *Field Dislocation Mechanics (FDM)*

Field dislocation mechanics pertains to a theoretical framework of continuum character that incorporates incompatibility of elastic and plastic deformation as an

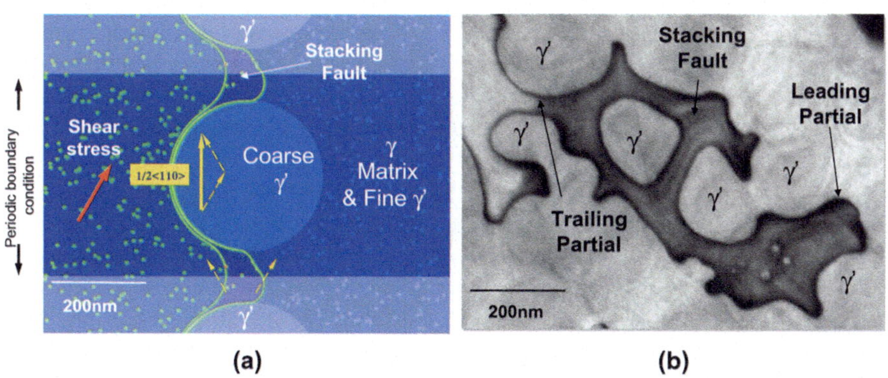


**Fig. 31** (**a**) MPF simulation of dissociated leading and trailing Shockley partial dislocations moving through the $\gamma'$ channel in a Ni-base superalloy with very fine tertiary $\gamma'$ precipitates and encountering larger secondary $\gamma'$ precipitates with size on the order of a few hundred nm, comparing with (**b**) TEM measurement. Reproduced with permission from Fig. 6 of Wang and Li (2010)

essential part of the field equations. FDM is framed in the context of compatibility of the total deformation gradient, with incompatibility of the plastic deformation gradient/distortion arising from geometrically necessary (net) Burgers vector at the scale of a Burgers circuit of interest, giving rise to lattice curvature (Nye 1953; Bilby and Smith 1956; Ashby 1970) and the excess dislocation density tensor. Accordingly, it offers a continuum coarse-graining methodology for incompatibility mechanics by considering a range of characteristic sizes of the Burgers circuit (defining the area through which dislocations cut). The theoretical construct of FDM was formalized by Acharya and co-workers over the past over a decade ago (Acharya 2001, 2003; Acharya et al. 2006, 2008; Roy and Acharya 2006; Roy et al. 2007), providing common ground with higher scale theories such as SCD and GC in treating incompatibility due to geometrically necessary dislocations (GNDs). Hence, FDM serves as a theoretical construct to consider the effect of size of Burgers circuit on response. For this reason, it has among the widest range of applicability among model classes shown in Fig. 4. Depending on its auxiliary constitutive relations to represent the evolution of dislocation density, its predictive character can be similar to that of DDD at lower scales and that of GC crystal plasticity at higher scales. Typically based on small strain/distortion assumptions to facilitate implementation, this class of theory explicitly incorporates lattice closure relations on incompatibility expressed either in terms of a single dislocation (e.g., approaching atomic scale, cf. Zhang 2015; Hartley and Mishin 2005; Mendis et al. 2006), or in terms of the Nye tensor evaluated over a finite Burgers circuit for a large number of dislocations (analogous to second gradient plasticity). In principle, one may consider FDM to encompass many other theories, with distinctions drawn based on the level of coarse-graining (i.e., scale of the Burgers circuit and representation of populations of signed dislocations) or on the reduction of model order for various elements of the constitutive framework beyond the incompatibility relations. Its ability to provide predictive bottom-up simulations is somewhat more advanced than that of typical mesoscale GC models owing to its flexibility to deal with different length scales of incompatibility and dislocation interactions.

Roy et al. (2007) outline a cascade of increasingly simplified forms of the theory with continued reduction of model order, ranging from a complete theory that models transport of excess (net) dislocation density to more simplified forms. This cascade embeds many previously proposed gradient plasticity theories, in principle. Similar to a regime for which DDD modeling is often employed, mesoscopic FDM has been used to model dislocation sources and size effects at initial yield in micron-scale specimens (e.g., Puri et al. 2009). We will discuss later the deep connection of FDM with SCD, with the latter in many cases representing a higher fidelity form of FDM that focuses on distributions of dislocation densities. In FDM, boundary conditions on the flow give rise to additional geometric configuration effects on the field problem. As a consequence, Bauschinger or internal stress effects emerge naturally with the constraint on flow and are strongly influenced by the nature of the boundary condition on the transport of excess dislocation density, similar to the case of DDD and some GC models. These internal stresses are critical to description of effects of boundary conditions and size and configuration of finite geometries. Boundary conditions may also be pursued to address slip transfer reactions at grain boundaries, although this involves complexity similar to that encountered by DDD and GC models. FDM has been more recently applied in quasi-static form to spatially periodic microstructures (Brenner et al. 2014). Acharya (2011) has discussed a thermodynamically consistent framework for FDM at the mesoscale. Das et al. (2016) have pointed out that non-convexity or non-quasi-convexity of the potential energy landscape is not necessary to realize dislocation patterning when wave propagation transport of polar dislocation density is considered, although such non-convexity has served as the basis for approaches to dislocation substructure pattern formation based on cooperative constrained energy minimization (cf. Kuhlmann-Wilsdorf 1989; Ortiz and Repetto 1999; Ortiz et al. 2000; Koslowski et al. 2004; Kochmann and Hackl 2011; Klusemann and Kochmann 2014).

FDM has been recently extended (Fressengeas et al. 2011; Upadhyay et al. 2013; Taupin et al. 2015) in principle to incorporate the kinematics of disclinations (Clayton et al. 2005), rotational lattice defects proposed originally by Volterra (1907). Disclination representations are useful in representation of coincidence site lattice grain boundaries, reflecting the role of interfacial dissociation of partial dislocations; disclinations are not incorporated into standard dislocation theory (e.g., DDD), but emerge naturally in atomistic simulations.

## 6.5 *Discrete Dislocation Dynamics (DDD) Models*

Implementations of discrete dislocation dynamics (DDD) models employ continuum linear elasticity solutions for dislocations to compute long range elastic interactions, representing the positions and velocities of all dislocations by discretizing the dislocation lines into a piecewise series of curves or line segments. This approach effectively regards dislocation plasticity as a non-equilibrium sequence of elastic states with evolving line defects. DDD methods typically assume a simple form of dislocation force-velocity relationship (governed by mobility) for screw and edge components of dislocations, and do not explicitly distinguish partial dislocations, opting instead for an approximation of modeling full dislocations. DDD models employ a fully nonlocal treatment to superimpose the stress fields associated with long range interactions of dislocations along with the applied stress (Lepinoux and Kubin 1987; Gulluoglu et al. 1989; Amodeo and Ghoniem 1988, 1990; Ghoniem 2005) to compute the driving force on each segment. An early and persistent limitation, the influence of nonlinear dislocation core effects (which are outside the singular theory of linear elasticity) on short range dislocation interactions, has been rather effectively addressed by non-singular formulations of DDD (Cai et al. 2006) that can replicate the influence of core spreading and attenuation of interaction energies. These approaches can be hierarchically informed using atomistic simulations (Martınez et al. 2008; Groh et al. 2009; Rudd et al.

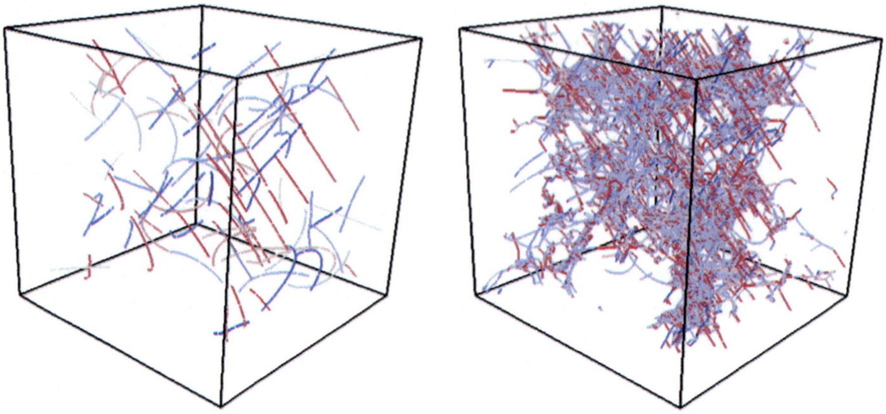

**Fig. 32** Initial (left) and final (right) dislocation microstructures in a 5 μm size crystal domain of FCC Ni with initial random distribution of Frank-Read sources with length of 600 nm, cycled under uniaxial strain-controlled tension-compression at a strain range of 0.4% and strain rate of 200 $s^{-1}$ in the [001] crystallographic direction (multislip). Dislocations are allowed to escape from all surfaces. Reproduced with permission from Fig. 5 of Hussein and El-Awady (2016)

2014). For example, work hardening can also be tackled with combined strategies of atomistically-informed DDD for certain dislocation-obstacle interactions (Arsenlis et al. 2007). Incorporation of short range interaction effects vary according to the perspective and expertise of the user.

The DDD method has received significant investment and is therefore maturing as a tool to support mesoscopic simulations (e.g., ParaDiS code, Arsenlis et al. 2007; Tang et al. 2011, and the microMegas code described by Devincre et al. 2011). Areas of recent focus include massive parallelization (Wang et al. 2006), fast multipole methods to address long range stress field interactions of segments, along with mesh adaptivity to segment the evolving dislocation lines, and employment of maximum power dissipation to assess dislocation junction node dissociation. Sills et al. (2016) have introduced advanced time integration algorithms for DDD simulations. While DDD has been extensively used to explore the problem of dislocation pattern formation and work hardening in single crystals, recent applications have considered the complexity of cyclic loading (Hussein and El-Awady 2016) for up to 80 cycles, including the influence of the domain size and considering the critical influence of cross slip processes in conjunction with formation of dislocation cell walls; Fig. 32 shows an example of initial and final dislocation microstructures formed in Ni after two loading cycles with a specified initial dislocation density of $10^{12}$ $m^{-2}$.

DDD has also been used to hierarchically inform GC or LC crystal plasticity formulations (cf. Zbib et al. 1998, 2002; Zbib and de la Rubia 2002; Khan et al. 2004; Wallin et al. 2008; Groh et al. 2009). It is interesting to note that accounting for continuum lattice rotation in DDD is a nontrivial multiscale issue, requiring and assignment of a characteristic volume element for this purpose (Zbib et al. 2002; Zbib and de la Rubia 2002; Khan et al. 2004). DDD is often applied with periodic boundary conditions to explore fundamentals of dislocation substructure formation and work-hardening for single crystals. It has also been applied to initial-boundary value problems for finite domains using implementations that couple with the finite element method to model responses of structures and devices (Yasin et al. 2001; Zbib et al. 2011). Groh and Zbib (2009) provide an overview of hierarchical versus concurrent coupling methods between DDD and continuum crystal plasticity.

A primary advantage of DDD simulations is the ability to quantify statistical details of emergent dislocation networks and response of submicron specimens up to the micron scale, in contrast to atomistic and most coarse-grained atomistic methods, as well as MPF and FDM models that resolve individual defects. DDD models aimed at exploring finite size specimen responses have been advanced to consider size effects, influence of free surfaces and image forces, intermittent/jerky flow behavior, and source limitation/starvation phenomena at submicron to micron scale specimens (Van der Giessen and Needleman 1995; Guruprasad and Benzerga 2008; Akarapu et al. 2010). Recent advances have considered highly parallel finite element or boundary element method implementations for arbitrary finite geometries (Crone et al. 2014) and for considering elastically anisotropic crystals (Liu et al. 2016), including associated fast multipole expansions for long range interactions (Yin et al. 2012). DDD has proven to be quite useful in modeling dislocation plasticity in confined geometries such as constrained thin films, multilayers, micropillars, and nanoindentation, providing key physical insights into scale dependent dislocation phenomena and interactions with surfaces and interfaces (cf. Zbib et al. 2011); this includes cases where kinetics dominates elastic interactions (Ryu et al. 2013). Moreover, DDD solutions can offer reference solutions for higher scale, lower fidelity models such as GC. SCM can be used to address similar regimes of cooperative dislocation behavior, but at a reduced level of fidelity in terms of describing individual junctions, reactions, and statistics of segment migration.

Since populations of dislocations are not decomposed into geometrically necessary and statistically stored components, DDD offers perhaps the clearest pathway to explicitly model effects on dislocation network irreversibility and accommodation of slip via the entire population of dislocations, in contrast to some implementations of FDM, SCM, and most applications of GC. DDD can address complex dislocation network formation, effects of strengths and distributions of dislocation sources, and dislocation junction formation (Zbib et al. 1998; Ghoniem et al. 2000; Bulatov et al. 1998, 2006; Devincre et al. 2008, 2011; Motz et al. 2009; Li et al. 2003; Arsenlis et al. 2007; Po et al. 2014). Modeling of dislocation pattern formation is actively being pursued, including important three-dimensional effects and contributions of cross slip (Hussein et al. 2015), as well as dislocation substructure formation under cyclic loading (Po et al. 2014). At the same time, 2D DDD methods (Van der Giessen and Needleman 1995; Benzerga et al. 2004; Keralavarma and Curtin 2016) have been advocated as a simplified but qualitatively correct framework to pursue understanding of work hardening and dislocation substructure formation in 3D; in this case, constitutive rules need to be prescribed to represent the effects of the corresponding phenomena in 3D. It is noted that dislocation mean paths in anisotropic crystals have directional character (Devincre et al. 2008), which has often been overlooked in discussion of most modeling results; this can have ramifications for mechanical properties that depend on mean free path, such as fatigue response.

As a reduced order model construct, DDD shares certain challenges with other continuum formulations, including representation of high angle grain boundaries and treatment of dislocation slip transfer reactions with interfaces. In particular, the modeling of dislocations in high angle grain boundaries is challenging since partial dislocations are not explicitly addressed. Since DDD explicitly models dislocation flux and storage, treatment of interface reactions requires additional and perhaps complex heuristics, or domain decomposition as per the CADD method to address slip transfer as accurately as other elements of the modeling framework. Moreover, since DDD models track migration of individual dislocation segments, updating lattice rotation involves multi-resolution complexity that is perhaps more easily resolved over a finite volume for models that employ statistical populations of dislocations (e.g., SCD, GC, LC). Representation of solute distribution, associated dislocation core interactions, and solute drag effects, as in dynamic strain aging, is challenging with DDD and may require two-way multiscale concurrency in cases where time scales of atomic level processes cannot be separated from scales of dislocation migration. Dislocation interactions with point defects (Cai et al. 2014) or interactions with diffusive vacancies at large strain or under cyclic loading at finite temperature requires extension of the framework that may vary according to modeler preference and are in early stages of development, whether involving DDD or atomistic methods that separate time scales to inform DDD (e.g., Keralavarma and Benzerga 2015; Baker and Curtin 2016; Matveev et al. 2016).

There are also certain challenges for DDD that owe to the nature of its discrete representation of dislocations in a continuous lattice. The potential energy landscape is coarse-grained to isolate the contributions of dislocations apart from lattice resistance or other obstacles. Certain assumptions made in DDD regarding interaction of multiple dislocations on the same slip system with an obstacle and its debris field may or may not accord with physical observations or atomistic simulations, so this is an area of uncertainty. Since dislocation sources (Frank-Read or pole sources) are explicitly represented, wait time heuristics must be introduced to model probability of source activation with time under an applied stress (Van der Giessen and Needleman 1995; Benzerga et al. 2004; Benzerga 2009). The resolution of DDD also places strong demands on kinetics and the associated range of time step sizes necessary to resolve individual dislocation interactions as well as cooperative processes of dislocation bypass of extended obstacles or sets of obstacles via thermal activation (rather than artificially overdriven kinetics), with time scales that can be many orders of magnitude higher. Setting aside consideration of these issues regarding thermally activated, dislocation configuration-dependent processes, large scale DDD simulations can be carried out to reasonable strain levels on the order of s (Ciorba et al. 2012) by exploiting massive parallelization. More efficient FFT formulations of DDD have recently been advanced that offer improved efficiency and scalability (cf. Bertin et al. 2015).

Certain thermally activated unit processes such as dislocation cross slip have been incorporated heuristically in DDD (cf. Groh and Zbib 2009; Groh et al. 2009; Deng and El-Azab 2010; Kang et al. 2014; Xia and El-Azab 2015; Xia et al. 2016; Hussein and El-Awady 2016). However, just as in coarse-grained atomistics, the ability to model thermally activated processes involving substantial activation volume (e.g., > 100 $b^3$) is limited by the ability to capture the probability of activating coordinated modes of thermal vibration of the dislocation line(s) that give rise to sufficient fluctuation to move the system through the saddle point of the reaction pathway for the barrier or set of barriers. There are two issues to resolve. In DDD, the stronger barriers are sessile junctions, unless finer scale precipitates or second phase particles are explicitly introduced as appropriate to the system considered (cf. Hiratani and Zbib 2003; Picu et al. 2009; Chen et al. 2010). Second, the time scales associated with coordinated thermally activated bypass of a field of obstacles can be much longer than those for a single obstacle. Recent work by Sobie et al. (2017a, b, c) has explored the use of the Nudged Elastic Band (NEB) method with DDD for bypass of nanoscale obstacles and obstacle fields to inform transition state theory approaches for probability of bypass; this includes an estimation of the entropic barrier strength associated with the thermal vibrational modes of sets of dislocation segments in the system (recall that the activation energy at finite temperature is the free energy). Using these kinds of NEB results along with entropic effects might then inform wait time heuristics more appropriately for such cases of extended obstacle systems, including thermally assisted activation of dislocation sources with continuing strain.

A rather fundamental issue in DDD is that of resolving the rate-limiting mechanisms for dislocation cross slip or bypass of barriers in the many body case for the interplay of a distribution of barrier strengths and/or strengthening mechanisms and size of activation volumes (cf. Dong and Curtin 2012). In statistical coarse-graining (e.g., kinetic Monte Carlo methods), GC, and LC this is accomplished by simply reflecting the kinetics of the statistically most probable mechanism(s) for thermally activated deformation in the overall population, which might require at most a few terms in a flow rule with different activation energies. Naturally, the approximation error in so doing may be significant if there are competing modes with their own statistical distributions, particularly if the distributions are wide and some involve extended interactions and relatively large activation energies and volumes (rare events). Selection of a dominant rate-limiting mechanism has been the standard textbook approach in materials science for many years, but modern alloys exhibit complexity of hierarchical structure that places increased demands on models. In DDD, the kinetics is particularly challenging for thermally activated dislocation bypass of obstacles involving a wide spectrum of transition times (or relaxation times) since the complete distribution of dislocation segments within an overall ensemble must be explicitly addressed, including the influence of internal (interaction) stresses. For this reason, using event probabilities to weight sub-populations of dislocation density evolution as in SCD, FDM or GC models, although a concession in reducing degrees of freedom, has a certain simplicity by comparison. If quasi-static simulations are undertaken in DDD at sufficiently high stress levels as necessary to overcome the strongest barriers in the system, the resulting distributions of internal stresses may be far from the case of thermally activated bypass corresponding to lower stress levels. In such cases, the trajectories of dislocations in the systems will likely not accord with realistic sequences of activated segment migration through a thermally activated process, which has important implications for modeling dislocation activity below the avalanche threshold pertaining to general yield conditions. This is particularly important for dislocation interactions with barriers of varying strength and character in the system, for example a grain boundary network. Another way of saying this is that dislocation evolution in standard DDD for such complex systems does not in general conform to a sequence of constrained equilibrium states associated with trajectories through the evolving energy landscape that mimic actual thermally activated processes, as discussed earlier in the context of quasi-static CAC.

Let us summarize a few key points. DDD, MPF and CGPF models that make use of eigenstrain methods or linear elastic solutions to represent dislocations have a significant advantage of rigorously accounting for long range interactions of dislocation segments. From the bottom-up, DFT and atomistic simulations are likely to connect more directly with MPF methods than with DDD. In terms of two-scale transitions to the continuum crystal plasticity level, the kinematics of DDD are logically addressed by concurrent schemes advanced by Zbib and coworkers (Zbib et al. 2011, 2002; Zbib and de la Rubia 2002), including coupling with FE methods for addressing finite geometries and consideration of free surfaces (Khraishi and Zbib 2002). However, the upscaling of kinetics will be limited in accuracy by the available stochastic treatment within DDD of distributions of barrier strengths (cf. Hiratani and Zbib 2003) and interfaces. Hierarchical methods to inform parameters of the flow and hardening rules in crystal plasticity using DDD constitute the prevalent approach, as shown in Fig. 33 (Groh et al. 2009; Groh and Zbib 2009).

Prospects for concurrent multiscale modeling across these particular model constructs have been advanced using spatial domain decomposition for purposes of modeling various regions in fracture and fatigue at crack tips (Curtin et al. 2010), as shown in Fig. 34. This includes atomistic DDD coupling (e.g., CADD method) and coupling of DDD and LC crystal plasticity models (Wallin et al. 2008). The atomistic region is preferred rather than a cohesive zone treatment (cf. Olarnrithinun et al. 2013 for DDD applied with a cohesive zone model) to admit the complexity of crack tip dislocation nucleation, emission, reconstructive reactions, and environmental effects, since cohesive zone methods focus on expressing idealized strength and fracture energy apart from these other phenomena. The large distinction of model form between atomistics, DDD, and crystal plasticity model constructs likely limits the efficacy of the full hierarchical multiscale schema shown in Fig. 33, and uncertainty quantification is essential. On the other hand, coarse-graining schemes within a given model construct (atomistics, DFM, SCM) that bridge sufficient scales might offer an alternative means to achieve greater concurrency, allow reduction of epistemic (reducible) uncertainty with advances in coarse-graining. It appears to be a fertile area of research.

## *6.6 Statistical Continuum Dislocation (SCD) Models*

Statistical continuum dislocation models are a class of reduced order continuum descriptions similar in spirit to DDD that formulate the governing equations in a manner that pertains to a statistical population/distribution of dislocations (cf. Groma 1997; Arsenlis and Parks 1999, 2002; El-Azab 2000, 2006; Zaiser 2001; LeSar and Rickman 2004; El-Azab et al. 2007; Deng and El-Azab 2010; Ispánovity et al. 2011; Xia and El-Azab 2015; Groma 2010; Kapetanou et al. 2015; Groma et al. 2016; Xia et al. 2016; Monavari et al. 2016). These models employ averaging procedures based on the statistical mechanics of interacting many-body systems, recognizing that the Nye tensor employed in FDM and numerous GC theories offers only a description of GNDs in an averaged sense over all the line directions in a volume element (or Burgers circuit) and not the overall distribution of dislocations. Accordingly, these latter models are more limited in nature than DDD or SCD. Statistical descriptions of dislocation densities and fluxes are nontrivial, as pointed out by Sandfeld et al. (2011). Limitations naturally arise associated with averaging assumptions that are made. Many of the early approaches were of 2D character addressing interactions of populations of edge dislocations for an infinite single crystal, for example, yielding important insights (Groma 1997, 2010; Groma et al. 2003). The description of dislocations goes beyond that of the classical dislocation density tensor (e.g., El-Azab 2000; Hochrainer et al. 2007) to provide more statistical information. Current trends are moving towards 3D applications and finite

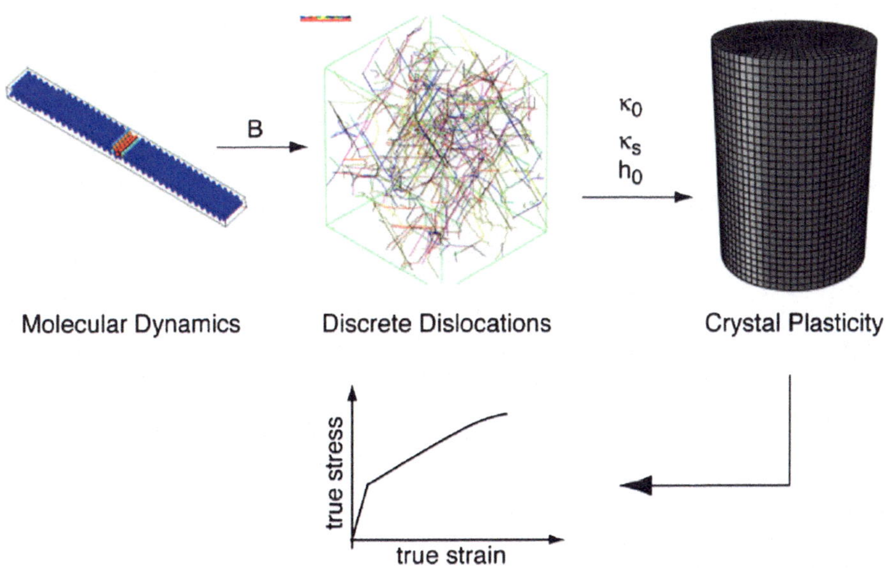


**Fig. 33** Hierarchical multiscale modeling with handoffs from MD to DDD to continuum crystal plasticity (GC or LC). Adapted with permission from Fig. 1 of Groh et al. (2009)

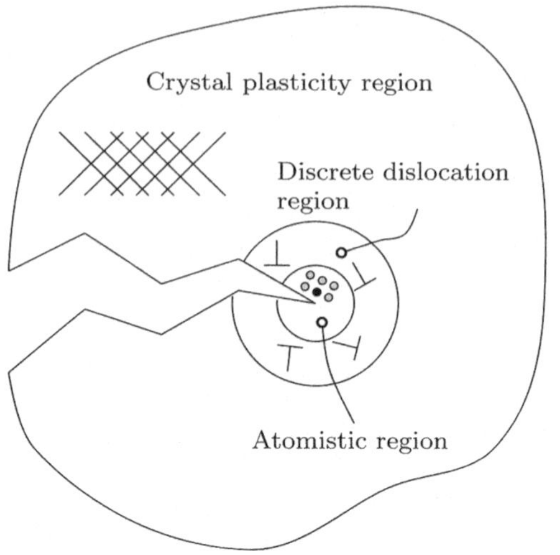


**Fig. 34** Concurrent multiscale modeling concept for fracture or fatigue that employs atomistics near the crack tip, abutting to a DDD region, then to an GC or LC crystal plasticity region in the far field. Modeling two-way transport of dislocation flux across various interfaces is challenging. Reproduced with permission from Fig. 5 of Curtin et al. (2010)

volumes, making contact with higher length scale GC and LC crystal plasticity. Approaches vary but share common elements of tracking geometrically necessary dislocation (GND) density evaluated using the Nye tensor at the scale of an integration point or solution node in addition to computing long range interactions between various statistical moments of the GND density distribution. In some cases, the total dislocation density is also tracked (i.e., statistically stored content as well as GND density) to convey more information. In 3D implementations (cf. Xia and El-Azab 2015), distributions of the dislocation segments of edge and screw character are each tracked at each integration point, along with mobility-dependent migration with wait times at barriers. As in DDD, appropriate boundary conditions are necessary for finite domains and free surfaces.

SCD models ostensibly bridge between DDD and GC models but have strong connectivity to the governing equations of FDM; they also consider statistical moments of dislocation ensembles and their interactions. Like FDM, SCD can resolve to the nanoscale (perhaps 20–30 nm scale), and like GC it can bridge upward to express incompatibility mechanics at scales of hundreds of nm to a μm. In terms of capabilities, certain formulations of SCD have much in common with coarse-grained implementations of FDM and GC in terms of capabilities. A principal advantage of SCD is its potential to shed light on scaling relations for dislocation substructure formation and to explore sensitivity of these relations to various physical parameters and mechanisms. Xia and El-Azab (2015) predicted arrangements of dislocations into cell structures by treating plastic deformation as an eigenstrain field, solving an incremental elastic boundary value problem for slip system driving forces and transport of dislocation densities, similar in nature to FDM models. Cross slip has been admitted at the unit process level of junction bypass (Xia and El-Azab 2015; Xia et al. 2016), and was found to play a key role in describing formation of dislocation cell structures, as was also found using DDD by Hussein and El-Awady (2016).

The implementation of SCD at high fidelity relies on a relatively small set of physical parameters in addition to solution scheme-related parameters. SCD models can accept information from DDD models as necessary to motivate model form and to validate and quantify assumptions regarding the evolution of spatial statistics of dislocation populations. Generally, most applications of SCD models are directed towards the challenging problem of predicting the formation and evolution of dislocation substructure in single crystals (similar to DDD with periodic boundary conditions). Although more computationally efficient and therefore offering greater access to modeling dislocation substructure formation at larger strains and for complex load histories, the predictive character of SCD does not seem to be as robust as DDD for a broad range of phenomena. Uncertainty in SCD model form is manifested in relations for junction formation and kinetics of thermally activated motion of dislocations, for example, among other things. There are similar issues in terms of representing dislocation-interface reactions, although thermal activation can perhaps be approached more effectively from a statistical perspective that considers dominant, rate limiting barriers among a complex distribution of barriers (Deng and El-Azab 2010; Xia et al. 2016), similar to the approaches taken by GC and LC models. The overlap with GC in this regard increases as the degree of coarse-graining of SCD models increases.

Another SCD framework has been developed in recent works of Hochrainer and co-workers (Hochrainer et al. 2007, 2014; Sandfeld et al. 2011), which they label as Continuum Dislocation Dynamics (CDD). This work emphasizes the need to go beyond local forms of dislocation density evolution equations to include dislocation transport processes, focusing on a consistent description of dislocation kinematics. According to Sandfeld et al. (2011), a particular advantage of their approach is the possibility of linking BCs of the gradient-dependent equations to the physics of dislocation motion near surfaces and interfaces. Recent work (Hochrainer et al. 2014) has introduced multipole expansions for the orientation dependence of the dislocations density functions to achieve a more tractable framework. Within the limits of its assumptions, CDD is a compact theory that offers a means of address dislocation behavior in small volumes/specimens, acknowledging that the theory will grow in complexity by inclusion of dislocation interactions on different slip systems, cross slip, and segment-specific dislocation mobility.

Building on the foundations of CDD, and starting with Hochrainer's (2015) approach based on multipole expansions, Monavari et al. (2016) introduced a methodology based on the maximum information entropy principle (MIEP) to derive closed-form evolution equations of curved and connected dislocation lines (loops) via a set of density measures. The authors compared results of dislocation bowing and loop interactions with walls within persistent slip band (PSB) type geometries and associated assumed dislocation velocity fields with results of DDD, with good agreement achieved using averaged information regarding edge and screw GND densities. Figure 35 shows a snapshot of dislocation configurations within the

PSB at left and comparison of the dislocation loop density and curvature versus applied strain from CDD with DDD at right.

SCD can serve as a tool, either mathematical or computational in nature, to assist in developing fundamental understanding of cooperative dislocation phenomena. The statistical physics community has garnered important insights from application of SCD to dislocation substructure formation and scaling relations based on 2D simulations of dislocations in infinitely extended single crystals. Implications for 3D structures are not always clear, but in this regard, we mention a few important contributions here. First, Ispánovity et al. (2011) have explored relaxation of dislocation segments in 2D systems of edge dislocation segments and find that the inherent quenched disorder of the position of slip planes, in a manner similar to systems exhibiting glass behavior with a wide spectrum of decay times, leads to a breakdown of the scaling relation for the dislocation velocity distribution, with cutoff time increasing with system size. Accordingly, the relaxation towards equilibrium increases with system size due to increasing complexity of the energy landscape. The system does not approach a global minimum energy configuration and relaxation depends on the initial state, with no intrinsic relaxation time. This reinforces the aforementioned shortcoming of local energy minimization schemes based on simplified collective kinetics of laminates to describe non-equilibrium dislocation substructures. Moreover, the notion of non-equilibrium trajectories along a sequence of constrained equilibrium states is reinforced; the energetic complexity of dislocation ensembles is akin to that of long chain polymer assemblies. This does not promote confidence in addressing thermally activated kinetics using explicit incremental strategies to advance individual dislocation segments within a large population based on local information of nearest neighbor junctions and segments, for example.

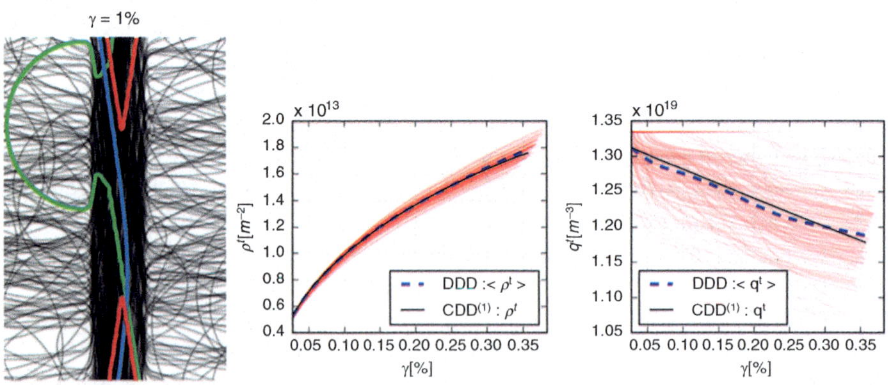


**Fig. 35** Arrangement of dislocations within a velocity field of a persistent slip band from (left) DDD at 1% applied strain and comparison of DDD with CDD based on plots of loop density (middle) and curvature (right), where the solid lines are DDD averages over 200 simulations, and the dashed lines are the results of CDD. The individual DDD results form a family of thin lines with variability. Reproduced with permission from Fig. 6 (DDD simulation) and Fig. B3 (loop density and curvature plots) of Monavari et al. (2016)

Second, Groma et al. (2015) have recently argued that dislocation-dislocation interaction is scale free (i.e., 1/r dependence), so dislocation substructure formation within an infinitely extended crystal has no relevant length scale parameter other than the average dislocation spacing. Dislocation configurations and associated interactions manifest scale effects that depend on finite geometries and boundary conditions, so there is no characteristic or intrinsic material length scale. By now this is reasonably well understood by those who work with GC crystal plasticity models, who typically regard introduction of normalizing length scales as a means to calibrate dependence of accommodational GNDs on specific sample configurations, boundary conditions, and sample sizes of interest (cf. Voyiadjis and Abu Al-Rub 2005; Mesarovic 2005). Of course, dislocation source distribution, second phase interactions, and free surface interactions at sub-micron scales can also impose scale effects. Intrinsic length scales may exert influence when interactions are not scale-free, for example being established by interactions with obstacles (Chakravarthy and Curtin 2010). Many practical alloy systems make use of dispersion or precipitate strengthening, which gives rise to natural length scales at scales much finer than typical grain size. Even in ultra-fine grain metals or nanocrystals, the grain boundary character distribution gives rise to characteristic scales that affect hardening and dislocation substructure formation.

Third, Ispánovity et al. (2014) have found that avalanche dynamics of 2D dislocation systems is scale free at every level of applied stress and cannot be considered as reflecting a depinning transition for which criticality is reached at a given level of applied stress associated with general yield. In fact, jamming effects may dominate, and the two scenarios are quite different. Depinning transition has been assumed to govern dislocation flow in some statistical theories (e.g., LBL theory of Langer et al. 2010) of ostensibly jammed dislocation networks in pure fcc metals. Such a depinning transition has previously been argued to apply (Papanikolaou et al. 2012) to the interpretation of yield via dislocation dynamics of infinite pure crystals in terms of quasi-periodic, self-organized avalanche bursts in the presence of dislocation relaxation, reaching a critical depinning point at yield and leading to dynamical quasi-periodic scale invariance. Ovaska et al. (2015) showed that introduction of randomly distributed quenched pinning points (e.g., immobile solute atoms) in large scale 2D DDD simulations can change the interpretation to that of strength governed by depinning transitions. Hence, it is very important to understand and incorporate the realistic nature of dislocation strengthening mechanisms (Chakravarthy and Curtin 2010) for practical alloy systems.

Finally, Zaiser and Sandfeld (2014) have presented a cogent and succinct argument that the proper mathematical structure of dislocation theory itself, with 1/r scaling of interaction stresses, apart from a arguments regarding avalanche dynamics and self-organized criticality of dislocation systems, gives rise to similitude relations for scaling; such relations include the Taylor relation between flow stress and dislocation density and the link between flow stress and the characteristic scale of periodic dislocation patterns. They approach this issue based on generic invariance properties of the dynamic equations employed in both 2D and 3D dislocation simulations. Fundamentally, if dislocation dynamics is controlled by elastic dislocation interactions, then characteristic lengths of substructure arrangements scale with the average dislocation spacing or inverse square root of dislocation density, while flow stress scales with the square root of dislocation density. They note that dislocation cross slip and annihilation reactions can lead to deviations from similitude. They also show that short-range self-interactions that depend on intrinsic scales break down similitude. These arguments have important implications for interpretation of size effects in DDD simulations and SCD models. They imply that any model (DDD or SCD) for which dynamics is governed by elastic dislocation interactions must produce patterning with wavelength that is inversely proportional to the stress. Moreover, the framing of the evolution equations for dislocation density in GC or LC crystal plasticity models that aspire to mimic fine scale substructure formation and similitude relations must adhere to the tenets of the underlying discrete dislocation theory. Of course, many GC and LC crystal plasticity theories address specific dislocation strengthening mechanisms in engineering alloys and are not framed to address scale free dislocation interactions in extended single crystals, nor do they address various elements of the construction of FDM, DDD, or SCD.

One clear opportunity to further advance SCD is to incorporate higher statistical moments of signed dislocation density in order to retain more information to express dislocation kinetics and interactions than is offered by reduced order descriptions of dislocation density (Sandfeld et al. 2011). This increases computational expense, but may offer improved predictive capability, particularly for applications involving finite geometries, strain gradients, polycrystals, or obstacle fields.

## *6.7 Generalized Continua (GC) Crystal Plasticity Models*

Generalized continua (GC) crystal plasticity models introduce dependence on specific dislocation configurations that are necessary to accommodate imposed strain as a function of specimen size and geometry, as reflected through scale dependent incompatibility of the elastic and plastic deformation gradients, in a manner that is analogous to FDM models. GC models for dislocation plasticity are necessarily of reduced order compared to DDD, FDM, SCD, or phase field models. They augment strengthening due to overall dislocation density with configuration-dependent strengthening expressed by dependence on the GND (or excess dislocation) density, differentiated in this regard by model assumptions. Advanced GC theories for dislocations in a lattice typically associate higher order stress quantities that conjugate with evolution of generalized kinematics to contribute to dissipation (e.g., second gradient theory, micropolar, and micromorphic models). They are intended to apply to length scales well above the average spacing of dislocations and to initial-boundary value problems with scales that are typically on the order of 500 nm and above, rather than to levels of structure hierarchy well below the micron scale. DDD, FDM and SCD are better choices for lower scales, and to more rigorously incorporate dislocation transport. Because they are reduced order continuum formulations and typically require introduction of normalizing length scale(s), most GC models do not offer predictive character in the same sense as the previously discussed models, requiring more model assumptions, phenomenology, increased number of parameters, and influence of associated calibration schemes. Still, they have considerable value since they are highly amenable to efficient finite element implementation and can span upwards through modeling at the length scale of polycrystals. A peculiar aspect of GC models is that the boundary conditions can be difficult to interpret and prescribe, owing to the inherent nature of reduced order models that attempt to incorporate more physics. We have also seen in the last section that care must be taken in prescription of dislocation density evolution in GC (or LC) crystal plasticity models to ensure that it conforms to the underlying physics of scale free dislocation systems as appropriate.

The origins of nonlocal GC models of elasticity and plasticity trace back at least over 45 years (cf. Eringen and Claus 1970; Eringen 1972, 1999; Dillon and Kratochvíl 1970). These early models were not framed in the context of crystalline plasticity per se. GC crystal plasticity models of strain gradient, micropolar and micromorphic type have advanced substantially over the past 15 years, to the point where they are viable and efficient reduced order alternatives to DDD or SCD for modeling mechanical response of domains with scales on the order of $\mu$m to tens or hundreds of $\mu$m, including polycrystals. As previously mentioned, these models employ the phenomenological notion of so-called 'characteristic length scale' parameters that normalize gradient or nonlocal terms to capture size effects. Of course, these length scale parameters are not true characteristic lengths but depend to varying degree on the initial conditions, boundary conditions, and level of inelastic deformation (Voyiadjis and Abu Al-Rub 2005).

Second gradient (plastic strain gradient or slip gradient) crystal plasticity models regularize the influence of strain gradients on the material work hardening behavior (Busso et al. 2000; Gurtin 2002; Evers et al. 2004; Cheong et al. 2005; Bayley et al. 2006; Gurtin and Anand 2007; Abu Al-Rub et al. 2007; Viatkina et al. 2007; Gerken and Dawson 2008; Dunne et al. 2012; Bargmann and Svendsen 2012). Back stress arises naturally in such models as a manifestation of GND gradients and manifest specimen size and boundary effects, which is a key advantage of these models relative to LC models, particularly with regard to modeling specimen configuration, size, and boundary condition effects. On the other hand, contributions to slip system back stress that emerge from intragranular dislocation substructure may not be addressed since these models typically don't consider dislocation substructure development in detail; these back stresses can be significant for low stacking fault energy crystals or precipitate-strengthened systems. These GC theories have emerged from initial conception by Ashby (1970) of the role of GNDs in size-dependent strengthening in plasticity, followed by early couple stress theories (Fleck and Hutchinson 1993; Fleck et al. 1994) and later by simplified strain gradient theory applied to indentation (Nix and Gao 1998). These early works moved GC theories more towards today's well-established second gradient crystal plasticity model structure, having much in common with FDM applied at coarse scale. GC models continue a trend towards extension to incorporate more salient physics.

For example, Hurtado and Ortiz (2013) have offered a novel GC crystal plasticity framework that includes both energy created by surface steps and self-energy of dislocations based on slip gradients (GNDs); their model is able to capture the significant influence of free surfaces on the stress-strain behavior of micro-pillars in compression.

Recent developments have cast GC crystal plasticity models into forms that express sensitivity to multi-phase microstructures and complex dislocation mechanisms (cf. Keshavarz and Ghosh 2013; Ghosh et al. 2016a, b; Shahba and Ghosh 2016), along with sophisticated mechanistic evolution equations at the slip system level and associated calibration schemes. It is very difficult to incorporate comparable levels of key dislocation mechanisms and material structure hierarchy in the other model constructs shown in Fig. 4 for complex alloy systems. For example, Keshavarz and Ghosh (2015) proposed a hierarchical GC crystal plasticity for Ni-base superalloys. Ghosh et al. (2016a, b) introduced a parametrically homogenized GC crystal plasticity model for $\gamma - \gamma'$ Ni-base superalloys that addresses transitions across three scales: sub-grain precipitates and dislocation-precipitate interactions, grain level response, and collective polycrystal behavior with realistic 3D microstructures. Their approach is at the frontier of model order reduction approaches via homogenization of GC descriptions from fine to coarse scales; it facilitates determination of the functional form of the macroscale constitutive parameters with respect to the sub-grain precipitate structure. The precipitate-channel length scale in $\gamma - \gamma'$ Ni-base superalloys typically lies within the mesoscale shown in Fig. 4, so there is uncertainty and non-uniqueness of models framed at this scale; for material systems with well-developed understanding of dislocation mechanisms, such approaches can be extremely useful to inform grain and polycrystal level continuum models.

Le (2018) has recently extended the Langer-Bouchbinder-Lookman thermodynamic theory of dislocation-enabled plasticity (cf. Langer et al. 2010; Langer 2017) to consider gradients of plastic deformation and associated dislocations of excess sign (i.e., GNDs). We note that application of this theory must take care to ensure that strain gradients are sufficiently weak such that the integration points can sample a sufficient population of dislocations to warrant the assignment of the configurational entropy of dislocations (i.e., well above the size of characteristic dislocation substructures expressing periodicity of the excess dislocation population). This requirement in fact holds for any gradient theory in which free energy functions for substructure are assigned.

Recognition of the role of stress gradient on dislocation nucleation was first addressed by Miller and Acharya (2004). Chakravarthy and Curtin (2011) contributed to understanding the effects of the stress gradient on yield stress for dislocation interaction with obstacles using discrete dislocation theory, and formalized the approach in GC plasticity. The concept was later applied to modeling of fracture using DDD (Olarnrithinun et al. 2013). Recent microstructure-sensitive GC crystal plasticity models that include both stress and strain gradient effects have been introduced (Taheri-Nassaj and Zbib 2015; Lyu et al. 2016). It is understood that stress gradients manifest size effects of initial yield stress due to pile-ups at barriers, while strain gradients manifest size effects in work hardening associated with dislocation substructure. Hence, stress and strain gradient effects are both necessary and in fact are complementary. DDD and SCD models naturally incorporate both types of effects only if strengthening obstacles are explicitly considered (which is not commonly done in SCD); framing of continuum theories of GC type requires intentional incorporation of both, not just strain gradient effects. Since incorporation of strengthening obstacles or barriers induces breakdown of similitude relations (discussed in the SCD section), we see that this is a critical consideration in modeling behavior of actual engineering alloy systems.

Typical GC crystal plasticity models embed all relevant elements of LC crystal plasticity and further add dependence of evolution equations on GND density, along with an analytic relation for back stress in terms of the plastic incompatibility (albeit typically limited to the average GND density within the volume of the integration point). This significantly extends capabilities to reflect the role of heterogeneity and compatibility near interfaces in polycrystals, as well as the influence of applied strain gradients, size effects on work hardening, etc.

The literature has emphasized strain gradient models. Other less conventional forms of GC crystal plasticity models have been advanced over the past 10–15 years that capture GNDs and other scale dependent effects. These include micropolar (Forest et al. 2000; Forest and Sievert 2003; Mayeur et al. 2011; Mayeur and McDowell 2013, 2014, 2015) and micromorphic (Aslan et al. 2011; Cordero et al. 2010, 2013) models framed as extensions of crystal plasticity that effectively admit rotation (micropolar, micromorphic) and deformation (micromorphic) of lattice director vectors as additional kinematic degrees of freedom to augment translational displacements. Mayeur and McDowell (2014) demonstrated close parallels between the structure of micropolar crystal plasticity and Gurtin's slip gradient model (2002). Implementation of the former does not require computation of gradient or curl operations on the plastic deformation. Additional degrees of freedom are introduced that effectively represent microrotation of the director vectors at each integration point. Relatively simple micropolar crystal plasticity models can capture sub-micron size effects in plasticity via incorporation of elastic torsion-curvature and an evolution equation for the plastic torsion curvature, expressing a yield potential or flow rule as a combined, single function of the resolved slip system shear stress and couple-stresses conjugate to GNDs of screw and edge type. Dislocation evolution distinguishes between GNDs and statistically stored dislocations (SSDs), and the slip system back stress is naturally related to the skew symmetric part of the Cauchy stress. Normalizing length scales are required for both for the elastic and plastic curvature effects (analogous to those employed in stress and strain gradient models mentioned in the foregoing). An example of application of micropolar crystal plasticity to shear of constrained thin films with planar double slip (edge dislocations only) is shown in Fig. 36 (Mayeur and McDowell 2013).

The ability to access sufficient length scales necessary to model polycrystalline material response is challenging for DDD or SCM models, or with FDM models framed at fine resolution. Polycrystalline mesoscales (see Fig. 4) are a logical scale for application of microstructure-sensitive GC crystal plasticity models; LC

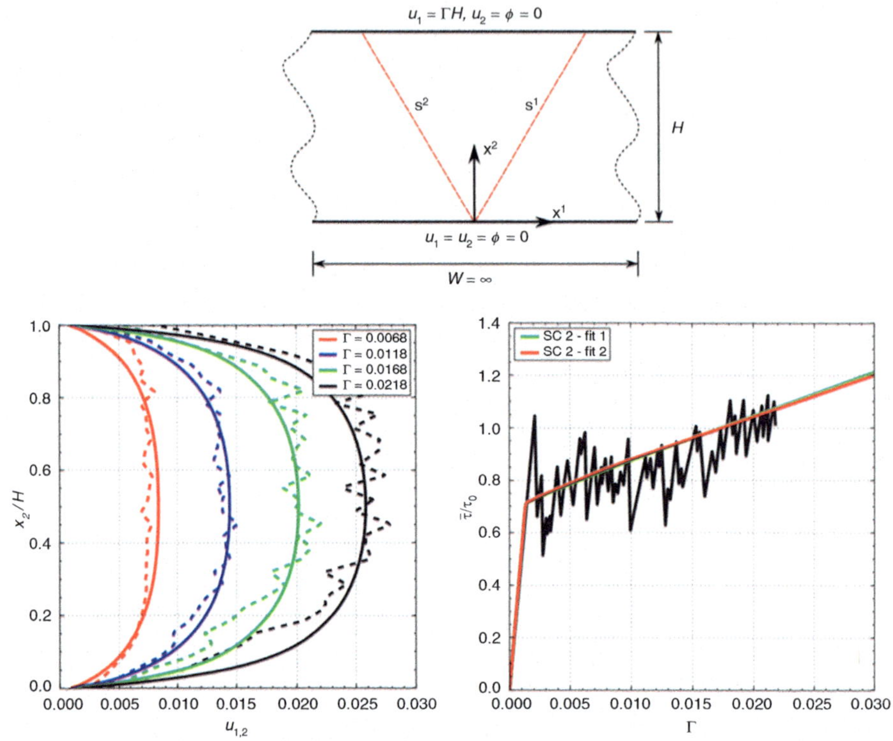


**Fig. 36** (Top) Geometry and boundary conditions for finite element analysis of planar double slip under imposed shear Γ of a constrained thin film using micropolar finite element simulations (simple single criterion strength model) for H = 1 μm. Results for simulated shear strain distributions in solid lines (bottom left) are compared with dashed line discrete dislocation modeling results (Shu et al. 2001), and (bottom right) the micropolar simulations of average shear stress-strain response (solid lines) are compared with the discrete dislocation results (jerky curve). Reproduced with permission from Fig. 1 (top), Fig. 18 (bottom right) and Fig. 19a (bottom left) of Mayeur and McDowell (2013)

crystal plasticity models are also quite applicable, especially those that incorporate size effects explicitly in the evolution equations for internal state variables related to material yield and work hardening. It should be noted that most applications have considered the error in simulating measured stress-strain responses of polycrystalline ensembles as a metric for convergence and accuracy of a given GC model framework. These frameworks do not commonly incorporate dislocation transport or slip transfer through grain or phase boundaries. Increasingly, GC crystal plasticity models are being employed with the goal of accurate prediction of local inelastic shear strain within grains and near interfaces to support modeling of local failure processes (cf. McDowell and Dunne 2010; Dunne et al. 2012). This is an unsettled issue since the predictive capability of GC crystal plasticity for modeling submicron scale processes involving dislocation substructure is limited. Moreover, understanding of dislocation slip transfer relations at grain boundaries is still in early stages (Bieler et al. 2009).

Microstructure-sensitive GC models that incorporate evolving dislocation density at the slip system level as internal variables (cf. Raabe et al. 2001; Ma et al. 2006; Roters et al. 2010; Dunne et al. 2012; Keshavarz and Ghosh 2013; Shahba and Ghosh 2016) offer enhanced correlative capabilities for complex loading conditions in many cases at the scale of individual grains. Still, the experimentally observed number of activated slip systems at the sub-grain scale is typically limited compared to simulations using most mesoscopic crystal plasticity models of GC and LC type (Lloyd 2010; Buchheit et al. 2005); this is an issue that requires increased focus on slip system hardening relations and comparison with *in situ* experimental data such as EBSD and digital image correlation measurements of surface deformation.

## 6.8 Local Continuum (LC) Crystal Plasticity

Local continuum (LC) crystal plasticity (cf. Asaro 1983a, b) emerged as a first order constitutive model construct based on characterization of slip in crystals pioneered by Taylor (1938) and others. The theory has a venerable history and is standard in commercial codes, employing a multiplicative decomposition of the total deformation gradient into elastic and plastic parts. Over the years, most implementations have gravitated towards the use of an isoclinic intermediate configuration in which the lattice director vectors are the same as in the reference configuration, serving as a natural configuration in which to analyze lattice hyperelasticity and frame evolution laws for lattice-based evolution of internal state variables. Historically, there have been a number of hypoelastic rate type formulations of crystal plasticity in the current configuration which can give rise to certain issues regarding unintended coupling with lattice rotation.

LC crystal plasticity models offer feasible computation of responses of sets of grains, subject to two primary limitations. First, they do not incorporate long range stresses due to dislocation interactions, and therefore have limited capability to express size effects. Sometimes, forms and parameters are introduced into the constitutive relations for work hardening based on the scale of microstructure (e.g., grain/phase size or spacing) as a surrogate for more complex dislocation interactions. Such local theories suffer from intrinsic mesh dependence and degeneracy of solutions with mesh refinement, so care must be taken in this regard. From a practical perspective, minimum mesh size should be limited to the order of a micron for such models in view of their spatial averaging approximation. For example, it is difficult to justify use of highly refined meshes on the order of 100 nm and below near microstructure heterogeneities such as grain or phase boundaries with local models, in spite of the desire for accurate geometric representation of microstructure. Second, only a few internal state variables are typically employed in local crystal plasticity models, such as drag stress, threshold stress and back stress, which renders them of more utility in correlating observed behavior, with only limited predictive capability. These local models can address scale effects of constraint on slip modes due to microstructure barriers to dislocation motion and effects of second phase particles (both coherent and incoherent) on mechanical response in approximate fashion by incorporating scale dependence in evolution of internal variables such as dislocation density (cf. McDowell 2008, 2010; Roters et al. 2010).

Some of these criticisms of LC crystal plasticity at the grain scale apply as well to most GC crystal plasticity models, including their largely correlative nature. Models that employ dislocation density as an evolving internal state variable tend to be preferred in correlations, including those that decompose geometrically necessary and statistically stored populations for a given size of Burgers circuit. Certainly, the issue of accounting for interface slip transfer reactions (reflection, absorption, desorption, or direct transmission) is a challenge faced not only by LC, but also GC, DDD, and SCD crystal plasticity approaches. At present, only coarse-grained atomistics and domain decomposition methods based on atomistics can offer predictive capabilities in this regard.

Like GC models, LC crystal plasticity models are phenomenological and lack predictive character at the slip system level in terms of kinetics and hardening relations, but they are computationally efficient relative to the other models listed in Fig. 4. Unlike GC models that employ incompatibility mechanics based on GND density or equivalent, LC models do not naturally lead to back stress, which must be incorporated as necessary based on physical mechanisms of dislocation-barrier interactions (cf. Hennessey et al. 2017). Incorporation of back stress has relied more on intuition and experience. The need for substantial intragranular back stress levels is overlooked in applications of LC models to precipitation strengthened alloy systems (Al alloys, Ni-base superalloys) or systems with relatively low stacking fault energy and high slip planarity. Even GC theories may understate the level of back stress, since they are not intrinsically predictive regarding underlying intragranular dislocation mechanisms by virtue of their coarse-grained representation of GNDs and lack of explicit consideration of dislocation substructure. LC models can be used to model responses of large numbers of grains in 3D polycrystal plasticity simulations and are therefore of utility in statistical exploration of the effects of both grain size and orientation distributions, as well as grain morphology and higher order spatial correlations of microstructure (e.g., nearest neighbor statistics), on the deformation response, texture evolution, and strain or failure processes such as fatigue (cf. Przybyla and McDowell 2010, 2011). Ghosh and Chakraborty (2013) build on the foundation of a wavelet transformation based technique for multi-time scaling (Joseph et al. 2010) to facilitate computation over the large numbers of cycles (including cyclic plastic strain and stress redistribution within polycrystals) necessary to evaluate fatigue lifetime (crack formation), based a non-local criterion for crack formation.

It can be said that even with their limitation of local action, microstructure-sensitive, dislocation mechanism-based LC models of crystal plasticity have received substantial attention in development (cf. Stainier et al. 2002; Arsenlis et al. 2004; Liu 2006; Shenoy et al. 2008; Bridier et al. 2009; Patra and McDowell

2012; Li et al. 2014; Castelluccio and McDowell 2017; Messner et al. 2017), including stochastic models (Askari et al. 2015). Time scaling via hTST (cf. Wang et al. 2017) can also be undertaken based on underlying rate-limiting deformation processes. Most of this work translates readily into a second gradient or micropolar/micromorphic GC crystal plasticity setting. Further, the physics of certain problems are amenable to this class of theory, e.g., modeling of short range nucleation phenomena (at scales below the element volume). Austin and McDowell (2011, 2012) introduced a dislocation density evolution LC model for shock plasticity that includes both heterogeneous and homogeneous dislocation nucleation, and considered both phonon drag and thermally activated flow regimes as well. Lloyd et al. (2014a, b) later cast this formulation within LC crystal plasticity for FCC systems and explored novel mesocopic phenomena in the interplay of shock wave propagation in polycrystals, including scattering and dispersion effects that arise from this underlying heterogeneity.

Substantial advances in computational efficiency for LC crystal plasticity have been achieved using FFT methods (Prakash and Lebensohn 2009; Lebensohn et al. 2012; Chen et al. 2014; Rollett et al. 2015; Lebensohn and Needleman 2016). Moreover, an inverse method has been developed recently (Paulson et al. 2017) for efficiently estimating local stress and strain states within polycrystalline microstructures by calibrating influence functions based on 2-point spatial statistics of microstructure to LC simulations on a single microstructure, and then predicting local states for a range of other microstructures (for the same ensemble boundary conditions); this method may be used to efficiently inform probability of activation of failure mechanisms as a function of a wide range of microstructure morphologies (grain size, shape, and orientation distributions).

The concept of "importance sampling" or weighting of responses in the process of time averaging relies on the expected probability of occurrence (cf. McDowell 1997, 1999, 2008, 2010; McDowell et al. 2011), and is a key aspect of homogenization of dislocation evolution processes in multiscale modeling. This sort of statistical volume averaging can be linked closely with GC and LC crystal plasticity modeling of microstructure ensembles focused on specific material responses. Along these lines, Alleman et al. (2015) have introduced a distribution-enhanced homogenization framework for continuum elasto-plasticity in which higher length scale models are formulated in a series expansion based on the microscale constitutive relations and moments of arbitrary order of the microscale field variables, with no a priori assumption of a homogeneous effective medium representation of the macroscale constitutive behavior. This approach weights the statistical moments of micro-responses (micro-scale density functions) differentially to inform evolution at the homogenized level, a form of statistical volume averaging based on relative importance of micro-responses. Results compare favorably with brute force direct numerical simulations of macroscopic responses for representative volume elements of heterogeneous systems. This was also described earlier as an objective of hierarchical multiscale modeling in the doctoral thesis of Luscher (2010).

From an engineering perspective, LC crystal plasticity models are clearly of value in enabling computation of materials over a suitably large number of grains/phases to begin to make contact with many practical applications. From an applied physics perspective, however, we may adopt the perspective that reduced order LC models are inherently correlative (GC models as well, in most respects), such that their role is more focused on accepting information from more predictive bottom-up simulations and making use of top-down routes (e.g., experiments) to inform and optimize model forms and parameters (cf. Tallman et al. 2017). Attempts to formulate *fundamental* theories of plasticity are more likely compatible with MPF or SCD model constructs that can incorporate statistical ensembles, and reflect characteristics of the primal, underlying atomistic and/or discrete dislocation behaviors.

## 7 Strengths and Weaknesses of Various Model Constructs

Table 2 summarizes how phenomena addressed by crystalline plasticity models listed in Table 1 map onto the various model constructs applicable to different scales shown in Fig. 4. This is an attempt to provide guidance in terms of strengths and limitations of various model constructs in addressing key physical phenomena of dislocation plasticity across length and time scales. It is obvious that no single model framework addresses all 22 phenomena listed in Table 1. Of course, the predictive nature of MD/MS results in greatest coverage of phenomena, but these models lack access to longer length scales and time scales for thermally activated processes, particularly those processes having large activation volume; MD focuses more on nanoscales. CG atomistics extends the access to larger spatial domains, but remains limited by solution time step size of the full atomistic region. MPF removes some of the time scale limitations of atomistics and employs few heuristics, but largely retains atomistic limitations in terms of spatial scales. DDD provides a balance of predictive access to some of the phenomena, including both short and long range spatial interactions, but is also limited in its ability to scale cooperative bypass of obstacles at longer time scales and involves considerably more heuristics. SCD has similar limitations, albeit lacking elastic anisotropy in most implementations to date, but provides a more direct coarse-graining pathway for modeling lattice rotation and temporal coarse-graining as necessary to model development of dislocation substructure in single crystals. FDM and GC have very similar characteristics and capabilities, although formal applications of FDM models have tended to employ fewer heuristics than GC models to date. We may anticipate that LC models address the lowest number of phenomena listed in Table 1. However, both GC and LC crystal plasticity models have well-established heuristics for coarse-grained representation of thermally activated processes, although very few such models have attempted to model competing kinetics of multiple mechanisms that may operate at different stress and temperature levels (spectrum of kinetics), requiring multiple activation energies and thresholds in the slip system flow rule relations (cf. Shenoy et al. 2008).

**Table 2** Mapping of phenomena listed in Table 1 onto various model constructs shown in Fig. 4

| MD or MS | CG atomistics | MPF | FDM | DDD | SCD | GC | LC |
|---|---|---|---|---|---|---|---|
| 1–2, 3[a], 4, 5[b], 6–13, 14[a], 15–16, 18[d], 19[d], 20, 21[b], 22 | 1–13, 14[a], 15–16, 17[a], 18[d], 19[a,d], 20–22 | 1–4, 5[b], 7–15, 17–18, 20, 21[b], 22 | 1–4, 5[f], 6, 8[c], 10[c], 11, 13, 17, 18[f], 19[f], 20[f] | 1–2, 3[e], 4, 5[f], 6[f], 7–13, 14[e], 15–16, 18[f], 19[f], 21, 22[e] | 1–2, 3[f], 4, 6[f], 8, 10[c], 11, 13, 14[f], 15–16, 18[f], 19[f], 21 | 1–9, 10[c], 11, 13, 16[f], 17[f], 18[f], 19[f], 20[f], 22[f] | 1–7, 16[f], 17[f], 18[f], 19[f], 20[f] |

[a]Nudged elastic band method, unit processes, limited size of activation volume
[b]Only nanocrystals and nanostructures
[c]Limited by representation of the net Burgers vector, depending on the scale of the description
[d]Limited accessible time scales
[e]Wait time heuristics
[f]Limited heuristics

# 8 Summary, Prospects and Gaps

This chapter highlights the need to generalize the conception of model constructs for dislocation behavior across a range of length scales from lattice spacing to those of structural applications. We have intentionally focused on modeling of slip processes; consideration of twinning and slip twin interactions (cf. Beyerlein et al. 2014; Ardeljan et al. 2017), particularly in low symmetry crystals, brings in another set of multiscale considerations and approximations with uncertainty that are quite relevant to the topic but beyond the scope of this introductory volume. Moreover, consideration of first or second order phase transformations (cf. Bartel et al. 2011) also bears relevance in coupling with dislocation evolution, and predictive model constructs to address these phenomena are even more limited than those for modeling dislocations alone. The most promising model constructs with regard to predictive character are atomistics, coarse-grained atomistics, and MPF. Phenomenology and attendant uncertainty tends to accumulate in posing reduced order models at higher scale for collective twinning, slip twin interactions and phase transformations. The caveat, of course, is that these latter processes have strongly mesoscopic character (beyond nucleation of individual twins or phases) in terms of growth and many body interactions. For these reasons, higher length and time scale model constructs (e.g., GGPF, GC, LC) have largely been used to explore them using inductive strategies (e.g., Ardeljan et al. 2017). Suffice it to say that joint consideration of slip and twinning or phase transformation at various length and time scales deserves substantial scrutiny and attention in future work in a manner similar to the route taken in this work for dislocation processes in crystals. Certainly, there is ample opportunity to reduce the epistemic uncertainty in modeling these coupled processes.

When uncertainty of model form, model parameters, operative mechanisms, and initial and/or boundary conditions are considered, it is clear that models should be selected that are specific to the scale(s) of interest and address the needs of each application. Two-scale hierarchical or concurrent transitions of models (either coarse-graining or reduction of model order) are most common owing to the vast range of length and time scales that exist in dislocation mechanics/dynamics. It is necessary to pursue statistical approaches, likely with increasing contributions of materials data science and informatics, to enhance the bridging of models and salient information across length and time scales. We are still in relatively early stages in this regard.

Some of the more challenging and longstanding mesoscale gaps for the hierarchy of crystalline plasticity models listed in Fig. 4 include:

- Distinguishing dislocation nucleation and growth processes that include source activation and migration of loops.
- Initial conditions and evolving structure of interfaces.
- Slip transfer at interfaces, including interface evolution and damage.
- More robust sub-micron slip system level evolution equations for internal variables that reflect evolving dislocation populations.

- Quantifying coarse-graining and model reduction errors.
- Uncertainty quantification of individual models, domain decomposition, and various two-scale transitions.
- Improving the predictive character of lower scale model descriptions in two-scale transitions that start at the mesoscale.
- Modeling dislocations in multi-component material systems (cf. Rao et al. 2017), including appropriate interatomic potentials for atomistic methods that include uncertainty quantification.
- Improving fidelity and accuracy of simulation of local states of slip system activation and degree of slip within individual grains to support material fatigue, fracture and ductility assessments.
- Coupling dislocation structure and behavior with point defect interactions, including climb, strain aging, and hydrogen, interstitial, and vacancy interactions.
- Time scaling across a spectrum of thermally activated processes over a wide range of activation volumes, including dislocation cross slip, obstacle interactions, etc., accurately reflecting a sequence of constrained equilibrium states that reflect the non-equilibrium nature of evolving dislocation fields.
- Consideration of both entropic and enthalpic effects in dislocation-obstacle interactions, as well as the complete description of free energy for dislocations in reduced order models.
- Uncertainty quantification in individual crystalline plasticity model constructs, as well as consideration of uncertainty propagation in concurrent or hierarchical multiscale modeling.

Although we recognize the large body of research in model order reduction approaches (e.g., homogenization) from the mesoscale upward, this Chapter has intentionally focused more on model constructs than on methods for multiscale model transitions. Examination of limitations of model constructs at various levels of hierarchy is critical to the issue of predictive multiscale modeling and mitigation of uncertainty propagation in multiscale model chains. In spite of significant limitations in time scaling and incorporation of finite temperature effects, atomistic coarse-graining methods presently have more predictive character than other scale transition schemes to support understanding and design of new material systems and microstructures affected by dislocation phenomena. DDD and SCD models hold great promise to predict scale-dependent dislocation strengthening, but operate with greater uncertainty at present in addressing realistic dislocation microstructure effects such as grain or phase boundaries, crystal anisotropy, etc. at substantial strain levels. On the other hand, higher length scale transitions (e.g., GC or LC crystal plasticity among scales) in systems with rather well-established understanding of dislocation mechanisms mitigate uncertainty by employing this knowledge base. The concern for GC and LC crystal plasticity models is their inability to serve as predictive tools for discovery and design of new alloy systems that do not have this extensive established knowledge base. To some extent, DDD relies on the prior knowledge base for materials (witness the large number of applications to pure metals in the literature), as well as uncertain information from lower scale models to inform parameters and heuristic algorithmic choices. In fact, the peak of the model form uncertainty at the mesoscale in Fig. 3 reflects this. For a new material system, the time required to reduce uncertainty in mesoscale descriptions relative to either atomistic scales or grain/polycrystal scales (the latter typically incorporates considerable top-down experimental information) can be expected to be long, particularly when multiple dislocation mechanisms operate with distinct driving forces and kinetics. Emerging data science techniques such as deep machine learning may be useful to reduce mesoscopic model form uncertainty, especially if properly constrained by training sets that involve inputs from physically-based models.

Regarding the last bullet point in the above list, time scales accessible to atomistic or coarse-grained atomistic models are limited by the requirement for full atomistic resolution of dislocation cores, coupled with fs time scale of atomic vibration for finite temperature applications. As emphasized in this Chapter, longer time scales associated with thermally activated dislocation processes are difficult to access for high fidelity and resolution models. Transition state theory can be employed to consider probability of bypass in temporal coarse-graining, based on activation energies for atomistic unit processes involving extended defects by employing tools such as the NEB method (Li 2007; Zhu et al. 2007, 2008). However, most atomistic studies have focused on dislocation phenomena having limited sizes of activation volumes. NEB methods can also be employed using MPF models (Wang and Li 2010) or even DDD (Sobie et al. 2017b, c) to inform upscaled continuum models regarding thermally activated dislocation processes. Consideration of the entropic part of the activation free energy is an important consideration in thermally activated dislocation bypass (Ryu et al. 2011; Sobie et al. 2017a). These thermally activated event probabilities can be cast in kinetic Monte Carlo schemes (Deo and Srolovitz 2002) and can handshake with DDD, FDM, MPF and various other continuum reduced order models or temporal coarse-graining approaches (Li et al. 2007) in terms of activation energies, attempt frequencies, and the bias of long range stresses, but the temporal coarse-graining should be grounded in statistical mechanics. The issue of time scaling of dislocation phenomena continues to present substantial challenges, in view of the constrained non-equilibrium character of dislocation networks, similar in many respects to challenges presented by the complex energy landscapes and the wide spectrum of relaxation times of partially cross-linked long chain polymer systems.

**Acknowledgment** Thet author is grateful for the support of the Carter N. Paden, Jr. Distinguished Chair in Metals Processing at Georgia Tech, as well as prior support in pursuit of various aspects of metal plasticity from AFOSR, ONR (N00014-05-C-024, N00014-17-1-2036), ARO, Eglin AFB, DARPA, NAVAIR, GE, Pratt & Whitney, Boeing, QuesTek, Simulia, the NSF-funded PSU-GT Center for Computational Materials Design (IIP-0541678, IIP-1034968), and NSF Grants CMMI-1232878, CMMI-0758265, CMMI-1030103, and CMMI-1333083.

# Bibliography


R.K. Abu Al-Rub, G.Z. Voyiadjis, D.J. Bammann, A thermodynamic based higher-order gradient theory for size dependent plasticity. Int. J. Solids Struct. **44**(9), 2888–2923 (2007)

A. Acharya, A model of crystal plasticity based on the theory of continuously distributed dislocations. J. Mech. Phys. Solids **49**, 761–784 (2001)

A. Acharya, Driving forces and boundary conditions in continuum dislocation mechanics. Proc. R. Soc. Lond. A **459**, 1343–1363 (2003)

A. Acharya, Microcanonical entropy and mesoscale dislocation mechanics and plasticity. J. Elast. **104**, 23–44 (2011)

A. Acharya, A. Roy, A. Sawant, Continuum theory and methods for coarse-grained, mesoscopic plasticity. Scr. Mater. **54**, 705–710 (2006)

A. Acharya, A.J. Beaudoin, R. Miller, New perspectives in plasticity theory: dislocation nucleation, waves and partial continuity of the plastic strain rate. Math. Mech. Solids **13**(3–4), 292–315 (2008)

S. Akarapu, H.M. Zbib, D.F. Bahr, Analysis of heterogeneous deformation and dislocation dynamics in single crystal micropillars under compression. Int. J. Plast. **16**, 239–257 (2010)

C. Alleman, D.J. Luscher, C. Bronkhorst, S. Ghosh, Distribution-enhanced homogenization framework for heterogeneous elasto-plastic problems. J. Mech. Phys. Solids **85**, 2015 (2015)

J.S. Amelang, G.N. Venturini, D.M. Kochmann, Microstructure evolution during nanoindentation by the quasicontinuum method. Proc. Appl. Math. Mech. **13**, 553–556 (2013)

J.S. Amelang, G.N. Venturini, D.M. Kochmann, Summation rules for a fully nonlocal energy-based quasicontinuum method. J. Mech. Phys. Solids **82**, 378–413 (2015)

R.J. Amodeo, N.M. Ghoniem, A review of experimental-observations and theoretical-models of dislocation cells and subgrains. Res. Mech. **23**(2–3), 137–160 (1988)

R.J. Amodeo, N.M. Ghoniem, Dislocation dynamics. I. A proposed methodology for micromechanics. Phys. Rev. B **41**, 6958 (1990)

M. Ardeljan, I.J. Beyerlein, M. Knezevic, Effect of dislocation density-twin interactions on twin growth in AZ31 as revealed by explicit crystal plasticity finite element modeling. Int. J. Plast. **99**, 81–101 (2017)

A. Arsenlis, D.M. Parks, Crystallographic aspects of geometrically-necessary and statistically-stored dislocation density. Acta Mater. **47**(5), 1597–1611 (1999)

A. Arsenlis, D.M. Parks, Modeling the evolution of crystallographic dislocation density in crystal plasticity. J. Mech. Phys. Solids **50**, 1979–2009 (2002)

A. Arsenlis, B.D. Wirth, M. Rhee, Dislocation density-based constitutive model for the mechanical behaviour of irradiated Cu. Philos. Mag. **84**(34), 3617–3635 (2004)

A. Arsenlis, W. Cai, M. Tang, M. Rhee, T. Oppelstrup, G. Hommes, T.G. Pierce, V.V. Bulatov, Enabling strain hardening simulations with dislocation dynamics. Model. Simul. Mater. Sci. Eng. **15**, 553–595 (2007)

R.J. Asaro, Crystal plasticity. ASME J. Appl. Mech. **50**, 921–934 (1983a)

R.J. Asaro, Micromechanics of crystals and polycrystals. Adv. Appl. Mech. **23**, 1–115 (1983b)

M.F. Ashby, The deformation of plastically non-homogeneous materials. Philos. Mag. **21**, 399–424 (1970)

H. Askari, M.R. Maughan, N. Abdolrahim, D. Sagapuram, D.F. Bahr, H.M. Zbib, A stochastic crystal plasticity framework for deformation of micro-scale polycrystalline materials. Int. J. Plast. **68**, 21–33 (2015)

O. Aslan, N.M. Cordero, A. Gaubert, S. Forest, Micromorphic approach to single crystal plasticity and damage. Int. J. Eng. Sci. **49**, 1311–1325 (2011)

R.A. Austin, D.L. McDowell, A viscoplastic constitutive model for polycrystalline FCC metals at very high rates of deformation. Int. J. Plast. **27**(1), 1–24 (2011)

R.A. Austin, D.L. McDowell, Parameterization of a rate-dependent model of shock-induced plasticity for copper, nickel and aluminum. Int. J. Plast. **32–33**, 134–154 (2012)

K.L. Baker, W.A. Curtin, Multiscale diffusion method for simulations of long-time defect evolution with application to dislocation climb. J. Mech. Phys. Solids **92**, 297–312 (2016)
D.J. Bammann, E.C. Aifantis, On a proposal for a continuum with microstructure. Acta Mech. **45**, 91–121 (1982)
D.J. Bammann, E.C. Aifantis, A model for finite-deformation plasticity. Acta Mech. **69**, 97–117 (1987)
S. Bargmann, B. Svendsen, Theoretical and algorithmic formulation of models for energetic GND-based hardening in single crystals. J. Multiscale Comput. Eng. **10**(6), 551–565 (2012)
T. Bartel, A. Menzel, B. Svendsen, Thermodynamic and relaxation-based modelling of the interaction between martensitic phase transformations and plasticity. J. Mech. Phys. Solids **59**, 1004 (2011)
M. Baskes, Modified embedded atom potentials for cubic materials and impurities. Phys. Rev. B **46**, 2727 (1992)
C.J. Bayley, W.A.M. Brekelmans, M.G.D. Geers, A comparison of dislocation-induced back stress formulations in strain gradient crystal plasticity. Int. J. Solids Struct. **43**, 7268–7286 (2006)
A.A. Benzerga, An analysis of exhaustion hardening in micron-scale plasticity. Int. J. Plast. **24**(7), 1128–1157 (2008)
A.A. Benzerga, Micro-pillar plasticity: 2.5D mesoscopic simulations. J. Mech. Phys. Solids **57**, 1459–1469 (2009)
A.A. Benzerga, Y. Brechet, A. Needleman, E. Van der Giessen, Incorporating three-dimensional mechanisms into two-dimensional dislocation dynamics. Model. Simul. Mater. Sci. Eng. **12**, 159–196 (2004)
V.L. Berdichevsky, On thermodynamics of crystal plasticity. Scr. Mater. **54**, 711–716 (2006)
V.L. Berdichevsky, Entropy of microstructure. J. Mech. Phys. Solids **56**, 742–771 (2008)
N. Bertin, M.V. Upadhyay, C. Pradalier, L. Capolungo, A FFT-based formulation for efficient mechanical fields computation in isotropic and anisotropic periodic discrete dislocation dynamics. Model. Simul. Mater. Sci. Eng. **23**(6), 065009 (2015)
I.J. Beyerlein, A. Misra, X. Zhang, Growth twins and deformation twins in metals. Annu. Rev. Mater. Res. **44**, 329–363 (2014)
T.R. Bieler, P. Eisenlohr, F. Roters, D. Kumar, D.E. Mason, M.A. Crimp, D. Raabe, The role of heterogeneous deformation on damage nucleation at grain boundaries in single phase metals. Int. J. Plast. **25**, 1655–1683 (2009)
B.A. Bilby, E. Smith, Continuous distributions of dislocations III. Proc. R. Soc. Lond. A **236**, 481–505 (1956)
A. Binder, M. Luskin, D. Perez, A.F. Voter, Analysis of transition state theory rates upon spatial coarse-graining. Multiscale Model. Simul. **13**, 890–915 (2015)
R. Brenner, A.J. Beaudoin, P. Suquet, A. Acharya, A. Acharya, Numerical implementation of static field dislocation mechanics theory for periodic media. Philos. Mag. **94**(16), 1764–1787 (2014)
F. Bridier, D.L. McDowell, P. Villechaise, J. Mendez, Crystal plasticity modeling of slip activity in Ti-6Al-4V under high cycle fatigue loading. Int. J. Plast. **25**(6), 1066–1082 (2009)
T.E. Buchheit, G.W. Wellman, C. Battaille, Investigating the limits of polycrystal plasticity modeling. Int. J. Plast. **21**(2), 221–249 (2005)
M.J. Buehler, *Atomistic Modeling of Materials Failure* (Springer, 2008). isbn:978-0-387-76426-9
M.J. Buehler, A. Hartmaier, M.A. Duchaineau, F.F. Abraham, H. Gao, The dynamical complexity of work-hardening: a large-scale molecular dynamics simulation. Acta Mech. Sinica **21**, 103–111 (2005)
M. Buehler, A. Van Duin, W.A. Goddard, Multiparadigm modeling of dynamical crack propagation in Silicon using a reactive force field. Phys. Rev. Lett. **96**(9), 095505 (2006)
V. Bulatov, F.F. Abraham, L. Kubin, B. Devincre, S. Yip, Connecting atomistic and mesoscale simulations of crystal plasticity. Nature **391**(6668), 669–672 (1998)
V.V. Bulatov, L.L. Hsiung, M. Tang, A. Arsenlis, M.C. Bartelt, W. Cai, J.N. Florando, M. Hiratani, M. Rhee, G. Hommes, T.G. Pierce, T.D. de la Rubia, Dislocation multi-junctions and strain hardening. Nature **44**, 1174–1178 (2006)

E.P. Busso, F.T. Meissonnier, N.P. O'Dowd, Gradient-dependent deformation of two-phase single crystals. J. Mech. Phys. Solids **48**, 2333–2361 (2000)

G.C. Butler, D.L. McDowell, Polycrystal constraint and grain subdivision. Int. J. Plast. **14**(8), 703–717 (1998)

W. Cai, A. Arsenlis, C.R. Weingberger, V.V. Bulatov, A non-singular continuum theory of dislocations. J. Mech. Phys. Solids **54**, 561–587 (2006)

W. Cai, R.B. Sills, D.M. Barnett, W.D. Nix, Modeling a distribution of point defects as misfitting inclusions in stressed solids. J. Mech. Phys. Solids **66**, 154–171 (2014)

C. Carstensen, K. Hackl, A. Mielke, Non-convex potentials and microstructures in finite-strain plasticity. Proc. R Soc Lond **A458**, 299–317 (2002)

G.M. Castelluccio, D.L. McDowell, Mesoscale cyclic crystal plasticity with dislocation substructures. Int. J. Plast. **98**, 1–26 (2017)

S.S. Chakravarthy, W.A. Curtin, Effect of source and obstacle strengths on yield stress: a discrete dislocation study. J. Mech. Phys. Solids **58**, 625–635 (2010)

S.S. Chakravarthy, W.A. Curtin, Stress-gradient plasticity. Proc. Natl. Acad. Sci. U S A **108**, 15716–15720 (2011)

L.Q. Chen, Phase-field models for microstructure evolution. Annu. Rev. Mater. Res. **32**, 113–140 (2002)

Y. Chen, Reformulation of microscopic balance equations for multiscale materials modeling. J. Chem. Phys. **130**, 134706 (2009)

Z. Chen, K.T. Chu, D.J. Srolovitz, J.M. Rickman, M.P. Haataja, Dislocation climb strengthening in systems with immobile obstacles: three-dimensional level-set simulation study. Phys. Rev. B **81**, 054104 (2010)

Y. Chen, J. Zimmerman, A. Krivtsov, D.L. McDowell, Assessment of atomistic coarse-graining methods. Int. J. Eng. Sci. **49**, 1137–1349 (2011)

L. Chen, J. Chen, R. Lebensohn, L.-Q. Chen, An integrated fast Fourier transform-based phase-field and crystal plasticity approach to model recrystallization of three dimensional polycrystals. Comput. Methods Appl. Mech. Eng. **285**, 829–848 (2014)

K. Cheong, E. Busso, A. Arsenlis, A study of microstructural length scale effects on the behavior of FCC polycrystals using strain gradient concepts. Int. J. Plast. **21**(9), 1797–1814 (2005)

J. Cho, T. Junge, F.-F. Molinari, G. Anciaux, Toward a 3D coupled atomistics and discrete dislocation dynamics simulation: dislocation core structures and Peierls stresses with several character angles in FCC aluminum. Adv. Model. Simul. Eng. Sci. **2**, 12 (2015). https://doi.org/10.1186/s40323-015-0028-6

F.M. Ciorba, S. Groh, M.F. Horstemeyer, Parallelizing discrete dislocation dynamics simulations on multi-core systems. Proc. Comp. Sci. **1**, 2135–2143 (2012)

J.D. Clayton, D.J. Bammann, D.L. McDowell, A geometric framework for the kinematics of crystals with defects. Philos. Mag. **85**(33–35), 3983–4010 (2005)

N.M. Cordero, A. Gaubert, S. Forest, E.P. Busso, F. Gallerneau, S. Kruch, Size effects in generalised continuum crystal plasticity for two-phase laminates. J. Mech. Phys. Solids **58**, 1963–1994 (2010)

N.M. Cordero, S. Forest, E.P. Busso, Micromorphic modelling of grain size effects in metal polycrystals. GAMM-Mitteilungen **36**(2), 186–202 (2013)

J.C. Crone, P.W. Chung, K.W. Leiter, J. Knap, S. Aubry, G. Hommes, A. Arsenlis, A multiply parallel implementation of finite element-based discrete dislocation dynamics for arbitrary geometries. Model. Simul. Mater. Sci. Eng. **22**, 035014–035041 (2014)

W.A. Curtin, V.S. Deshpande, A. Needleman, E. Van der Giessen, M. Wallin, Hybrid discrete dislocation models for fatigue crack growth. Int. J. Fatigue **32**, 1511–1520 (2010)

A. Das, A. Acharya, P. Suquet, Microstructure in plasticity without nonconvexity. Comput. Mech. **57**(3), 387–403 (2016)

M.S. Daw, M.I. Baskes, Embedded-atom method: derivation and application to impurities, surfaces, and other defects in metals. Phys. Rev. B **29**, 6443 (1984)

J. Deng, A. El-Azab, Temporal statistics and coarse graining of dislocation ensembles. Philos. Mag. **90**(27–28), 3651–3678 (2010)

C.S. Deo, D.J. Srolovitz, First passage time Markov chain analysis of rare events for kinetic Monte Carlo: double kink nucleation during dislocation glide. Model. Simul. Mater. Sci. Eng. **10**, 581–596 (2002)

B. Devincre, T. Hoc, L. Kubin, Dislocation mean free paths and strain hardening of crystals. Science **320**(5884), 1745–1748 (2008)

B. Devincre, R. Madec, G. Monnet, S. Queyreau, R. Gatti, L. Kubin, *Modeling crystal plasticity with dislocation dynamics simulations: the 'microMegas' code. Mechanics of nano-objects*. (Presses del'Ecole des Mines de Paris, Paris, 2011), pp. 81–100. http://zig.onera.fr/mm_home_page/doc/Article_mM_2011.pdf. Accessed 20 Dec 2017

M.P. Dewald, W.A. Curtin, Multiscale modelling of dislocation/grain boundary interactions. II. Screw dislocations impinging on tilt boundaries in Al. Philos. Mag. **87**, 4615–4641 (2007a)

M.P. Dewald, W.A. Curtin, Multiscale modeling of dislocation/grain boundary interactions: I. Edge dislocations impinging on S11 (113) tilt boundary in Al. Model. Simul. Mater. Sci. Eng. **15**, S193–S215 (2007b)

M.P. Dewald, W.A. Curtin, Multiscale modeling of dislocation/grain-boundary interactions: III. 60° dislocations impinging on $\Sigma3{:}\Sigma9$ and $\Sigma11$ tilt boundaries in Al. Model. Simul. Mater. Sci. Eng. **19**, 055002 (2011)

O.W. Dillon, J. Kratochvíl, A strain gradient theory of plasticity. Int. J. Solids Struct. **6**, 1513–1533 (1970)

Y. Dong, W.A. Curtin, Thermally activated plastic flow in the presence of multiple obstacle types. Model. Simul. Mater. Sci. Eng. **20**, 075006 (2012)

F.P.E. Dunne, R. Kiwanuka, A.J. Wilkinson, Crystal plasticity analysis of micro-deformation, lattice rotation and geometrically necessary dislocation density. Proc. R. Soc. A **468**, 2509–2531 (2012)

L.M. Dupuy, E.B. Tadmor, R.E. Miller, R. Phillips, Finite-temperature quasicontinuum: molecular dynamics without all the atoms. Phys. Rev. Lett. **95**, Art. No. 060202 (2005)

B. Eidel, A. Stukowski, A variational formulation of the quasicontinuum method based on energy sampling in clusters. J. Mech. Phys. Solids **57**, 87–108 (2009)

A. El-Azab, Statistical mechanics treatment of the evolution of dislocation distributions in single crystals. Phys. Rev. B **61**(18), 11956–11966 (2000)

A. El-Azab, Statistical mechanics of dislocation systems. Scr. Mater. **54**, 723–727 (2006)

A. El-Azab, J. Deng, M. Tang, Statistical characterization of dislocation ensembles. Philos. Mag. **87**(8–9), 1201–1223 (2007)

K.R. Elder, M. Grant, Modeling elastic and plastic deformations in non-equilibrium processing using phase field crystals. Phys. Rev. E **70**, 051605 (2004)

K.R. Elder, M. Katakowski, M. Haataja, M. Grant, Modeling elasticity in crystal growth. Phys. Rev. Lett. **88**, 245701 (2002)

A.C. Eringen, Nonlocal polar elastic continua. Int. J. Eng. Sci. **10**, 1–16 (1972)

A.C. Eringen, *Microcontinuum Field Theories. I. Foundations and Solids* (Springer, New York, 1999)

A.C. Eringen, W.D. Claus Jr, A micromorphic approach to dislocation theory and its relation to several existing theories, in *Fundamental Aspects of Dislocation Theory*, vol. 2, ed. by J.A. Simmons, R. de Wit, R. Bullough, Library of Congress Catalog Number 70-602416, Conference Proceedings (National Bureau of Standards, U.S. Department of Commerce, 1970), pp. 1023–1040

M.I. Espanol, D.M. Kochmann, S. Conti, M. Ortiz, A $\Gamma$-convergence analysis of the quasicontinuum method. Multscale Model. Simul. (SIAM) **11**(3), 766–794 (2013)

L.P. Evers, W.A.M. Brekelmans, M.G.D. Geers, Non-local crystal plasticity model with intrinsic SSC and GND effects. J. Mech. Phys. Solids **52**, 2379–2401 (2004)

J. Fan, *Multiscale Analysis of Deformation and Failure of Materials* (Wiley, West Sussex, 2011)

R. Fan, Z. Yuan, J. Fish, Adaptive two-scale nonlinear homogenization. Int. J. Comput. Methods Eng. Sci. Mech. **11**, 27–36 (2010)

Fish J, *Multiscale Methods: Bridging the Scales in Science and Engineering*, 1st edn. (Oxford University Press, Oxford, 2009). isbn: 978-0-19-923385-4

J.C. Fisher, Application of Cottrell's theory of yielding to delayed yielding in steel. Trans. Am. Soc. Met. **47**, 451 (1955)

N.A. Fleck, J.W. Hutchinson, A phenomenological theory for strain gradient effects in plasticity. J. Mech. Phys. Solids **41**, 1825–1857 (1993)

N.A. Fleck, G.M. Muller, M.F. Ashby, J.W. Hutchinson, Strain gradient plasticity: theory and experiments. Acta Metall. Mater. **42**, 475–487 (1994)

S. Forest, R. Sievert, Elastoviscoplastic constitutive frameworks for generalized continua. Acta Mech. **160**, 71–111 (2003)

S. Forest, D.K. Trinh, Generalized continua and non-homogeneous boundary conditions in homogenization methods. Z. Angew. Math. Mech. **91**(2), 90–109 (2011)

S. Forest, F. Barbe, G. Cailletaud, Cosserat modelling of size effects in the mechanical behaviour of polycrystals and multi-phase materials. Int. J. Solids Struct. **37**, 7105–7126 (2000)

C. Fressengeas, V. Taupin, L. Capolungo, An elasto-plastic theory of dislocation and disclination fields. Int. J. Solids Struct. **48**, 3499–3509 (2011)

J. Friedel, in *Electron Microscopy and Strength of Crystals*, ed. by G. Thomas, J. Washbum (Interscience, Wiley, New York, NY, 1962), pp. 605–651

T.S. Gates, J.A. Hinkley, *Computational materials: modeling and simulation of nanostructured materials and systems*. NASA/TM-2003-212163, 2003

M.G.D. Geers, V.G. Kouznetsova, W.A.M. Brekelmans, Multiscale first-order and second-order computational homogenization of microstructures towards continua. Int. J. Multiscale Comput. Eng. **1**(4), 371–386 (2003)

M.G.D. Geers, V.G. Kouznetsova, W.A.M. Brekelmans, Multi-scale computational homogenization: trends and challenges. J. Comput. Appl. Math. **234**(7), 2175–2182 (2010)

J.M. Gerken, P.R. Dawson, A crystal plasticity model that incorporates stresses and strains due to slip gradients. J. Mech. Phys. Solids **56**(4), 1651–1672 (2008)

N.M. Ghoniem, A perspective on dislocation dynamics, in *Handbook of Materials Modeling, vol. 1: Methods and Models*, ed. by S. Yip (Springer, Netherlands, 2005), pp. 1–7

N.M. Ghoniem, S.-H. Tong, L.Z. Sun, Parametric dislocation dynamics: a thermodynamics-based approach to investigations of mesoscopic plastic deformation. Phys. Rev. B **61**(1), 1–15 (2000)

S. Ghosh, P. Chakraborty, Microstructure and load sensitive fatigue crack nucleation in Ti-6242 using accelerated crystal plasticity FEM simulations. Int. J. Fatigue **48**, 231–246 (2013)

S. Ghosh, K. Lee, P. Raghavan, A multi-level computational model for multi-scale damage analysis in composite and porous materials. Int. J. Solids Struct. **38**(14), 2335–2385 (2001)

S. Ghosh, J. Bai, P. Raghavan, Concurrent multi-level model for damage evolution in micro structurally debonding composites. Mech. Mater. **39**(3), 241–266 (2007)

S. Ghosh, A. Shahba, X. Tu, E.L. Huskins, B.E. Schuster, Crystal plasticity FE modeling of Ti alloys for a range of strain-rates. Part II: Image-based model with experimental validation. Int. J. Plast. **87**, 69–85 (2016a)

S. Ghosh, G. Weber, S. Keshavarz, Multiscale modeling of polycrystalline Nickel-based superalloys accounting for subgrain microstructures. Mech. Res. Commun. **78**, 34–46 (2016b)

H. Gleiter, Nanostructured materials: basic concepts and microstructure. Acta Mater. **48**(1), 1–29 (2000)

R. Gracie, T. Belytschko, An adaptive concurrent multiscale method for the dynamic simulation of dislocations. Int. J. Numer. Methods Eng. **86**(4–5), 575–597 (2011)

A. Granato, K. Lucke, J. Schlipf, L. Teutonico, Entropy factors for thermally activated unpinning of dislocations. J. Appl. Phys. **35**(9), 2732–2745 (1964)

M.S. Green, Markoff random processes and the statistical mechanics of time-dependent phenomena. II. Irreversible processes in fluids. J. Chem. Phys. **22**, 398 (1954)

S. Groh, H.M. Zbib, Advances in discrete dislocations dynamics and multiscale modeling. J. Eng. Mater. Technol. **131**, 041209-1–041209-10 (2009)

S. Groh, E.B. Marin, M.F. Horstemeyer, H.M. Zbib, Multiscale modeling of the plasticity in an aluminum single crystal. Int. J. Plast. **25**, 1456–1473 (2009)

I. Groma, Link between the microscopic and mesocopic length-scale description of the collective behavior of dislocations. Phys. Rev. B **56**(10), 5807–5813 (1997)

I. Groma, Statistical physical approach to describe the collective properties of dislocations, in *Multiscale Modelling of Plasticity and Fracture by Means of Dislocation Mechanics*, ed. by P. Gumbsch, R. Pippan (Eds), (CISM International Centre for Mechanical Science, 2010), pp. 213–270. isbn:978-3-7091-0283-1
I. Groma, F.F. Csikor, M. Zaiser, Spatial correlations and higher-order gradient terms in a continuum description of dislocation dynamics. Acta Mater. **51**, 1271 (2003)
I. Groma, Z. Vandrus, P.D. Ispanovity, Scale-free phase field theory of dislocations. Phys. Rev. Lett. **114**, 015503 (2015)
I. Groma, M. Zaiser, P.D. Ispanovity, *Dislocation Patterning in a 2D Continuum Theory of Dislocations*. arXiv:1601.07831 [cond-mat.mtrl-sci] (2016)
R.K. Guduru, K.A. Darling, R.O. Scattergood, C.C. Koch, K.L. Murty, Mechanical properties of electrodeposited nanocrystalline copper using tensile and shear punch tests. J. Mater. Sci. **42**(14), 5581–5588 (2007)
A.N. Gulluoglu, D.J. Srolovitz, R. Lesar, P.S. Lomdahl, Dislocation distributions in two dimensions. Scr. Metall. **23**, 1347–1352 (1989)
P. Gumbsch, An atomistic study of brittle fracture: toward explicit failure criteria from atomistic modeling. J. Mater. Res. **10**(11), 2897–2907 (1995)
B. Gurrutxaga-Lerma, D.S. Balint, D. Dini, D.A. Eakins, A.P. Sutton, A dynamic discrete dislocation plasticity method for the simulation of plastic relaxation under shock loading. Proc. R. Soc. A **469**, 20130141 (2013)
M.E. Gurtin, A gradient theory of single-crystal viscoplasticity that accounts for geometrically necessary dislocations. J. Mech. Phys. Solids **50**(1), 5–32 (2002)
M.E. Gurtin, L. Anand, A gradient theory for single-crystal plasticity. Model. Simul. Mater. Sci. Eng. **15**, S263–S270 (2007)
P.J. Guruprasad, A.A. Benzerga, Size effects under homogeneous deformation of single crystals: a discrete dislocation analysis. J. Mech. Phys. Solids **56**, 132–156 (2008)
K. Hackl, U. Hoppe, D.M. Kochmann, Generation and evolution of inelastic microstructures – an overview. GAMM Mitteilungen **35**, 91–106 (2012)
C.S. Hartley, Y. Mishin, Representation of dislocation cores using Nye tensor distributions. Mater. Sci. Eng. A **400–401**, 18–21 (2005)
G. Henkelman, H. Jonsson, Long time scale kinetic Monte Carlo simulations without lattice approximation and predefined event table. J. Chem. Phys. **115**, 9657–9666 (2001)
G. Henkelman, B.P. Uberuaga, H. Jonsson, A climbing image nudged elastic band method for finding saddle points and minimum energy paths. J. Chem. Phys. **113**(22), 9901–9904 (2000)
C. Hennessey, G.M. Castelluccio, D.L. McDowell, Sensitivity of polycrystal plasticity to slip system kinematic hardening laws for Al 7075-T6. Mater. Sci. Eng. A **687**, 241–248 (2017)
M. Hiratani, H.M. Zbib, On dislocation–defect interactions and patterning: stochastic discrete dislocation dynamics (SDD). J. Nucl. Mater. **323**, 290–303 (2003)
M. Hiratani, H.M. Zbib, M.A. Khaleel, Modeling of thermally activated dislocation glide and plastic flow through local obstacles. Int. J. Plast. **19**(9), 1271–1296 (2003)
J.P. Hirth, A brief history of dislocation theory. Metall. Trans. A. **16**(12), 2085–2090 (1985)
T. Hochrainer, Multipole expansion of continuum dislocation dynamics in terms of alignment tensors. Philos. Mag. **95**, 1321–1367 (2015)
T. Hochrainer, M. Zaiser, P. Gumbsch, A three-dimensional continuum theory of dislocation systems: kinematics and mean-field formulation. Philos. Mag. **87**, 1261–1282 (2007)
T. Hochrainer, S. Sandfeld, M. Zaiser, P. Gumbsch, Continuum dislocation dynamics: towards a physical theory of crystal plasticity. J. Mech. Phys. Solids **63**, 167–178 (2014)
M.F. Horstemeyer, D.L. McDowell, Modeling effects of dislocation substructure in polycrystal elastoviscoplasticity. Mech. Mater. **27**, 145–163 (1998)
M.F. Horstemeyer, M.I. Baskes, V.C. Prantil, J. Philliber, S. Vonderheide, A multiscale analysis of fixed-endsimple shear using molecular dynamics, crystal plasticity, and a macroscopic internal state variable theory. Model. Simul. Mater. Sci. Eng. **11**, 265–286 (2003)
D.A. Hughes, N. Hansen, High angle boundaries and orientation distributions at large strains. Scr. Metall. Mater. **33**(2), 315–321 (1995)

D.A. Hughes, Q. Liu, D.C. Chrzan, N. Hansen, Scaling of microstructural parameters: misorientations of deformation induced boundaries. Acta Mater. **45**(1), 105–112 (1997)
D. Hull, D.J. Bacon, *Introduction to Dislocations*, 5th edn. (Elsevier, Oxford, 2011)
A. Hunter, F. Saied, C. Le, M. Koslowski, Large-scale 3D phase field dislocation dynamics simulations on high-performance architecture. Int. J. High Perform. Comput. Appl. **25**(2), 223–235 (2010)
D.E. Hurtado, M. Ortiz, Finite element analysis of geometrically necessary dislocations in crystal plasticity. Int. J. Numer. Methods Eng. **93**, 66–79 (2013)
A.M. Hussein, J.A. El-Awady, Quantifying dislocation microstructure evolution and cyclic hardening in fatigued face-centered cubic single crystals. J. Mech. Phys. Solids **91**, 126–144 (2016)
A.M. Hussein, S.I. Rao, M.D. Uchic, D.M. Dimiduk, J.A. El-Awady, Microstructurally based cross-slip mechanisms and their effects on dislocation microstructure evolution in FCC crystals. Acta Mater. **85**, 180–190 (2015)
J. Irving, J. Kirkwood, The statistical mechanical theory of transport processes. IV. The equations of hydrodynamics. J. Chem. Phys. **8**, 817–829 (1950)
P.D. Ispánovity, I. Groma, G. Györgyi, P. Szabo, W. Hoffelner, Criticality of relaxation in dislocation systems. Phys. Rev. Lett. **107**, 085506 (2011)
P.D. Ispánovity, L. Laurson, M. Zaiser, I. Groma, S. Zapperi, M.J. Alava, Avalanches in 2D dislocation systems: plastic yielding is not depinning. Phys. Rev. Lett. **112**, 235501 (2014)
D.S. Joseph, P. Chakraborty, S. Ghosh, Wavelet transformation based multi-time scaling method for crystal plasticity FE simulations under cyclic loading. Comput. Methods Appl. Mech. Eng. **199**, 2177–2194 (2010)
K. Kang, J. Yin, W. Cai, Stress dependence of cross slip energy barrier for face-centered cubic nickel. J. Mech. Phys. Solids **62**, 181–193 (2014)
O. Kapetanou, V. Koutsos, E. Theotokoglou, D. Weygand, M. Zaiser, Statistical analysis and stochastic dislocation based modeling of microplasticity. J. Mech. Behav. Mater. **24**(3-4), 105–113 (2015)
C. Kelchner, S.J. Plimpton, J.C. Hamilton, Dislocation nucleation and defect structure during surface-indentation. Phys. Rev. B **58**, 11085–11088 (1998)
S.M. Keralavarma, A.A. Benzerga, High-temperature discrete dislocation plasticity. J. Mech. Phys. Solids **82**, 1–22 (2015)
S.M. Keralavarma, W.A. Curtin, Strain hardening in 2D discrete dislocation dynamics simulations: a new '2.5D' algorithm. J. Mech. Phys. Solids **95**, 132–146 (2016)
S. Keshavarz, S. Ghosh, Multi-scale crystal plasticity finite element model approach to modeling nickel-based superalloys. Acta Mater. **61**, 6549–6561 (2013)
S. Keshavarz, S. Ghosh, Hierarchical crystal plasticity FE model for Nickel-based superalloys: sub-grain microstructures to polycrystalline aggregates. Int. J. Solids Struct. **55**, 17–31 (2015)
Kestin J, Rice JR, in *A Critical Review of Thermodynamics,* ed. by E. B. Stuart et al. (Mono Book Corporation, Baltimore, 1970)
S.M.A. Khan, H.M. Zbib, D.A. Hughes, Modeling planar dislocation boundaries using multi-scale dislocation dynamics plasticity. Int. J. Plast. **20**, 1059–1092 (2004)
T. Khraishi, H.M. Zbib, Free-surface effects in 3D dislocation dynamics: formulation and modeling. ASME J. Eng. Mater. Technol. **124**(3), 342–351 (2002)
W.K. Kim, E.B. Tadmor, Entropically stabilized dislocations. Phys. Rev. Lett. **112**, 105501 (2014)
W.K. Kim, M. Luskin, D. Perez, A.F. Voter, E.B. Tadmor, Hyper-QC: an accelerated finite-temperature quasicontinuum method using hyperdynamics. J. Mech. Phys. Solids **63**, 94–112 (2014)
J.G. Kirkwood, The statistical mechanical theory of transport processes. I. General theory. J. Chem. Phys. **14**, 180 (1946)
B. Klusemann, D.M. Kochmann, Microstructural pattern formation in finite-deformation single-slip crystal plasticity under cyclic loading: relaxation vs. gradient plasticity. Comput. Methods Appl. Mech. Eng. **278**, 765–793 (2014)
J. Knap, M. Ortiz, An analysis of the quasicontinuum method. J. Mech. Phys. Solids **49**, 1899–1923 (2001)

D.M. Kochmann, K. Hackl, The evolution of laminates in finite crystal plasticity: a variational approach. Contin. Mech. Thermodyn. **23**, 63–85 (2011)

D.M. Kochmann, G.N. Venturini, A meshless quasicontinuum method based on local maximum-entropy interpolation. Model. Simul. Mater. Sci. Eng. **22**, 034007–034035 (2014)

U.F. Kocks, H. Mecking, Physics and phenomenology of strain hardening: the FCC case. Prog. Mater. Sci. **48**(2003), 171–273 (2003)

U.F. Kocks, A.S. Argon, M.F. Ashby, Thermodynamics and kinetics of slip. Prog. Mater. Sci. **19**, 1 (1975)

S. Kohlhoff, P. Gumbsch, H.F. Fischmeister, Crack propagation in b.c.c. crystals studied with a combined finite-element and atomistic model. Philos. Mag. A **64**, 851–878 (1991)

M. Koslowski, A. Cuitino, M. Ortiz, R. LeSar, R. Thomson, *Dislocation patterns and the deformation of metals*. 2004 TMS Annual Meeting and Exhibition, Charlotte, NC, March 17, 2004. http://www.ortiz.caltech.edu/presentations/pdf/tms-04.pdf. Accessed 16 Jan 2018

B. Kouchmeshky, N. Zabaras, Microstructure model reduction and uncertainty quantification in multiscale deformation processes. Comput. Mater. Sci. **48**, 213–227 (2010)

V. Kouznetsova, M.G.D. Geers, W.A.M. Brekelmans, Multi-scale constitutive modelling of heterogeneous materials with a gradient-enhanced computational homogenization scheme. Int. J. Numer. Methods Eng. **54**(8), 1235–1260 (2002)

V.G. Kouznetsova, M.G.D. Geers, W.A.M. Brekelmans, Multi-scale second-order computational homogenization of multi-phase materials: a nested finite element solution strategy. Comput. Methods Appl. Mech. Eng. **193**(48–51), 5525–5550 (2004)

J. Kratochvil, R. Sedlacek, Pattern formation in the framework of the continuum theory of dislocations. Phys. Rev. B **67**(9), 094105 (2003)

J. Kratochvil, M. Kruzik, R. Sedlacek, Statistically based continuum model of misoriented dislocation cell structure formation. Phys. Rev. B **75**, 064104 (2007)

R. Kubo, M. Yokota, S. Nakajima, Statistical-mechanical theory of irreversible processes. II. Response to thermal disturbance. J. Phys. Soc. Jpn. **12**, 1203 (1957)

D. Kuhlmann-Wilsdorf, Theory of plastic deformation: properties of low energy dislocation structures. Mater. Sci. Eng. A **113**, 1–41 (1989)

Y. Kulkarni, J. Knap, M. Ortiz, A variational approach to coarse-graining of equilibrium and non-equilibrium atomistic description at finite temperature. J. Mech. Phys. Solids **56**(4), 1417–1449 (2008)

K.S. Kumar, H. Van Swygenhoven, S. Suresh, Mechanical behavior of nanocrystalline metals and alloys. Acta Mater. **51**(19), 5743–5774 (2003)

S. Kwon, Y. Lee, J.Y. Park, D. Sohn, J.H. Lim, S. Im, An efficient three-dimensional adaptive quasicontinuum method using variable-node elements. J. Comput. Phys. **228**, 4789–4810 (2009)

J.S. Langer, Statistical thermodynamics of strain hardening in polycrystalline solids. Phys. Rev. E **92**, 032125 (2015)

J.S. Langer, Thermal effects in dislocation theory. Phys. Rev. E **94**, 063004 (2016)

J.S. Langer, Thermodynamic theory of dislocation-enabled plasticity. Phys. Rev. E **96**, 053005 (2017)

J.S. Langer, E. Bouchbinder, T. Lookman, Thermodynamic theory of dislocation-mediated plasticity. Acta Mater. **58**(10), 3718–3732 (2010)

K.C. Le, Thermodynamic dislocation theory for non-uniform plastic deformations. J. Mech. Phys. Solids **111**, 157–169 (2018)

K.C. Le, T.M. Tran, Dislocation mediated plastic flow in aluminum: comparison between theory and experiment. Int. J. Eng. Sci. **119**, 50–54 (2017)

K.C. Le, T.M. Tran, J.S. Langer, Thermodynamic dislocation theory of high-temperature deformation in aluminum and steel. Phys. Rev. E **96**, 013004 (2017)

R.A. Lebensohn, A. Needleman, Numerical implementation of non-local polycrystal plasticity using fast Fourier transforms. J. Mech. Phys. Solids **97**, 333–351 (2016)

R.A. Lebensohn, K.A. Kanjarla, P. Eisenlohr, An elasto-viscoplastic formulation based on fast Fourier transforms for the prediction of micromechanical fields in polycrystalline materials. Int. J. Plast. **32–33**, 59–69 (2012)
J. Lemaitre, J.L. Chaboche, *Mechanics of Solid Materials* (Cambridge University Press, Cambridge, 1990)
J. Lepinoux, L.P. Kubin, The dynamic organization of dislocation structures – a simulation. Scr. Metall. **21**, 833–838 (1987)
R. LeSar, J.M. Rickman, Incorporation of local structure in continuous theory of dislocations. Phys. Rev. B **69**, 172105 (2004)
J. Li, The mechanics and physics of defect nucleation. MRS Bull. **32**, 151–159 (2007)
J. Li, A.H.W. Ngan, P. Gumbsch, Atomistic modeling of mechanical behavior. Acta Mater. **51**(19), 5711–5742 (2003)
J. Li, P.G. Kevrekidis, C.W. Gear, I.G. Kevrekidis, Deciding the nature of the coarse equation through microscopic simulations: the baby-bathwater scheme. SIAM Rev. **49**(3), 469–487 (2007)
D. Li, H. Zbib, X. Sun, M. Khaleel, Predicting plastic flow and irradiation hardening of iron single crystal with mechanism-based continuum dislocation dynamics. Int. J. Plast. **52**, 3–17 (2014)
C. Liu, *Dislocation-Based Crystal Plasticity Finite Element Modeling of Polycrystalline Material Deformation*. PhD Thesis, University of California-Los Angeles, 2006
W.K. Liu, H.S. Park, D. Qian, E.G. Karpov, H. Kadowaki, G.J. Wagner, Bridging scale methods for nanomechanics and materials. Comput. Methods Appl. Mech. Eng. **195**(13–16), 1407–1421 (2006)
B. Liu, A. Arsenlis, S. Aubry, Computing forces on interface elements exerted by dislocations in an elastically anisotropic crystalline material. Model. Simul. Mater. Sci. Eng. **24**, 055013 (2016)
J.T. Lloyd, *Implications of Limited Slip in Crystal Plasticity*. M.S. Thesis, Woodruff School of Mechanical Engineering, Georgia Institute of Technology, 2010
J.T. Lloyd, J.D. Clayton, R.A. Austin, D.L. McDowell, Plane wave simulation of elastic-viscoplastic single crystals. J. Mech. Phys. Solids **69**, 14–32 (2014a)
J.T. Lloyd, J.D. Clayton, R.C. Becker, D.L. McDowell, Simulation of shock wave propagation in single crystal and polycrystalline Aluminum. Int. J. Plast. **60**, 118–144 (2014b)
M.-A. Louchez, L. Thuinet, R. Besson, A. Legris, Microscopic phase-field modeling of HCP/FCC interfaces. Comput. Mater. Sci. **132**, 62–73 (2017)
L. Lu, Y.F. Shen, X.H. Chen, L.H. Qian, K. Lu, Ultrahigh strength and high electrical conductivity in copper. Science **304**(5669), 422–426 (2004)
D.J. Luscher, *A Hierarchical Framework for the Multiscale Modeling of Microstructure Evolution in Heterogeneous Materials*, PhD Thesis, Georgia Tech, 2010
D.J. Luscher, D.L. McDowell, C.A. Bronkhorst, A second gradient theoretical framework for hierarchical multiscale modeling of materials. Int. J. Plast. **26**(8), 1248–1275 (2010)
D.J. Luscher, D.L. McDowell, C.A. Bronkhorst, Essential features of fine scale boundary conditions for second gradient multiscale homogenization of statistical volume elements. J. Multiscale Comput. Eng. **10**(5), 461–486 (2012)
H. Lyu, N. Taheri-Nassaj, H.M. Zbib, A multiscale gradient-dependent plasticity model for size effects. Philos. Mag. **96**(18), 1883–1908 (2016)
X. Ma, N. Zabaras, Kernal principal component analysis for stochastic input model generation. J. Comput. Phys. **230**, 7311–7331 (2011)
A. Ma, F. Roters, D. Raabe, A dislocation density based constitutive model for crystal plasticity FEM including geometrically necessary dislocations. Acta Mater. **54**, 2169–2179 (2006)
J. Marian, G. Venturini, B.L. Hansen, J. Knap, M. Ortiz, G.H. Campbell, Finite-temperature extension of the quasicontinuum method using Langevin dynamics: entropy losses and analysis of errors. Model. Simul. Mater. Sci. Eng. **18**(1), 015003 (2010)
X. Markenscoff, R.J. Clifton, The nonuniformly moving edge dislocation. J. Mech. Phys. Solids **29**, 253–262 (1981)
E. Martınez, J. Mariana, A. Arsenlis, M. Victoria, J.M. Perlado, Atomistically informed dislocation dynamics in FCC crystals. J. Mech. Phys. Solids **56**, 869–895 (2008)

M.V. Matveev, O.V. Selivanikova, D.N. Cherepanov, Formation of deformation substructures in FCC crystals under the influence of point defect fluxes. Mater. Sci. Eng. **124**, 012129 (2016)
J.R. Mayeur, D.L. McDowell, Bending of single crystal thin films as predicted by micropolar crystal plasticity. Int. J. Eng. Sci. **49**, 1357–1366 (2011)
J.R. Mayeur, D.L. McDowell, An evaluation of higher-order single crystal strength models for constrained thin films subjected to simple shear. J. Mech. Phys. Solids **61**(9), 1935–1954 (2013)
J.R. Mayeur, D.L. McDowell, A comparison of Gurtin-type and micropolar single crystal plasticity with generalized stresses. Int. J. Plast. **57**, 29–51 (2014)
J.R. Mayeur, D.L. McDowell, Micropolar crystal plasticity simulations of particle strengthening. Model. Simul. Mater. Sci. Eng. **23**(6), 065007 (2015)
J.R. Mayeur, D.L. McDowell, D.J. Bammann, Dislocation-based micropolar single crystal plasticity: comparison of multi- and single-criterion theories. J. Mech. Phys. Solids **59**(2), 398–422 (2011)
D.L. McDowell, *Evolving Structure and Internal State Variables*. Nadai Award Lecture, ASME Materials Division, ASME IMECE, Dallas, TX, November 20 (1997)
D.L. McDowell, Non-associative aspects of multiscale evolutionary phenomena, in *Proceedings of the 4th International Conference on Constitutive Laws for Engineering Materials*, 1999, ed. by R.C. Picu, E. Krempl, pp. 54–57
McDowell DL, Internal state variable theory, in *Handbook of Materials Modeling, Part A: Methods*, ed. by S. Yip, M.F. Horstemeyer (Springer, The Netherlands, 2005), pp. 1151–1170
D.L. McDowell, Viscoplasticity of heterogeneous metallic materials. Mater. Sci. Eng. R. Rep. **62**(3), 67–123 (2008)
D.L. McDowell, A perspective on trends in multiscale plasticity. Int. J. Plast. **26**(9), 1280–1309 (2010)
D.L. McDowell, F.P.E. Dunne, Microstructure-sensitive computational modeling of fatigue crack formation. Int. J. Fatigue **32**(9), 1521–1542 (2010)
D.L. McDowell, J.H. Panchal, H.-J. Choi, C.C. Seepersad, J.K. Allen, F. Mistree, *Integrated Design of Multiscale, Multifunctional Materials and Products* (Elsevier, 2009, 392 pages), ISBN-13: 978-1-85617-662-0
D.L. McDowell, S. Ghosh, S.R. Kalidindi, Representation and computational structure-property relations of random media. JOM **63**(3), 45–51 (2011)
B.G. Mendis, Y. Mishin, C.S. Hartley, K.J. Hemker, Use of the Nye tensor in analyzing HREM images of bcc screw dislocations. Philos. Mag. **86**(29–31), 4607–4640 (2006)
S.D. Mesarovic, Energy, configurational forces and characteristic lengths associated with the continuum description of geometrically necessary dislocations. Int. J. Plast. **21**, 1855–1889 (2005)
M.C. Messner, M. Rhee, A. Arsenlis, N.R. Barton, A crystal plasticity model for slip in hexagonal close packed metals based on discrete dislocation simulations. Model. Simul. Mater. Sci. Eng. **25**, 044001 (2017)
W. Meyer, H. Neldel, Uber die Beziehungen zwischen der Energiekonstanten e und der Mengenkonstanten a der Leitwerts-Temperaturformel bei oxydischen Halbleitern. Z. Tech. Phys. **12**, 588–593 (1937)
J.R. Mianroodi, B. Svendsen, Atomistically determined phase-field modeling of dislocation dissociation, stacking fault formation, dislocation slip, and reactions in fcc systems. J. Mech. Phys. Solids **77**, 109–122 (2015)
R.E. Miller, A. Acharya, A stress-gradient based criterion for dislocation nucleation in crystals. J. Mech. Phys. Solids **42**(7), 1507–1525 (2004)
R. Miller, E.B. Tadmor, R. Phillips, M. Ortiz, Quasicontinuum simulation of fracture at the atomic scale. Model. Simul. Mater. Sci. Eng. **6**(5), 607–638 (1998a)
R. Miller, M. Ortiz, R. Phillips, V. Shenoy, E.B. Tadmor, Quasicontinuum models of fracture and plasticity. Eng. Fract. Mech. **61**(3-4), 427–444 (1998b)
Y. Mishin, D. Farkas, M.J. Mehl, D.A. Papaconstantopoulos, Interatomic potentials for monoatomic metals from experimental data and ab initio calculations. Phys. Rev. B **59**, 3393–3407 (1999)

Y. Mishin, M.J. Mehl, D.A. Papaconstantopoulos, A.F. Voter, J.D. Kress, Structural stability and lattice defects in copper: Ab initio, tight-binding, and embedded-atom calculations. Phys. Rev. B **63**, 224106 (2001)
A. Misra, X. Zhang, D. Hammon, R.G. Hoagland, Work hardening in rolled nanolayered metallic composites. Acta Mater. **53**(1), 221–226 (2005)
M. Monavari, S. Sandfeld, M. Zaiser, Continuum representation of systems of dislocation lines: a general method for deriving closed-form evolution equations. J. Mech. Phys. Solids **95**, 575–601 (2016)
A.P. Moore, C. Deo, M.I. Baskes, M. Okuniewski, D.L. McDowell, Understanding the uncertainty of interatomic potentials' parameters and formalism. Comput. Mater. Sci. **126**, 308–320 (2017)
C. Motz, D. Weygan, J. Senger, P. Gumbsch, Initial dislocation structures in 3-D discrete dislocation dynamics and their influence on microscale plasticity. Acta Mater. **57**(6), 1744–1754 (2009)
W. Muschik, *Non-equilibrium Thermodynamics with Application to Solids* (Springer, New York, 1993)
S. Narayanan, D.L. McDowell, T. Zhu, Crystal plasticity model for BCC iron atomistically informed by kinetics of correlated kinkpair nucleation on screw dislocation. J. Mech. Phys. Solids **65**, 54–68 (2014)
A.A. Nazarov, A.E. Romanov, R.Z. Valiev, Random disclination ensembles in ultrafine-grained materials produced by severe plastic deformation. Scr. Mater. **34**(5), 729–734 (1996)
C.F.A. Negre, S. Plimpton, M.J. Cawkwell, R. Perriot, N. Aguirre, M.N. Niklasson, D. Perez, T. Germann, A. Thompson, B. Uberuaga, R.J. Zamora, M. Wood, A. Voter, *Performing Density-Functional Tight-Binding Calculations with LAMMPS and the LATTE Library*, Los Alamos National Laboratory (2017). http://lammps.sandia.gov/workshops/Aug17/pdf/negre.pdf. Accessed 20 Dec 2017
L.D. Nguyen, K.L. Baker, D.H. Warner, Atomistic predictions of dislocation nucleation with transition state theory. Phys. Rev. B **84**, 024118 (2011)
W.D. Nix, H.J. Gao, Indentation size effects in crystalline materials: A law for strain gradient plasticity. J. Mech. Phys. Solids **46**(3), 411–425 (1998)
J.F. Nye, Some geometrical relations in dislocated crystals. Acta Metall. **1**, 153–162 (1953)
S. Olarnrithinun, S.S. Chakravarthy, W.A. Curtin, Discrete dislocation modeling of fracture in plastically anisotropic metals. J. Mech. Phys. Solids **61**, 1391–1406 (2013)
OpenKIM online suite of open source tools for molecular simulation of materials. https://openkim.org/. Accessed 21 June 2016
M. Ortiz, E.A. Repetto, Nonconvex energy minimization and dislocation structures in ductile single crystals. J. Mech. Phys. Solids **47**, 397–462 (1999)
M. Ortiz, E.A. Repetto, L. Stainier, A theory of subgrain dislocation structures. J. Mech. Phys. Solids **48**, 2077–2114 (2000)
M. Ovaska, L. Laurson, M.J. Alava, Quenched pinning and collective dislocation dynamics. Sci. Rep. **5**, 10580 (2015)
J.H. Panchal, S.R. Kalidindi, D.L. McDowell, Key computational modeling issues in ICME. Comput. Aided Des. **45**(1), 4–25 (2013)
W. Pantleon, Formation of disorientations in dislocation structures during plastic deformation. Solid State Phenom. **87**, 73–92 (2002)
S. Papanikolaou, D.M. Dimiduk, W. Choi, J.P. Sethna, M.D. Uchic, C.F. Woodward, S. Zapperi, Quasi-periodic events in crystal plasticity and the self-organized avalanche oscillator. Nature **490**, 517–521 (2012)
A. Patra, D.L. McDowell, Crystal plasticity-based constitutive modeling of irradiated bcc structures. Philos. Mag. **92**(7), 861–887 (2012)
N.H. Paulson, M.W. Priddy, D.L. McDowell, S.R. Kalidindi, Reduced-order structure-property linkages for polycrystalline microstructures based on 2-Point statistics. Acta Mater. **129**, 428–438 (2017)
F. Pavia, W.A. Curtin, Parallel algorithm for multiscale atomistic/continuum simulations using LAMMPS. Model. Simul. Mater. Sci. Eng. **23**, 055002 (2015). (23p)

R. Peierls, The size of a dislocation. Proc. Phys. Soc. Lond. **52**, 34–37 (1940)
R.C. Picu, R. Li, Z. Xu, Strain rate sensitivity of thermally activated dislocation motion across fields of obstacles of different kind. Mater. Sci. Eng. A **502**(1–2), 164–171 (2009)
P.A. Pluchino, X. Chen, M. Garcia, L. Xiong, D.L. McDowell, Y. Chen, Dislocation migration across coherent phase interfaces in SiGe superlattices. Comput. Mater. Sci. **111**, 1–6 (2016)
G. Po, M.S. Mohamed, T. Crosby, C. Erel, A. El-Azab, N. Ghoniem, Recent progress in discrete dislocation dynamics and its applications to micro plasticity. JOM **66**(10), 2108–2120 (2014)
A. Prakash, R.A. Lebensohn, Simulation of micromechanical behavior of polycrystals: finite elements vs. fast Fourier transforms. Model. Simul. Mater. Sci. Eng. **17**, 064010 (2009)
C.P. Przybyla, D.L. McDowell, Microstructure-sensitive extreme value probabilities for high cycle fatigue of Ni-base superalloy IN100. Int. J. Plast. **26**(3), 372–394 (2010)
C.P. Przybyla, D.L. McDowell, Simulated microstructure-sensitive extreme value probabilities for high cycle fatigue of duplex Ti-6Al-4V. Int. J. Plast. **27**, 1871–1895 (2011)
S. Puri, A. Roy, A. Acharya, D. Dimiduk, Modeling dislocation sources and size effects at initial yield in continuum plasticity. J. Mech. Mater. Struct. **4**(9), 1603–1618 (2009)
S. Qu, V. Shastry, W.A. Curtin, R.E. Miller, A finite-temperature dynamic coupled atomistic/discrete dislocation method. Model. Simul. Mater. Sci. Eng. **13**(7), 1101–1118 (2005)
D. Raabe, M. Sachtleber, Z. Zhao, F. Roters, S. Zaefferer, Micromechanical and macromechanical effects in grain scale polycrystal plasticity experimentation and simulation. Acta Mater. **49**(17), 3433–3441 (2001)
V. Racherla, J.L. Bassani, Strain burst phenomena in the necking of a sheet that deforms by non-associated plastic flow. Model. Simul. Mater. Sci. Eng. **15**, S297–S311 (2007)
H. Rafii-Tabar, L. Hua, M. Cross, A multi-scale atomistic-continuum modelling of crack propagation in a two-dimensional macroscopic plate. J. Phys. Condens. Matter **10**(11), 2375–2387 (1998)
S.I. Rao, C. Woodward, T.A. Parthasarathy, O. Senkov, Atomistic simulations of dislocation behavior in a model FCC multicomponent concentrated solid solution alloy. Acta Mater. **134**, 188–194 (2017)
J.R. Rice, Inelastic constitutive relations for solids: an internal variable theory and its application to metal plasticity. J. Mech. Phys. Solids **19**, 433–455 (1971)
A.D.T. Rollett, R. Pokharel, R.A. Lebensohn, R.M. Suter, Comparison in 3D of experiments on, and simulations of plastic deformation of polycrystals. Microsc. Microanal. **21**(S3), 2371–2372 (2015)
F. Roters, P. Eisenlohr, L. Hantcherli, D.D. Tjahjanto, T.R. Bieler, D. Raabe, Overview of constitutive laws, kinematics, homogenization and multiscale methods in crystal plasticity finite-element modeling: theory, experiments, applications. Acta Mater. **58**(4), 1152–1211 (2010)
A. Roy, A. Acharya, Size effects and idealized dislocation microstructure at small scales: predictions of a phenomenological model of mesoscopic field dislocation mechanics. J. Mech. Phys. Solids **54**, 1711–1743 (2006)
A. Roy, S. Puri, A. Acharya, Phenomenological mesoscopic dislocation mechanics, lower-order gradient plasticity and transport of mean excess dislocation density. Model. Simul. Mater. Sci. Eng. **15**, S167–S180 (2007)
R.E. Rudd, J.Q. Broughton, Coarse-grained molecular dynamics and the atomic limit of finite elements. Phys. Rev. B **58**(10), R5893–R5896 (1998)
R.E. Rudd, J.Q. Broughton, Concurrent coupling of length scales in solid state systems. Phys. Status Solidi B **217**(1), 251–291 (2000)
R.E. Rudd, A. Arsenlis, N.R. Barton, R.M. Cavallo, A.J. Comley, B.R. Maddox, J. Marian, H.-S. Park, S.T. Prisbrey, C.E. Wehrenberg, L. Zepeda-Ruiz, B.A. Remington, Multiscale strength (MS) models: their foundation, their successes, and their challenges. 18th APS-SCCM and 24th AIRAPT. J. Phys. Conf. Ser. **500**, 112055 (2014)
S. Ryu, K. Kang, W. Cai, Predicting the dislocation nucleation rate as a function of temperature and stress. J. Mater. Res. **26**(18), 2335–2354 (2011)

I. Ryu, W.D. Nix, W. Cai, Plasticity of bcc micropillars controlled by competition between dislocation multiplication and depletion. Acta Mater. **61**, 3233–3241 (2013)

S. Sandfeld, T. Hochrainer, M. Zaiser, P. Gumbsch, Continuum modeling of dislocation plasticity: theory, numerical implementation and validation by discrete dislocation simulations. J. Mater. Res. **26**, 623–632 (2011)

S. Saroukhani, L.D. Nguyen, K.W.K. Leung, C.V. Singh, D.H. Warner, Harnessing atomistic simulations to predict the rate at which dislocations overcome obstacles. J. Mech. Phys. Solids **90**, 203–214 (2016)

M. Schöberl, N. Zabaras, P.-S. Koutsourelakis, Predictive coarse-graining. J. Comput. Phys. **333**, 49–77 (2017)

R. Sedlacek, J. Kratochvil, Variational approach to subgrain formation. Z. Metallkd. **96**, 602–607 (2005)

M. Seefeldt, Disclinations in large-strain plastic deformation and work-Hardening. Rev. Adv. Mater. Sci. **2**, 44–79 (2001)

M. Seefeldt, P. Klimanek, Modeling of plastic deformation by means of dislocation-disclination dynamics. Solid State Phenom. **87**, 93–112 (2002)

M. Seefeldt, L. Delannay, B. Peeters, S.R. Kalidindi, P. Van Houtte, A disclination-based model for grain subdivision. Mater. Sci. Eng. A **319–321**, 192–196 (2001a)

M. Seefeldt, L. Delannay, B. Peeters, E. Aernoudt, P. Van Houtte, Modeling the initial stage of grain subdivision with the help of a coupled substructure and texture evolution algorithm. Acta Mater. **49**, 2129–2143 (2001b)

A. Shahba, S. Ghosh, Crystal plasticity FE modeling of Ti alloys for a range of strain-rates. Part I: a unified constitutive model and flow rule. Int. J. Plast. **87**, 48–68 (2016)

C. Shen, Y. Wang, Modeling dislocation network and dislocation–precipitate interaction at mesoscopic scale using phase field method. Int. J. Multiscale Comput. Eng. **1**(1), 91–104 (2003)

C. Shen, J. Li, Y. Wang, Predicting structure and energy of dislocations and grain boundaries. Acta Mater. **74**, 125–131 (2014)

V.B. Shenoy, R. Miller, E.B. Tadmor, R. Phillips, M. Ortiz, Quasicontinuum models of interfacial structure and deformation. Phys. Rev. Lett. **80**(4), 742–745 (1998)

V.B. Shenoy, R. Miller, E. Tadmor, D. Rodney, R. Phillips, M. Ortiz, An adaptive finite element approach to atomic-scale mechanics – the quasicontinuum method. J. Mech. Phys. Solids **47**(3), 611–642 (1999)

M. Shenoy, Y. Tjiptowidjojo, D.L. McDowell, Microstructure-sensitive modeling of polycrystalline IN 100. Int. J. Plast. **24**, 1694–1730 (2008)

B. Shiari, R.E. Miller, W.A. Curtin, Coupled atomistic/discrete dislocation simulations of nanoindentation at finite temperature. ASME J. Eng. Mater. Technol. **127**(4), 358–368 (2005)

L.E. Shilkrot, W.A. Curtin, R.E. Miller, A coupled atomistic/continuum model of defects in solids. J. Mech. Phys. Solids **50**, 2085–2106 (2002a)

L.E. Shilkrot, R.E. Miller, W.A. Curtin, Coupled atomistic and discrete dislocation plasticity. Phys. Rev. Lett. **89**, 025501–025501 (2002b)

L.E. Shilkrot, R.E. Miller, W.A. Curtin, Multiscale plasticity modeling: coupled atomistics and discrete dislocation mechanics. J. Mech. Phys. Solids **52**, 755–787 (2004)

T. Shimokawa, T. Kinari, S. Shintaku, Dislocation-grain boundary interactions by the Quasicontinuum method. Key Eng. Mater. **340–341**, 973–978 (2007a)

T. Shimokawa, T. Kinari, S. Shintaku, Interaction mechanism between edge dislocations and asymmetrical tilt grain boundaries investigated via quasicontinuum simulations. Phys. Rev. B **75**, 144108 (2007b)

J.Y. Shu, N.A. Fleck, E. Van der Giessen, A. Needleman, Boundary layers in constrained plastic flow: comparison of nonlocal and discrete dislocation plasticity. J. Mech. Phys. Solids **49**(6), 1361–1395 (2001)

S. Silling, M. Epton, O. Weckner, J. Xu, E. Askari, Peridynamic states and constitutive modeling. J. Elast. **88**, 151–184 (2007)

R.B. Sills, A. Aghaei, W. Cai, Advanced time integration algorithms for dislocation dynamics simulations of work hardening. Model. Simul. Mater. Sci. Eng. **24**, 045019 (2016). (17pp)

C. Sobie, L. Capolungo, D.L. McDowell, E. Martinez, Modal analysis of dislocation vibration and reaction attempt frequency. Acta Mater. **134**, 203–210 (2017a)

C. Sobie, L. Capolungo, D.L. McDowell, E. Martinez, Scale transition using dislocation dynamics and the nudged elastic band method. J. Mech. Phys. Solids **105**, 161–178 (2017b)

C. Sobie, D.L. McDowell, E. Martinez, L. Capolungo, Thermal activation of dislocations in large scale obstacle bypass. J. Mech. Phys. Solids **105**, 150–160 (2017c)

J. Song, W.A. Curtin, T.K. Bhandakkar, H.J. Gao, Dislocation shielding and crack tip decohesion at the atomic scale. Acta Mater. **58**, 5933–5940 (2010)

M.R. Sørensen, A.F. Voter, Temperature-accelerated dynamics for simulation of infrequent events. J. Chem. Phys. **112**, 9599–9606 (2000)

D.E. Spearot, *Atomistic Calculations of Nanoscale Interface Behavior in FCC Metals*, PhD thesis, Woodruff School of Mechanical Engineering, Georgia Institute of Technology, Atlanta, GA, 2005

D.E. Spearot, K.I. Jacob, D.L. McDowell, Dislocation nucleation from bicrystal interfaces with dissociated structure. Int. J. Plast. **23**, 143–160 (2007a)

D.E. Spearot, M.A. Tschopp, K.I. Jacob, D.L. McDowell, Tensile strength of h100i and h110i tilt bicrystal copper interfaces. Acta Mater. **55**(2), 705–714 (2007b)

L. Stainier, A.M. Cuitino, M. Ortiz, A micromechanical model of hardening, rate sensitivity and thermal softening in bcc single crystals. J. Mech. Phys. Solids **50**, 1511–1545 (2002)

S. Sun, V. Sundararaghavan, A peridynamic implementation of crystal plasticity. Int. J. Solids Struct. **51**, 3350–3360 (2014)

E.B. Tadmor, R.E. Miller, *Modeling Materials: Continuum, Atomistic and Multiscale Techniques* (Cambridge University Press, New York, 2011)

E.B. Tadmor, M. Ortiz, R. Phillips, Quasicontinuum analysis of defects in solids. Philos. Mag. A **73**(6), 1529–1563 (1996a)

E.B. Tadmor, R. Phillips, M. Ortiz, Mixed atomistic and continuum models of deformation in solids. Langmuir **12**(19), 4529–4534 (1996b)

E.B. Tadmor, F. Legoll, W.K. Kim, L.M. Dupuy, R.E. Miller, Finite-temperature quasi-continuum. Appl. Mech. Rev. **65**, 010803-1–010803-27 (2013)

N. Taheri-Nassaj, H.M. Zbib, On dislocation pileups and stress-gradient dependent plastic flow. Int. J. Plast. **74**, 1–16 (2015)

A. Tallman, L.P. Swiler, Y. Wang, D.L. McDowell, Reconciled top-down and bottom-up hierarchical multiscale calibration of bcc Fe crystal plasticity. Int. J. Comput. Methods Eng. **15**(6), 1–19 (2017)

M. Tang, G. Hommes, S. Aubry, A. Arsenlis, *ParaDiS-FEM Dislocation Dynamics Simulation Code Primer*, 2011. LLNL-TR-501662:https://e-reports-ext.llnl.gov/pdf/519124.pdf. Accessed 21 June 2016

V. Taupin, L. Capolungo, C. Fressengeas, M. Upadhyay, B. Beausir, A mesoscopic theory of dislocation and disclination fields for grain boundary-mediated crystal plasticity. Int. J. Solids Struct. **71**, 277–290 (2015)

G.I. Taylor, Plastic strain in metals. J. Inst. Met. **62**(1), 307–324 (1938)

I. Tembhekar, J.S. Amelang, L. Munk, D.M. Kochmann, Automatic adaptivity in the fully nonlocal quasicontinuum method for coarse-grained atomistic simulations. Int. J. Numer. Methods Eng. **110**, 878–900 (2017)

S. Tiwari, *Methods of Atomistic Input into the Initial Yield and Plastic Flow Criteria for Nanoscrystalline Materials*. PhD Thesis, School of Materials Science & Engineering, Georgia Institute of Technology, Atlanta, GA, 2014

S. Tiwari, G.J. Tucker, D.L. McDowell, Simulated defect growth avalanches during deformation of nanocrystalline copper. Philos. Mag. **93**(5), 478–498 (2013)

M.A. Tschopp, D.L. McDowell, Structures and energies of Σ3 asymmetric tilt grain boundaries in copper and aluminum. Philos. Mag. **87**(22), 3147–3173 (2007a)

M.A. Tschopp, D.L. McDowell, Tension-compression asymmetry in homogeneous dislocation nucleation in single crystal copper. Appl. Phys. Lett. **90**, 121911–121916 (2007b)

M.A. Tschopp, D.E. Spearot, D.L. McDowell (2008a) Influence of grain boundary structure on dislocation nucleation in FCC metals, in *Dislocations in Solids*, A tribute to F.R.N. Nabarro, J.P. Hirth, Elsevier, **14**, 43–139.

M.A. Tschopp, G.J. Tucker, D.L. McDowell, Atomistic simulations of dislocation nucleation in copper grain boundaries under uniaxial tension and compression. Comput. Mater. Sci. **44**(2), 351–362 (2008b)

M.A. Tschopp, S.P. Coleman, D.L. McDowell, Symmetric and asymmetric tilt grain boundary structure and energy in Cu and Al (and transferability to other FCC metals). Integr. Mater. Manuf. Innov. **4**, 11 (2015). https://doi.org/10.1186/s40192-015-0040-1

G.J. Tucker, J.A. Zimmerman, D.L. McDowell, Shear deformation kinematics of bicrystalline grain boundaries in atomistic simulations. Model. Simul. Mater. Sci. Eng. **18**, 015002 (2010)

G.J. Tucker, J.A. Zimmerman, M.D. DL, Continuum metrics for deformation and microrotation from atomistic simulations: Application to grain boundaries. Int. J. Eng. Sci. **49**, 1424–1434 (2011)

G.J. Tucker, S. Tiwari, J.A. Zimmerman, D.L. McDowell, Investigating the deformation of nanocrystalline copper with microscale kinematic metrics and molecular dynamics. J. Mech. Phys. Solids **60**, 471–486 (2012)

M.V. Upadhyay, L. Capolungo, V. Taupin, C. Fressengeas, Elastic constitutive laws for incompatible crystalline media: the contributions of dislocations, disclinations and G-disclinations. Philos. Mag. **93**(7), 794–832 (2013)

E. Van der Giessen, A. Needleman, Discrete dislocation plasticity: a simple planar model. Model. Simul. Mater. Sci. Eng. **3**, 689–735 (1995)

E.M. Viatkina, W.A.M. Brekelmans, M.G.D. Geers, Modelling of the internal stress in dislocation cell structures. Eur. J. Mech. A/Solids **26**, 982–998 (2007)

V. Volterra, Sur l'equilibre des carps elastiques multiplement connexes. Ann. Ecole Norm. Sup. Paris **24**, 401–517 (1907)

A.F. Voter, A method for accelerating the molecular dynamics simulation of infrequent events. J. Chem. Phys. **106**, 4665–4667 (1997)

A.F. Voter, Parallel replica method for dynamics of infrequent events. Phys. Rev. B **57**, 13985–13988 (1998)

G.Z. Voyiadjis, R.K. Abu Al-Rub, Gradient plasticity theory with a variable length scale parameter. Int. J. Solids Struct. **42**(14), 3998–4029 (2005)

G.J. Wagner, W.K. Liu, Coupling of atomistic and continuum simulations using a bridging scale decomposition. J. Comput. Phys. **190**, 249–274 (2003)

D. Walgraef, E.C. Aifantis, On the formation and stability of dislocation patterns, I-III. Int. J. Eng. Sci. **12**, 1351–1372 (1985)

M. Wallin, W.A. Curtin, M. Ristinmaa, A. Needleman, Multi-scale plasticity modeling: Coupled discrete dislocation and continuum crystal plasticity. J. Mech. Phys. Solids **56**, 3167–3180 (2008)

Y.Z. Wang, L.Q. Chen, Simulation of microstructural evolution using the field method, in *Methods in Materials Research*, ed. by E.N. Kaufmann (Wiley, New York, (2000), Chapters 2a.3.1–2a.3.23

Y. Wang, J. Li, Phase field modeling of defects and deformation. Acta Mater. **58**(4), 1212–1235 (2010)

J. Wang, A. Misra, An overview of interface-dominated deformation mechanisms in metallic multilayers. Curr. Opinion Solid State Mater. Sci. **15**(1), 20–28 (2011)

Y.U. Wang, Y.M. Jin, A.M. Cuitino, A.G. Khachaturyan, Nanoscale phase field microelasticity theory of dislocations: Model and 3D simulations. Acta Mater. **49**(10), 1847–1857 (2001)

Y.U. Wang, Y.M. Jin, A.G. Khachaturyan, Dislocation dynamics – phase field, in *Handbook of Materials Modeling* S. Yip) (2005, Chapter 7.12, pp. 2287–2305)

W. Wang, N. Ghoniem, L.S.R. Swaminaryan, A parallel algorithm for 3D dislocation dynamics. J. Comput. Phys. **219**, 608–621 (2006)

H. Wang, L. Capolungo, B. Clausen, C.N. Tome, A crystal plasticity model based on transition state theory. Int. J. Plast. **93**, 251–268 (2017)

E. Weinan, Z. Huang, Matching conditions in atomistic-continuum modeling of materials. Phys. Rev. Lett. **8713**(13), 135501 (2001)

S.P. Xia, T. Belytschko, A bridging domain method for coupling continua with molecular dynamics. Comput. Methods Appl. Mech. Eng. **193**, 1645–1669 (2004)

S. Xia, A. El-Azab, Computational modelling of mesoscale dislocation patterning and plastic deformation of single crystals. Model. Simul. Mater. Sci. Eng. **23**(5), 55009–55034 (2015)

S. Xia, J. Belak, A. El-Azab, The discrete-continuum connection in dislocation dynamics: I. Time coarse graining of cross slip. Model. Simul. Mater. Sci. Eng. **24**, 075007 (2016). 22p

L. Xiong, G.J. Tucker, D.L. McDowell, Y. Chen, Coarse-grained atomistic simulation of dislocations. J. Mech. Phys. Solids **59**, 160–177 (2011)

L. Xiong, Q. Deng, G.J. Tucker, D.L. McDowell, Y. Chen, A concurrent scheme for passing dislocations from atomistic to continuum regions. Acta Mater. **60**(3), 899–913 (2012a)

L. Xiong, Q. Deng, G.J. Tucker, D.L. McDowell, Y. Chen, Coarse-grained atomistic simulations of dislocations in Al, Ni and Cu crystals. Int. J. Plast. **38**, 86–101 (2012b)

L. Xiong, D.L. McDowell, Y. Chen, Sub-THz phonon drag on dislocations by coarse-grained atomistic simulations. Int. J. Plast. **55**, 268–278 (2014a)

L. Xiong, X. Chen, D.L. McDowell, Y. Chen, Predicting phonon properties of 1D polyatomic crystals through the concurrent atomistic-continuum simulations. Arch. Appl. Mech., special issue in Honor of Professor G. Maugin **84**, 1665–1675 (2014b)

L. Xiong, S. Xu, D.L. McDowell, Y. Chen, Concurrent atomistic-continuum simulations of dislocation-void interactions in FCC crystals. Int. J. Plast. **65**, 33–42 (2015)

L. Xiong, J. Rigelesaiyin, X. Chen, S. Xu, D.L. McDowell, Y. Chen, Coarse-grained elastodynamics of fast moving dislocations. Acta Mater. **104**, 143–155 (2016)

S. Xu, R. Che, L. Xiong, Y. Chen, D.L. McDowell, A quasistatic implementation of the concurrent atomistic-continuum method for FCC crystals. Int. J. Plast. **72**, 91–126 (2015)

S. Xu, L. Xiong, Y. Chen, D.L. McDowell, Sequential slip transfer of mixed character dislocations across $\Sigma 3$ coherent twin boundary in FCC metals: a concurrent atomistic-continuum study. npg Comput. Mater. **2**, 15016 (2016a). https://doi.org/10.1038/npjcompumats.2015.16

S. Xu, L. Xiong, Q. Deng, D.L. McDowell, Mesh refinement schemes for the concurrent atomistic-continuum method. Int. J. Solids Struct. **90**, 144–152 (2016b)

S. Xu, L. Xiong, Y. Chen, D.L. McDowell, Shear stress- and line length-dependent screw dislocation cross-slip in FCC Ni. Acta Mater. **122**, 412–419 (2017a)

S. Xu, L. Xiong, Y. Chen, D.L. McDowell, Comparing EAM potentials to model slip transfer of sequential mixed character dislocations across two symmetric tilt grain boundaries in Ni. JOM **69**(5), 814–821 (2017b)

H. Yasin, H.M. Zbib, M.A. Khaleel, Size and boundary effects in discrete dislocation dynamics: coupling with continuum finite element. Mater. Sci. Eng. A **309–310**, 294–299 (2001)

J. Yin, D.M. Barnett, S.P. Fitzgerald, W. Cai, Computing dislocation stress fields in anisotropic elastic media using fast multipole expansions. Model. Simul. Mater. Sci. Eng. **20**, 045015 (2012)

W. Yu, Z. Wang, Interactions between edge lattice dislocations and $\Sigma 11$ symmetrical tilt grain boundaries in copper: a quasi-continuum method study. Acta Mater. **60**(13–14), 5010–5021 (2012)

W.S. Yu, Z.Q. Wang, Interactions between edge lattice dislocations and Sigma 11 symmetrical tilt grain boundary: comparisons among several FCC metals and interatomic potentials. Philos. Mag. **94**, 2224–2246 (2014)

M. Zaiser, Statistical modeling of dislocation systems. Mater. Sci. Eng. A **309–310**, 304–315 (2001)

M. Zaiser, S. Sandfeld, Scaling properties of dislocation simulations in the similitude regime. Model. Simul. Mater. Sci. Eng. **22**, 065012 (2014)

H.M. Zbib, T.D. de la Rubia, A multiscale model of plasticity. Int. J. Plast. **18**(9), 1133–1163 (2002)

H.M. Zbib, M. Rhee, J.P. Hirth, On plastic deformation and the dynamics of 3d dislocations. Int. J. Mech. Sci. **40**(2), 113–127 (1998)

H.M. Zbib, T.D. de la Rubia, V. Bulatov, A multiscale model of plasticity based on discrete dislocation dynamics. ASME J. Eng. Mater. Technol. **124**(1), 78–87 (2002)

H.M. Zbib, C.T. Overman, F. Akasheh, D. Bahr, Analysis of plastic deformation in nanoscale metallic multilayers with coherent and incoherent interfaces. Int. J. Plast. **27**, 1618–1639 (2011)

X. Zhang, *Field Dislocation Mechanics with Applications in Atomic, Mesoscopic and Tectonic Scale Problems*. Dissertations, Carnegie Mellon University Paper 585, 2015. http://repository.cmu.edu/cgi/viewcontent.cgi?article=1624&context=dissertations

J. Zhang, S. Chakraborty, S. Ghosh, Concurrent atomistic-continuum model for developing self-consistent modeling of crystalline solids with cracks. Int. J. Multiscale Comput. Eng. **15**(2), 99–119 (2017)

T. Zhu, J. Li, Ultra-strength Materials. Prog. Mater. Sci. **55**(7), 710–757 (2010)

T. Zhu, D.L. McDowell, *Coarse graining: time scaling from unit processes*. IIMEC Summer School on Computational Materials Science Across Scales, Texas A&M, July 29–30, 2015

T. Zhu, J. Li, A. Samanta, H.G. Kim, S. Suresh, Interfacial plasticity governs strain rate sensitivity and ductility in nanostructured metals. Proc. Natl. Acad. Sci. U S A **104**, 3031–3036 (2007)

T. Zhu, L. Li, A. Samanta, A. Leach, K. Gall, Temperature and strain-rate dependence of surface dislocation nucleation. Phys. Rev. Lett. **100**, 025502 (2008)

T. Zhu, J. Li, S. Yip, *Atomistic Reaction Pathway Sampling: The Nudged Elastic Band Method and Nanomechanics Applications. Nano and Cell Mechanics: Fundamentals and Frontiers*, vol. 1, ed. by H.D. Espinosa, G. Bao (Wiley, Chichester, West Sussex, (2013), Chapter 12

J.A. Zimmerman, H. Gao, F.F. Abraham, Generalized stacking fault energies for embedded atom FCC metals. Model. Simul. Mater. Sci. Eng. **8**(2), 103–116 (2000)

J.A. Zimmerman, C.L. Kelchner, P.A. Klein, J.C. Hamilton, S.M. Foiles, Surface step effects on nanoindentation. Phys. Rev. Lett. **8716**, 165507 (2001)

# Generalized Continua and Phase-Field Models: Application to Crystal Plasticity

**Samuel Forest, Kais Ammar, Benoit Appolaire, Victor de Rancourt, and Stephan Wulfinghoff**

**Abstract** Three continuum field theories are presented that account for the size-dependent behaviour of materials. The micromorphic medium is endowed with microdeformation degrees of freedom that describe the rotation and distortion of a triad of microstructural directions, like crystallographic lattice directions. It is a very general framework that can be specialized to strain gradient plasticity theory dedicated to the modelling of plastic events in metals and alloys. Both frameworks are developed here in the special case of crystal plasticity as a complete example of transition from micro-physical phenomena to continuum macro-modelling. Finally the phase field method is introduced in this landscape as a continuum modelling approach to the motion of phase boundaries and interfaces driven by thermodynamics and mechanics.

## 1 Introduction

There are several ways of modelling the evolution of microstructures under mechanical loading. Classical homogenization theory for instance replaces micro-heterogenous Cauchy continua by a homogeneous equivalent medium endowed with effective linear or nonlinear properties (François et al. 2012). This methodology is very effective in capturing morphological mixture effects but usually does not predict size-dependent microstructural properties (grain, inclusion, precipitate size effects or those arising from the discrete nature of dislocation driven plasticity,

S. Forest (✉) · K. Ammar · V. de Rancourt
Centre des Matériaux, Mines ParisTech CNRS, Paris, France
e-mail: samuel.forest@mines-paristech.fr; kais.ammar@ensmp.fr

B. Appolaire
Institut Jean Lamour, University of Lorraine, Nancy, France
e-mail: benoit.appolaire@onera.fr

S. Wulfinghoff
Institute of Applied Mechanics, RWTH Aachen University, Aachen, Germany

 
S. Mesarovic et al. (eds.), *Mesoscale Models*, CISM International Centre for Mechanical Sciences 587, https://doi.org/10.1007/978-3-319-94186-8_6

see chapters “Multiscale Dislocation-Based Plasticity”, “Statistical Theory of Dislocation” and “Multiscale Modeling of Interfaces, Dislocations, and Dislocation Field Plasticity”).

Size-dependent continuum modelling of materials requires the consideration of microstructure related additional degrees of freedom and their gradient, or, alternatively, of some nonlocal material responses represented by strain gradient or nonlocal constitutive laws (Maugin and Metrikine 2010; Altenbach et al. 2011). One of the most general and versatile higher order continuum is the micromorphic theory conceived by Eringen and Suhubi (1964) and Mindlin (1964). Each material point is endowed with translational degrees of freedom (described by the usual displacement vector) and a microdeformation tensor accounting for rotation and distortion of a triad of microstructural directions like lattice directions in a crystal. The gradient of microdeformation introduces curvature and strain gradient effects that are essentially size-dependent. A recent account of the micromorphic model can be found in the CISM course (Forest 2012b). The first section of this chapter deals with the presentation of a reduced micromorphic model specialized for crystal plasticity. In contrast to the full micromorphic model which relies on the gradient of the microdeformation tensor, only the curl part of the microdeformation gradient is introduced in the free energy function. The reason is the connection of the theory to crystal plasticity in which Kröner’s dislocation density tensor defined as the curl of the plastic deformation field plays an essential role. The dislocation density tensor is identical with the concept of geometrically necessary dislocations (GND) introduced by Ashby (1970). A detailed description of the dislocation density tensor can be found in the CISM courses (Forest 2012a; Forest et al. 2014) and in the original work (Cermelli and Gurtin 2001).

When the microdeformation tensor is constrained to coincide with the deformation gradient itself, Eringen’s micromorphic theory reduces to the strain gradient model initially introduced by Mindlin (1965). Neglecting the effect of elastic deformation gradients then leads to the realm of strain gradient plasticity (Aifantis 1987; Fleck and Hutchinson 1997). Such a reduction of the *microcurl* model to strain gradient plasticity involving the dislocation density tensor in the constitutive modelling is presented in the second section of this chapter. The attention is focused on the choice of the macroscopic constitutive law taken as a function of the dislocation density tensor. The merits of quadratic and non-quadratic potentials are analyzed with respect to physical phenomena occurring in particular in cyclic plasticity.

The third and final section of this chapter considers another field theory, the so-called phase field approach, and its coupling to mechanics. This is a currently open field of nonlinear continuum physics which combines the motion of phase boundaries and the plasticity occurring in the single phases whose volume fractions evolve in the process. Phase field models are closely related to strain gradient theories as put forward by Gurtin (1996) with a general formalism adopted in this chapter. The assessment of such theories combining phase transformation and crystal plasticity is still difficult today due to the lack of in situ experimental data which are still hard to obtain. The formulation of the theory relies on the use of homogenization methods to describe the behaviour of diffuse interfaces which are the main feature of phase field approaches.

## 2 Micromorphic Continuum Theory for Crystal Plasticity

The micromorphic theory applied to crystal plasticity is called the *microcurl* model by Cordero et al. (2010). It is presented here within the finite deformation framework allowing for large rotations and large distortion of the lattice, particularly suited for prediction of crystallographic texture evolution and grain fragmentation during material processing.

### *2.1 Model Formulation*

#### 2.1.1 Balance Equations

The degrees of freedom of the proposed theory are the displacement vector $\boldsymbol{u}$ and the microdeformation variable $\hat{\boldsymbol{\chi}}^p$, a generally non-symmetric second rank tensor. The field $\hat{\boldsymbol{\chi}}^p(\boldsymbol{X})$ is generally not compatible, meaning that it does not derive from a vector field. The exponent $^p$ indicates, in advance, that this variable will eventually be constitutively related to plastic deformation occurring at the material point. In particular, the microdeformation $\hat{\boldsymbol{\chi}}^p$ is treated as an invariant quantity with respect to rigid body motion. A first gradient theory is considered with respect to the degrees of freedom. However, the influence of the microdeformation gradient is limited to its curl part because of the intended relation to the dislocation density tensor associated with the curl of plastic distortion. The following sets of degrees of freedom and of their gradients are therefore defined:

$$DOF = \{\boldsymbol{u}\,, \quad \hat{\boldsymbol{\chi}}^p\}, \quad GRAD = \{\boldsymbol{F} := \boldsymbol{1} + \boldsymbol{u} \otimes \boldsymbol{\nabla}_0, \quad \boldsymbol{K} := \text{Curl}\,\hat{\boldsymbol{\chi}}^p\} \tag{1}$$

The $\boldsymbol{\nabla}_0$ operator denotes the gradient of the field with respect to Lagrangian coordinates $\boldsymbol{X}$. The initial (resp. current) configuration of the body is called $V_0$ (resp. $V$). The following definition of the Curl operator is adopted:

$$\text{Curl}\,\hat{\boldsymbol{\chi}}^p := \frac{\partial \hat{\boldsymbol{\chi}}^p}{\partial X_k} \times \boldsymbol{e}_k, \quad K_{ij} := \epsilon_{jkl} \frac{\partial \hat{\chi}^p_{ik}}{\partial X_l} \tag{2}$$

where $\epsilon_{ijk}$ is the permutation tensor.

The method of virtual power is used to derive the balance and boundary conditions, following (Germain 1973). For that purpose, the power density of internal forces is defined as a linear form with respect to the velocity fields and their Eulerian gradients:

$$p^{(i)} = \boldsymbol{\sigma} : (\dot{\boldsymbol{u}} \otimes \boldsymbol{\nabla}) + \boldsymbol{s} : \dot{\hat{\boldsymbol{\chi}}}^p + \boldsymbol{M} : \text{curl}\,\dot{\hat{\boldsymbol{\chi}}}^p, \quad \forall \boldsymbol{x} \in V \tag{3}$$

Here, the conjugate quantities are the Cauchy stress tensor $\boldsymbol{\sigma}$, which is symmetric in the absence of volume couples, the microstress tensor, $\boldsymbol{s}$, and the generalized couple stress tensor $\boldsymbol{M}$. The curl of the microdeformation rate is defined as

$$\operatorname{curl} \dot{\hat{\boldsymbol{\chi}}}^{p} := \epsilon_{jkl} \frac{\partial \dot{\hat{\chi}}_{ik}^{p}}{\partial x_l} \boldsymbol{e}_i \otimes \boldsymbol{e}_j = \dot{\boldsymbol{K}}\, \boldsymbol{F}^{-1} \tag{4}$$

The form of the power density of internal forces dictates the form of the power density of contact forces:

$$p^{(c)} = \boldsymbol{t} \cdot \dot{\boldsymbol{u}} + \boldsymbol{m} : \dot{\hat{\boldsymbol{\chi}}}^{p}, \quad \forall \boldsymbol{x} \in \partial V \tag{5}$$

where $\boldsymbol{t}$ is the usual simple traction vector and $\boldsymbol{m}$ the double traction tensor. The principle of virtual power is stated in the static case and in the absence of volume forces for the sake of brevity:

$$-\int_D p^{(i)}\, dV + \int_{\partial D} p^{(c)}\, dS = 0 \tag{6}$$

for all virtual fields $\dot{\boldsymbol{u}}$, $\dot{\hat{\boldsymbol{\chi}}}^{p}$, and any subdomain $D \subset V$. By application of Gauss divergence theorem, assuming sufficient regularity of the fields, this statement expands into

$$\int_V \frac{\partial \sigma_{ij}}{\partial x_j} \dot{u}_i\, dV + \int_V \left( \epsilon_{kjl} \frac{\partial M_{ik}}{\partial x_l} - s_{ij} \right) \dot{\hat{\chi}}_{ij}^{p}\, dV$$
$$+ \int_{\partial V} \left(t_i - \sigma_{ij} n_j\right) \dot{u}_i\, dS + \int_{\partial V} \left(m_{ik} - \epsilon_{jkl} M_{ij} n_l\right) \dot{\hat{\chi}}_{ik}^{p}\, dS = 0, \quad \forall \dot{u}_i, \forall \dot{\hat{\chi}}_{ij}^{p}$$

which leads to the two field equations of balance of momentum and generalized balance of moment of momentum:

$$\operatorname{div} \boldsymbol{\sigma} = 0, \quad \operatorname{curl} \boldsymbol{M} + \boldsymbol{s} = 0, \quad \forall \boldsymbol{x} \in V \tag{7}$$

and two boundary conditions

$$\boldsymbol{t} = \boldsymbol{\sigma} \cdot \boldsymbol{n}, \quad \boldsymbol{m} = \boldsymbol{M} \cdot \boldsymbol{\epsilon} \cdot \boldsymbol{n}, \quad \forall \boldsymbol{x} \in \partial V \tag{8}$$

the index representation of the latter relation being $m_{ij} = M_{ik} \epsilon_{kjl} n_l$.

### 2.1.2 Constitutive Equations

The deformation gradient is decomposed into elastic and plastic parts in the form,

$$\boldsymbol{F} = \boldsymbol{F}^e\, \boldsymbol{F}^p \tag{9}$$

The isoclinic intermediate configuration is defined in a unique way by keeping the crystal orientation unchanged from the initial to the intermediate configuration following (Mandel 1973). The plastic distortion $\boldsymbol{F}^p$ is invariant with respect to rigid body motions that are carried by $\boldsymbol{F}^e$. The current mass density is $\rho$ whereas the mass density of the material element in the intermediate configuration is $\tilde{\rho}$, such that $\tilde{\rho}/\rho = J_e := \det(\boldsymbol{F}^e)$. The elastic strain is defined as,

$$\boldsymbol{E}^e := \frac{1}{2}(\boldsymbol{F}^{eT}\,\boldsymbol{F}^e - \boldsymbol{1}) \tag{10}$$

The microdeformation is linked to the plastic deformation via the introduction of a relative deformation measure, defined as

$$\boldsymbol{e}^p := \boldsymbol{F}^{p-1}\,\hat{\boldsymbol{\chi}}^p - \boldsymbol{1} \tag{11}$$

This tensor $\boldsymbol{e}^p$ measures the departure of the microdeformation from the plastic deformation. The state variables are assumed to be the elastic strain, the relative deformation, the curl of microdeformation and some internal variables, $\alpha$:

$$STATE := \{\boldsymbol{E}^e, \quad \boldsymbol{e}^p, \quad \boldsymbol{K}, \quad \alpha\} \tag{12}$$

The specific Helmholtz free energy density, $\psi$, is assumed to be a function of this set of state variables. In particular, in this simple version of the model, the curl of microdeformation is assumed to contribute entirely to the stored energy. In more sophisticated models, as proposed in Forest and Sievert (2003, 2006), Forest (2009), and Gurtin and Anand (2009), the relative deformation, the microdeformation and its gradient can be split into elastic and plastic parts.

When the internal constraint $\boldsymbol{e}^p \equiv 0$ is enforced, the plastic microdeformation coincides with the plastic deformation so that the curl of the plastic microdeformation is directly related to the dislocation density tensor, $\boldsymbol{A}$, defined by

$$\boldsymbol{K} := \text{Curl}\,\hat{\boldsymbol{\chi}}^p \equiv \text{Curl}\,\boldsymbol{F}^p = J\,\boldsymbol{A}\,\boldsymbol{F}^{-T} \tag{13}$$

The micromorphic model then reduces to strain gradient plasticity according to Gurtin (2002).

The dissipation rate density is the difference

$$\Delta^{intr} := p^{(i)} - \rho\dot{\psi} \geq 0 \tag{14}$$

which must be positive according to the second principle of thermodynamics. When the previous strain measures are introduced, the power density of internal forces takes the following form:

$$\begin{aligned} p^{(i)} &= \boldsymbol{\sigma} : \dot{\boldsymbol{F}}^e\,\boldsymbol{F}^{e-1} + \boldsymbol{\sigma} : (\boldsymbol{F}^e\,\dot{\boldsymbol{F}}^p\,\boldsymbol{F}^{p-1}\,\boldsymbol{F}^{e-1}) + \boldsymbol{s} : (\boldsymbol{F}^p\,\dot{\boldsymbol{e}}^p + \dot{\boldsymbol{F}}^p\,\boldsymbol{e}^p) + \boldsymbol{M} : \dot{\boldsymbol{K}}\,\boldsymbol{F}^{-1} \\ &= \frac{\rho}{\tilde{\rho}}\boldsymbol{\Pi}^e : \dot{\boldsymbol{E}}^e + \frac{\rho}{\tilde{\rho}}\boldsymbol{\Pi}^M : \dot{\boldsymbol{F}}^p\,\boldsymbol{F}^{p-1} + \boldsymbol{s} : (\boldsymbol{F}^p\,\dot{\boldsymbol{e}}^p + \dot{\boldsymbol{F}}^p\,\boldsymbol{e}^p) + \boldsymbol{M} : \dot{\boldsymbol{K}}\,\boldsymbol{F}^{-1} \end{aligned} \tag{15}$$

where $\mathbf{\Pi}^e$ is the Piola stress tensor with respect to the intermediate configuration and $\mathbf{\Pi}^M$ is the Mandel stress tensor:

$$\mathbf{\Pi}^e := J_e \boldsymbol{F}^{e-1}\, \boldsymbol{\sigma}\, \boldsymbol{F}^{e-T}, \quad \mathbf{\Pi}^M := J_e \boldsymbol{F}^{eT}\, \boldsymbol{\sigma}\, \boldsymbol{F}^{e-T} = \boldsymbol{F}^{eT}\, \boldsymbol{F}^{e}\, \mathbf{\Pi}^e \tag{16}$$

On the other hand,

$$\rho\dot{\psi} = \rho\frac{\partial \psi}{\partial \boldsymbol{E}^e} : \dot{\boldsymbol{E}}^e + \rho\frac{\partial \psi}{\partial \boldsymbol{e}^p} : \dot{\boldsymbol{e}}^p + \rho\frac{\partial \psi}{\partial \boldsymbol{K}} : \dot{\boldsymbol{K}} + \rho\frac{\partial \psi}{\partial \alpha}\dot{\alpha} \tag{17}$$

We compute

$$\begin{aligned} J_e \Delta^{intr} &= (\mathbf{\Pi}^e - \tilde{\rho}\frac{\partial \psi}{\partial \boldsymbol{E}^e}) : \dot{\boldsymbol{E}}^e + (J_e \boldsymbol{F}^{pT}\, \boldsymbol{s} - \tilde{\rho}\frac{\partial \psi}{\partial \boldsymbol{e}^p}) : \dot{\boldsymbol{e}}^p \\ &+ (J_e \boldsymbol{M}\, \boldsymbol{F}^{-T} - \tilde{\rho}\frac{\partial \psi}{\partial \boldsymbol{K}}) : \dot{\boldsymbol{K}} \\ &+ (\mathbf{\Pi}^M + J_e \boldsymbol{s}\, \hat{\boldsymbol{\chi}}^{pT}) : \dot{\boldsymbol{F}}^p\, \boldsymbol{F}^{p-1} - \tilde{\rho}\frac{\partial \psi}{\partial \alpha}\dot{\alpha} \geq 0 \end{aligned} \tag{18}$$

Assuming that the processes associated with $\dot{\boldsymbol{E}}^e$, $\dot{\boldsymbol{e}}^p$ and $\dot{\boldsymbol{K}}$ are non-dissipative, the state laws are obtained:

$$\mathbf{\Pi}^e = \tilde{\rho}\frac{\partial \psi}{\partial \boldsymbol{E}^e}, \quad \boldsymbol{s} = J_e^{-1} \boldsymbol{F}^{p-T}\, \tilde{\rho}\frac{\partial \psi}{\partial \boldsymbol{e}^p}, \quad \boldsymbol{M} = J_e^{-1}\tilde{\rho}\frac{\partial \psi}{\partial \boldsymbol{K}}\, \boldsymbol{F}^T \tag{19}$$

The residual dissipation rate is

$$J_e D = (\mathbf{\Pi}^M + J_e \boldsymbol{s}\, \hat{\boldsymbol{\chi}}^{pT}) : \dot{\boldsymbol{F}}^p\, \boldsymbol{F}^{p-1} - R\dot{\alpha} \geq 0, \quad \text{with} \quad R := \tilde{\rho}\frac{\partial \psi}{\partial \alpha} \tag{20}$$

At this stage, a dissipation potential that depends on stress measures, $\Omega(\boldsymbol{\mathcal{S}}, R)$, is introduced in order to formulate the evolution equations for plastic flow and internal variables:

$$\dot{\boldsymbol{F}}^p\, \boldsymbol{F}^{p-1} = \frac{\partial \Omega}{\partial \boldsymbol{\mathcal{S}}}, \quad \text{with} \quad \boldsymbol{\mathcal{S}} := \mathbf{\Pi}^M + J_e \boldsymbol{s}\, \hat{\boldsymbol{\chi}}^{pT} \tag{21}$$

$$\dot{\alpha} = -\frac{\partial \Omega}{\partial R} \tag{22}$$

where $R$ is the thermodynamic force associated with the internal variable $\alpha$, and $\boldsymbol{\mathcal{S}}$ is the effective stress conjugate to plastic strain rate, the driving force for plastic flow.

In the case of crystal plasticity, a generalized Schmid law is adopted for each slip system $s$ in the form:

$$f^s(\boldsymbol{\mathcal{S}}, \tau_c^s) = |\boldsymbol{\mathcal{S}} : \boldsymbol{P}^s| - \tau_c^s \geq 0, \quad \text{with} \quad \boldsymbol{P}^s = \boldsymbol{l}^s \otimes \boldsymbol{n}^s \tag{23}$$

for activation of slip system $s$ with slip direction, $\boldsymbol{l}^s$, and normal to the slip plane, $\boldsymbol{n}^s$. We call $\boldsymbol{P}^s$ the orientation tensor. The critical resolved shear stress is $\tau_c^s$ which may be a function of $R$ in the presence of isotropic hardening. The kinematics of plastic slip follows from the choice of a dissipation potential, $\Omega(f^s)$, that depends on the stress variables through the yield function itself, $f^s$:

$$\dot{\boldsymbol{F}}^p \boldsymbol{F}^{p-1} = \sum_{s=1}^{N} \frac{\partial \Omega}{\partial f^s} \frac{\partial f^s}{\partial \boldsymbol{\mathcal{S}}} = \sum_{s=1}^{N} \dot{\gamma}^s \boldsymbol{P}^s, \quad \text{with} \quad \dot{\gamma}^s = \frac{\partial \Omega}{\partial f^s} \text{sign}(\boldsymbol{\mathcal{S}} : \boldsymbol{P}^s) \tag{24}$$

A possible viscoplastic potential is then:

$$\Omega(f^s) = \frac{K}{n+1} < \frac{f^s}{K} >^{n+1} \tag{25}$$

where $K, n$ are viscosity parameters associated with viscoplastic slip, and the brackets stand for $< \cdot >= \text{Max}(0, \cdot)$. The generalized resolved shear stress can be decomposed into two contributions:

$$\boldsymbol{\mathcal{S}} : \boldsymbol{P}^s = \tau^s - x^s, \quad \text{with} \quad \tau^s = \boldsymbol{\Pi}^M : \boldsymbol{P}^s \quad \text{and} \quad x^s = -\boldsymbol{s}\, \hat{\boldsymbol{\chi}}^{pT} : \boldsymbol{P}^s \tag{26}$$

The usual resolved shear stress is $\tau^s$ whereas $x^s$ can be interpreted as an internal stress or back-stress leading to kinematic hardening. The fact that the introduction of the effect of the dislocation density tensor or, more generally, of gradient of plastic strain tensor, leads to the existence of internal stresses induced by higher order stresses has already been noticed by Steinmann (1996), see also Forest (2008). The back-stress component is induced by the microstress $\boldsymbol{s}$ or, equivalently, by the curl of the generalized couple stress tensor, $\boldsymbol{M}$, via the balance Eq. (7).

## 2.2 Geometrically Linearized Model

When deformations and rotations remain sufficiently small, the previous equations can be linearized as follows:

$$\boldsymbol{F} = \boldsymbol{1} + \boldsymbol{u} \otimes \nabla_0 \simeq \boldsymbol{1} + \boldsymbol{H}^e + \boldsymbol{H}^p, \quad \boldsymbol{H}^e = \boldsymbol{\varepsilon}^e + \boldsymbol{\omega}^e, \quad \boldsymbol{H}^p = \boldsymbol{\varepsilon}^p + \boldsymbol{\omega}^p \tag{27}$$

where $\boldsymbol{\varepsilon}^e, \boldsymbol{\omega}^e$ (resp. $\boldsymbol{\varepsilon}^p, \boldsymbol{\omega}^p$) are the symmetric and skew-symmetric parts of $\boldsymbol{F}^e - \boldsymbol{1}$ (resp. $\boldsymbol{F}^p - \boldsymbol{1}$). When microdeformation is small, the relative deformation is linearized as

$$\boldsymbol{e}^p = (\boldsymbol{1} + \boldsymbol{H}^p)^{-1} (\boldsymbol{1} + \boldsymbol{\chi}^p) - \boldsymbol{1} \simeq \boldsymbol{\chi}^p - \boldsymbol{H}^p, \quad \text{with} \quad \boldsymbol{\chi}^p = \hat{\boldsymbol{\chi}}^p - \boldsymbol{1} \tag{28}$$

When linearized, the state laws (19) become:

$$\boldsymbol{\sigma} = \rho \frac{\partial \psi}{\partial \boldsymbol{\varepsilon}^e}, \quad \boldsymbol{s} = \rho \frac{\partial \psi}{\partial \boldsymbol{e}^p}, \quad \boldsymbol{M} = \rho \frac{\partial \psi}{\partial \boldsymbol{K}} \tag{29}$$

The evolution equations read then:

$$\dot{\boldsymbol{\varepsilon}}^p = \frac{\partial \Omega}{\partial (\boldsymbol{\sigma} + \boldsymbol{s})}, \quad \dot{\alpha} = -\frac{\partial \Omega}{\partial R} \tag{30}$$

The most simple case of a quadratic free energy potential is first considered:

$$\rho \psi(\boldsymbol{\varepsilon}^e, \boldsymbol{e}^p, \boldsymbol{K}) = \frac{1}{2} \boldsymbol{\varepsilon}^e : \mathbb{C} : \boldsymbol{\varepsilon}^e + \frac{1}{2} H_\chi \boldsymbol{e}^p : \boldsymbol{e}^p + \frac{1}{2} A \boldsymbol{K} : \boldsymbol{K} \tag{31}$$

The usual four-rank tensor of elastic moduli is denoted by $\mathbb{C}$. The higher order moduli have been limited to only two additional parameters: $H_\chi$ (unit MPa) and $A$ (unit MPa.mm$^2$). Their essential impact on the prediction of size effects will be analyzed in the next section. It follows that:

$$\boldsymbol{\sigma} = \mathbb{C} : \boldsymbol{\varepsilon}^e, \quad \boldsymbol{s} = H_\chi \boldsymbol{e}^p, \quad \boldsymbol{M} = A \boldsymbol{K} \tag{32}$$

Large values of $H_\chi$ ensure that $\boldsymbol{e}^p$ remains small so that $\boldsymbol{\chi}^p$ remains close to $\boldsymbol{H}^p$ and $\boldsymbol{K}$ is close to the dislocation density tensor. The yield condition for each slip system becomes:

$$f^s = |\tau^s - x^s| - \tau_c^s \tag{33}$$

with

$$x^s = -\boldsymbol{s} : \boldsymbol{P}^s = (\operatorname{curl} \boldsymbol{M}) : \boldsymbol{P}^s = A(\operatorname{curl} \operatorname{curl} \boldsymbol{\chi}^p) : \boldsymbol{P}^s \tag{34}$$

It follows that the micromorphic back-stress is related to the second spatial derivative of plastic deformation, in the spirit of Aifantis celebrated gradient plasticity model (Aifantis 1987).

## 2.3 Dislocation Density Tensor vs. Lattice Curvature

Experimental techniques like EBSD provide the field of lattice orientation and, consequently, of lattice rotation $\boldsymbol{R}^e$ during deformation. The rotation $\boldsymbol{R}^e$ appears in the polar decomposition of the elastic deformation $\boldsymbol{F}^e = \boldsymbol{R}^e \boldsymbol{U}^e$, where $\boldsymbol{U}^e$ is the lattice stretch tensor. Since

$$\boldsymbol{A} = -\operatorname{curl} \boldsymbol{F}^{e-1} = -\operatorname{curl} (\boldsymbol{U}^{e-1} \cdot \boldsymbol{R}^{eT}) \tag{35}$$

the hypothesis of small elastic strain implies

$$\boldsymbol{A} \simeq -\text{curl}\, \boldsymbol{R}^{eT} \tag{36}$$

This approximation also requires that the gradient of elastic strain is also small, which is not ensured even if the elastic strain is small. If, in addition, elastic rotations are small, we have

$$\boldsymbol{A} \simeq -\text{curl}\,(\boldsymbol{1} - \boldsymbol{\omega}^e) = \text{curl}\, \boldsymbol{\omega}^e \tag{37}$$

The small rotation axial vector is defined as

$$\overset{\times}{\boldsymbol{\omega}}{}^e = -\frac{1}{2}\boldsymbol{\epsilon} : \boldsymbol{\omega}^e, \quad \boldsymbol{\omega}^e = -\boldsymbol{\epsilon} \cdot \overset{\times}{\boldsymbol{\omega}}{}^e \tag{38}$$

or, in matrix notations,

$$[\boldsymbol{\omega}^e] = \begin{bmatrix} 0 & \omega^e_{12} & -\omega^e_{31} \\ -\omega^e_{12} & 0 & \omega^e_{23} \\ \omega^e_{31} & -\omega^e_{23} & 0 \end{bmatrix} = \begin{bmatrix} 0 & -\overset{\times}{\omega}{}^e_3 & \overset{\times}{\omega}{}^e_2 \\ \overset{\times}{\omega}{}^e_3 & 0 & -\overset{\times}{\omega}{}^e_1 \\ -\overset{\times}{\omega}{}^e_2 & \overset{\times}{\omega}{}^e_1 & 0 \end{bmatrix} \tag{39}$$

The gradient of the lattice rotation field delivers the lattice curvature tensor. In the small deformation context, the gradient of the rotation tensor is represented by the gradient of the axial vector:

$$\boldsymbol{\kappa} := \overset{\times}{\boldsymbol{\omega}}{}^e \otimes \boldsymbol{\nabla} \tag{40}$$

One can establish a direct link between $\text{curl}\, \boldsymbol{\omega}^e$ and the gradient of the axial vector associated with $\boldsymbol{\omega}$. For that purpose, the matrix form of $\text{curl}\, \boldsymbol{\omega}^e$ is derived according to:

$$[\text{curl}\, \boldsymbol{\omega}^e] = \begin{bmatrix} \omega^e_{12,3} + \omega^e_{31,2} & -\omega^e_{31,1} & -\omega^e_{12,1} \\ -\omega^e_{23,2} & \omega^e_{12,3} + \omega^e_{23,1} & -\omega^e_{12,2} \\ -\omega^e_{23,3} & -\omega^e_{31,3} & \omega^e_{23,1} + \omega^e_{31,2} \end{bmatrix} \tag{41}$$

or equivalently

$$[\text{curl}\, \boldsymbol{\omega}^e] = \begin{bmatrix} -\overset{\times}{\omega}{}^e_{3,3} - \overset{\times}{\omega}{}^e_{2,2} & \overset{\times}{\omega}{}^e_{2,1} & \overset{\times}{\omega}_{3,1} \\ \overset{\times}{\omega}{}^e_{1,2} & -\overset{\times}{\omega}{}^e_{3,3} - \overset{\times}{\omega}{}^e_{1,1} & \overset{\times e}{\omega}_{3,2} \\ \overset{\times}{\omega}{}^e_{1,3} & \overset{\times}{\omega}{}^e_{2,3} & -\overset{\times}{\omega}{}^e_{1,1} - \overset{\times}{\omega}{}^e_{2,2} \end{bmatrix} \tag{42}$$

from which it becomes apparent that

$$\boldsymbol{A} = \boldsymbol{\kappa}^T - (\text{trace}\,\boldsymbol{\kappa})\mathbf{1}, \quad \boldsymbol{\kappa} = \boldsymbol{A}^T - \frac{1}{2}(\text{trace}\,\boldsymbol{A})\mathbf{1} \tag{43}$$

This is a remarkable relation linking, with the context of small elastic strains[1] and rotations, the dislocation density tensor to lattice curvature. It is known as Nye's formula (Nye 1953).

Examples of applications dealing with size-effects associated with dislocation glide in narrow channels can be found in the CISM course (Forest et al. 2014) and in the references Cordero et al. (2010), and Chang et al. (2016). Instead, examples are provided in the next section in the special case of strain gradient plasticity regarded as a special case of the *microcurl* model.

## 3 Strain Gradient Crystal Plasticity

From a physical point of view, the introduction of additional energy density terms involving the dislocation density tensor can be motivated by the incomplete nature of the continuum theory. Clearly, the continuum description does not contain the full information on the discrete dislocation microstructure. In particular, single dislocations are not resolved. Instead, the continuum representation may be interpreted as a smoothed version of the real system, where information is lost deliberately. There is no reason to assume that the total elastic energy of the continuum representation coincides with the elastic energy of the real system including discrete dislocations. This is due to the loss of information as a result of the smoothing procedure (Mesarovic et al. 2010). Additional energy terms in gradient plasticity may therefore be interpreted as an attempt to partially compensate the error in the continuum elastic energy. This is done by taking into account available kinematical information on the dislocation microstructure as additional argument of the energy.

The optimal form of the energy is subject of current research. Most applications are based on a pragmatic quadratic approach like in the formula (31), see also (e.g. Cordero et al. 2012; Reddy et al. 2012; Wulfinghoff and Böhlke 2012; Miehe et al. 2014; Wulfinghoff et al. 2013; Mesarovic et al. 2015).

Instead, a more physical approach seems to be based on a variable internal length scale as a function of the dislocation state (Groma et al. 2003; Forest 2008; Mesarovic et al. 2010). The quadratic form (31) was shown to provide physically unrealistic scaling in the size-dependent response of laminate microstructures under shear (Cordero et al. 2010; Forest and Guéninchault 2013). Since quadratic forms are unusual in classical dislocation theory, alternative free energy potentials were proposed in the past 10 years. Rank-one energies that are linear with respect to

[1]and in fact of small gradient of elastic strain.

the GND densities, have been shown to lead to a size-dependent yield stress in certain situations. Additionally motivated by line tension (and more elaborate) arguments, they are used by several authors (Ortiz and Repetto 1999; Conti and Ortiz 2005; Ohno and Okumura 2007; Kametani et al. 2012; Hurtado and Ortiz 2013). Asymptotic methods can be used to derive alternative effective potentials for distributions of edge dislocations. Systematic derivations of back stress distributions were performed in Geers et al. (2013) by means of asymptotic methods.

The choice of a logarithmic energy is inspired by the statistical theory of dislocations of Groma et al. (2003), Berdichevsky (2006), and Groma et al. (2007). Here, the internal length scale of the back stress is determined by the dislocation microstructure (see also Svendsen and Bargmann 2010; Forest and Guéninchault 2013). In the latter reference, the rank one and logarithmic formulations were applied to strain gradient plasticity theories involving the full dislocation density tensor instead of the individual GND densities.

This chapter follows the lines of the contribution by Wulfinghoff et al. (2015).

## 3.1 *Formulation of Two Free Energy Potentials*

It is assumed that the volumetric stored energy density has the form

$$\rho\psi = W = W_e + W_g + W_h, \tag{44}$$

with $W_e = (\boldsymbol{\varepsilon} - \boldsymbol{\varepsilon}^p) : \mathbb{C} : (\boldsymbol{\varepsilon} - \boldsymbol{\varepsilon}^p)/2$. The expressions $W_h$ and $W_g$ are assumed to be functions of internal (hardening) variables $\alpha$ and the dislocation density tensor $\boldsymbol{A}$, respectively. Size-independent isotropic hardening is accounted for by $W_h$, while $W_g$ models size effects.

The following defect energies are investigated:

$$W_g^1 = cGb\|\boldsymbol{A}\|, \quad W_g^{ln} = c_0\|\boldsymbol{A}\| \ln \frac{\|\boldsymbol{A}\|}{A_0}, \tag{45}$$

where $c$ is a constant of order unity, $G$ is the macroscopic shear modulus, $b$ is the Burgers vector, $A_0$ is a constant and $c_0$ is given by

$$c_0 = \frac{Gb\beta}{2\pi(1-\nu)}, \tag{46}$$

where $\nu$ is Poisson's ratio and $\beta$ is of order unity. The Euclidean norm of the dislocation density tensor is defined as: $\|\boldsymbol{A}\| = \sqrt{\boldsymbol{A} \cdot \boldsymbol{A}}$.

The rank-one energy $W_g^1$ can be motivated by simple line tension arguments, see Ortiz and Repetto (1999) and Hurtado and Ortiz (2013).

The logarithmic energy $W_g^{ln}$ (Eq. (45)) is motivated by the form of the associated back stress (Forest and Guéninchault 2013). It turns out that the approach $W_g^{ln}$ leads to a back stress which is formally close to the one derived in the statistical theory of Groma et al. (2003), given in 1D by

$$-\frac{Gc_1}{2\pi(1-\upsilon)\rho}\partial^2_{x_1}\gamma \tag{47}$$

for a single slip situation with slip direction $\boldsymbol{e}_1$. Here, $\rho$ denotes the total dislocation density. In the two-dimensional single slip regime, the back stress involves the second derivative of the plastic slip, as postulated by Aifantis (1987). However, the internal length scale is not interpreted as a material constant but determined by the dislocation microstructure, if $W_g^{ln}$ is applied.

The stresses $\boldsymbol{\sigma}$ and $\boldsymbol{M}$ are assumed to be energetic, i.e.

$$\boldsymbol{\sigma} = \frac{\partial W}{\partial \boldsymbol{\varepsilon}^e}, \quad \boldsymbol{M} = \frac{\partial W}{\partial \boldsymbol{A}} \tag{48}$$

If the stored energy is not differentiable at $\boldsymbol{A} = \boldsymbol{0}$, the symbol $\partial$ in Eq. $(48)_2$ is interpreted as a sub-differential operator (see, e.g., Han and Reddy 2013), i.e.

$$\boldsymbol{M}\Big|_{\boldsymbol{\alpha}=\boldsymbol{0}} \in \{\boldsymbol{M} : W_g(\boldsymbol{A}) - \boldsymbol{M}\cdot\boldsymbol{A} \geq 0 \,\forall\, \boldsymbol{A}\}. \tag{49}$$

This can be interpreted as follows. If the stress $\boldsymbol{M}$ is applied at a material point, $\boldsymbol{A}$ will take a value which minimizes the expression $W_g(\boldsymbol{A}) - \boldsymbol{M}\cdot\boldsymbol{A}$. For small values of $\boldsymbol{M}$, the minimum is given by $\boldsymbol{A} = \boldsymbol{0}$. However, for sufficiently large values of $\boldsymbol{M}$, the value of $\boldsymbol{A}$ can be determined from the stationarity condition $\boldsymbol{M} = \partial_{\boldsymbol{\alpha}} W_g$.

For example, if $W_g$ is given by $W_g^1 = cGb\|\boldsymbol{A}\|$, it follows that $\boldsymbol{A} = \boldsymbol{0}$ if

$$\boldsymbol{M} : \boldsymbol{A} \leq W_g^1(\boldsymbol{A}) = cGb\|\boldsymbol{A}\| \quad \forall \boldsymbol{A} \Leftrightarrow \boldsymbol{M} : \boldsymbol{A} \leq \|\boldsymbol{M}\|\|\boldsymbol{A}\| \leq cGb\|\boldsymbol{A}\| \quad \forall \boldsymbol{A}. \tag{50}$$

Hence, it is found that

$$\boldsymbol{M} \in \{\boldsymbol{M} : \varphi(\boldsymbol{M}) \leq 0\}, \text{ if } \boldsymbol{A} = \boldsymbol{0} \tag{51}$$

$$\boldsymbol{M} = cGb\frac{\boldsymbol{A}}{\|\boldsymbol{A}\|}, \text{ else} \tag{52}$$

with $\varphi(\boldsymbol{M}) = \|\boldsymbol{M}\| - cGb$.

Note that the generalized stress $\boldsymbol{M}$ can be computed uniquely from $\boldsymbol{A}$ only if $\boldsymbol{A} \neq \boldsymbol{0}$. This makes analytical solutions as well as the numerical implementation difficult. Possible numerical strategies concerning this problem are discussed in Kametani et al. (2012) as well as Hurtado and Ortiz (2013).

## 3.2 Application to the Shearing of the Periodic Laminate

In this section, the two new potentials are applied to the elasto-plastic laminate microstructure considered in Cordero et al. (2010), and Wulfinghoff et al. (2015), see Fig. 1. The two promising candidates of the defect energy function $W_g$ are investigated concerning their effect on the overall size effects as well as the dislocation pile-up structures building up at impenetrable boundaries. The dislocation density tensor can be expressed in terms of the edge density $\rho_\perp = -\partial_{x_1}\gamma$

$$\boldsymbol{A} = -\rho_\perp \boldsymbol{e}_1 \otimes \boldsymbol{e}_3. \tag{53}$$

The quantity $\rho_\perp$ represents the total Burgers vector amount per unit area of edge dislocations. Note, that its unit ($\mu\text{m}^{-1}$) differs from the unit of the total line length per unit volume $\rho$, given by $\mu\text{m}^{-2}$.

Assuming the defect energy $W_g$ to be a function of $\|\boldsymbol{A}\|$, the generalized stress $\boldsymbol{M}$ reads

$$\boldsymbol{M} = \partial_{\boldsymbol{A}} W_g = \partial_{\|\boldsymbol{A}\|} W_g \frac{\boldsymbol{A}}{\|\boldsymbol{A}\|} = M(x_1)\boldsymbol{e}_1 \otimes \boldsymbol{e}_3. \tag{54}$$

From the balance Eq. $(7)_2$, it follows that

$$s_{12} - M' = 0. \tag{55}$$

Throughout this section, the isotropic hardening contribution will be neglected, i.e. $W_h = 0$.

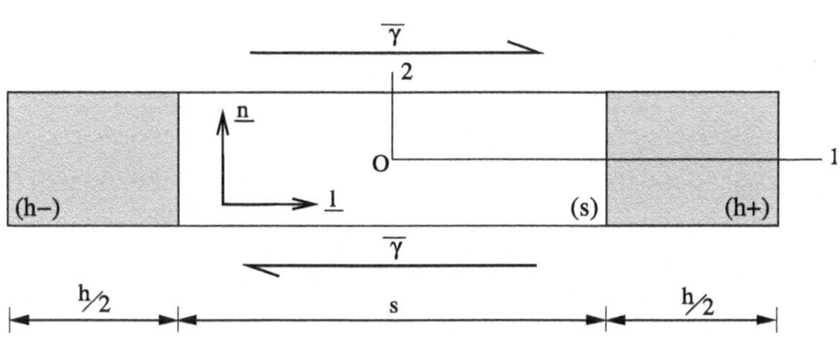


**Fig. 1** Single slip in a periodic two-phase single crystal laminate under simple shear: the grey phase (h) displays a purely linear elastic behavior whereas the inelastic deformation of the white elasto-plastic phase (s) is controlled by a single slip system ($\boldsymbol{n}$ , $\boldsymbol{l}$)

### 3.3 Rank-One Defect Energy

For the laminate, the following energy is adopted

$$W_g^1 = cGb\,\|\boldsymbol{A}\| = cGb\,|\rho_\perp|, \tag{56}$$

where $c$ is of order unity (Ortiz and Repetto 1999). According to Eq. (54), the generalized stress $M$ reads

$$M = -\frac{\rho_\perp}{|\rho_\perp|}cGb = -\text{sign}\,\rho_\perp cGb, \quad \text{if } |\rho_\perp| > 0 \tag{57}$$

$$|M| \leq cGb, \quad \text{if } |\rho_\perp| = 0. \tag{58}$$

where the second line follows from Eq. (51).

Subsequently, a monotonic shear deformation in the positive direction is prescribed such that the following relations hold in the soft phase:
$\tau^{eff} \geq \tau^C, \quad \dot{\gamma} \geq 0.$

The principle of virtual power is written for the real field on the laminate unit cell $V$:

$$\int_V \boldsymbol{\sigma} : \dot{\boldsymbol{\varepsilon}} + \boldsymbol{s} : \dot{\boldsymbol{H}}^p + \boldsymbol{M} : \text{curl}\,\dot{\boldsymbol{H}}^p\, dV = \int_{\partial V} \boldsymbol{t} \cdot \dot{\boldsymbol{u}}\, dS + \boldsymbol{m} : \dot{\boldsymbol{H}}^p\, dS \tag{59}$$

The last term of the right-hand side vanishes due to the fact that $\boldsymbol{H}^p$ is periodic whereas $\boldsymbol{m}$ is anti-periodic. The first term of the right-hand side coincides with the first term in the left-hand side, as can be shown by means of Gauss theorem. As a result we obtain,

$$\int_{\partial V} \boldsymbol{s} : \dot{\boldsymbol{H}}^p + \boldsymbol{M} : \text{curl}\,\dot{\boldsymbol{H}}^p\, dV = 0 \tag{60}$$

For the laminate under single slip, this gives:

$$\int_{-s/2}^{s/2+h} s_{12}\dot{\gamma} + (-cGb\,\text{sign}\,\gamma_{,1})(-\dot{\gamma}_{,1})\, dx_1 = 0 \tag{61}$$

Under monotonic loading, sign $\gamma_{,1}$ = sign $\dot{\gamma}_{,1}$, so that

$$\int_{-s/2}^{s/2+h} s_{12}\dot{\gamma} + cGb|\dot{\gamma}_{,1}|\, dx_1 = 0 \tag{62}$$

According to Schmid's law, $\sigma_{12} + s_{12} = \tau_c$ where the fields $\sigma_{12}, s_{12}$ are uniform. The solution is such that the plastic slip field $\gamma(x_1)$ is uniform in $]-s/2, s/2[$ at each instant. So does $\dot{\gamma}(x_1) = \dot{\gamma}(0)$ in $]-s/2, s/2[$. It vanishes in $]s/2, s/2+h[$. It is therefore discontinuous at $\pm s/2$. As a result, the derivative of the plastic slip rate is the sum of two Dirac functions:

$$\dot{\gamma}_{,1}(x_1) = \dot{\gamma}(0)(\delta(x_1 + \frac{s}{2}) - \delta(x_1 - \frac{s}{2})) \tag{63}$$

The integration of these Dirac functions (in fact the absolute values due to (62)) on the interval $[-s/2, s/2 + h]$ finally gives

$$s(\tau_c c - \sigma_{12})\dot{\gamma}(0) + 2cGb\dot{\gamma}(0) = 0 \tag{64}$$

The scaling law follows:

$$\sigma_{12} = \tau_c + \frac{2cGb}{s}. \tag{65}$$

This equation holds in the plastic regime. Clearly, the application of the rank-one energy increases the macroscopic yield stress by $2cGb/s$, i.e., the increase scales inversely with the size of the soft phase (see Fig. 2). The same scaling behavior has been found by Ohno and Okumura (2007) for a spherical grain, also using a rank-one energy. As illustrated in Fig. 2, the dislocations localize in dislocation walls at the elasto-plastic interface.

For the material parameters of aluminum ($G = 26.12$ GPa and $b = 0.286$ nm) and $c = 1$, the size effect becomes important when the system size is below ~10 µm. The plastic shear strain is constant in the bulk, i.e., the dislocations form singularities (walls) at the boundaries. The back stress is constant (w.r.t. space) in the bulk. During the first period, it increases and thereby impedes any plastic deformation. Therefore, the overall deformation is purely elastic during this period. At a certain point, the plastic deformation starts and the back stress remains constant afterwards. Its value is given by $2cGb/s$. It has the same scaling behaviour as Orowan's precipitate hardening law in physical metallurgy and essentially accounts for the same type of dislocation driven size effects.

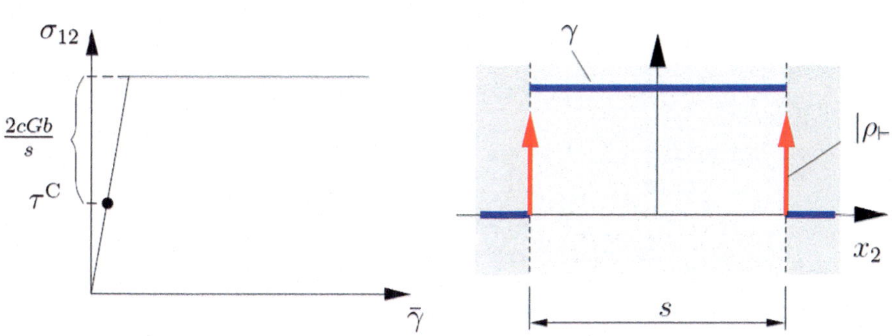


**Fig. 2** Macroscopic shear stress strain curve for the rank-one energy. The increase of the overall yield point scales inversely with the size of the soft phase, after (Wulfinghoff et al. 2015)

### 3.4 Logarithmic Energy

This section investigates the following defect energy

$$W_g^{ln} = c_0 \|\boldsymbol{A}\| \ln \frac{\|\boldsymbol{A}\|}{A_0}, \tag{66}$$

with the constant $c_0$ as defined in Eq. (46). The energy is motivated by the statistical theory of dislocations by Groma et al. (2003). The authors derive a back stress term which involves the second gradient of slip as postulated by Aifantis (1987). However, their theory involves an internal length scale which is given by $1/\sqrt{\rho}$, where $\rho$ denotes the total dislocation density.

In pure metals, the geometrical characteristics of the microstructure are essentially determined by the dislocation arrangement. This is a strong argument for a (variable) internal length scale, which is determined by the available dislocation field variables (instead of a constant length scale parameter, see also Forest and Sedláček (2003) where this dependency is derived from a dislocation line tension model).

It is demonstrated subsequently, that the approach (66) leads to a back stress which is similar to that of Groma et al. (2003). However, it should be mentioned that this energy is neither convex nor smooth with respect to the dislocation density tensor (a regularization is necessary for computational applications). For the laminate problem, the generalized stress $M$ reads (see Eq. (54))

$$M = -\text{sign}\, \rho_\perp\, c_0 \left( \ln \frac{|\rho_\perp|}{A_0} + 1 \right). \tag{67}$$

In this section, rate-independent plasticity will be considered based on the yield criterion

$$f = |\tau^{eff}| - \tau^C \leq 0. \tag{68}$$

Here, the effective stress reads

$$\tau^{eff} = (\boldsymbol{\sigma} + \boldsymbol{s}) \cdot (\boldsymbol{l} \otimes \boldsymbol{n}) = \sigma_{12} + s_{12} \overset{(55)}{=} \sigma_{12} + M'. \tag{69}$$

With Eq. (67) and $M' = (\partial_{\rho_\perp} M)\, (\partial_{x_1} \rho_\perp)$, it follows that

$$\tau^{eff} = \tau - \frac{c_0}{|\rho_\perp|} \partial_{x_1} \rho_\perp = \tau + \frac{G\beta}{2\pi(1-\nu)} \frac{b}{|\rho_\perp|} \partial_{x_1}^2 \gamma. \tag{70}$$

Here, the second term can be interpreted as a back stress. Note that the back stress involves no internal length scale parameter. Instead, the internal length scale, $\sqrt{b/\rho_\perp}$, is determined by the dislocation microstructure. In contrast to the back stress of Groma et al. (2003), the internal length scale is determined by the GND-density $\rho_\perp$ instead of the total density $\rho$. Hence, the influence of statistically stored dislocations (SSDs) is ignored. Therefore, a homogeneous initial GND-density $|\rho_\perp| = A_0$ will be assumed to be given. In addition, it is assumed that the SSD-density is equal or less than $A_0$. The soft phase is assumed to be under plastic loading, with $\tau^{eff} = \tau_c$ in the soft phase. In this case,

$$M' \overset{(55)}{=} s_{12} = -(\sigma_{12} - \tau_c) = \text{const.} \Rightarrow M = -(\sigma_{12} - \tau_c)x_1, \tag{71}$$

where, again, the constant of integration vanishes due to the symmetry requirement $|M(-s/2)| = |M(s/2)|$. The plastic slip $\gamma$ can be derived from the equality of Eqs. (67) and (71), which yields a differential equation for $\gamma$. The solution reads

$$\gamma = \frac{A_0 L}{e}\left(\exp\left(\frac{s}{2L}\right) - \exp\left(-a\frac{x_1}{L}\right)\right) \quad \text{with } L = \frac{c_0}{\sigma_{12} - \tau_C}, \tag{72}$$

where the matching conditions $\gamma(-s/2) = \gamma(s/2) = 0$ have been exploited and where $e = \exp(1)$. The variable $a$ is defined by $a = \operatorname{sign}\gamma'$ and is assumed positive in $(-s/2, 0)$ and negative in $(0, s/2)$.

The macroscopic stress strain relation follows

$$\bar{\gamma} = \frac{A_0 L}{e(s+h)}\left(\exp\left(\frac{s}{2L}\right)(s - 2L) + 2L\right) + \frac{\sigma_{12}}{G}. \tag{73}$$

The solution is evaluated for the following material parameters

| $E$[GPa] | $\nu$ | $\tau^C$[MPa] | $b$[nm] | $\beta$ | $A_0/b$[$\mu$m$^{-2}$] |
|---|---|---|---|---|---|
| 70 | 0.34 | 10 | 0.286 | 1 | 1 |

A very thin hard phase with negligible width $h$ is considered ($h/s = 10^{-6}$ for the analytical solution).

The macroscopic stress-strain curve (73) is illustrated in Fig. 3. A clear size effect is visible. Apparently, mainly the overall yield stress is affected. The hardening shows less size dependence. It is remarkable that the model provides a size-dependent yield stress and non-linear kinematic hardening.

Since there is no distinct yield stress, the evaluation of the scaling behavior is based on the offset yield stress at 0.2% plastic strain. The offset yield stress as a function of the inverse of the size $1/s$ exhibits the same behavior as in the results obtained from the rank-one energy. It scales inversely with channel size, see Wulfinghoff et al. (2015).

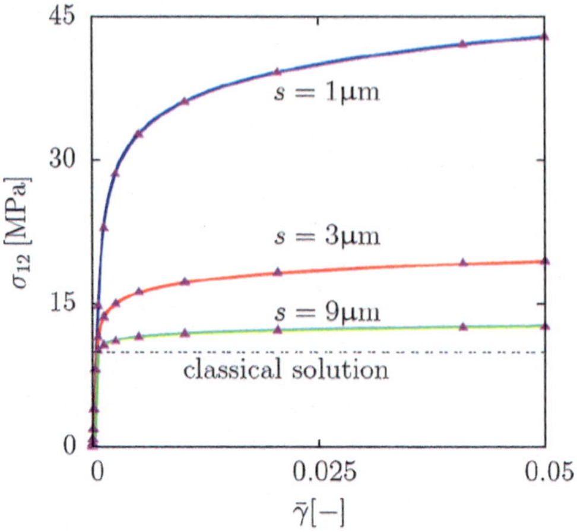


**Fig. 3** Macroscopic stress-strain diagram for three different sizes. Analytical (lines) and regularized, numerical (triangles) solution for the logarithmic potential, after (Wulfinghoff et al. 2015)

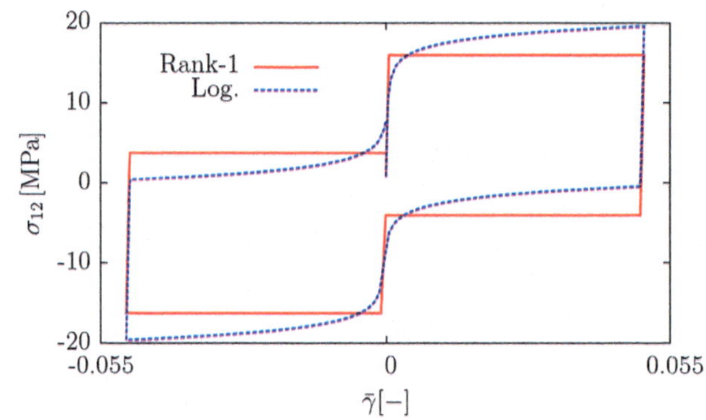


**Fig. 4** Cyclic loading for $s = 3\,\mu$m in a laminate under single slip according strain gradient plasticity with two distinct potentials

## *3.5 Cyclic Behavior of the Laminate*

The laminate is now submitted to one full cycle $\bar{\gamma} = \pm 0.05$. The hysteresis loops $\sigma_{12}$ vs. $\bar{\gamma}$ for both rank one and logarithmic potentials are represented in Fig. 4 for $s = 3\,\mu$m. In the absence of isotropic hardening, the loops are stabilized after one full cycle. They are characterized by pure kinematic hardening. The influence of the back stress is clearly observable. The One striking feature of the results is the non-convexity of the obtained loops. According to the rank one model the first unloading stage is characterized by reverse plasticity at a constant negative shear stress. When $\bar{\gamma}$ goes through zero again, the overall shear stress experiences a jump of the same magnitude as computed analytically for monotonic loading in Sects. 3.3

and 3.4. The non convex loop obtained for the logarithmic potential is similar but smoother and displays smooth nonlinear kinematic hardening. A similar non convex hysteresis loop was obtained by Ohno and Okumura (2008) for the rank one model.

The type of non-linear kinematic hardening observed for both models can be identified with Asaro's type KIII model, corresponding to a *first in / last out* sequence of dislocation motion (Asaro 1975). It is considered by Asaro as the most perfect form of recovery of plastic memory. Such stress-strain loops display inflection points that are observed in some materials, see Asaro (1975) for a Nimonic alloy, but such observations have also been made in several Nickel based superalloys. It is usually attributed to substructural recovery on the microscale, for instance pile-up formation and destruction at $\gamma'$ precipitates. In the present simple single crystal model, it is the single active hardening mechanism induced by strain gradient plasticity and the presence of the hard phase in the laminate. It represents an accurate continuum description of dislocation piling-up and unpiling-up phenomena.

The experimental evidence of such non convex loops is illustrated in Figs. 5 and 6 in the case of polycrystalline Fe-Cr and Al-Cu-Mg alloys, respectively. The first loop in Fig. 5 (left) exhibits two inflection points but the convexity is restored after a few cycles and the usual shape with still a strong Bauschinger effect is retrieved in Fig. 5 (right). The Fig. 6 shows that the amount of plastic recoverability is controlled by the annealing degree of the dislocation microstructure. Further evidence of non-convex loops in the cyclic behavior of FCC alloys can be found in the recent contribution by Proudhon et al. (2008) dealing with aluminium alloys. The common characteristics of these FCC alloys is that they all contain a population of nonshearable intragranular precipitates. This distribution of particles represents the first series of obstacles to be overcome by dislocations for the plasticity to start. The distance between precipitates presents a small scatter and the average value is the characteristic length responsible for the size-dependent yield limit. This distance is comparable to the width $s$ in our ideal laminate model. As illustrated by the TEM observations by Stoltz and Pelloux (1974, 1976), Taillard and Pineau (1982) and Proudhon et al. (2008), dislocation loops multiply around precipitates and can

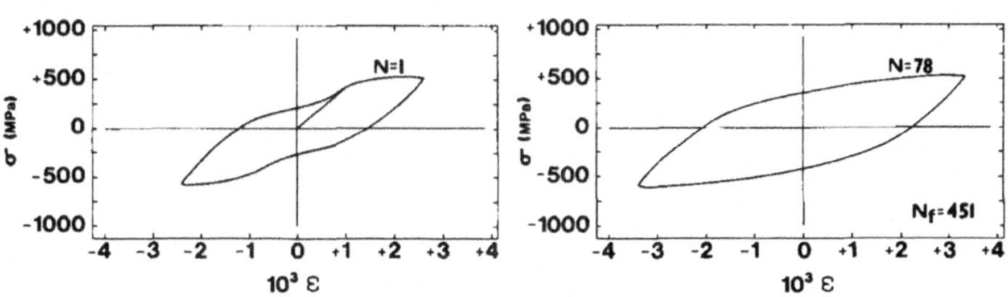


**Fig. 5** Shape of the stress-strain hysteresis loop as a function of the number $N$ of cycles for a Fe-19wt.%Cr alloy aged at 923 K for 72 h and mechanically tested at room temperature: $N = 1$ (left), $N = 48$ (right, $N_f$ indicates the number of cycles to failure), after (Taillard and Pineau 1982)

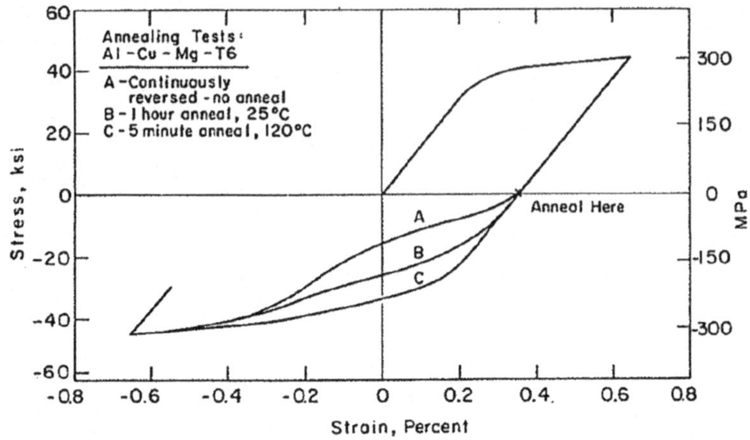


**Fig. 6** Interrupted annealing hysteresis curves for an Al–Cu–Mg–T6 alloy tested at room temperature: A–continuously reverse load (no anneal), B–annealed 1 h/25 °C and C–annealed 5 min/120 °C, after (Stoltz and Pelloux 1976)

be destroyed after reverse loading unless the material is annealed before reversing the load, see Fig. 6, or unless the multiplication of forest dislocations or cross-slip effects limit the recoverability of cyclic plasticity. The effect has also been observed in nickel-base single crystal superalloys for tension along $< 111 >$, see Fig. 7. The simulations based on the logarithmic potential provide smooth loops that are closer to the experimental shapes. Our simulations deal with ideal single crystal laminates and simulations for polycrystals remain to be done. However, as shown by the two-dimensional strain gradient plasticity simulations performed by Ohno and Okumura (2008) based on the rank-one potential the effect pertains for polycrystals.

The two non-quadratic energy potentials represent continuum models of a discrete phenomenon which can be illustrated for a single dislocation source, as shown in Fig. 8. The cyclic response of a Frank–Read source, simulated by discrete dislocation dynamics (DDD) (Déprés et al. 2004; Chang et al. 2016), provides a non-convex loop which is identical to the one predicted by the rank-one continuum model. This is related to the fact that an instability of the loop behavior is observed for a critical stress that is inversely proportional to the length of the source. The scenario of dislocation source bowing and sudden propagation and multiplication can be reversed entirely in the absence of strong interaction with the dislocation forest and in the absence of cross-slip. This explains the concave shape of the stress-strain loop predicted by the DDD which is accurately translated by the continuum model, see Fig. 4. Statistical effects of the collective behavior of dislocations finally destroy the recoverability of plastic deformation and the associated transmission of the single source behavior to the macroscopic response.

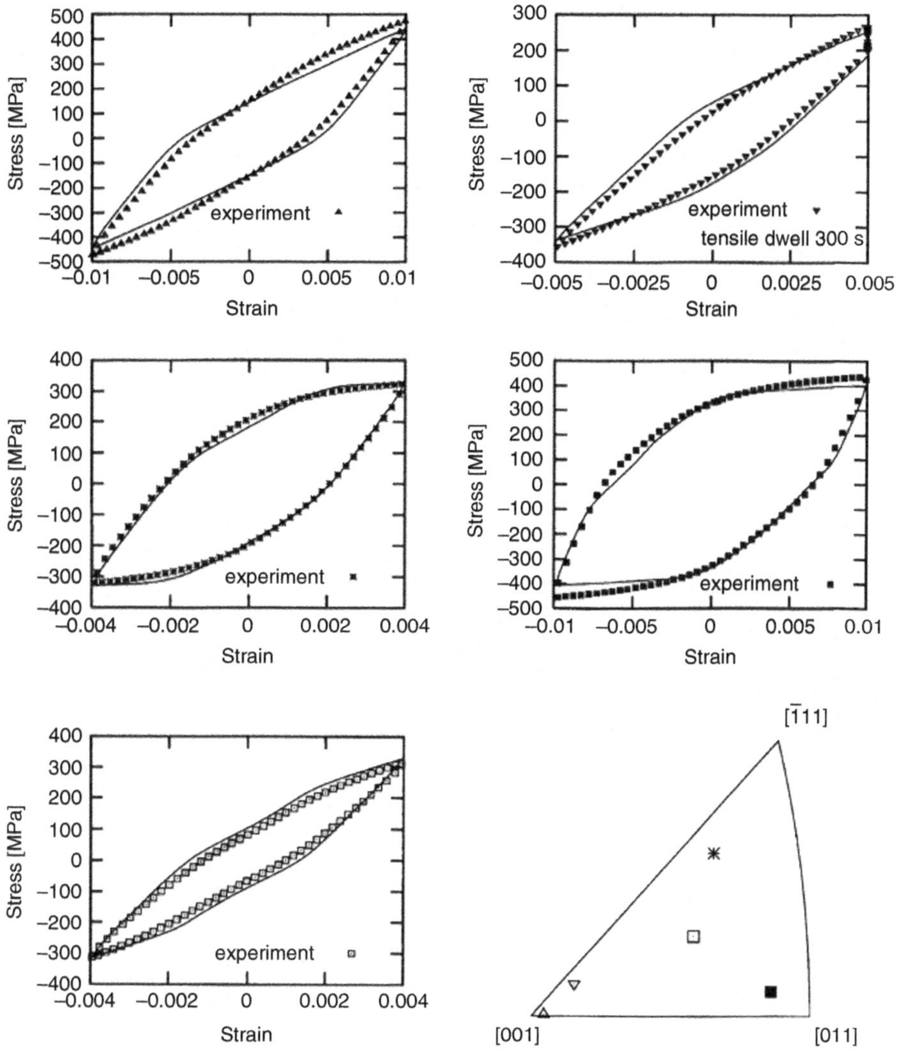


**Fig. 7** Stabilized stress-strain loops for nickel-base single crystal superalloy SC16 at 950 °C, experiment vs. simulation after (Fedelich 2002)

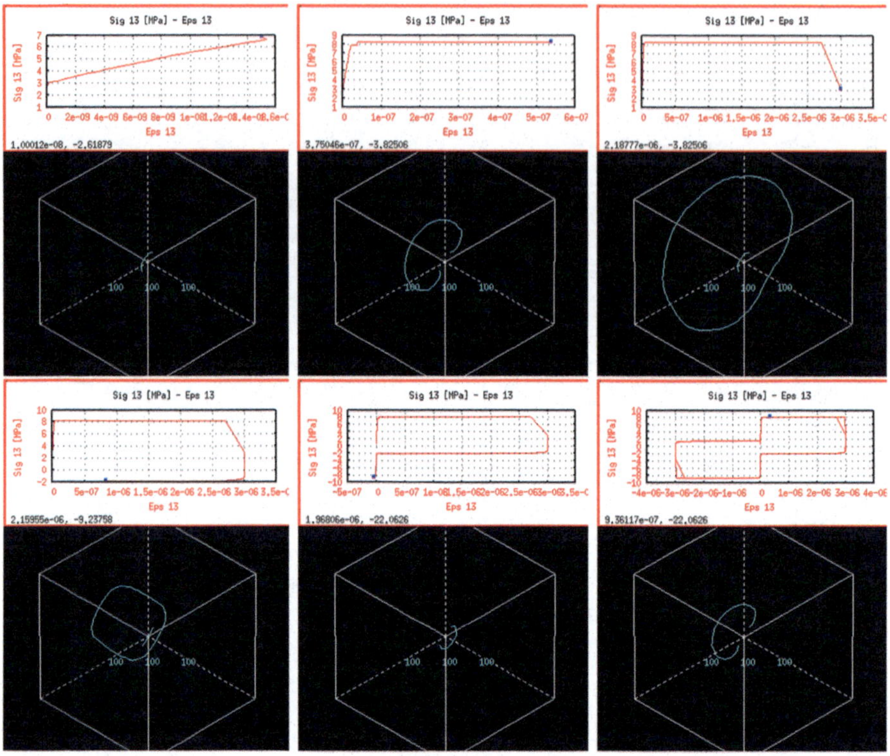


**Fig. 8** Cyclic behavior of a single Frank–Read dislocation source simulated by discrete dislocation density. Courtesy of Dr. M. Fivel

# 4 Phase Field Modelling of Microstructure Evolution in Crystals

The phase field approach has become a ubiquitous tool to simulate microstructure evolution in materials undergoing phase transformations (Steinbach 2009; Finel et al. 2010). Phases are described by continuous fields where each one is endowed with a transport equation, or phase field equation. Such fields vary rapidly inside the phase boundaries over a given width, which is also called the interface thickness. The interface is then said to be diffuse, which can have a physical justification when the thickness goes beyond a few inter-atomic distances. The transport equations associated to the phase field variables allow to predict the evolving morphology of new phases, and naturally embed the phase transformation's kinetics. For many diffusion-controlled phase transformations, stresses can play a significant role in both the transformation kinetics and the morphology evolution of the growing phases. The transport equations for phase fields were firstly coupled to Cauchy's first law of motion restricted to the elastic behavior of the phases in Wang et al.

(1993), and Dreyer and Müller (2000). More recently, the attention was focused on the extension to plasticity and viscoplasticity, as done by Guo et al. (2005), Ubachs et al. (2005), Uehara et al. (2007), Gaubert et al. (2008, 2010), and Ammar et al. (2009).

Two main approaches have been proposed to determine the effective behavior of the interface, namely the interpolation and homogenization approaches according to the terminology by Ammar et al. (2009, 2011, 2014). In the latter approach, the diffuse interface is seen as a mixture of two phases that each retain their macroscopic properties, and for which standard homogenization methods coming for instance from the mechanics of heterogeneous materials (Zaoui 2002; Qu and Cherkaoui 2006; François et al. 2012), can be used to derive the effective response. In this case, the phase field parameter then plays a role akin to the volume fraction of one phase within the diffuse interface. In contrast, a single constitutive behavior is used for both phases in the interpolation approach for which the corresponding material parameters are interpolated between the bounding values reached in each phase.

The choice of the interpolation for the interface effective behavior originated from the work of Tiaden et al. (1998) who introduced the solute partitioning on the basis of the mixture theory. Such a solute partitioning was then reinterpreted shortly after by Kim et al. (1999) who used a Reuss like assumption for the diffusion behavior throughout the interface, which they called the pseudo-equilibrium condition. Steinbach and Apel (2006) then used the Reuss assumption, by analogy to the work of Kim et al. (1999), to account for the elastic behavior of phases. Ammar et al. (2009, 2011, 2014) also applied the Voigt homogenization scheme that relies on identical strain values taken by both phases at each material point. Mixed schemes as they arise in laminate microstructures were then employed by Durga et al. (2013), and Mosler et al. (2014) in order to reduce spurious stress concentration within the interface. Nevertheless, the popularity of the interpolation method—known as Katchaturyan's approach—was such that the first attempts to introduce nonlinear material behavior of phases relied on the interpolation method, see for instance (Guo et al. 2005; Gaubert et al. 2008, 2010; Cottura et al. 2012).

The choice of a suitable homogenization scheme generally depends on the morphology of the mixture of phases within the diffuse interface. As such, a mixture very often does not correspond to a physical reality for actual interfaces. That is why various schemes have been applied without reference to the real structure of the interface. But the homogenization and interpolation schemes can be of practical importance with respect to the convergence of the diffuse interface problem towards the sharp interface problem. It is now admitted that the Reuss-like assumption done in Kim et al. (1999) improves such a convergence w.r.t. to interpolation methods. Finally, Ammar et al. (2014) highlighted the paradigm of the interpolation and homogenization methods though the concept of inheritance. The interpolation method embeds inheritance of the internal variables during the phase transformation. On the contrary, the homogenization method makes the internal variables of each phase independent from each other. Nevertheless, inheritance can be added or discarded by the use of suitable evolutionary laws for the internal variables.

Durga et al. (2013) and Mosler et al. (2014) successfully applied the combination of homogenization and phase field methods to the elastic behavior of phases. These frameworks were extended to the consideration of elastoviscoplastic deformation modes of the individual phases by Rancourt et al. (2016). This approach is reported in the present section following the same lines. The role of thermodynamics, and especially the second law or entropy principle, will be shown to be instrumental in selecting suitable homogenization schemes in the phase field models. The constitutive setting extends the continuum thermodynamic framework used in nonlinear mechanics of materials to the phase field approach (Germain et al. 1983; Maugin 1992, 1999). This represents an alternative to the use of variational derivatives of the free energy function as done classically in phase field approaches since the pioneering work by Cahn and Hilliard (1958). It has the advantage that the theory can be readily applied to finite bodies with specific and unambiguous boundary conditions, in contrast to usual periodic or infinite microstructures and corresponding vanishing boundary conditions used in the latter reference. The space of state variables is extended to include internal variables accounting for material work-hardening. Mechanical dissipation then results from the competition between energy storage due to the multiplication of dislocations and to dissipation induced by dislocation motion. The theory is suitable for the material evolution at a mesoscale where individual dislocations are not resolved but rather represented by suitable densities. It typically applies at the grain scale in a polycrystal where the continuum crystal plasticity has proved in many cases to provide a reliable description of slip processes. The introduction of crystal plasticity into phase field models goes back to works by Gaubert et al. (2008, 2010), and Cottura et al. (2012) for single crystals and by Abrivard et al. (2012a,b) for polycrystals, including grain boundary migration.

For the sake of simplicity, the theory is exposed within the isothermal small perturbation framework although extensions to anisothermal and finite deformation models are possible and necessary for many applications, e.g. Mosler et al. (2014).

## *4.1 Phase Field Modelling Using the Homogenization Approach*

The present phase field framework is based on the introduction of a phase field parameter $\phi$ accounting for a two-phase system $\alpha$-$\beta$, with constant values $\phi = 1$ and $\phi = 0$, respectively in $\alpha$ and $\beta$. The phase field varies rapidly and continuously from 0 to 1 in the diffuse interface zone over a thickness $\delta$ as defined in Kim et al. (1999). The physical degrees of freedom attributed to each material point in the continuum are the concentration $c_i$ of a given chemical element $i$, and the displacement vector $\boldsymbol{u}$ with the corresponding strain tensor $\boldsymbol{\varepsilon}$ defined as the symmetric part of the gradient of the displacement field.

Following the homogenization method, all of the chosen energy potentials along with their respective state variables, fluxes and conjugates are interpolated as follows:

$$\Xi_{i} = h(\phi)\, \Xi_{i}^{\alpha} + \bar{h}(\phi)\, \Xi_{i}^{\beta}\,, \tag{74}$$

where $\Xi$ can be the Helmholtz free energy density $f$ or the Gibbs free energy density $g$, the concentration $c_i$ of component $i$ with the conjugate diffusion potential $\mu_i$ along with mass flux $\boldsymbol{J}_i$, or the strain $\boldsymbol{\varepsilon}$ and stress $\boldsymbol{\sigma}$. Accordingly, the quantities $\Xi^\alpha$ and $\Xi^\beta$: $c_{i}^{\alpha}$, $c_{i}^{\beta}$, $\boldsymbol{\varepsilon}^\alpha$, etc., can be regarded as fictitious or auxiliary variables in the theory according to Kim et al. (1999), and Ammar et al. (2009). The weighting function $h$, analogous to a phase fraction, fulfils the following requirements:

$$h(1) = 1, \quad h(0) = 0, \qquad h'(\phi) = \begin{cases} 0, & \text{if } \phi = 1 \text{ or } \phi = 0, \\ > 0, & \text{if } 0 < \phi < 1\,. \end{cases} \tag{75}$$

The complementing function $\bar{h}$ is defined as $\bar{h}(\phi) = 1 - h(\phi)$. The conditions on the derivative of $h$ exclude the linear function $h(\phi) = \phi$ classically required by volume averaging in homogenization theory. In phase field models, they aim at localising the mixture of phases within a narrow diffuse interface zone. A standard choice for $h$ is the following third order polynomial function:

$$h(\phi) = \phi^2\,(3 - 2\phi)\,. \tag{76}$$

Finally, the total strain tensor attributed to each phase at each material point is split into three distinct mechanical contributions as follows:

$$\boldsymbol{\varepsilon}^{\psi} = \boldsymbol{\varepsilon}_{e}^{\psi} + \boldsymbol{\varepsilon}_{p}^{\psi} + \boldsymbol{\varepsilon}_{\star}^{\psi}\,, \tag{77}$$

where $\psi$ stands for either $\alpha$ or $\beta$ (the same convention will be adopted in the sequel), and where the subscripts $e$ stands for elasticity, $p$ for (visco)plasticity and $\star$ for a constant and uniform eigenstrain, the latter being assumed, in this work, to be independent of concentration and temperature, for the sake of brevity.

### 4.1.1 Balance Equations

The total concentrations are subjected to the local balance of mass in the form:

$$\dot{c}_{i} + \operatorname{div} \boldsymbol{J}_{i} = 0, \tag{78}$$

for each component $i$, in the absence of source terms. The mass balance equation is associated with Dirichlet or Neumann conditions at the boundary of the body where the concentration or the normal flux $j = \boldsymbol{J} \cdot \boldsymbol{n}$ are prescribed, $\boldsymbol{n}$ denoting the normal surface vector.

The stress tensor fulfils the static balance of momentum equation:

$$\operatorname{div} \boldsymbol{\sigma} = \mathbf{0}\,, \tag{79}$$

in the absence of body forces for simplicity. The Cauchy balance equation is complemented by displacement or stress vector, $\boldsymbol{t} = \boldsymbol{\sigma} \cdot \boldsymbol{n}$, controlled boundary conditions.

The internal energy density enters the local balance of energy equation

$$\dot{e} = \boldsymbol{\sigma} : \dot{\boldsymbol{\varepsilon}} - \pi \dot{\phi} + \boldsymbol{\xi} \cdot \boldsymbol{\nabla} \dot{\phi}\,, \tag{80}$$

where generalized stresses $\boldsymbol{\xi}$ and $\pi$ are introduced as dual quantities to the change in phase field parameter $\phi$ and its gradient, following (Gurtin 1996; Ammar et al. 2009). The first term in the right-hand side is the classical mechanical power of internal forces whereas the second and third terms account for additional internal energy change associated with boundary motion. The generalized stresses are postulated to fulfil an additional balance equation in the form:

$$\pi + \operatorname{div} \boldsymbol{\xi} = 0. \tag{81}$$

The balance of generalized stresses is associated with Dirichlet or Neumann boundary conditions where the phase field parameter or the generalized stress vector, $\boldsymbol{\xi} \cdot \boldsymbol{n}$, are prescribed at the boundary.

In the spirit of homogenization theory, the partial quantities $c_i^{\psi}, \boldsymbol{\sigma}^{\psi}$ are *not* subjected to similar balance equations. This is in contrast to mixture theories as discussed by Müller (2001).

The variational formulation of balance Eqs. (79) and (81) constitute the generalized principle of virtual power

$$\int_V \left(\boldsymbol{\sigma} : \operatorname{grad} \boldsymbol{u}^* + \pi \phi^* - \boldsymbol{\xi} \cdot \operatorname{grad} \phi^*\right) dV = \int_S \left(\boldsymbol{n} \cdot \boldsymbol{\sigma} \cdot \boldsymbol{u}^* - \boldsymbol{\xi} \cdot \boldsymbol{n}\, \phi^*\right) dS\,, \tag{82}$$

using the virtual velocity fields $\{\boldsymbol{u}^*, \phi^*\}$, as discussed in Ammar et al. (2009). The generalized principle of virtual power is complemented by the variational formulation of the balance of mass equation (78):

$$\int_V \left(\dot{c}_i\, c^* - \boldsymbol{J}_i \cdot \operatorname{grad} c^*\right) dV = \int_S (\boldsymbol{J}_i \cdot \boldsymbol{n})\, c^*\, dS\,, \tag{83}$$

for the field of virtual concentration rates $c^*$.

The previous variational formulations are directly applicable to the finite element method (Ammar et al. 2009).

## 4.2 Constitutive Framework Based on the Helmholtz Free Energy Potential

Thermodynamic principles are inoperative as long as the state and internal variables of the theory are not defined. That is why the setting up of a constitutive theory starts with the proper definition of the state and internal variables. The internal energy, $e$, and Helmholtz free energy, $f$, functions differ by one argument, namely the entropy variable, $s$, entering the internal energy function and the temperature variable, $T$, arising in the free energy. They are related by the corresponding Legendre transform:

$$f = e - s\,T\,. \tag{84}$$

These functions are defined per unit volume in the present work.

The approach is exposed in the case of a two-phase multicomponent system made of solid crystalline materials, as defined in the previous section. To render the model more transparent, its presentation is specialized to the case of a single crystal phase $\beta$ undergoing plastic slip according to crystal plasticity theory, see Asaro and Lubarda (2006), and Besson et al. (2009), coexisting with phase $\alpha$ which exhibits an isotropic elasto-viscoplastic behavior. This does not limit the approach to such a configuration, and it is straightforward to adopt any combination of nonlinear constitutive laws, as will be demonstrated in Sect. 4.8.

## 4.3 State and Internal Variables

The considered space of state and internal variables for the two-phase $\alpha$-$\beta$ system is chosen as:

$$\text{STATE} = \left\{\phi, \operatorname{grad}\phi,\ c_{\mathrm{i}}^{\alpha}, \boldsymbol{\varepsilon}_e^{\alpha},\ c_{\mathrm{i}}^{\beta}, \boldsymbol{\varepsilon}_e^{\beta}, \rho_s^{\beta}, x_s^{\beta}\right\}, \tag{85}$$

where $\phi$ is a smooth phase indicator field, $c_{\mathrm{i}}^{\psi}$ are site fractions occupied by the chemical species $i$, $\boldsymbol{\varepsilon}_e^{\psi}$ is the elastic strain tensor attributed to each phase at each material point. The internal variables of the model are represented by the hardening variables, $\rho_s^{\beta}$ and $x_s^{\beta}$, where the subscript $s$ indicates the slip system number according to the crystal plasticity model. The variable $\rho_s^{\beta}$ is akin to a dislocation density measure directly related to the critical resolved shear stress for the activation of slip system $s$ according to Schmid's law, whereas $x_s$ accounts for kinematic hardening needed for cyclic plasticity. Let $N$ denote the total number of slip systems characterising the single crystalline phase $\beta$, typically $N = 12$ in f.c.c. crystals. For simplicity, no hardening variable is introduced in the phase $\alpha$, without restriction to the generality of the constitutive formulation that allows for distinct behavior laws for both phases. Under general anisothermal conditions, the previous set should be complemented by the temperature variable.

The dual set is composed of the thermodynamical forces associated to the state and internal variables following the framework of continuum thermodynamics:

$$\text{FORCES} = \left\{ -\pi, \boldsymbol{\xi}\,,\; \mu_{\text{i}}^{\alpha}, \boldsymbol{\sigma}^{\alpha},\; \mu_{\text{i}}^{\beta}, \boldsymbol{\sigma}^{\beta}, R_{s}^{\beta}, X_{s}^{\beta} \right\}, \tag{86}$$

where $\pi$ and $\boldsymbol{\xi}$ are the generalized stresses introduced by Gurtin (1996), and Ammar et al. (2009), $\mu_i^{\psi}$ and $\boldsymbol{\sigma}^{\psi}$ are respectively the diffusion potentials and stress tensors of phase $\psi$, $R_s^{\beta}$ is the slip resistance and $X_s^{\beta}$ the back-stress as described in Busso and Cailletaud (2005), and Besson et al. (2009) for each slip system $s$ in phase $\beta$.

### 4.4 Decomposition of the Free Energy Density Function

The volumetric free energy density $f$ of the two-phase material is decomposed at each material point into diffuse interface, chemical and mechanical parts, respectively called $f_{\phi}$, $f_c$, $f_{\varepsilon}$:

$$f = f_{\phi} + \sum_{i} f_{c_i} + f_{\varepsilon}\,. \tag{87}$$

The interface contribution of the free energy density is defined by Kim et al. (1999) as:

$$f_{\phi}(\phi, \operatorname{grad}\phi) = 3\gamma\left(\frac{1}{\lambda} W(\phi) + \lambda \parallel \operatorname{grad}\phi \parallel^2\right), \tag{88}$$

with $\gamma$ the interface energy, $\lambda = \delta/(2z)$ where $\delta$ is the diffuse interface thickness defined between $\phi^{-} = 0.05$ and $\phi^{+} = 0.95$, $z$ a parameter defined as $z = \log\left(\phi^{+}/\phi^{-}\right)$ and $W$ the double well function defined as:

$$W(\phi) = \phi^2\,(1-\phi)^2\,. \tag{89}$$

Bulk free energies are coupled by means of the phase field variable using the same averaging relations as in Sect. 4.1 for the state variables, thermodynamical forces and fluxes (74):

$$f_{c_i}(\phi, c_{\text{i}}^{\alpha}, c_{\text{i}}^{\beta}) = h(\phi)\, f_{c_i}^{\alpha}(c_{\text{i}}^{\alpha}) + \bar{h}(\phi)\, f_{c_i}^{\beta}(c_{\text{i}}^{\beta})\,, \tag{90}$$

$$f_{\varepsilon}(\phi, \boldsymbol{\varepsilon}_{e}^{\alpha}, \boldsymbol{\varepsilon}_{e}^{\beta}, \rho_{s}^{\beta}, x_{s}^{\beta}) = h(\phi)\, f_{\varepsilon}^{\alpha}(\boldsymbol{\varepsilon}_{e}^{\alpha}) + \bar{h}(\phi)\, f_{\varepsilon}^{\beta}(\boldsymbol{\varepsilon}_{e}^{\beta}, \rho_{s}^{\beta}, x_{s}^{\beta})\,. \tag{91}$$

Possible dependence of $f_{\varepsilon}^{\psi}$ on concentrations can be added in the latter equation but is not considered here for simplicity, without restriction. As a result of the coupling with the phase field variable $\phi$ shown in Eq. (91), the minimisation of the free energy may lead to two phases in equilibrium and, thus, to the construction of diagrams corresponding to chemical equilibrium under mechanical stress, as studied extensively in the references Cahn and Larché (1984), Johnson (1987), and Frolov and Mishin (2012).

## 4.5 State Laws Relating Thermodynamic Variables and Associated Forces

The local form of the second principle of thermodynamics is written as:

$$\dot{s} \geq \operatorname{div} \boldsymbol{\Phi} \,, \tag{92}$$

where $s$ is the volumetric entropy density and $\boldsymbol{\Phi}$ the entropy density flux. The time derivative of (84) in the isothermal case allows us to rewrite (92) into:

$$\dot{e} - \dot{f} \geq T \operatorname{div} \boldsymbol{\Phi} \,. \tag{93}$$

The free energy density is a function of the state variables (85). Its time derivative can be expanded according to the chain rule:

$$\dot{f} = \frac{\partial f}{\partial \phi} \dot{\phi} + \frac{\partial f}{\partial \operatorname{grad} \phi} \cdot \operatorname{grad} \dot{\phi} + \sum_i \frac{\partial f}{\partial c_{\mathrm{i}}^{\alpha}} + \sum_i \frac{\partial f}{\partial c_{\mathrm{i}}^{\beta}} \dot{c}_{\mathrm{i}}^{\beta} + \frac{\partial f}{\partial \boldsymbol{\varepsilon}_e^{\alpha}} : \dot{\boldsymbol{\varepsilon}}_e^{\alpha} + \frac{\partial f}{\partial \boldsymbol{\varepsilon}_e^{\beta}} : \dot{\boldsymbol{\varepsilon}}_e^{\beta}$$

$$+ \sum_s \frac{\partial f}{\partial \rho_s^{\beta}} \dot{\rho}_s^{\beta} + \sum_s \frac{\partial f}{\partial x_s^{\beta}} \dot{x}_s^{\beta} \,. \tag{94}$$

The entropy flux is related to the mass entropy flux and the diffusion potential, following (de Groot and Mazur 1962–1984; Coleman and Noll 1963; Ammar et al. 2009; Villani et al. 2014):

$$\boldsymbol{\Phi} = \frac{1}{T} \sum_i \mu_{\mathrm{i}} \boldsymbol{J}_{\mathrm{i}} \,. \tag{95}$$

Expanding the Eqs. (93)–(95) and using the mass and energy balance Eqs. (78) and (80) combined with the averaging relations (91) lead to the Clausius-Duhem inequality:

$$- \left( \pi + \frac{\partial f}{\partial \phi} \right) \dot{\phi} + \left( \boldsymbol{\xi} - \frac{\partial f}{\partial \operatorname{grad} \phi} \right) \cdot \operatorname{grad} \dot{\phi} + \sum_i \mu_{\mathrm{i}} \dot{c}_{\mathrm{i}} - h \frac{\partial f^{\alpha}}{\partial c_{\mathrm{i}}^{\alpha}} \dot{c}_{\mathrm{i}}^{\alpha} - \bar{h} \frac{\partial f^{\beta}}{\partial c_{\mathrm{i}}^{\beta}} \dot{c}_{\mathrm{i}}^{\beta}$$

$$- \sum_i \boldsymbol{J}_{\mathrm{i}} \cdot \operatorname{grad} \mu_{\mathrm{i}} + \boldsymbol{\sigma} : \dot{\boldsymbol{\varepsilon}} - h \frac{\partial f^{\alpha}}{\partial \boldsymbol{\varepsilon}_e^{\alpha}} : \dot{\boldsymbol{\varepsilon}}_e^{\alpha}$$

$$- \bar{h} \frac{\partial f^{\beta}}{\partial \boldsymbol{\varepsilon}_e^{\beta}} : \dot{\boldsymbol{\varepsilon}}_e^{\beta} - \sum_s \bar{h} \frac{\partial f^{\beta}}{\partial \rho_s^{\beta}} \dot{\rho}_s^{\beta} - \sum_s \bar{h} \frac{\partial f^{\beta}}{\partial x_s^{\beta}} \dot{x}_s^{\beta} \geq 0 \,. \tag{96}$$

The previous inequality is first exploited for the individual phases. The particular case $\phi = 1$ corresponds to the pure phase $\alpha$:

$$\sum_i \left(\mu_i^\alpha - \frac{\partial f^\alpha}{\partial c_i^\alpha}\right)\dot{c}_i^\alpha - \sum_i \boldsymbol{J}_i^\alpha \cdot \operatorname{grad} \mu_i^\alpha + \left(\boldsymbol{\sigma}^\alpha - \frac{\partial f^\alpha}{\partial \boldsymbol{\varepsilon}_e^\alpha}\right) : \dot{\boldsymbol{\varepsilon}}_e^\alpha + \boldsymbol{\sigma}^\alpha : \dot{\boldsymbol{\varepsilon}}_p^\alpha \geq 0\,, \tag{97}$$

whereas the particular case for the phase $\beta$, $\phi = 0$, reads:

$$\sum_i \left(\mu_i^\beta - \frac{\partial f^\beta}{\partial c_i^\beta}\right)\dot{c}_i^\beta - \sum_i \boldsymbol{J}_i^\beta \cdot \operatorname{grad} \mu_i^\beta + \left(\boldsymbol{\sigma}^\beta - \frac{\partial f^\beta}{\partial \boldsymbol{\varepsilon}_e^\beta}\right) : \dot{\boldsymbol{\varepsilon}}_e^\beta + \boldsymbol{\sigma}^\beta : \dot{\boldsymbol{\varepsilon}}_p^\beta$$
$$- \sum_s \frac{\partial f^\beta}{\partial \rho_s^\beta} \dot{\rho}_s^\beta - \sum_s \frac{\partial f^\beta}{\partial x_s^\beta} \dot{x}_s^\beta \geq 0\,. \tag{98}$$

Assuming that the diffusion potentials and the stress tensors do not depend explicitly on $\dot{c}_i^\alpha$, $\dot{c}_i^\beta$, $\dot{\boldsymbol{\varepsilon}}_e^\alpha$ and $\dot{\boldsymbol{\varepsilon}}_e^\beta$, the left-hand sides of the two previous inequalities are linear forms with respect to the concentration and elastic strain increments. Following (Coleman and Noll 1963), their positivity requires that the cofactors must vanish:

$$\mu_i^\alpha = \frac{\partial f^\alpha}{\partial c_i^\alpha}, \qquad \mu_i^\beta = \frac{\partial f^\beta}{\partial c_i^\beta}, \qquad \boldsymbol{\sigma}^\alpha = \frac{\partial f^\alpha}{\partial \boldsymbol{\varepsilon}_e^\alpha}, \qquad \boldsymbol{\sigma}^\beta = \frac{\partial f^\beta}{\partial \boldsymbol{\varepsilon}_e^\beta}\,. \tag{99}$$

The thermodynamic forces associated with the hardening variables are then defined as:

$$R_s^\beta = \frac{\partial f^\beta}{\partial \rho_s^\beta}, \qquad X_s^\beta = \frac{\partial f^\beta}{\partial x_s^\beta}\,. \tag{100}$$

The residual dissipation rate inequality is obtained after combining (96) with (99), (100) and (74):

$$-\pi^\sharp\,\dot{\phi} + \left(\boldsymbol{\xi} - \frac{\partial f}{\partial \operatorname{grad} \phi}\right) \cdot \operatorname{grad} \dot{\phi} + \sum_i h\,(\mu_i - \mu_i^\alpha)\,\dot{c}_i^\alpha + \sum_i \bar{h}\,(\mu_i - \mu_i^\beta)\,\dot{c}_i^\beta$$
$$- \sum_i \boldsymbol{J}_i \cdot \operatorname{grad} \mu_i + h\,(\boldsymbol{\sigma} - \boldsymbol{\sigma}^\alpha) : \dot{\boldsymbol{\varepsilon}}_e^\alpha + \bar{h}\,(\boldsymbol{\sigma} - \boldsymbol{\sigma}^\beta) : \dot{\boldsymbol{\varepsilon}}_e^\beta$$
$$+ h\boldsymbol{\sigma} : \dot{\boldsymbol{\varepsilon}}_p^\alpha + \bar{h}\boldsymbol{\sigma} : \dot{\boldsymbol{\varepsilon}}_p^\beta - \bar{h} \sum_s R_s^\beta \dot{\rho}_s^\beta - \bar{h} \sum_s X_s^\beta \dot{x}_s^\beta \geq 0\,, \tag{101}$$

where $\pi^\sharp$ is defined as

$$\pi^\sharp = \pi + \frac{\partial f}{\partial \phi} - h' \Big[ \sum_i \mu_i \,(c_i^\alpha - c_i^\beta) + \boldsymbol{\sigma} : (\boldsymbol{\varepsilon}^\alpha - \boldsymbol{\varepsilon}^\beta) \Big] \tag{102}$$

and is similar to the dissipative microforce introduced by Gurtin (1996).

At this point, it is still assumed that the diffusion potential and stress tensor functions do not explicitly depend on the rates $\dot{c}_{\rm i}^{\alpha}$, $\dot{c}_{\rm i}^{\beta}$ and $\dot{\boldsymbol{\varepsilon}}_e^{\alpha}$ and $\dot{\boldsymbol{\varepsilon}}_e^{\beta}$. In that case, it follows that the following conditions must hold to ensure positivity of the dissipation rate:

$$\mu_{\rm i} = \mu_{\rm i}^{\beta} = \mu_{\rm i}^{\alpha}, \quad \boldsymbol{\sigma} = \boldsymbol{\sigma}^{\alpha} = \boldsymbol{\sigma}^{\beta}. \tag{103}$$

The first condition of equal partial diffusion potentials was adopted as a starting point by Kim et al. (1999). The equality of partial stresses associated with each phase at a material point corresponds to the well-known Reuss approximation in the mechanical homogenization theory and was first applied by Steinbach and Apel (2006) and discussed in Ammar et al. (2009), Durga et al. (2013), and Mosler et al. (2014).

Finally, following (Gurtin 1996), the generalized stress vector $\boldsymbol{\xi}$ is assumed to be independent of grad $\dot{\phi}$, which provides the last state law:

$$\boldsymbol{\xi} = \frac{\partial f}{\partial \operatorname{grad} \phi}. \tag{104}$$

As a result, the residual dissipation inequality reads:

$$-\pi^{\sharp}\,\dot{\phi} - \boldsymbol{J}_{\rm i} \cdot \operatorname{grad} \mu_{\rm i} + h\boldsymbol{\sigma} : \dot{\boldsymbol{\varepsilon}}_p^{\alpha} + \bar{h}\Big(\boldsymbol{\sigma} : \dot{\boldsymbol{\varepsilon}}_p^{\beta} - \sum_s R_s^{\beta}\dot{\rho}_s^{\beta} - \sum_s X_s^{\beta}\dot{x}_s^{\beta}\Big) \geq 0. \tag{105}$$

The mechanical contributions in the previous inequality contain the plastic power and the possible energy storage by means of increase of internal variables.

At this stage, specific choices for the bulk volumetric free energy densities can be introduced in the form of quadratic potentials for the purpose of illustration. The chemical contributions

$$f_{c_i}^{\psi}(c_{\rm i}^{\psi}) = \frac{1}{2} k_i^{\psi}\left(c_{\rm i}^{\psi} - \widehat{c}_{\rm i}^{\psi}\right)^2, \tag{106}$$

correspond to elliptic paraboloids with minimum at $\widehat{c}_i^{\psi}$ and curvature given by $k_i^{\psi}$. The elasticity contribution for the $\alpha$ phase

$$f_{\varepsilon}^{\alpha}(\boldsymbol{\varepsilon}_e^{\alpha}) = \frac{1}{2}\boldsymbol{\varepsilon}_e^{\alpha} : \mathbb{C}^{\alpha} : \boldsymbol{\varepsilon}_e^{\alpha}, \tag{107}$$

involves the fourth order tensor of elastic moduli of the bulk phase, $\mathbb{C}^{\alpha}$. The energy stored by elasticity and hardening for the phase $\beta$ is taken from Busso and Cailletaud (2005), and Besson et al. (2009):

$$f_{\varepsilon}^{\beta}(\boldsymbol{\varepsilon}_e^{\beta}) = \frac{1}{2}\boldsymbol{\varepsilon}_e^{\beta} : \mathbb{C}^{\beta} : \boldsymbol{\varepsilon}_e^{\beta} + \frac{1}{2} b^{\beta} Q^{\beta} \sum_{s,r} H^{sr} \rho_s^{\beta} \rho_r^{\beta} + \frac{1}{2} C^{\beta} \sum_s \left(x_s^{\beta}\right)^2. \tag{108}$$

The hardening parameters are $Q^\beta$, $b^\beta$ and $C^\beta$. The interaction matrix $H^{sr}$ accounts for self and latent hardening induced by dislocation interactions.

Based on the previous quadratic potentials, the state laws (99), (100) and (104) combined with (103) provide:

$$\mu_i^\psi = k_i^\psi\left(c_i^\psi - \hat{c}_i^\psi\right) = \mu_i\,, \tag{109}$$

along with

$$\boldsymbol{\sigma}^\psi = \mathbb{C}^\psi : \boldsymbol{\varepsilon}_e^\psi = \boldsymbol{\sigma}\,, \tag{110}$$

$$R_s^\beta = b^\beta Q^\beta \sum_r H^{sr}\rho_r^\beta\,, \qquad X_s^\beta = C^\beta x_s^\beta\,. \tag{111}$$

The generalized stress vector is proportional to the gradient of the phase field variable, following (88) and (104):

$$\boldsymbol{\xi} = 6\gamma\lambda\,\mathrm{grad}\,\phi\,. \tag{112}$$

## 4.6 Constitutive Laws from the Dissipation Potential

The positivity of the dissipation rate in (105) can be identically fulfilled by introducing a convex dissipation potential depending on the generalized stresses, conjugate to the flow of phase field, concentration, plastic strain and hardening variables:

$$\Omega = \Omega(\pi^\sharp,\ \mathrm{grad}\,\mu_i^\alpha, \mathrm{grad}\,\mu_i^\beta,\ \boldsymbol{\sigma}^\alpha, \boldsymbol{\sigma}^\beta, R_s^\beta, X_s^\beta;\ \phi)\,, \tag{113}$$

The dissipation potential $\Omega$ is a function of thermodynamical forces and may also depend on the thermodynamical variables that have to be considered as parameters, according to Germain et al. (1983). Here $\phi$ is used in the decomposition of the dissipation potential in weighted contributions of the individual phases, as done for the free energy density function:

$$\Omega = \Omega_\pi(\pi^\sharp) + h(\phi)\,\Omega^\alpha(\mathrm{grad}\,\mu_i^\alpha, \boldsymbol{\sigma}^\alpha) + \bar{h}(\phi)\,\Omega^\beta(\mathrm{grad}\,\mu_i^\beta, \boldsymbol{\sigma}^\beta, R_s^\beta, X_s^\beta)\,. \tag{114}$$

Looking at (97), (98), (99) and (100), quadratic dissipation potentials can be chosen in the form:

$$\Omega_\pi(\pi^\sharp) = M_\phi \pi^{\sharp 2} \tag{115}$$

$$\Omega^\alpha(\text{grad}\,\mu_{\text{i}}^\alpha, \boldsymbol{\sigma}^\alpha) = \frac{1}{2} M_i^\alpha |\text{grad}\,\mu_{\text{i}}^\alpha|^2 + \frac{K^\alpha}{n^\alpha + 1}\left\langle \frac{\sqrt{3/2\,\boldsymbol{\sigma}_{\text{dev}}^\alpha : \boldsymbol{\sigma}_{\text{dev}}^\alpha} - R_0^\alpha}{K^\alpha} \right\rangle^{n^\alpha+1} \tag{116}$$

$$\begin{aligned}\Omega^\beta(\text{grad}\,\mu_i^\beta, \boldsymbol{\sigma}^\beta, R_s^\beta, X_s^\beta) &= \frac{1}{2} M_i^\beta |\text{grad}\,\mu_{\text{i}}^\beta|^2 \\ &+ \frac{K^\beta}{n^\beta + 1} \sum_s \left\langle \frac{|\boldsymbol{\sigma}^\beta : \boldsymbol{P}_s^\beta - X_s^\beta| - R_0^\beta - R_s^\beta}{K^\beta} \right\rangle^{n^\beta+1}.\end{aligned} \tag{117}$$

$M_\phi$ and $M_i^\psi$ are the mobility parameters respectively for the phase-field and diffusion in phase $\psi$. For simplicity, cross terms between the different species are not considered (the matrix of mobility is diagonal) and the mobilities are taken isotropic. The brackets $\langle . \rangle$ denote the Macaulay brackets, they deliver the positive part of the concerned quantity. A von Mises yield criterion with constant threshold $R_0^\alpha$ is chosen for the $\alpha$ phase whereas the Schmid law is adopted in the $\beta$ phase, $R_0^\beta$ being the initial critical resolved shear stress and the $R_s^\beta$ the hardening for each slip system. The deviatoric stress tensors are called $\boldsymbol{\sigma}_{\text{dev}}^\psi$. The crystallographic orientation tensor $\boldsymbol{P}_s^\beta$ is the tensor product of the slip direction and normal to the slip plane for slip system $s$, see Eq. (23). The material parameters $K^\psi$ and $n^\psi$ account for the generally nonlinear viscous properties of both phases.

The dissipation potential is used to derive the flow rules for diffusion and viscoplasticity. The mass flux vectors present in the Clausius-Duhem inequalities obtained in (97) and (98) are derived from the dissipation potential for $\phi = 1$ and $\phi = 0$ as follows:

$$\boldsymbol{J}_{\text{i}}^\alpha = -\frac{\partial \Omega^\alpha}{\partial \text{grad}\,\mu_{\text{i}}^\alpha} = -M_i^\alpha \,\text{grad}\,\mu_{\text{i}}^\alpha\,, \quad \boldsymbol{J}_{\text{i}}^\beta = -\frac{\partial \Omega^\beta}{\partial \text{grad}\,\mu_{\text{i}}^\beta} = -M_i^\beta \,\text{grad}\,\mu_{\text{i}}^\beta \tag{118}$$

where the diffusion mobilities $L_i^\psi$ can be related to the diffusivities $D_i^\psi$ as follows, in order to retrieve the generalized Fick law of diffusion:

$$L_i^\psi = D_i^\psi \left( \frac{\partial^2 f^\psi}{\partial c_{\text{i}}^{\psi\,2}} \right)^{-1}. \tag{119}$$

Using then the flux averaging relations (74) and (103), one obtains:

$$\boldsymbol{J}_{\mathrm{i}} = -\Big(h(\phi)\, L_i^{\alpha} + \bar{h}(\phi)\, L_i^{\beta}\Big)\mathrm{grad}\, \mu_{\mathrm{i}}\,, \tag{120}$$

so that effective mobilities can be defined as the average of the bulk diffusivities of the phases:

$$L_i(\phi) = h(\phi)\, L_i^{\alpha} + \bar{h}(\phi)\, L_i^{\beta}\,. \tag{121}$$

The plastic flow and hardening rules can also be derived from the dissipation potential following the theory of standard generalized materials (Besson et al. 2009):

$$\dot{\boldsymbol{\varepsilon}}_p^{\alpha} = \frac{\partial \Omega_p^{\alpha}}{\partial \boldsymbol{\sigma}^{\alpha}} = \left\langle \frac{\sqrt{3/2\, \boldsymbol{\sigma}^{\alpha}_{\mathrm{dev}} : \boldsymbol{\sigma}^{\alpha}_{\mathrm{dev}}} - R_0^{\alpha}}{K^{\alpha}} \right\rangle^{n^{\alpha}} \frac{\boldsymbol{\sigma}^{\alpha}_{\mathrm{dev}}}{\sqrt{\boldsymbol{\sigma}^{\alpha}_{\mathrm{dev}} : \boldsymbol{\sigma}^{\alpha}_{\mathrm{dev}}}}\,. \tag{122}$$

This corresponds to usual von Mises power-law viscoplasticity with threshold $R_0^{\alpha}$. As for the phase $\beta$:

$$\dot{\boldsymbol{\varepsilon}}_p^{\beta} = \frac{\partial \Omega_p^{\beta}}{\partial \boldsymbol{\sigma}^{\beta}} = \sum_s \left\langle \frac{|\boldsymbol{\sigma}^{\beta} : \boldsymbol{P}_s^{\beta} - X_s^{\beta}| - R_s^{\beta}}{K} \right\rangle^{n^{\beta}} \boldsymbol{P}_s^{\beta}\, \mathrm{sign}\left(\tau_s^{\beta} - X_s^{\beta}\right) = \sum_s \dot{\gamma}_s \boldsymbol{P}_s\,, \tag{123}$$

where $\dot{\gamma}_{\alpha}$ is the slip rate for the slip system number $s$. The double contraction of the stress tensor $\boldsymbol{\sigma}^{\beta}$ and the orientation $\boldsymbol{P}_s^{\alpha}$ is called the resolved shear stress, $\tau_s^{\beta}$, on slip system $s$ and is the driving force for slip activation in the presence of a back-stress $X_s^{\beta}$.

The evolution laws for nonlinear hardening are defined following (Busso and Cailletaud 2005; Besson et al. 2009):

$$\dot{\rho}_s^{\beta} = \Big(1 - b^{\beta}\rho_s^{\beta}\Big)|\dot{\gamma}_s|, \qquad \dot{x}_s^{\beta} = \Big(\mathrm{sign}\,(\boldsymbol{\sigma}^{\beta} : \boldsymbol{P}_s^{\beta} - X_s^{\beta}) - d^{\beta}x_s^{\beta}\Big)|\dot{\gamma}_s|\,. \tag{124}$$

Finally, the evolution of the phase field variable is derived from the quadratic part of the dissipation potential depending on $\pi^{\sharp}$ as follows:

$$\dot{\phi} = -\frac{\partial \Omega}{\partial \pi^{\sharp}} = -M_{\phi}\pi^{\sharp}\,. \tag{125}$$

The latter equation can be reformulated into the well-known Allen-Cahn equation by introducing the generalized stress balance equation (81) and the state law (104) combined with (102):

$$\frac{\dot{\phi}}{M_{\phi}} = 6\gamma\lambda\, \Delta\phi - \frac{\partial f}{\partial \phi} + h'\Big[\sum_i \mu_{\mathrm{i}}(c_{\mathrm{i}}^{\alpha} - c_{\mathrm{i}}^{\beta}) + \boldsymbol{\sigma} : (\boldsymbol{\varepsilon}^{\alpha} - \boldsymbol{\varepsilon}^{\beta})\Big]. \tag{126}$$

It is convenient to rewrite the enhanced Allen-Cahn equation (126) by introducing $F_\phi^R$ as follows:

$$\frac{\dot{\phi}}{M_\phi} = 3\gamma\Big(2\lambda\,\Delta\phi - \frac{1}{\lambda}W'\Big) - h' F_\phi^R\,, \tag{127}$$

where

$$F_\phi^R = f^\alpha - f^\beta - \sum_i \mu_{\rm i}\left(c_{\rm i}^\alpha - c_{\rm i}^\beta\right) - \boldsymbol{\sigma} : \left(\boldsymbol{\varepsilon}^\alpha - \boldsymbol{\varepsilon}^\beta\right)\,. \tag{128}$$

It can be noted that when $F_\phi^R = 0$ the equilibrium condition in Frolov and Mishin (2012, eq. 29) is recovered, see also Spatschek and Eidel (2013). The above constitutive laws are then consistent with a positive dissipation rate at each material point and at each time. Note that Eq. (128) is valid for any form of the free energy function. In the particular case of quadratic elastic and chemical potentials, Eq. (128) becomes:

$$F_\phi^R = -\frac{1}{2}\sum_i \mu_{\rm i}\left(c_{\rm i}^\alpha - c_{\rm i}^\beta\right) - \frac{1}{2}\boldsymbol{\sigma} : \left(\boldsymbol{\varepsilon}^\alpha - \boldsymbol{\varepsilon}^\beta\right)\,. \tag{129}$$

### *4.7 Constitutive Framework Based on the Gibbs Free Energy Potential*

It has been shown that the restrictions implied by the entropy principle on the form of the constitutive and state laws based on the homogenization approach from the volumetric free energy density have lead us to select the Reuss homogenization scheme regarding stresses and diffusion potentials attached to each phase at each material point. Following a similar procedure, an alternative formulation can be proposed based on the use of the Gibbs free energy or free enthalpy. The detailed derivation can be found in Rancourt et al. (2016). It leads to an alternative homogenization scheme for the mechanical part of the model. The volumetric Gibbs energy density function $g$ is defined as the double Legendre transform of the volumetric internal energy density $e$:

$$g = e - T\,s - \boldsymbol{\sigma} : \boldsymbol{\varepsilon}\,. \tag{130}$$

The exploitation of the entropy principle based on the Gibbs free energy leads to sufficient conditions for positive dissipation in the form:

$$\mu_{\rm i} = \mu_{\rm i}^\beta = \mu_{\rm i}^\alpha, \qquad \boldsymbol{\varepsilon} = \boldsymbol{\varepsilon}^\alpha = \boldsymbol{\varepsilon}^\beta\,. \tag{131}$$

The latter strain relations correspond to the Voigt approximation in homogenization theory. This is in contrast to the Reuss approximation used previously.

A complete constitutive framework can then be proposed in the same way as in the case of the Helmholtz free energy based on the Voigt interface conditions. The reader is referred to Rancourt et al. (2016) for the corresponding detailed constitutive laws including the reversible and dissipative parts.

It is worth noting that prior to the choice of a specific homogenization scheme, the equivalence of both models is not ensured. Indeed, if we introduce the Legendre transformation from the Helmholtz to the Gibbs free energy densities for any two phased system:

$$g = f - \boldsymbol{\sigma} : \boldsymbol{\varepsilon}\,, \tag{132}$$

Applying the mixture rules (74) to the strain and stress tensors in the previous expression gives:

$$g = f - \left[h(\phi)\boldsymbol{\sigma}^{\alpha} + \bar{h}(\phi)\boldsymbol{\sigma}^{\beta}\right] : \left[h(\phi)\boldsymbol{\varepsilon}^{\alpha} + \bar{h}(\phi)\boldsymbol{\varepsilon}^{\beta}\right]. \tag{133}$$

On the other hand, for each phase, the bulk free enthalpy can also be expressed as:

$$g^{\psi} = f^{\psi} - \boldsymbol{\sigma}^{\psi} : \boldsymbol{\varepsilon}^{\psi}\,. \tag{134}$$

If we apply (134) into the second model based on the interpolation of the bulk Gibbs energies, it comes that:

$$g = h(\phi)g^{\alpha} + \bar{h}(\phi)g^{\beta} \tag{135}$$

$$= h(\phi)\,f^{\alpha} + \bar{h}(\phi)\,f^{\beta} - \left[h(\phi)\,\boldsymbol{\sigma}^{\alpha} : \boldsymbol{\varepsilon}^{\alpha} + \bar{h}(\phi)\,\boldsymbol{\sigma}^{\beta} : \boldsymbol{\varepsilon}^{\beta}\right] \tag{136}$$

$$= f - \left[h(\phi)\,\boldsymbol{\sigma}^{\alpha} : \boldsymbol{\varepsilon}^{\alpha} + \bar{h}(\phi)\,\boldsymbol{\sigma}^{\beta} : \boldsymbol{\varepsilon}^{\beta}\right], \tag{137}$$

which is *a priori* not equivalent to (133).

While writing $\langle x \rangle$ the average value, or here the interpolated value, of $x$ according to Eq. (74), the equivalence of (133) and (137) can be reformulated as follows:

$$\langle \boldsymbol{\sigma} \rangle : \langle \boldsymbol{\varepsilon} \rangle = \langle \boldsymbol{\sigma} : \boldsymbol{\varepsilon} \rangle\,, \tag{138}$$

which is known as the Hill-Mandel principle of macro-homogeneity (Zaoui 2002). In the framework of continuum mechanics, this condition is used to ensure the scale transition. The homogenization schemes to be used within the present framework are required to fulfill this condition.

## 4.8 Growth of a Precipitate in a Viscoplastic Matrix

The relevance of the proposed framework is illustrated by investigating the diffusion-controlled growth of misfitting precipitates in supersaturated matrices. This section examines how viscoplasticity impacts the kinetics and the shape evolution of growing precipitates.

For the sake of simplicity, a binary alloy is considered with materials parameters in Table 1. So as to draw general trends relevant for any micrometer-sized microstructures in metallic materials evolving at high temperature, they are reported in their dimensionless forms (denoted with tildes), where the length, time and energy scales are defined respectively by $L$, commensurate with the system size,

**Table 1** Dimensionless materials parameters for the phase field model coupled with viscoplasticity

| | | |
|---|---|---|
| *Chemical equilibrium* | | |
| | Equilibrium concentration $\tilde{c}^{\alpha}$ | 1 |
| | Equilibrium concentration $\tilde{c}^{\beta}$ | 0 |
| | Curvature $\tilde{k}^{\alpha}$ | 100 |
| | Curvature $\tilde{k}^{\beta}$ | 1 |
| *Interface properties* | | |
| Interface thickness $\tilde{\delta}$ | Interface energy $\tilde{\gamma}$ | 0.02 |
| | Planar precipitates | [0.005; 0.025] |
| | Circular precipitates | 0.025 |
| *Mobilities* | | |
| | Diffusivity $\tilde{D}^{\alpha}$ | $8 \times 10^{-3}$ |
| | Diffusivity $\tilde{D}^{\beta}$ | 1 |
| | Phase field mobility $\tilde{M}_{\phi}$ | $6.17 \times 10^{6}$ |
| *Elasticity* | | |
| | $\varepsilon^{\star}$ | 0.01 |
| | $\tilde{E}^{\alpha} = \tilde{E}^{\beta} = \tilde{E}$ | 210 |
| | $\nu^{\alpha} = \nu^{\beta} = \nu$ | 0.3 |
| *Isotropic viscoplasticity* $\tilde{R}_0$ | | |
| | Planar precipitates | 0 |
| | Circular precipitates | 0.2 |
| | $\tilde{K}$ | $0.01 - 100$ |
| | $n$ | 5 |
| *Crystal plasticity* | | |
| | $\tilde{R}_0^{\beta}$ | 0.065 |
| | $\tilde{R}_s^{\beta}$ | $\tilde{Q} = 0.45, b = 3$ |
| | $\tilde{X}^{\beta}$ | $\tilde{C} = 0.6, d = 20$ |
| | $\tilde{K}^{\beta}$ | 0.1 |
| | $n^{\beta}$ | 5 |

the diffusion characteristic time $\tau = L^2/D$ with $D$ the chemical diffusivity of the alloying species, and the ratio $E = \gamma/\delta$ of the interface energy $\gamma$ to the interface thickness $\delta$. Moreover, the concentrations are scaled by the difference in concentrations $\hat{c}^\alpha - \hat{c}^\beta$ (Table 1) between the coexisting phases as given by the phase diagram. For the sake of illustration, Norton parameters are recast such as to rescale viscosities $K$ by the energy scale, assuming that the characteristic time entering $K$ is $L^2/D$ (this has no consequence because there is no intrinsic time scale in viscoplasticity Norton type laws). Nonetheless, particular values have been chosen to comply with the growth of oxide layers at the surface of stainless steels (de Rancourt 2015). In that case, typical physical values of the characteristic scales are $L = 200\,\mathrm{nm}$, $\tau = 700\,\mathrm{s}$ and $E = 10^9\,\mathrm{Jm}^{-3}$.

The expression of the mobility of the phase field proposed by Kim et al. (1999) is used to achieve local equilibrium at the interface rapidly, i.e. on times much smaller than the duration of the diffusion-controlled process. $\tilde{M}_\phi$ reads:

$$\tilde{M}_\phi = \tilde{D}^\beta / (6\tilde{\lambda}^2 \zeta)\,, \tag{139}$$

with

$$\zeta = \tilde{k}^\alpha \tilde{k}^\beta (\hat{c}^\alpha - \hat{c}^\beta)^2 \int_0^1 \frac{h\bar{h}}{\bar{h}\tilde{k}^\alpha + h\tilde{k}^\beta} \frac{d\phi}{\phi(1-\phi)}.$$

In all the subsequent calculations, the initial microstructure is set such that the phase field features an equilibrium profile in tanh in the direction normal to the interface. The initial concentration field $c$ is defined such as to display the chemical equilibrium value in the precipitate, i.e. $c^\alpha = \hat{c}^\alpha$, and a supersaturated constant value in the matrix, i.e. $c^\beta > \hat{c}^\beta$, sufficient to promote growth.

Calculations involving viscoplasticity are performed so as to assess its possible impact on the kinetics of diffusion-controlled transformations, by comparing to purely chemical and elastic kinetics. A simple Norton constitutive law is considered in either $\alpha$ or $\beta$ phase with parameters in Table 1. It is assumed that all phases are initially elastic, i.e. $\boldsymbol{\varepsilon}^p(\boldsymbol{x}) = 0,\ \forall \boldsymbol{x}$. Moreover, it is worth mentioning that within the present framework, the growing phase does not inherit plasticity from the mother phase, as discussed by Ammar et al. (2014). Detailed results on the impact of the homogenization scheme and of viscoplasticity on the kinetics are to be found in Rancourt et al. (2016).

The interplay between viscoplasticity and morphological evolutions during the diffusion-controlled growth of precipitates is investigated. In contrast with recent works addressing issues concerning specific materials and involving complex situations, such as rafting in Ni-base superalloys (Gaubert et al. 2010; Cottura et al. 2012) or Widmanstätten plate growth (Cottura et al. 2015), we focus on a simple generic situation by examining the destabilization of a circular precipitate growing in a supersaturated matrix, which is known to undergo a Mullins-Sekerka morphological destabilization (Mullins and Sekerka 1963). It has been shown in Leo and Sekerka (1989) that elasticity may change the critical radius for destabilization, depending on the ratio of shear moduli between the precipitate and the matrix: the destabilization of a hard precipitate growing inside a soft matrix is delayed whereas the inverse situation is predicted for a soft precipitate in a hard matrix. Misfitting precipitates of radius $\tilde{r}_0 = 0.15$ are located at the center of a $4 \times 4$ square matrix supersaturated with $c^0 = 0.05$ so as to promote growth. Different combinations of constitutive laws are considered: (1) an elastic precipitate embedded in a matrix with isotropic viscoplasticity; (2) a precipitate with isotropic viscoplasticity embedded in an elastic matrix; and (3) an elastic precipitate embedded in a matrix endowed with crystal plasticity behavior.

The evolution of the precipitate in the first case is shown in Fig. 9 where the cumulative plastic strain ranges from 0 (light yellow) to 2% (orange). The precipitate remains circular for the whole growth duration. The plastic strain is localized around the precipitate in the matrix and is not inherited by the precipitate. Its maximum value, located at the interface, remains constant, and the radius of the plastic zone increases with the precipitate radius in agreement with the analytical solution and phase field calculations considering time-independent plasticity (Ammar et al. 2011).

In the second case (Fig. 10), the viscoplastic precipitate growing in an elastic matrix undergoes a very different evolution: the circular shape is rapidly destabilized to give rise to a star shape with four branches aligned with the discretization grid. A shell of plastic strain develops inside the precipitate close to the interface very early during the growth process. During the transient stage of destabilization, the increase in plastic strain is slightly lower along the directions of destabilization (lighter regions in Fig. 10) than in other directions. But rapidly the tips grow at a constant rate along the horizontal and vertical directions. This is very similar to dendrite growth, observed also at the solid state in alloys with moderate isotropic misfits partially relaxed by misfit dislocations at the interface, e.g. Husain et al. (1999), mimicked crudely by viscoplasticity in the present work, see also Appolaire et al. (2010). Note that the unstable modes are selected by the slight anisotropy of interface energy induced by the discretization grid, e.g. Bösch et al. (1995): indeed,

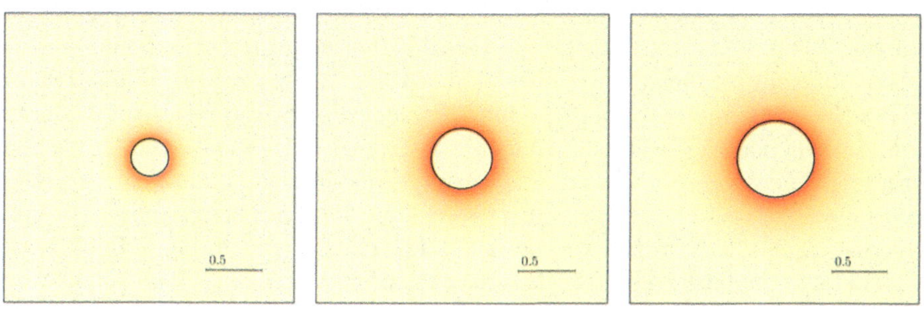


**Fig. 9** Snapshots (enlargements of the $4 \times 4$ system) of the cumulative equivalent plastic strain (red maximum value = 0.01) at times ($1.3 \times 10^{-6}$, 0.5 and 1.1) during the diffusion-controlled growth an elastic precipitate in a supersaturated matrix with isotropic viscoplastic behavior

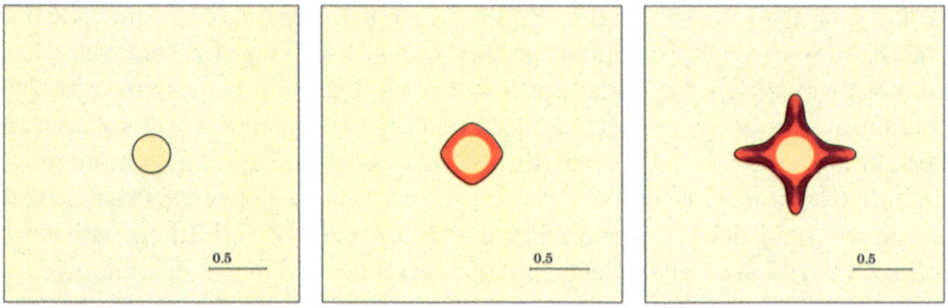


**Fig. 10** Snapshots (enlargements of the 4 × 4 system) of the cumulative equivalent plastic strain (red maximum value = 0.03) at times (1.3× $10^{-6}$, 0.5 and 1.1) during the diffusion-controlled growth a precipitate with isotropic viscoplastic behavior in an elastic supersaturated matrix

by twisting the square mesh with respect to the horizontal axis, we have obtained star shaped precipitates with branches no longer aligned horizontally and vertically but aligned with the mesh (not shown).

The differences between both cases are similar to the differences found by Leo and Sekerka (1989) in the pure elastic case between hard precipitate in soft matrix and soft precipitate in hard matrix. Indeed, qualitatively, viscoplastic relaxation by a $J_2$ theory amounts qualitatively to decrease the secant shear modulus, above the yield stress. Then, an elastic precipitate in a viscoplastic matrix behaves similarly to a hard precipitate in a soft matrix and displays a larger critical radius for morphological destabilization, whereas a viscoplastic circular precipitate behaves like a soft precipitate in an elastic matrix and destabilizes at a smaller radius. This interpretation is supported by additional calculations, not shown, where strain-hardening is introduced in the viscoplastic precipitate, and where the destabilization occurs later than without hardening.

Finally, the growth of an elastic precipitate is computed for a matrix that displays anisotropic viscoplasticity along a single slip system (indicated by letter A in Fig. 11). Very early in the growth process, plastic strain is localized near the interface in the matrix at 45° with respect to the square box with short horizontal and vertical bands respectively aligned and perpendicular to the slip direction (slip and kink bands). The resolved shear stress is maximal at this angle around a circular precipitate with isotropic misfit, e.g. Geslin et al. (2014). The heterogeneous distribution caused by the anisotropy of plastic relaxation induces an anisotropic shape evolution with smooth corners developing along the horizontal and vertical directions, where there is no plastic relaxation. This case demonstrates that the symmetries of plastic relaxation can interfere with the other symmetries that drive the morphology of the microstructures, as already shown in phase field calculations of rafting in Ni-based superalloys (Gaubert et al. 2010; Cottura et al. 2012). Moreover, it demonstrates that the interplay of plasticity and diffusion-controlled phase transformations cannot be predicted with simple qualitative arguments: indeed, whereas plasticity promotes growth of a planar interface with respect to elasticity, it can hinder growth as in Fig. 11. This complexity which discards often oversimplified arguments based on

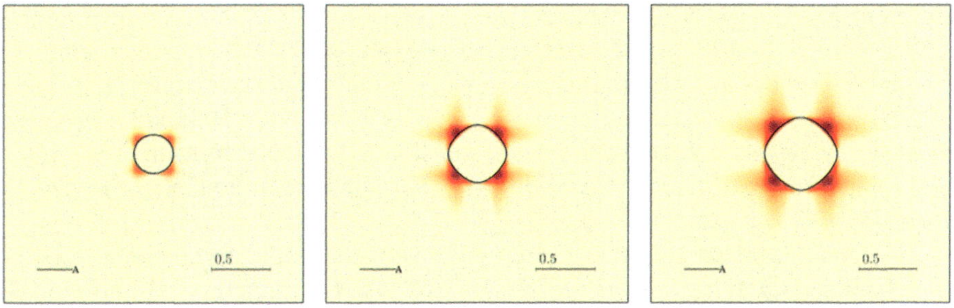


**Fig. 11** Snapshots (enlargements of the 4 × 4 system) of the cumulative equivalent plastic strain (red maximum value = 0.02) at times ($2.8 \times 10^{-4}$, 0.6 and 1.3) during the diffusion-controlled growth an elastic precipitate in a supersaturated matrix with anisotropic viscoplastic behavior (slip system denoted by A)

overall balances, e.g. Embury et al. (2003), can be related to the indirect influence of mechanics on the morphologies through the boundary conditions of diffusion driving the evolution processes.

## 5 Conclusions

Three continuum field approaches have been sketched and illustrated in the case of crystal plasticity in metals. All three frameworks are sensitive to the size of the microstructure and therefore represent enhancements of classical continuum mechanics. This is due to the fact that they involve gradients of dimensionless quantities (strain or phase field variables) and thus necessarily include some internal lengths in the constitutive relations. Strain gradient plasticity has emerged as a special case of the general micromorphic theory initiated by Mindling and Eringen. The formulation of higher order constitutive laws remains a largely open question in current research. Phenomenological Ansaetze were presented and compared to physical arguments coming from dislocation theories making use of the dislocation density tensor. Recent results by Mesarovic et al. (2015) show the limitations in the use of the dislocation density tensor as a whole as a constitutive variable. More sophisticated laws involve the individual GND densities for all slip systems. Intrinsic lengths can be identified from experimental field measurements and from computational or theoretical results depicted in the chapters "Multiscale Dislocation-Based Plasticity", "Statistical Theory of Dislocation" and "Multiscale Modeling of Interfaces, Dislocations, and Dislocation Field Plasticity".

A general framework has been established describing diffusion driven phase transformation in elastoviscoplastic multicomponent materials, based on the combination of homogenization and phase field approaches. In contrast to existing frameworks applied to nonlinear materials, mainly relying on parameter interpolation, it has the advantage that arbitrary types of material behavior for the different phases can be combined within a consistent thermomechanical theory. The exploitation of the second law of thermodynamics has been used to motivate two privileged homogenization schemes, namely the Voigt/Taylor and Reuss/Sachs models, respectively obtained from the consideration of the Helmholtz or Gibbs free energy density potentials. It extends the strategy initiated by Kim et al. (1999) and limited to the diffusion part of the model, to the case of nonlinear mechanical behavior of the phases.

The computational analysis of the model shows that the competition between the characteristic times for mass diffusion and viscoplastic stress relaxation leads to complex kinetic behavior compared to the purely chemical and elastic reference cases. Viscoplasticity also controls the morphology of a growing circular precipitate in a matrix. It is responsible for possible loss of stability characterized by deviation from the circular form during growth. Crystal plasticity of the matrix was shown to induce anisotropic growth of the precipitate due to the formation of slip and kink bands at preferred locations of the precipitate's boundary. The question of inheritance of plasticity and hardening should also be considered within a homogenization based phase field approach, as initiated in the reference Ammar et al. (2014). This should however be supported by relevant experimental data including field measurements of concentration, strain and crystal orientation.

Combination of strain gradient plasticity and phase field models is necessary since the interaction between phase boundaries and matrix mechanical response takes place at low scales involving dislocations. The continuum approach can still be used at least up to a certain extent as demonstrated by recent computations for the creep behaviour of single nickel-base superalloys, see Cottura et al. (2012).

## References

G. Abrivard, E.P. Busso, S. Forest, B. Appolaire, Phase field modelling of grain boundary motion driven by curvature and stored energy gradients. Part I: theory and numerical implementation. Philos. Mag. **92**, 3618–3642 (2012a)

G. Abrivard, E.P. Busso, S. Forest, B. Appolaire, Phase field modelling of grain boundary motion driven by curvature and stored energy gradients. Part II: application to recrystallisation. Philos. Mag. **92** 3643–3664 (2012b)

E.C. Aifantis, The physics of plastic deformation. Int. J. Plast. **3**, 211–248 (1987)

H. Altenbach, G.A. Maugin, V. Erofeev, *Mechanics of Generalized Continua*. Advanced Structured Materials, vol. 7 (Springer, Berlin, 2011)

K. Ammar, B. Appolaire, G. Cailletaud, S. Forest, Combining phase field approach and homogenization methods for modelling phase transformation in elastoplastic media. Eur. J. Comput. Mech. **18**, 485–523 (2009)

K. Ammar, B. Appolaire, G. Cailletaud, S. Forest, Phase field modeling of elasto-plastic deformation induced by diffusion controlled growth of a misfitting spherical precipitate. Philos. Mag. Lett. **91**, 164–172 (2011)

K. Ammar, B. Appolaire, S. Forest, M. Cottura, Y. Le Bouar, A. Finel, Modelling inheritance of plastic deformation during migration of phase boundaries using a phase field method. Meccanica **49**, 2699–2717 (2014). https://doi.org/10.1007/s11012-014-0011-1

B. Appolaire, E. Aeby-Gautier, J.D. Teixeira, M. Dehmas, S. Denis, Non-coherent interfaces in diffuse interface models. Philos. Mag. **90**, 461–483 (2010)
R.J. Asaro, Elastic–plastic memory and kinematic hardening. Acta Metall. **23**, 1255–1265 (1975)
R.J. Asaro, V.A. Lubarda, *Mechanics of Solids and Materials* (University Press, Cambridge, 2006)
M.F. Ashby, The deformation of plastically non-homogeneous materials. Philos. Mag. **21**, 399–424 (1970)
V.L. Berdichevsky, On thermodynamics of crystal plasticity. Scr. Mater. **54**, 711–716 (2006)
J. Besson, G. Cailletaud, J.-L. Chaboche, S. Forest, M. Blétry, *Non-linear Mechanics of Materials*. Series: Solid Mechanics and Its Applications, vol. 167 (Springer, New York, 2009), 433 p. ISBN: 978-90-481-3355-0
A. Bösch, H. Müller-Krumbhaar, O. Shochet, Phase-field models for moving boundary problems: controlling metastability and anisotropyi. Z. Phys. **97**, 367–377 (1995)
E.P. Busso, G. Cailletaud, On the selection of active slip systems in crystal plasticity. Int. J. Plast. **21**, 2212–2231 (2005)
J.W. Cahn, J.E. Hilliard, Free energy of a nonuniform system. I. interfacial free energy. J. Chem. Phys. **28**, 258–267 (1958)
J.W. Cahn, F. Larché, A simple model for coherent equilibrium. Acta Metall. **32**, 1915–1923 (1984)
P. Cermelli, M.E. Gurtin, On the characterization of geometrically necessary dislocations in finite plasticity. J. Mech. Phys. Solids **49**, 1539–1568 (2001)
H.J. Chang, N.M. Cordero, C. Déprés, M. Fivel, S. Forest, Micromorphic crystal plasticity versus discrete dislocation dynamics analysis of multilayer pile-up hardening in a narrow channel. Arch. Appl. Mech. **86**, 21–38 (2016). https://doi.org/10.1007/s00419-015-1099-z
B.D. Coleman, W. Noll, The thermodynamics of elastic materials with heat conduction and viscosity. Arch. Ration. Mech. Anal. **13**, 167–178 (1963)
S. Conti, M. Ortiz, Dislocation microstructures and the effective behavior of single crystals. Arch. Ration. Mech. Anal. **176**, 103–147 (2005)
N.M. Cordero, A. Gaubert, S. Forest, E.P. Busso, F. Gallerneau, S. Kruch, Size effects in generalised continuum crystal plasticity for two-phase laminates. J. Mech. Phys. Solids **58**, 1963–1994 (2010)
N.M. Cordero, S. Forest, E.P. Busso, S. Berbenni, M. Cherkaoui, Grain size effects on plastic strain and dislocation density tensor fields in metal polycrystals. Commun. Math. Sci. **52**, 7–13 (2012)
M. Cottura, Y. Le Bouar, A. Finel, B. Appolaire, K. Ammar, S. Forest, A phase field model incorporating strain gradient viscoplasticity: application to rafting in Ni-base superalloys. J. Mech. Phys. Solids **60**, 1243–1256 (2012)
M. Cottura, B. Appolaire, A. Finel, Y. Le Bouar, Plastic relaxation during diffusion-controlled growth of widmanstätten plates. Scr. Mater. **108**, 117–121 (2015). https://doi.org/10.1016/j.scriptamat.2015.06.032
S.R. de Groot, P. Mazur, *Non-equilibrium Thermodynamics* (Dover, North Holland, 1962–1984)
V. de Rancourt, Odelling the oxidation of polycristalline austenitic stainless steels using a phase field approach coupled with mechanics. PhD, Mines ParisTech (2015)
V. de Rancourt, B. Appolaire, S. Forest, K. Ammar, Homogenization of viscoplastic constitutive laws within a phase field approach. J. Mech. Phys. Solids **88**, 35–48 (2016). https://doi.org/10.1016/j.jmps.2015.12.026
C. Déprés, C.F. Robertson, M.C. Fivel, Low-strain fatigue in aisi 316l steel surface grains: a three-dimensional discrete dislocation dynamics modelling of the early cycles i. dislocation microstructures and mechanical behaviour. Philos. Mag. **84**(22), 2257–2275 (2004). https://doi.org/10.1080/14786430410001690051
W. Dreyer, W.H. Müller, A study of the coarsening in tin/lead solders. Int. J. Solids Struct. **37**, 3841–3871 (2000)
A. Durga, P. Wollants, N. Moelans, Evaluation of interfacial excess contributions in different phase-field models for elastically inhomogeneous systems. Model. Simul. Mater. Sci. Eng. **21**, 055018 (2013)

J.D. Embury, A. Deschamps, Y. Brechet, The interaction of plasticity and diffusion controlled precipitation reactions. Scr. Mater. **49**, 927–932 (2003)
A.C. Eringen, E.S. Suhubi, Nonlinear theory of simple microelastic solids. Int. J. Eng. Sci. **2**, 189–203, 389–404 (1964)
B. Fedelich, A microstructural model for the monotonic and the cyclic mechanical behavior of single crystals of superalloys at high temperatures. Int. J. Mech. Sci. **18**, 1–49 (2002)
A. Finel, Y. Le Bouar, A. Gaubert, U. Salman, Phase field methods: Microstructures, mechanical properties and complexity. C. R. Phys. **11**, 245–256 (2010)
N.A. Fleck, J.W. Hutchinson, Strain gradient plasticity. Adv. Appl. Mech. **33**, 295–361 (1997)
S. Forest, Some links between cosserat, strain gradient crystal plasticity and the statistical theory of dislocations. Philos. Mag. **88**, 3549–3563 (2008)
S. Forest, The micromorphic approach for gradient elasticity, viscoplasticity and damage. ASCE J. Eng. Mech. **135**, 117–131 (2009)
S. Forest, Generalized continuum modelling of crystal plasticity, in ed. by C. Sansour, S. Skatulla. *Generalized Continua and Dislocation Theory*. CISM International Centre for Mechanical Sciences, Courses and Lectures, vol. 537 (Springer, Berlin, 2012a), pp. 181–287
S. Forest, Micromorphic media, in ed. by H. Altenbach, V. Eremeyev, *Generalized Continua – from the Theory to Engineering Applications*. CISM International Centre for Mechanical Sciences, Courses and Lectures, vol. 541 (Springer, Berlin, 2012b), pp. 249–300
S. Forest, N. Guéninchault, Inspection of free energy functions in gradient crystal plasticity. Acta Mech. Sinica **29**, 763–772 (2013)
S. Forest, R. Sedláček, Plastic slip distribution in two–phase laminate microstructures: dislocation–based vs. generalized–continuum approaches. Philos. Mag. A **83**, 245–276 (2003)
S. Forest, R. Sievert, Elastoviscoplastic constitutive frameworks for generalized continua. Acta Mech. **160**, 71–111 (2003)
S. Forest, R. Sievert, Nonlinear microstrain theories. Int. J. Solids Struct. **43**, 7224–7245 (2006)
S. Forest, K. Ammar, B. Appolaire, N.M. Cordero, A. Gaubert, Micromorphic approach to crystal plasticity and phase transformation, in J. Schroeder, K. Hackl, *Plasticity and Beyond*. CISM International Centre for Mechanical Sciences, Courses and Lectures, vol. 550 (Springer, New York, 2014), pp. 131–198
F. François, A. Pineau, A. Zaoui, *Mechanical Behaviour of Materials. Volume 1: Micro and Macroscopic Constitutive Behaviour*. Solid Mechanics and its Applications, vol. 180 (Springer, New York, 2012)
T. Frolov, Y. Mishin, Thermodynamics of coherent interfaces under mechanical stresses. I. Theory. Phys. Rev. B **85**, 224106 (2012). https://doi.org/10.1103/PhysRevB.85.224106
A. Gaubert, A. Finel, Y. Le Bouar, G. Boussinot, Viscoplastic phase field modellling of rafting in ni base superalloys, in *Continuum Models and Discrete Systems CMDS11* (Mines Paris Les Presses, 2008), pp. 161–166
A. Gaubert, Y. Le Bouar, A. Finel, Coupling phase field and viscoplasticity to study rafting in Ni-based superalloys. Philos. Mag. **90**, 375–404 (2010)
M.G.D. Geers, R.H.J. Peerlings, M.A. Peletier, L. Scardia, Asymptotic behaviour of a pile–up of infinite walls of edge dislocations. Arch. Ration. Mech. Anal. **209**, 495–539 (2013)
P. Germain, The method of virtual power in continuum mechanics. Part 2 : Microstructure. SIAM J. Appl. Math. **25**, 556–575 (1973)
P. Germain, Q.S. Nguyen, P. Suquet, Continuum thermodynamics. J. Appl. Mech. **50**, 1010–1020 (1983)
P.A. Geslin, B. Appolaire, A. Finel, Investigation of coherency loss by prismatic punching with a nonlinear elastic model. Acta Mater. **71**, 80–88 (2014)
I. Groma, F.F. Csikor, M. Zaiser, Spatial correlations and higher-order gradient terms in a continuum description of dislocation dynamics. Acta Mater. **51**, 1271–1281 (2003)
I. Groma, G. Györgyi, B. Kocsis, Dynamics of coarse grain grained dislocation densities from an effective free energy. Philos. Mag. **87**, 1185–1199 (2007)
X.H. Guo, S.Q. Shi, X.Q. Ma, Elastoplastic phase field model for microstructure evolution. Appl. Phys. Lett. **87**(22), 221910 (2005). https://doi.org/10.1063/1.2138358

M.E. Gurtin, Generalized Ginzburg–Landau and Cahn–Hilliard equations based on a microforce balance. Physica D **92**, 178–192 (1996)
M.E. Gurtin, A gradient theory of single–crystal viscoplasticity that accounts for geometrically necessary dislocations. J. Mech. Phys. Solids **50**, 5–32 (2002)
M.E. Gurtin, L. Anand, Thermodynamics applied to gradient theories involving the accumulated plastic strain: the theories of Aifantis and Fleck and Hutchinson and their generalization. J. Mech. Phys. Solids **57**, 405–421 (2009)
W. Han, B.D. Reddy, *Plasticity: Mathematical Theory and Numerical Analysis* (Springer, New York, 2013)
D.E. Hurtado, M. Ortiz, Finite element analysis of geometrically necessary dislocations in crystal plasticity. Int. J. Numer. Methods Eng. **93**(1), 66–79 (2013)
S.W. Husain, M.S. Ahmed, I. Qamar, Dendritic morphology observed in the solid-state precipitation in binary alloys. Metal. Mater. Trans. A **30**, 1529–1534 (1999)
W.C. Johnson, On the inapplicability of Gibbs phase rule to coherent solids. Metall. Trans. A **18**, 1093–1097 (1987)
R. Kametani, K. Kodera, D. Okumura, N. Ohno, Implicit iterative finite element scheme for a strain gradient crystal plasticity model based on self-energy of geometrically necessary dislocations. Commun. Math. Sci. **53**(1), 53–59 (2012)
S.G. Kim, W.T. Kim, T. Suzuki, Phase-field model for binary alloys. Phys. Rev. E **60**, 7186–7197 (1999)
P.H. Leo, R.F. Sekerka, The effect of elastic fields on the morphological stability of a precipitate grown from solid solution. Acta Metal. **37**, 3139–3149 (1989)
J. Mandel, Equations constitutives et directeurs dans les milieux plastiques et viscoplastiques. Int. J. Solids Struct. **9**, 725–740 (1973)
G.A. Maugin, *Thermomechanics of Plasticity and Fracture* (Cambridge University Press, Cambridge, 1992)
G.A. Maugin, *Thermomechanics of Nonlinear Irreversible Behaviors* (World Scientific, Singapore, 1999)
G.A. Maugin, A.V. Metrikine, *Mechanics of Generalized Continua, One Hundred Years After the Cosserats*. Advances in Mechanics and Mathematics, vol. 21 (Springer, New York, 2010)
S.D. Mesarovic, R. Baskaran, A. Panchenko, Thermodynamic coarsening of dislocation mechanics and the size-dependent continuum crystal plasticity. J. Mech. Phys. Solids **58**(3), 311–329 (2010)
S.D. Mesarovic, S. Forest, J.P. Jaric, Size-dependent energy in crystal plasticity and continuum dislocation models. Proc. R. Soc. A **471**, 20140868 (2015). https://doi.org/10.1098/rspa.2014.0868
C. Miehe, S. Mauthe, F.E. Hildebrand, Variational gradient plasticity at finite strains. Part III: local-global updates and regularization techniques in multiplicative plasticity for single crystals. Comput. Methods Appl. Mech. Eng. **268**, 735–762 (2014)
R.D. Mindlin, Micro–structure in linear elasticity. Arch. Rat. Mech. Anal. **16**, 51–78 (1964)
R.D. Mindlin, Second gradient of strain and surface–tension in linear elasticity. Int. J. Solids Struct. **1**, 417–438 (1965)
J. Mosler, O. Shchyglo, H. Montazer Hojjat, A novel homogenization method for phase field approaches based on partial rank-one relaxation. J. Mech. Phys. Solids **68**, 251–266 (2014). https://doi.org/10.1016/j.jmps.2014.04.002
I. Müller, Thermodynamics of mixtures and phase field theory. Int. J. Solids Struct. **38**, 1105–1113 (2001)
W.W. Mullins, R.F. Sekerka, Morphological stability of a particle growing by diffusion or heat flow. J. Appl. Phys. **34**, 323–329 (1963). https://doi.org/10.1063/1.1702607
J.F. Nye, Some geometrical relations in dislocated crystals. Acta Metall. **1**, 153–162 (1953)
N. Ohno, D. Okumura, Higher-order stress and grain size effects due to self energy of geometrically necessary dislocations. J. Mech. Phys. Solids **55**, 1879–1898 (2007)
N. Ohno, D. Okumura, Grain–size dependent yield behavior under loading, unloading and reverse loading. Int. J. Mod. Phys. B **22**, 5937–5942 (2008)

M. Ortiz, E.A. Repetto, Nonconvex energy minimization and dislocation structures in ductile single crystals. J. Mech. Phys. Solids **47**(2), 397–462 (1999)
H. Proudhon, W.J. Poole, X. Wang, Y. Bréchet, The role of internal stresses on the plastic deformation of the Al–Mg–Si–Cu alloy AA611. Philos. Mag. **88**, 621–640 (2008)
J. Qu, M. Cherkaoui, *Fundamentals of Micromechanics of Solids* (John Wiley and Sons Inc, Hoboken, 2006)
B.D. Reddy, C. Wieners, B. Wohlmuth, Finite element analysis and algorithms for single-crystal strain-gradient plasticity. Int. J. Numer. Methods Eng. **90**(6), 784–804 (2012)
R. Spatschek, B. Eidel, Driving forces for interface kinetics and phase field models. Int. J. Solids Struct. **50**, 2424–2436 (2013)
I. Steinbach, Phase-field models in materials science. Model. Simul. Mater. Sci. Eng. **17**(7), 073001 (2009)
I. Steinbach, M. Apel, Multi phase field model for solid state transformation with elastic strain. Physica D **217**(2), 153–160 (2006)
P. Steinmann, Views on multiplicative elastoplasticity and the continuum theory of dislocations. Int. J. Eng. Sci. **34**, 1717–1735 (1996)
R.E. Stoltz, R.M. Pelloux, Cyclic deformation and Bauschinger effect in Al–Cu–Mg alloys. Scr. Metall. **8**, 269–276 (1974)
R.E. Stoltz, R.M. Pelloux, The Bauschinger effect in precipitation strengthened aluminum alloys. Metall. Trans. A **7**, 1295–1306 (1976)
B. Svendsen, S. Bargmann, On the continuum thermodynamic rate variational formulation of models for extended crystal plasticity at large deformation. J. Mech. Phys. Solids **58**(9), 1253–1271 (2010)
R. Taillard, A. Pineau, Room temperature tensile properties of Fe-19wt.% Cr alloys precipitation hardened by the intermetallic compound NiAl. Mater. Sci. Eng. **56**, 219–231 (1982)
J. Tiaden, B. Nestler, H.J. Diepers, I. Steinbach, The multiphase-field model with an integrated concept for modelling solute diffusion. Physica D **115**, 73–86 (1998)
R.L.J.M. Ubachs, P.J.G. Schreurs, M.G.D. Geers, Phase field dependent viscoplastic behaviour of solder alloys. Int. J. Solids Struct. **42**, 2533–2558 (2005)
T.T. Uehara, T. Tsujino, N. Ohno, Elasto-plastic simulation of stress evolution during grain growth using a phase field model. J. Cryst. Growth **300**, 530–537 (2007)
A. Villani, E.P. Busso, K. Ammar, S. Forest, M.G.D. Geers, A fully coupled diffusional-mechanical formulation: numerical implementation, analytical validation, and effects of plasticity on equilibrium. Arch. Appl. Mech. **84**, 1647–1664 (2014)
Y. Wang, A.G. Khachaturyan, J.W. Jr Morris, Kinetics of strain-induced morphological transformation in cubic alloys with a miscibility gap. Acta Metall. Mater. **41**, 279–296 (1993)
S. Wulfinghoff, T. Böhlke, Equivalent plastic strain gradient enhancement of single crystal plasticity: theory and numerics. Proc. R. Soc. A: Math. Phys. Eng. Sci. **468**(2145), 2682–2703 (2012)
S. Wulfinghoff, E. Bayerschen, T. Böhlke, A gradient plasticity grain boundary yield theory. Int. J. Plast. **51**, 33–46 (2013)
S. Wulfinghoff, S. Forest, T. Böhlke, Strain gradient plasticity modeling of the cyclic behavior of laminate microstructures. J. Mech. Phys. Solids **79**, 1–20 (2015). https://doi.org/10.1016/j.jmps.2015.02.008
A. Zaoui, Continuum micromechanics: survey. ASCE J. Eng. Mech. **128**, 808–816 (2002)

Printed by Printforce, the Netherlands